Subcellular Biochemistry

Volume 12

Immunological Aspects

A Continuation Order Plan is available for this series. A continuation order will bring delivery of each new volume immediately upon publication. Volumes are billed only upon actual shipment. For further information please contact the publisher.

Subcellular Biochemistry

Volume 12

Immunological Aspects

Edited by

J. R. Harris

North East Thames Regional Transfusion Centre
Brentwood, Essex, England

Springer Science+Business Media, LLC

The Library of Congress cataloged the first volume of this title as follows:

Sub-cellular biochemistry.

London, New York, Plenum Press.
v. illus. 23 cm. quarterly.
Began with Sept. 1971 issue. Cf. New serial titles.
1. Cytochemistry – Periodicals. 2. Cell organelles – Periodicals.
QH611.S84 574.8'76 73-643479

ISBN 978-1-4899-1683-9 ISBN 978-1-4899-1681-5 (eBook)
DOI 10.1007/978-1-4899-1681-5

This series is a continuation of the journal *Sub-Cellular Biochemistry,* Volumes 1 to 4 of which were published quarterly from 1972 to 1975

Originally published by Plenum, New York in 1988.
Softcover reprint of the hardcover 1st edition 1988

Contributors

Humberto de Assis Paiva Department of Immunology, Middlesex Hospital Medical School, London W1P 9PG, United Kingdom

Gopa Biswas-Hughes Department of Immunology, Middlesex Hospital Medical School, London W1P 9PG, United Kingdom

Stephen Brimijoin Department of Pharmacology, Mayo Clinic, Rochester, Minnesota 55905

Peter Byfield Department of Endocrinology, MRC Clinical Research Centre, Harrow HA1 3UJ, United Kingdom

Brian R. Champion Department of Immunology, Middlesex Hospital Medical School, London W1P 9PG, United Kingdom; *present address:* Immunobiology Department, Glaxo Group Research Ltd, Greenford UB6 0HE, United Kingdom

Yves Chicheportiche Department of Pathology, University Medical Center, 1211 Geneva 4, Switzerland

Anne Cooke Department of Immunology, Middlesex Hospital Medical School, London W1P 9PG, United Kingdom

L. Dickson MRC Human Biochemical Genetics Unit, University College, London NW1 2HE, United Kingdom

Carl G. Gahmberg Department of Biochemistry, University of Helsinki, SF-00170 Helsinki, Finland

H. C. Gooi Regional Blood Transfusion Centre, Leeds LS15 7TW, United Kingdom

F. R. Green MRC Human Biochemical Genetics Unit, University College, London NW1 2HE, United Kingdom

P. Greenwell Division of Immunochemical Genetics, MRC Clinical Research Centre, Harrow HA1 3UJ, United Kingdom

B. Griffiths MRC Human Biochemical Genetics Unit, University College, London NW1 2HE, United Kingdom

Hans-Peter Hauri Department of Pharmacology, Biocenter of the University of Basel, CH-4056 Basel, Switzerland

Patricia Hutchings Department of Immunology, Middlesex Hospital Medical School, London W1P 9PG, United Kingdom

J. Paul Luzio Department of Clinical Biochemistry, University of Cambridge, Cambridge CB2 2QR, United Kingdom

David Male Institute of Psychiatry, Denmark Hill, London SE5 1JK, United Kingdom

S. K. Malhotra Department of Zoology, University of Alberta, Edmonton, Alberta, Canada T6G 2E9

A. H. Merry Blood Group Reference Laboratory, Radcliffe Infirmary, Oxford OX2 6HE, United Kingdom

J. Noades MRC Human Biochemical Genetics Unit, University College, London NW1 2HE, United Kingdom

Kevin Page Department of Immunology, Middlesex Hospital Medical School, London W1P 9PG, United Kingdom

Michael R. Price Cancer Research Campaign Laboratories, University of Nottingham, University Park, Nottingham NG7 2RD, United Kingdom

Zoltán Rakonczay Central Research Laboratory, Medical University, H-6720 Szeged, Hungary

David C. Rayner Department of Pathology, University of Manitoba, Winnipeg, Canada R3G 0W3

Peter J. Richardson Department of Clinical Biochemistry, University of Cambridge, Cambridge CB2 2QR, United Kingdom

D. M. Swallow MRC Human Biochemical Genetics Unit, University College, London NW1 2HE, United Kingdom

Alan Michael Tartakoff Institute of Pathology, Case Western Reserve University, Cleveland, Ohio 44106

D. Voak Regional Transfusion and Immuno-Haematology Centre, Cambridge CB2 2PT, United Kingdom

Preface

This volume and future ones in the series, which is now under new editorship, depart from the pattern established by previous volumes in containing chapters loosely related to one another. It is thus appropriate that each volume should have a relevant subtitle. In the present volume the underlying theme is immunology and its application to the study of a number of different biochemical systems.

Almost inevitably, the study of the cell-surface antigens and glycoproteins of a number of different tissues tends to dominate the contents, but considerable emphasis is also placed on subcellular organelles and enzymes. Undoubtedly, it is the impact of readily available monoclonal antibodies that dominates work in the field. The epitope specificity provided by monoclonal antibodies gives the biochemist a powerful lever with which to probe the location, structure, and function of cellular protein and glycoprotein antigens. A number of the chapters included in this volume have a strong biomedical emphasis. This is indeed intentional, as it is now the policy of the series to cover the more applied aspects of the subject, while in no way neglecting more fundamental topics in subcellular biochemistry.

Thus, the current plans for the series expand rather than diminish those developed by the founding editor, the late Donald Roodyn. Dr. Roodyn started the quarterly journal *Sub-Cellular Biochemistry* in 1971 and molded it into book format, as *Subcellular Biochemistry,* in 1978. This series title is to be retained along with the varying subject-oriented subtitles. A theme close to Dr. Roodyn's heart was the unification of cell biology and classical biochemistry. Time has certainly shown that the two do indeed walk hand-in-glove and that strict academic disciplines are at times quite artificial. Furthermore, it is now clear that the whole of biology—together with medical, agricultural, and technological biochemistry in all their diverse aspects—must now fall within the range of the series. Previously, the discipline of physiology was considered to be the "experimental basis of medicine," but this designation must surely now be at least shared with, if not actually handed over to, biochemistry. At the time of Claude Bernard, physiological and biological chemistry had not evolved fully into the discipline of biochemistry as we know it today. Yet it is the continuing penetrative influence of chemistry and physics in the natural and applied biological sciences that has led to the current strength of the experimental approaches of modern biochemical science, hopefully leading ultimately to Donald Roodyn's "integrated view of the cell."

While demonstrating the application of immunology in subcellular biochemistry, the chapters in the present volume fall into two discrete groups: those that dwell primarily on the study of cell-surface antigens and those that deal with a range of other cellular, subcellular, and enzymic applications.

Chapter 1, by Michael R. Price, presents a discussion of breast cancer antigens defined by monoclonal antibodies and immediately sets the "applied" character of the book. This chapter contains a detailed presentation of the epitope analysis of breast-carcinoma-associated antigens and the diagnostic and prognostic applications of monoclonal antibodies. Chapter 2, by A. H. Merry, H. C. Gooi, and D. Voak, deals with the immunochemistry and serology of the major human blood group antigens. This is followed by a chapter on the molecular characterization of the $Rh_0(D)$ blood group by Carl G. Gahmberg. The fourth chapter, on the expression of ABH, Lewis, and related blood group antigens on the glycoproteins of the human jejunal brush border, by F. R. Green and colleagues, integrates very well with the two previous chapters on blood groups. In Chapter 5, a more fundamental account of the biogenesis and intracellular transport of intestinal brush border membrane hydrolases, investigated using antibody probes and tissue culture, is provided by Hans-Peter Hauri.

Leading into the second group of chapters, Chapter 6 presents a discussion of the immunoaffinity purification of membrane fractions from mammalian cells, contributed by Peter J. Richardson and J. Paul Luzio. There follows a survey of the use of antibodies for the analysis of secretory and endocytic paths of eukaryotic cells, by Yves Chicheportiche and Alan Michael Tartakoff. In Chapter 8, S. K. Malhotra discusses the use of monoclonal antibodies for investigating astrocyte structure and function. This is followed by a detailed account of current molecular approaches to experimental thyroid autoimmunity, written by Brian R. Champion and his colleagues. Finally, Chapter 10, by Zoltán Rakonczay and Stephen Brimijoin, contains a thorough description of the biochemistry and pathophysiology of the molecular forms of the cholinesterases.

J. R. Harris

Brentwood, Essex, United Kingdom

Contents

Chapter 1

Breast-Cancer-Associated Antigens Defined by Monoclonal Antibodies

Michael R. Price

Chapter 4

Expression of the ABH, Lewis, and Related Antigens on the Glycoproteins of the Human Jejunal Brush Border

F. R. Green, P. Greenwell, L. Dickson, B. Griffiths, J. Noades, and D. M. Swallow

Chapter 5

Biogenesis and Intracellular Transport of Intestinal Brush Border Membrane Hydrolases: Use of Antibody Probes and Tissue Culture

Hans-Peter Hauri

Chapter 6

Immunoaffinity Purification of Membrane Fractions from Mammalian Cells

Peter J. Richardson and J. Paul Luzio

Chapter 7

The Use of Antibodies for Analysis of the Secretory and Endocytic Paths of Eukaryotic Cells

Yves Chicheportiche and Alan Michael Tartakoff

Chapter 8

Monoclonal Antibodies in Investigations on Astrocytes

S. K. Malhotra

Chapter 9

Current Molecular Approaches to Experimental Thyroid Autoimmunity

Brian R. Champion, David C. Rayner, Patricia Hutchings, Anne Cooke, Kevin Page, Gopa Biswas-Hughes, Humberto de Assis Paiva, David Male, and Peter Byfield

Chapter 10

Biochemistry and Pathophysiology of the Molecular Forms of Cholinesterases

Zoltán Rakonczay and Stephen Brimijoin

Chapter 1

Breast-Cancer-Associated Antigens Defined by Monoclonal Antibodies

Michael R. Price

1. INTRODUCTION

The development of monoclonal antibodies against malignant cells is projected to have a number of significant applications related to the clinical management of breast cancer. These include the early diagnosis of tumors, monitoring patients for the detection of recurrence or effects of therapy, and therapeutic modalities, whereby antibodies are employed to target cytotoxic agents (drugs, toxins, or radionuclides) to the site of the tumor. Since in Western countries, breast cancer will afflict in the region of one out of every 14 women, it is essential that the potential of these agents should be explored fully. Conventional wisdom indicates that aids to earlier diagnoses would have an important impact on prognosis. Present methods of diagnosis by palpation and mammography permit detection of tumor masses with diameters as small as 1 cm. However, even at this apparently early stage, tumors may be sufficiently established for the dissemination of tumor cells to have already occurred (reviewed by Tagnon, 1986). Thus, if breast cancer is only rarely a localized disease when it is clinically evident, there is clearly a need for assays that might provide earlier, accurate diagnoses and that reflect tumor burden. Given the sensitivity and specificity that can be obtained with monoclonal antibodies, their use in immunoassays offers the possibilities

Michael R. Price Cancer Research Campaign Laboratories, University of Nottingham, University Park, Nottingham NG7 2RD, United Kingdom.

for the measurement of these clinical parameters and for optimizing therapy.

The development of monoclonal antibodies against breast tumors will shed further light on more fundamental aspects of the biology of malignant disease of the breast. Identification of antigens or epitopes unique to tumors, or the detection of components of ectopic or fetal origin, or products related to cellular differentiation or dedifferentiation would all help in our understanding of the biology of breast cancer. Any subclassification of this disease, which is characterized by its heterogeneity (both at the cellular level and with respect to the response to treatment regimes), would be highly desirable and may provide a guide to more accurate prognosis.

2. DEVELOPMENT OF MONOCLONAL ANTIBODIES AGAINST BREAST CANCER

Three main approaches have been adopted for the development of monoclonal antibodies reactive with breast carcinoma cells. The techniques are based on those developed and described by Köhler and Milstein (1975) and by which individual clones of antibody-secreting cells are immortalized by fusing them with cultured cells of a mouse myeloma.

With the first approach, mice are immunized with breast carcinoma cells or subcellular fractions or extracts derived from breast tumors, and these animals are employed as donors of sensitized spleen cells that are then fused with the mouse myeloma cells. The aim here is to select fusion products (hybridomas) secreting antibodies that show at least preferential reactivity with tumors. In the second approach, mice are immunized with normal materials isolated from human milk (i.e., milk fat globule membranes or skim milk), this being a convenient source of proteins and glycoproteins derived from terminally differentiated breast epithelia (Ceriani *et al.*, 1977). The impetus for pursuing this particular line of research stems from the study of normal differentiation in the hemopoietic system, which has aided the clinical management of many leukemias (Janossy *et al.*, 1980). Finally, human monoclonal antibodies have been produced using patients' lymphocytes as the source of sensitized cells to be fused with mouse myeloma cells (Schlom *et al.*, 1980). A major limitation with such interspecies hybridomas is their relative chromosomal instability with preferential deletion of human chromosomes and concomitant loss of antibody production or viability. In addition, this approach presumes the existence of tumor antigens capable of inducing a significant immune response in the autologous host and this, at the present stage of development, remains to be proved. Recent evidence suggests that specificity for tumors can be demonstrated in antibodies produced from clones of Epstein–Barr virus-transformed B lymphocytes obtained from breast cancer patients (Campbell *et al.*, 1986). The ability to produce

Table I
Summary of Monoclonal Antibodies Used in the Present Study

Monoclonal antibody	Type/ isotype	Immunogen	Reference
NCRC-11	IgM	Dissociated breast carcinoma cells	Ellis et al. (1984)
472/18	IgM	Purified NCRC-11-defined antigen	S. Edwards and E. Jacobs (unpublished observation)
HMFG-1	IgG1	Human milk fat globule	Burchell *et al.* (1983)
HMFG-2	IgG1		
EMA	IgG2a	Human milk fat globule	Cordell *et al.* (1985); Heyderman *et al.* (1985)
M8	IgG1	Human milk fat globule	Edwards and Brooks (1984)
M18	IgM		
M24	IgM		
115D8	IgG1	Human milk fat globule	Hilkens *et al.* (1984)
115F5	IgG2		
115G2	IgG2		
Ca1	IgM	Glycoproteins from cultured laryngeal carcinoma H.Ep2 cells	Ashall *et al.* (1982)
Ca2	IgG1	Purified Ca1-defined antigen	Bramwell *et al.* (1985)
Ca3	IgG1		

antitumor antibodies of human origin would represent an important development since these would be more acceptable for use in therapeutic modalities.

Table I summarizes details of a number of murine monoclonal antibodies that have been employed in the present investigation. A common feature of these antibodies is that their target antigens are glycoproteins, although it should be pointed out that several antibodies reactive with glycolipids have been developed (e.g., the MBr1 antibody described by Canevari *et al.*, 1983), and some antibodies define epitopes coexpressed on both glycoproteins and glycolipids (McIlhinney *et al.*, 1985).

3. ANALYSIS OF THE REACTIVITY OF MONOCLONAL ANTIBODIES AGAINST BREAST CANCER

3.1. Reactions of Monoclonal Antibodies with Tumors

One of the original aims of investigations on the development of monoclonal antibodies against breast cancer was to produce reagents that show specificity for malignant cells. However, despite many attempts, it is now clear that most, if not all, antibodies available display some level of reactivity against cells derived from normal tissues. This would be expected to impose serious limitations on their clinical application, but these antibodies have been employed successfully for the radiodiagnostic imaging of breast, ovarian, and gastrointestinal tumors, where tumor specificity would be expected to be an overriding prerequisite (Epenetos *et al.*, 1982a; Rainsbury *et al.*, 1983).

Many of the monoclonal antibodies produced, including those in Table I, recognize antigens widely expressed on tumors of epithelial origin and in some forms of epithelial metaplasia. The NCRC-11 antibody, for example, when tested against a series of 50 breast carcinomas, reacted with all tumors as assessed by immunocytochemical staining of tissue sections (Ellis *et al.*, 1984). Staining, however, was variable between samples and heterogeneous within individual tumors. Similar patterns of NCRC-11 antigen expression were found in adenocarcinomas from other sites (prostate, kidney, colon, pancreas, and endometrium) and with transitional cell carcinomas and squamous cell carcinomas, although some of the latter failed to express this antigen. None of a series of lymphomas, sarcomas of various types, or melanomas showed positive staining (Ellis *et al.*, 1984).

A major characteristic of the antigens defined by the antibodies detailed in Table I is that their expression in breast, and other carcinomas, is heterogeneous. The antigens involved here are usually membrane antigens, although cytoplasmic staining by antibodies is also found. Moreover, the pattern of staining using antibodies prepared against milk fat globules, for example, may be variable among cells of apparently the same morphology and this is neither experimental artifact nor related to the mitotic cell cycle (Edwards, 1985). Antigen heterogeneity among and within tumors, like other manifestations of tumor cell heterogeneity (e.g., morphology and response to cytotoxic drugs), may impose some limitations on the development of antibody-directed therapy, although it may provide an important means for the subclassification of tumors and guide to prognosis.

3.2. Reactions of Monoclonal Antibodies with Normal Tissues

The reactivity of six well-characterized antibodies with normal tissues is presented in summary in Table II. The antibodies include the antibreast carcinoma antibody, NCRC-11 (Ellis *et al.*, 1984), three monoclonal antibodies against human milk fat globules, HMFG-1, HMFG-2 (Arklie *et al.*, 1981), and 115D8 (Hilkens *et al.*, 1984), the antilaryngeal carcinoma antibody, Ca1 (McGee *et al.*, 1982), and the antibody termed EMA. The results with this latter reagent were initially obtained with an extensively adsorbed rabbit antiserum prepared against defatted human cream and reactive with the so-called epithelial membrane antigen (EMA) (Sloane and Ormerod, 1981) and the findings have

Table II
Reactivity of Monoclonal Antibodies with Normal Tissues

	NCRC-11	HMFG-1	HMFG-2	EMA	115D8	Ca1
Alimentary System						
Salivary gland						
Serous acini		+	+		+	
Mucinous acini		–	–		–	
Ducts	+			+	+	
Esophagus						–
Squamous epithelium	–			–	+	
Submucosa					–	
Stomach		–	–			–
Mucus-secreting cells					–	
Parietal cells	+				+	
Peptic cells					+	
Lamina propria					–	
Submucosa					–	
Small intestine	–	–	–		–	–
Large intestine	–	–	–		–	–
Liver						–
Parenchyma	–	–	–	–	–	
Bile ducts	+	+	+	+	–	
Pancreas						
Acini	+	+	+	+	+	–
Ducts	+				+	
Nervous System						
Brain					–	–
Cerebral cortex	–			–		
Cerebellum	–			–		
Spinal cord	–			–		
Peripheral nerves	–			–	–	
Endocrine System						
Adrenals	–					–
Medula and cortex				–	–	
Pancreas						
Islets of Langerhans		–	–	–	–	–
Thyroid	–	–	–			–
Follicle epithelium					+	
Parafollicular cells					–	
Pituitary gland				–		
Posterior pituitary					–	
Pars intermedia (follicle epithelium)					+	
Anterior pituitary					–	
Lymphoreticular System						
Spleen and lymph nodes	–			–	–	–

(*continued*)

Table II (*Continued*)

	NCRC-11	HMFG-1	HMFG-2	EMA	115D8	Ca1
Thymus		–	–			
Cortex and medulla				–	–	
Hassal's corpuscles					+	
Circulating cells and bone marrow cells	–			–		
Generative System						
Breast						
Resting acini and ducts	+	+	+	+	+	+
Lactating acini and ducts	+	+	+	+	+	+
Cervix	–			–		–
Uterus						
Endometrium	+	+	+	+	+	–
Myometrium					–	
Fallopian tube	+	–	–			+
Ovary				–		–
Follicular epithelium					+	
Surface epithelium					+	
Stroma					–	
Prostate					–	
Testis	–	–	–	–	–	–
Epididymus						
Ductal epithelium		–	+		+	
Skin						
Epidermis	–	–	–	–	–	–
Sweat glands	+			+		
Acini and ducts		–	+		+	
Sebaceous glands	+	+	+	+	+	–
Musculoskeletal System						
Muscle (striated, smooth, and cardiac)	–				–	
Connective tissues	–					–
Urinary System						
Kidney						
Glomeruli	–				–	
Proximal tubules	–			–	–	
Distal tubules	+			+	+	
Collecting tubules	+	+	+	+	+	–
Bladder						
Urothelium	+	–	–	+	+	+
Respiratory System						
Lung	+					–
Respiratory epithelium		–	+	+	+	
Ciliated epithelium		–	+	+	+	
Serous bronchial glands		+	+	+	+	
Mucous bronchial glands					–	

essentially been reproduced using a monoclonal reagent termed E29 (Cordell *et al.*, 1985; Heyderman *et al.*, 1985). The main feature of the results detailed in this table is the remarkable similarity in the profiles of reactivity obtained with these reagents despite the fact that they originated from five different laboratories using different immunogens to prepare the antibodies. The antigen defined by NCRC-11, for example, like the determinants defined by the other antibodies, has a wide but highly specific distribution in normal tissues, being virtually confined to the surface of certain specialized epithelia. The antigen was predominantly expressed on the luminal surface of the cuboidal and nonmucinous columnar epithelium of acini and ducts. Surface staining was also found in urothelium, mesothelium, and ciliated columnar epithelium of the respiratory tract, fallopian tube, and vas deferens.

Despite the marked similarities in the activity of these antibodies (Table II), there are specific differences in reactivity with normal tissues that cannot be accounted for solely by variabilities of interpretation in different laboratories. With HMFG-1 and HMFG-2, for example, HMFG-1 bound only to the luminal surfaces of classically defined exocrine glands, to their associated ducts, and to the collecting tubules of kidney and bronchioles of the lung (Arklie *et al.*, 1981). HMFG-2 showed a similar pattern of staining but also bound to sweat glands, the alveolar epithelium of the lung, and the luminal epithelium of the ductuli efferentes of the epididymis.

4. BIOCHEMICAL CHARACTERIZATION OF BREAST-CARCINOMA-ASSOCIATED ANTIGENS

In the preceding section, attention was directed toward the marked similarity in reactivity of several monoclonal antibodies with normal and malignant tissues. The reaction of six antibodies with normal tissues is presented in some detail in Table II, and from these considerations, it might be anticipated that molecular targets defined by these various antibodies are comparable or indeed the same moieties. A brief survey of the literature indicates that many of the monoclonal antibodies presently available react with complex glycoproteins of high molecular weight. Since these are apparently heavily glycosylated, anomalous behavior on gel filtration or using sodium dodecyl sulfate–polyacrylamide gel electrophoresis (SDS–PAGE) prevents accurate determination of molecular weight so that values quoted must be regarded as approximations.

The NCRC-11-defined antigen has been isolated from detergent-solubilized extracts of breast carcinoma subcellular membranes and purified by immunoadsorbent chromatography using Sepharose-linked NCRC-11 antibodies (Price *et al.*, 1985b). The antigen, when radiolabeled with ^{125}I, was examined by SDS–

PAGE followed by autoradiographic analysis of the dried gels and this migrated into the gels as a single but diffuse band with an apparent molecular weight in the region of 400 kD or above. When adsorbed to the wells of plastic microtest plates and treated with proteolytic enzymes (papain at 1 μg/mL, or much higher concentrations of pronase), the antigen was degraded or fragmented in situ so that the subsequent binding of NCRC-11 antibodies was totally abolished. NCRC-11 antigen was effectively adsorbed by Sepharose-linked wheat germ agglutinin, whereas the antibody-binding activity of antigen preparations was only partially reduced by adsorption with immobilized lentil lectin or *Helix pomatia* lectin, suggesting heterogeneity in the expression or accessibility of the appropriate carbohydrate residues reactive with the latter two lectins. NCRC-11 antibody-binding activity was fully retained after boiling antigen preparations for up to 5 min, suggesting that the NCRC-11-defined epitope was likely to be expressed in oligosaccharide side chains. However, antibody binding was unaffected by neuraminidase treatment of antigen preparations, indicating that sialic acid residues were not involved in the epitope (Price *et al.*, 1986a). Many of these findings have been reproduced with NCRC-11 antigen preparations isolated from breast carcinoma tissues and MCF-7 tissue culture cells as well as NCRC-11-defined antigens from ovarian mucinous and serous cyst adenocarcinomas (Price *et al.*, 1985b,1986a). Very similar results have been obtained in an analysis of antigens defined by the monoclonal antibody 115D8 (Hilkens *et al.*, 1985), although in this case the antigen, which has been termed the MAM-6 antigen, was isolated from human skim milk rather than from tumor-derived materials. Comparably, the monoclonal antibodies HMFG-1 and HMFG-2 react with oligosaccharide determinants present on a large molecular weight component (>400 kD) of the human milk fat globule (Burchell *et al.*, 1983), which, if not identical to the MAM-6 antigen, is almost certainly closely related to it. The HMFG-1-defined determinants were apparently more frequently expressed on this molecule than the HMFG-2 binding sites. These determinants, however, were more strongly expressed on carcinoma cells and could be carried on a range of glycoproteins with molecular weights from 80 to 300 kD. It was suggested that the HMFG-1 antibody may recognize an oligosaccharide determinant present on a large, complex carbohydrate side chain, which is more representative of normal differentiation, while the HMFG-2 antibody defines an epitope on a simpler carbohydrate, which may represent an abortive attempt at correct glycosylation by the malignant cell (Burchell and Taylor-Papadimitriou, 1985; Burchell *et al.*, 1983).

Attempts to characterize the EMA (defined by an adsorbed rabbit antiserum against defatted human cream) indicated that the antigenic activity in milk was spread over a wide range of molecular weights, and although purification produced a general reduction in size, the antigen remained heterogeneous (Ormerod *et al.*, 1983). A monoclonal antibody against EMA (termed E29) also reacted with material in milk fat globule membranes covering a wide range of molecular weights from 265 to 400 kD (Cordell *et al.*, 1985).

EMA preparations are isolated from human skimmed milk, and EMA activity during purification is monitored by immunochemical assay (Ormerod *et al.*, 1983). In a totally separate investigation, Shimizu and Yamauchi (1982) described the isolation and characterization of a high molecular weight mucinlike glycoprotein that had been extracted with detergent from milk fat globule membranes. This component, termed PAS-O, was purified to homogeneity using conventional biochemical techniques to give a single band on analytical ultracentrifugation. Subsequently, it was recognized that PAS-O was likely to be closely related to EMA since it was the membrane-associated counterpart of EMA. The marked similarities between these materials were revealed in a compositional study involving comparison of their amino acid analyses and also their carbohydrate compositions (Ormerod *et al.*, 1985).

As already indicated, NCRC-11 antigens (i.e., antigens isolated by their binding to and elution from immobilized NCRC-11 antibodies) have been obtained from both breast and ovarian tumors (Price *et al.*, 1986a). Monoclonal antibodies produced specifically against ovarian carcinoma cells also appear to define high molecular weight glycoproteins similar to those found in breast carcinomas and human milk. De Kretser *et al.* (1986) have developed one such antibody defining an ovarian-associated antigen of 360 kD, and this was considered likely to be similar to that defined by HMFG-2. Similarly, the antihuman laryngeal carcinoma antibody Ca1 reacted with high molecular weight glycoproteins of 350 and 390 kD, although in this case, normal urine was found to be a convenient source of antigen for purification (Bramwell *et al.*, 1983).

The implication from these various biochemical studies characterizing antigens associated with human milk and various carcinomas is that it is likely that there is a restricted family of related high molecular weight glycoproteins associated with epithelial cells and their malignancies, and that these bear diverse epitopes expressed in complex oligosaccharide side chains that are strongly immunogenic in the mice originally employed for the production of monoclonal antibodies. While this may indeed be the case, there are a variety of other macromolecules in breast tumors that may be exploited as breast-cancer-associated antigens defined by monoclonal antibodies. With antibodies prepared by immunization against the human breast tumor cell line MCF-7, for example, a number of antigens have been defined. Clearly, this cell line expresses the characteristic high molecular weight glycoprotein(s) already described (and against which antibodies have been produced directly by immunization with MCF-7 cells; Papsidero and Johnson, 1986; Papsidero *et al.*, 1983). In addition, smaller protein antigens have been defined including a 43,000 molecular weight membrane-associated glycoprotein (Edwards *et al.*, 1986) and an estrogen-regulated glycoprotein of 52,000 molecular weight (Garcia *et al.*, 1985). This latter component, being a cytoplasmic protein, is an inappropriate target for antibody-directed therapy or tumor imaging, but since it is secreted by exocytosis, it is feasible that it has applications in immunodiagnosis. Ménard *et al.* (1983) have also employed MCF-7 cells for the production of monoclonal antibodies reacting with breast tumors, although here antigenic activity was recovered in neutral glycolipid fractions (Canevari *et al.*, 1983; Ménard *et al.*, 1983).

5. EPITOPE ANALYSIS OF BREAST-CARCINOMA-ASSOCIATED ANTIGENS

5.1. Immunoassay Procedures

The high molecular weight, mucinlike glycoproteins associated with carcinoma of the breast and other tissues, and reactive with monoclonal antibodies against tumors and the human milk fat globule, were subjected to further immunochemical analysis, particularly with respect to examining the epitopes defined by the antibody NCRC-11 and to comparing them with those defined by other available monoclonal antibodies. For this purpose, NCRC-11-defined antigen was purified from breast carcinomas by immunoadsorbent chromatography using Sepharose-linked NCRC-11 antibodies. Three types of solid-phase radioimmunoassay were then employed to investigate the reactivity of a panel of antibodies with the NCRC-11-defined antigen. These assays are illustrated diagrammatically in Figure 1, and in each case, the reactions were performed in the 10-μL wells of Terasaki tissue culture microtest plates. With the first procedure—the microradioisotopic antiglobulin assay—antigen is coated onto the surface of the wells either by overnight incubation at 5°C or by drying the antigen onto the wells. Remaining nonspecific binding sites in the well are then blocked by incubation with excess protein (0.1% bovine serum albumin and 0.1% rabbit serum in phosphate buffered saline) and monoclonal antibodies are then dispensed into the wells. After incubation to allow antibody binding (1–2 h), the wells are washed and ^{125}I-labeled, affinity-purified $F(ab')_2$ fragments of a rabbit antimouse Ig antiserum are added. After incubation (again 1–2 h), the wells are washed to remove excess labeled antiglobulin, and the radioactivity remaining in each well is determined. In this way, the levels of reactivity of a number of antibodies with a particular antigen may be compared.

With the second assay—the competitive inhibition of antibody-binding assay—antigen is again adsorbed to the wells of the microtest plate and remaining nonspecific binding sites are blocked with excess protein. Radioiodinated antibody is then added to each well either in admixture with control buffer solution or together with an excess amount of unlabeled monoclonal antibody. Clearly, if

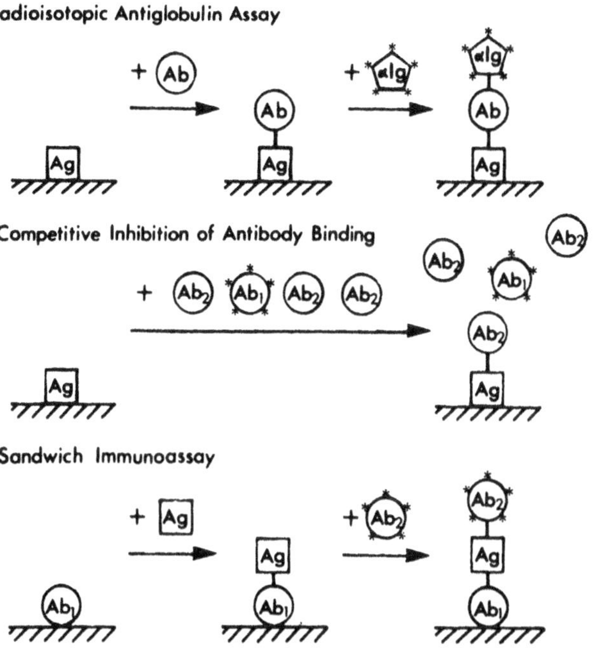

FIGURE 1. Solid-phase immunoassays employed in the analysis of monoclonal antibody-defined epitopes on breast-carcinoma-associated antigens.

the labeled and unlabeled antibodies are reactive with the same or closely related binding sites, the binding of the excess unlabeled antibody will prevent the binding of the labeled component (Figure 1).

Using the sandwich immunoassay, antibody is first adsorbed to the wells of the microtest plate. This is followed by antigen, which binds to the antibody on the solid phase. Bound antigen is then revealed by the addition of a second radiolabeled antibody to complete the "sandwich." Thus, if both antibodies are the same, the binding of the labeled antibody only occurs if the epitope is a repeated structure of the antigen. Alternatively, if the two antibodies are different, it is only possible for the labeled antibody to bind if epitopes defined by both antibodies are coexpressed on the same molecule (Figure 1). Further details concerning the technical aspects of these assays have been presented elsewhere (Price *et al.*, 1985a,b, 1986a,b).

5.2. Epitope Analysis of Tumor-Associated NCRC-11 Antigens

Table III illustrates the binding of a large panel of antibodies with a NCRC-11 antibody and a second-generation monoclonal antibody, prepared by

Table III
Binding of Monoclonal Antibodies to NCRC-11-Defined Antigen Isolated from Breast Carcinoma

Monoclonal antibody	Mean cpm ± SD bound to NCRC-11-defined antigen	Monoclonal antibody	Mean cpm ± SD bound to NCRC-11-defined antigen
NCRC-11	6886 ± 244	MBr1	173 ± 65
472/18	8874 ± 255	MBr8	159 ± 88
HMFG-1	5501 ± 316	MOv2	171 ± 81
HMFG-2	6623 ± 206	MOv8	128 ± 89
EMA	8176 ± 212	MOv15	75 ± 15
M8	6656 ± 90	C154	61 ± 43
M18	40 ± 65	C161	39 ± 109
M24	9 ± 283	11.285.14	−42 ± 86
115D8	12906 ± 387	791T/36	18 ± 30
115F5	6080 ± 372	791T/48	−41 ± 98
115G2	251 ± 24	W6/32	140 ± 238
Ca1	3950 ± 294	IgG	52 ± 127
Ca2	6483 ± 500	IgM	−53 ± 75
Ca3	7243 ± 290	NMS	148 ± 66

immunization with purified NCRC-11 antigen and termed 472/18 (Table I), both reacted well with this antigen. High levels of reactivity were also displayed in tests with the anti-milk-fat-globule antibodies HMFG-1, HMFG-2, EMA, M8, 115D8, and 115F5 and also with antibodies of the Ca series. These were produced against laryngeal carcinoma cells (i.e., the Ca1 antibody; Ashall *et al.*, 1982) or using purified Ca1-defined antigen as the immunogen (i.e., antibodies Ca2 and Ca3; Bramwell *et al.*, 1985) (Table I). Other antibodies against milk fat globules (M18, M24, 115G2, MBr1, and MBr8) failed to react with the NCRC-11 antigen. The remaining antibodies that were unreactive with this antigen included a series of antibodies against ovarian carcinomas (MOv2, MOv8, and MOv15; Miotti *et al.*, 1985) and antibodies against colorectal carcinomas (C154, C161, and 11.285.14) of which C161 reacts with the normal cross-reacting antigen, NCA (Price *et al.*, 1985a), and 11.285.14 reacts specifically with carcinoembryonic antigen, CEA (Corvalan *et al.*, 1984). The antisarcoma antibodies 791T/36 and 791T/48 and anti-HLA antibody W6/32 were all unreactive with the NCRC-11 antigen, as was normal IgG, IgM, and control mouse serum (Table III).

By immunocytochemical staining of tumors, the NCRC-11 antibody was shown to be reactive with a number of epithelial malignancies, including ovarian carcinomas (Ellis *et al.*, 1984). To explore the relation between NCRC-11 antigens from different tumor sources, immunoadsorbent purified preparations of NCRC-11-defined antigens were isolated from breast carcinomas, ovarian mucinous carcinomas, and ovarian serous cyst carcinomas (Price *et al.*, 1986a). These preparations each consisted of high molecular weight glycoproteins as described for NCRC-11 antigen from breast carcinomas. Using the micro-radioisotopic antiglobulin assay, a panel of 13 monoclonal antibodies were tested against each of the three antigen preparations and, as shown in Figure 2, their profiles of reactivity with each antigen were remarkably similar. This would suggest that the relative concentrations of each monoclonal antibody-defined epitope in each antigen were highly comparable. The major differences between the antigen preparations were that both ovarian carcinoma antigens expressed epitopes for the anti-milk-fat-globule membrane antibodies M18 and 115G2, which were apparently not expressed in this or other breast cancer antigen preparations (Figure 1; Price *et al.*, 1986a). The M18 antibody defines the I(Ma) blood group antigen, Galβ1→4GlcNAcβ1→6-; previously, Foster and Neville (1984) determined that desialylation of breast tumor tissue sections with neuraminidase exposes this oligosaccharide structure. The three NCRC-11 antigen preparations (Figure 2) were treated with neuraminidase and then examined for their capacity

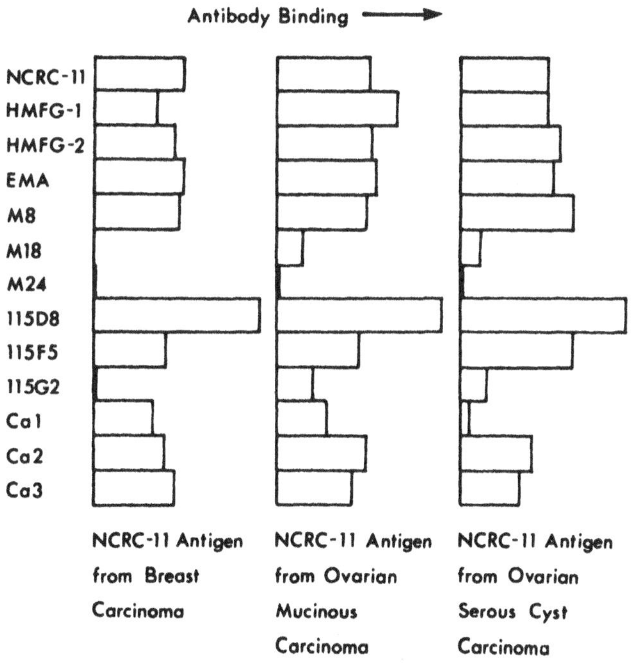

FIGURE 2. Profile of monoclonal antibody binding to immunoadsorbent purified NCRC-11-defined antigens isolated from breast carcinoma, ovarian mucinous carcinoma, and ovarian serous cyst carcinoma.

to bind the M18 antibody. Untreated and treated antigens were also examined for their capacity to bind the antibodies NCRC-11 and Ca1. Measurement of the effect of neuraminidase on Ca1 binding was included as a positive control for the enzyme digestion, since the epitope for Ca1 involves sialic acid (Bramwell *et al.*, 1985). As shown in Table IV, neuraminidase treatment of NCRC-11 antigens from ovarian or breast tumors failed to modify their capacity to bind NCRC-11 antibody, whereas the binding of Ca1 was in each case virtually abolished. Conversely, the binding of the M18 antibody was considerably increased with the treated antigens. Clearly, these tests indicate that excess sialylation of the I(Ma) blood group antigen, defined by the M18 antibody, leaves this determinant in a cryptic or masked form, particularly in the NCRC-11 antigen from breast carcinoma although this determinant is at least partially exposed in the ovarian carcinoma antigens (Table IV).

To probe further the distribution of monoclonal antibody-defined epitopes of NCRC-11 antigens, the antibodies shown in Figure 2 were assayed for their capacity to inhibit the binding of ^{125}I-labeled NCRC-11 antibody to NCRC-11 antigen preparations. Table V shows an example of data obtained in testing the capacity of three antibodies (in ascitic fluids) to inhibit the binding of labeled NCRC-11 antibody to NCRC-11 antigen from breast carcinoma. Clearly, ascitic fluids containing NCRC-11 antibody or M8 antibody competed with the labeled NCRC-11 antibody for binding to antigen. Conversely, the 115D8 antibody

Table IV
Binding of Monoclonal Antibodies to Neuraminidase-Treated NCRC-11-Defined Antigens

NCRC-11-defined antigen isolated from:	Monoclonal antibody	Mean cpm ± SD bound to antigen treated with: —	Neuraminidase
Breast carcinoma	P3NS1	50 ± 34	51 ± 18
	NCRC-11	5899 ± 4	4874 ± 223
	Ca1	3920 ± 162	181 ± 40
	M18	110 ± 32	2736 ± 17
Ovarian mucinous carcinoma	P3NS1	18 ± 53	6 ± 15
	NCRC-11	6805 ± 283	6202 ± 225
	Ca1	4491 ± 390	936 ± 33
	M18	2275 ± 94	3571 ± 93
Ovarian serous cyst carcinoma	P3NS1	51 ± 9	7 ± 31
	NCRC-11	2927 ± 132	2696 ± 127
	Ca1	451 ± 13	46 ± 22
	M18	857 ± 43	2475 ± 127

Table V
Competitive Inhibition of Binding of ^{125}I-Labeled NCRC-11 Antibody to NCRC-11-Defined Antigen Isolated from Breast Carcinoma

Competing ascitic fluid	Dilution of competing ascitic fluid	Binding of ^{125}I-labeled NCRC-11 antibody to NCRC-11-defined antigen	
		Mean cpm ± SD	Percentage binding
—	—	20149 ± 566	100
NCRC-11	10^{-3}	441 ± 66	2
	10^{-4}	4855 ± 849	24
	10^{-5}	17574 ± 81	87
	10^{-6}	19067 ± 374	95
M8	10^{-3}	1853 ± 202	9
	10^{-4}	7283 ± 139	36
	10^{-5}	14972 ± 191	74
	10^{-6}	17861 ± 97	97
115D8	10^{-3}	14473 ± 244	72
	10^{-4}	16802 ± 534	84
	10^{-5}	18884 ± 453	94
	10^{-6}	19673 ± 225	98

failed to inhibit labeled antibody binding over the whole concentration range tested. This finding is of interest since the previous studies demonstrated that 115D8 antibody-defined epitopes were very well represented in the NCRC-11 antigen preparations (Table III and Figure 2). Thus, from Table V, it can be deduced that the M8 antibody defines an epitope similar or identical to that defined by the NCRC-11 antibody, whereas that defined by 115D8 appears to be quite separately expressed on the NCRC-11 antigen. The findings from this type of approach have been extended further in a comparative analysis of NCRC-11 antigens from breast and ovarian tumors (Price *et al.*, 1986a). As summarized in Table VI, the profiles of inhibitory activity of 13 antibodies for three NCRC-11 antigens were virtually identical, indicating that the topographical distribution of epitopes defined by these antibodies was remarkably similar within each antigen preparation. This finding enhances the previous result that the three antigens are virtually equivalent with respect to their relative levels of epitope expression (Figure 2), and that the major differences between these NCRC-11 antigens can be accounted for by differences in the degrees of sialylation in the three types of tumor (Table IV; Price *et al.*, 1986a).

Epitope expression within the three NCRC-11 antigen preparations from breast, ovarian mucinous, and ovarian serous cyst carcinomas was examined further using the sandwich immunoassay (Figure 1). As shown in Table VII,

Table VI
Competitive Inhibition of Binding of ^{125}I-Labeled NCRC-11 Antibody to Various NCRC-11-Defined Antigen Preparations—A Summary

Competing monoclonal antibody	Capacity of monoclonal antibodies to inhibit the binding of ^{125}I-labeled NCRC-11 antibody to NCRC-11-defined antigens isolated from: Breast carcinoma	Ovarian mucinous carcinoma	Ovarian serous cyst carcinoma
NCRC-11	++	++	++
HMFG-1	+	+	+
HMFG-2	+	+	+
EMA	++	++	++
M8	++	++	++
M18	−	−	−
M24	−	−	−
115D8	−	−	−
115F5	−	−	−
115G2	−	−	−
Ca1	−	−	−
Ca2	−	−	+
Ca3	−	−	−

Table VII
Analysis of Epitope Expression on NCRC-11-Defined Antigen Preparations Using a Sandwich Immunoassay

Wells coated with ascitic fluid(1/500)	Binding of ^{125}I-labeled NCRC-11 antibody to antibody-coated wells treated with NCRC-11 antigen isolated from: Breast carcinoma	Ovarian mucinous adenocarcinoma	Ovarian serous cyst adenocarcinoma
NCRC-11	7425 ± 464	6687 ± 94	3362 ± 105
	(76 ± 46)[a]	(−93 ± 43)	(80 ± 23)
115D8	8617 ± 489	7762 ± 179	3155 ± 47
	(163 ± 23)	(−31 ± 8)	(135 ± 13)
M8	1859 ± 86	1781 ± 243	719 ± 43
	(129 ± 13)	(−86 ± 24)	(94 ± 13)

[a]Figures in parentheses represent the binding of ^{125}I-labeled NCRC-11 antibody in the absence of added antigen.

^{125}I-labeled NCRC-11 antibody bound to wells coated with NCRC-11, 115D8, and M8 antibodies to which antigens had been added. Thus, for ^{125}I-labeled NCRC-11 antibody to bind to antigen that had been captured by the adsorbed NCRC-11 epitopes must represent repeated structures on the antigens. Also, for binding of labeled antibody to antigen-treated wells that were coated with 115D8 or M8 antibodies, the 115D8- and M8-defined epitopes must be coexpressed with NCRC-11 epitopes on at least a proportion of the NCRC-11 antigenic molecules isolated from the three tumor sources.

The major conclusion from these epitope mapping studies is that NCRC-11-defined antigens express a number of important epitopes defined by several well-characterized monoclonal antibodies. The relative levels of expression and topographical distribution of these epitopes in NCRC-11 antigens from breast and ovarian tumors are highly comparable, and differences in the expression of some epitopes may be accounted for by varying sialylation of NCRC-11 antigens in different tissues.

5.3. Epitope Analysis of NCRC-11 Antigens from Normal Sources

Since the NCRC-11 antibody reacted with bladder urothelium, it was reasoned that the NCRC-11 antigen should be present in urine from normal individuals as a component shed or secreted from viable cells. Similarly, since many anti-milk-fat-globule antibodies reacted with the purified NCRC-11 antigen, it was also considered likely that this antigen should be in human milk. Preliminary tests indicated that these proposals were correct, and NCRC-11 antigens were purified from both normal skimmed milk and urine using immunoadsorbent chromatography as applied to the isolation of antigen from tumors: again, the antigens isolated were high molecular weight glycoproteins (>400 kD) as judged by SDS–PAGE. The results of testing a panel of monoclonal antibodies against these antigens using the microradiosotopic antiglobulin assay are presented in Figure 3, and the profiles of antibody reactivity are compared with that obtained with NCRC-11 antigen from breast tumor. Essentially, the profiles were very similar but the reactivity of the NCRC-11 antibody with the milk antigen was reduced somewhat in comparison to its reaction with the other two antigens (Figure 3). The antigen from milk also bore epitopes for the antibodies M18, M24, and 115G2, which were absent (or masked, as with the M18 epitope) in the antigen from breast tumors. The antigens from urine and breast cancer were very similar with respect to their profile of epitopes.

5.4. Implications of the Expression of Multiple Epitopes on NCRC-11-Defined Antigens

Clearly, the finding that NCRC-11-defined antigens express an array of epitopes for a variety of monoclonal antibodies produced by different research

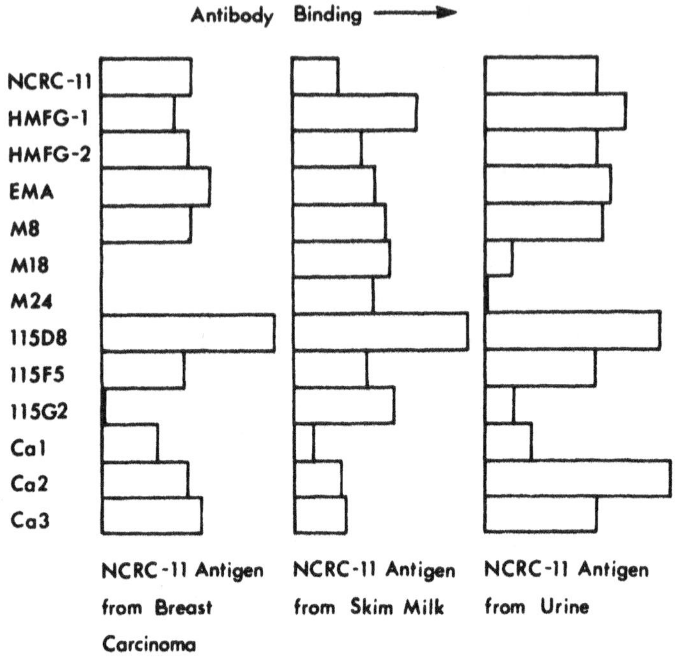

FIGURE 3. Profile of monoclonal antibody binding to immunoadsorbent purified NCRC-11-defined antigens isolated from breast carcinoma, skim milk, and urine.

groups reflects the powerful immunogenicity of this large glycoprotein or, more appropriately, family of high molecular weight glycoproteins. The NCRC-11-defined epitope appears most similar to that defined by the antibodies EMA and M8, although careful analysis of the distributions of the epitopes defined by these three antibodies, both on isolated antigens and with regard to antibody reactions with tissue sections, would suggest that they are not identical. Be that as it may, NCRC-11-defined antigens express epitopes that have been exploited for diverse purposes, which will be discussed in more detail in Section 7. Most of the antibodies are considered to react with carbohydrate epitopes (an exception being the M8 antibody; McIlhinney *et al.*, 1985), the evidence for this coming mainly from tests showing stability to heat and to the denaturing solvents employed for immunocytochemical staining of tissue sections, as well as in tests exploring their sensitivity to glycolytic enzymes or chemical treatments such as periodate oxidation. This would be entirely in accord with studies on the biochemical characterization of these high molecular weight glycoproteins, which have drawn attention to the extensive glycosylation of these antigens.

The present findings are also in accord with the proposal that NCRC-11 antigens from tumors contain epitopes that are associated with the state of differentiation of the tissue of origin, whereas other epitopes may represent determinants generated as a consequence of malignant change (Price *et al.*, 1985b, 1986a,b). For example, HMFG-1 epitopes are considered to be more representative of normal differentiation, while it has been suggested that HMFG-2 epitopes are more simple carbohydrates produced by the malignant cell in an abortive attempt at correct glycosylation (Burchell and Taylor-Papadimitriou, 1985). Obviously, antibodies produced against normal products such as milk fat globules are more likely to be representative of normal differentiation and antibodies reflecting malignant change or modification will be favored by immunization with tumor cells or materials derived from them.

6. FUNCTIONAL AND GENETIC ASPECTS OF THE EXPRESSION OF BREAST-CARCINOMA-ASSOCIATED ANTIGENS

6.1. Functional Aspects

There is passing reference in the literature to the possible biological function of the high molecular weight, mucinlike glycoproteins defined by a number of monoclonal antibodies and described in this chapter. Burchell *et al.* (1984) have speculated that the component of milk fat globule membranes that bears epitopes for HMFG-1 and HMFG-2 probably has a protective function when secreted onto the cell surface. Similarly, Shimizu and Yamauchi (1982) considered that the PAS-O molecule associated with milk fat globule membranes (and which is the membrane-associated counterpart of the high molecular weight antigen EMA; Ormerod *et al.*, 1985) also has a protective role. Since PAS-O was comparatively resistant to proteolytic enzymes such as pepsin, trypsin, and chymotrypsin, it was argued that it might be involved in protecting the surface of milk fat globule membranes against attack by proteases in milk and in the digestive tract. It was further suggested that PAS-O could be involved in the stabilization of milk fat emulsions in the digestive tract until milk fat is hydrolysed by lipases and absorbed in the intestine (Shimizu and Yamauchi, 1982).

Bramwell *et al.* (1983) attempted to resolve the function(s) of the high molecular weight mucins that have been defined by the Ca1 monoclonal antibody. It was reasoned that since the Ca1 antigen was a component of normal urothelium its function could be that of a classical mucin glycoprotein, which is to shield the epithelium that produces it from the toxic agents that would otherwise be destructive. From experiments involving the culture of HeLa cells in the presence of added urine, it was concluded that one function of the Ca1 antigen was to protect epithelial cells from the extremes of pH, particularly since the pH of shed urine can be as low as 4.5. It was further proposed that lactic acid, rather than hydrogen ion concentration alone, was more likely to be the inducer of Ca1 antigen synthesis since apparently lactate is generated by the glycolytic activity of the epithelial cells themselves (Mastroianni *et al.*, 1958). Of several cell lines grown in the presence of sodium lactate (10 mg/mL), a 20-fold increase in Ca1 antigen expression was induced in the bladder carcinoma line RT112/84, which normally expresses only low levels of the Ca1 antigen. Thus, with regard to the expression of the Ca1 antigen in tumors, it was concluded that while some malignant cells clearly synthesize the Ca1 antigen in vitro, others can be induced to do so by high lactate concentrations, and so Ca1 antigen in some malignant tumors may be a reflection of the metabolic conditions prevailing in those tumors (Bramwell *et al.*, 1983).

6.2. Genetic Aspects

Antibodies like Ca1 and NCRC-11 clearly react with high molecular weight components in normal urine so that urine has been used as a convenient source for antigen isolation (Bramwell *et al.*, 1983; Figure 3). Similar high molecular weight urinary glycoconjugates have previously been characterized by their reaction with peanut agglutinin as assessed by SDS–PAGE separation of urine followed by electrophoretic transfer onto nitrocellulose paper and detection with radioiodinated peanut lectin (Karlsson *et al.*, 1983). At least four components of different mobility were identified with each individual giving one or two bands. Figure 4 is a diagrammatic representation of the possible patterns showing the different phenotypes of lectin-binding urinary mucins, and the numbers identifying the bands and describing the phenotypes, as proposed by Karlsson *et al.* (1983), are shown. The results of family studies demonstrated that the patterns display Mendelian inheritance and that they are consistent with the notion that the phenotypes are determined by four codominant alleles at an autosomal locus. Furthermore, from population studies of a sample of 118 individuals, the observed distribution of phenotypes lies close to that expected for a population in a Hardy–Weinberg equilibrium, assuming an autosomal codominant inheritance.

Now the relevance of this genetically determined polymorphism to breast carcinoma antigens is that the monoclonal antibodies HMFG-1, HMFG-2, Ca1,

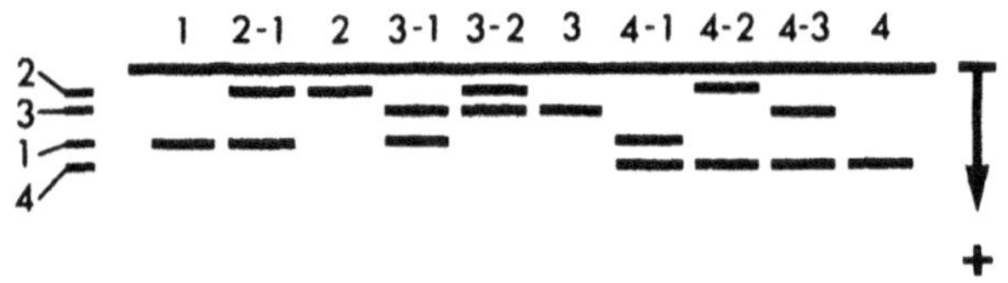

FIGURE 4. Diagrammatic representation of the possible phenotypes of peanut lectin binding glycoproteins in urine detected following SDS–PAGE and Western blotting (Karlsson *et al.*, 1983).

Ca2, and Ca3 all reacted with this family of high molecular weight urinary mucins (Swallow *et al.*, 1986). In addition, using a panel of urines of defined phenotype, kindly provided by Dr. D. Swallow, it was possible to show that the NCRC-11 antibody both reacted with the high molecular weight urinary components and had the capacity to distinguish between the various phenotypes. This observation provided some explanation for the slight differences in apparent molecular weight observed for the large glycoprotein antigens defined by various monoclonal antibodies and also for the fact that the antigens involved were often represented by a doublet as resolved by Western blotting techniques.

These findings begged the question whether individuals of a specific phenotype might be more susceptible to cancer. This is feasible since, if this class of glycoprotein has a "protective" function (as discussed above), the different molecular phenotypes expressed in the normal tissues with which these molecules are associated (Table II) might confer preferential resistance or susceptibility within those tissues to etiologic agents of cancer. Experimental investigation of these proposals is at a very early stage of development but two points are emerging from an analysis of the phenotypes of urines from a small sample of 35 breast cancer patients. As shown in Table VIII, the observed distribution of NCRC-11-defined glycoprotein phenotypes lies close to that expected and calculated for a population in a Hardy–Weinberg equilibrium, assuming codominant

Table VIII
Distribution of NCRC-11-Defined Glycoprotein Phenotypes in Urine Samples from 35 Breast Cancer Patients

Phenotype	1	2 − 1	2	3 − 1	3 − 2	3	4 − 1	4 − 2	4 − 3	4	Total
Observed number (O)	6	7	6	6	2	1	4	1	2	0	35
Expected number (E)	5.9	8.9	3.4	4.9	3.7	1.0	2.9	2.2	1.2	0.4	35
$\frac{(O-E)^2}{E}$	0.001	0.400	1.988	0.247	0.781	0.000	0.417	0.655	0.533	0.400	

$\chi^2 = 5.42;\ p > 0.05$

The expected distribution was calculated assuming Hardy–Weinberg equilibrium and the observed frequencies of alleles for the NCRC-11-defined urinary glycoproteins (bands 1–4) being:

1 = 0.41; 2 = 0.31; 3 = 0.17; 4 = 0.10

PUM (peanut-reactive urinary mucin) gene frequencies in a population of 118 normal individuals are as follows[a]:

1 = 0.42; 2 = 0.33; 3 = 0.20; 4 = 0.06

[a]From Karlsson *et al.* (1983).

inheritance. In addition, it should be noted that the frequencies of alleles for the NCRC-11-defined urinary glycoproteins (bands 1–4) were virtually identical to those described for the peanut-reactive urinary mucins in a population of normal individuals (Table VIII; Karlsson *et al.*, 1983).

Further studies are required to assess the biological significance of this novel class of genetically determined polymorphic components demonstrable by their reaction with particular lectins and breast cancer monoclonal antibodies.

7. DIAGNOSTIC, PROGNOSTIC, AND THERAPEUTIC APPLICATIONS OF MONOCLONAL ANTIBODIES AGAINST BREAST-CARCINOMA-ASSOCIATED ANTIGENS

7.1. Immunocytochemistry

In a study of 126 women with primary breast carcinoma, a highly significant relation was found between the proportion of cells staining with the NCRC-11 antibody on paraffin sections of tumors and patient survival (with a minimum follow-up time of 5 years or to death) (Ellis *et al.*, 1985). Staining related weakly to histological grade but not significantly to estrogen receptor status or pathological lymph node stage. Furthermore, a Cox multivariate analysis indicated that the relation of staining to survival was independent of other known prognostic factors (Ellis *et al.*, 1985). This investigation has now been extended to include a total of 435 patients and the relation between NCRC-11 antibody staining and survival has remained significant (I. O. Ellis, unpublished observations).

The most appealing interpretation of these observations is that the expression of NCRC-11 epitopes on tumors reflects a normal differentiation event, the retention of which is indicative of a favorable prognosis whereas its loss is associated with a poor prognosis. Similar proposals have been made in studies on the reactivity of anti-milk-fat-globule membrane antibodies with breast carcinomas, although since antigen expression with epithelial tumors is often heterogeneous (Edwards, 1985; Edwards and Brooks, 1984), it is still uncertain to what extent particular epitopes are associated with normal differentiation. For example, a high expression of the HMFG-1 determinant as an extracellular component in primary breast tumors probably reflects the ability of the tumors to secrete surface mucinlike glycoproteins with completed carbohydrate chains (Wilkinson *et al.*, 1984). Thus, positive extracellular straining with HMFG-1 was associated with a favorable prognosis and complete absence of staining was indicative of a poor prognosis. These findings were not confirmed in another study with HMFG-1, although different criteria were adopted for assessing immunohistochemical staining (Berry *et al.*, 1985).

The expression of HMFG-2 epitopes is not considered to be associated with differentiation but instead represents abortive processing of carbohydrate chains that may be related to invasive potential (Burchell *et al.*, 1983, 1984; Wilkinson *et al.*, 1984). Thus, HMFG-2 staining of tumors is apparently without prognostic significance (Berry *et al.*, 1985; Wilkinson *et al.*, 1984).

Since antibodies (monoclonal and polyclonal) against milk fat globule membrane components react characteristically with cells of epithelial origin, they have a number of important applications in diagnostic pathology. These include the detection of bone marrow and lymph node matastases from breast carcinoma (Dearnaley *et al.*, 1981; Heyderman *et al.*, 1979; Sloane and Ormerod, 1981), for the identification of malignant cells in serous effusions (Epenetos *et al.*, 1982b), and for the differential diagnosis of anaplastic tumors (Gatter *et al.*, 1984). In addition, monoclonal antibodies against metastatic breast carcinoma have been employed for evaluating the nature of breast nodules using fine-needle aspirate cytology (Nuti *et al.*, 1986). Finally, the use of "cocktails" of antibodies to increase staining intensities is only now beginning to be investigated (Heyderman *et al.*, 1985) and may offer a means for improved discrimination in diagnostic histopathology.

7.2. Serum Immunoassays

Since epitopes on the high molecular weight glycoprotein antigens associated with breast carcinomas represent repeated structures of those antigens, it is feasible to employ "sandwich" immunoassays (Figure 1) for the detection of those antigens in body fluids and, in particular, in the sera of cancer patients. Both HMFG-1 and HMFG-2 have been assayed in patients' sera, and 30% of sera from advanced breast cancer patients contained elevated levels of HMFG-1 antigen as compared with 6% of sera from healthy control women, whereas 53% of patients showed elevated levels of HMFG-2 antigen compared with 17% of controls (Burchell *et al.*, 1984). By immune blotting techniques, it was found that both antibodies react with similar high molecular weight components in tumor tissue and serum, and the variations in the banding patterns shown in the immune blots presented closely resemble those obtained in studies on the polymorphic expression of high molecular weight urinary glycoproteins in that single bands and doublets, with slight variations in mobility in the gel, were apparent (Burchell *et al.*, 1984).

When the NCRC-11 antibody was employed in sandwich immunoassays to detect circulating antigens in breast cancer patients, over 40% (28/69) of samples displayed elevated levels of NCRC-11 antigen as compared with 0% of sera from the healthy control group (0/60) (Figure 5). Again, high molecular weight polymorphic forms of the NCRC-11 antigen were detected in sera from cancer patients but not from normal individuals by SDS–PAGE, Western blotting, and

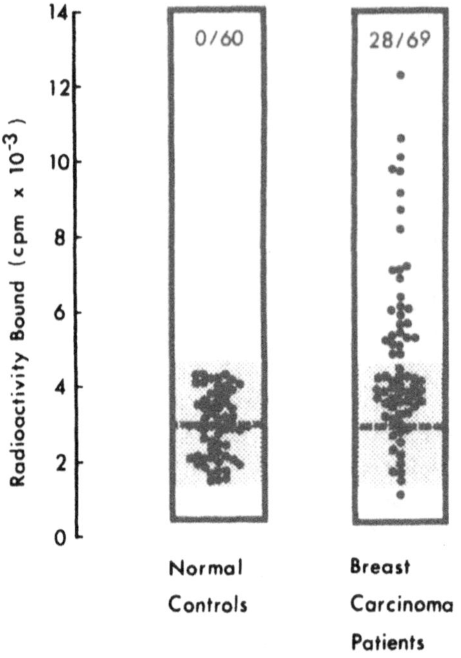

FIGURE 5. Levels of NCRC-11 antibody binding activity in the serum of normal healthy control women and patients with breast cancer as assessed using the solid-phase sandwich immunoassay.

NCRC-11 staining of the nitrocellulose sheets. At the present stage of development, the immunoassay for NCRC-11 antigen in breast cancer patients' sera displays excellent specificity for tumors, but clearly the sensitivity of the assay needs improvement.

Sandwich immunoassays, based on the use of 115D8, have also been employed to detect the so-called MAM-6 antigen in the sera of breast cancer patients (Hilkens *et al.*, 1985, 1986). Elevated levels of the antigen were found with increasing frequency according to the staging of the disease so that 79% of sera from patients with advanced breast cancer (Stage IV) were positive in this assay. The potential value of this serum marker was illustrated by showing that increasing or decreasing MAM-6 levels correlated with breast cancer progression or regression, indicating that the assay can be used to monitor the course of the disease during therapy (Hilkens *et al.*, 1986).

7.3. Immunoscintigraphy and Antibody-Directed Therapy

It would appear that the developing and invasive tumor disrupts normal tissue architecture sufficiently such that tumor products may gain access to the circulation. Thus, investigations have established that antigens defined by monoclonal antibodies may be present in the sera of breast cancer patients (Section

7.2). Conversely, it follows that products such as antitumor antibodies, administered systemically, should have access to the tumor, with the proviso that they are not rapidly eliminated. In the healthy individual, many of the "tumor" antigens discussed have a specific distribution in normal tissues being confined predominantly to the luminal surface of gland epithelium (Table II). However, it should be noted that in these locations, secreted or shed antigen is not directed into the circulatory system but, as already described, into fluids such as urine or milk. Therefore, these components should not interfere with the tumor-localizing capacity of systemically administered antibodies, and antibodies against milk fat globule membranes, for example, should have the capacity to localize in tumors.

These proposals have already been evaluated in clinical studies. Thus, Rainsbury *et al.* (1983) have successfully imaged metastatic breast carcinomas using ^{111}In-labeled M8 antibody. In this investigation, labeled antibody localized to the primary tumor and skeletal metastases were identified, although soft tissue metastases were not detected.

The HMFG-2 antibody has been used effectively for the localization of tumors of ovarian origin and that express antigens related to breast differentiation (Epenetos *et al.*, 1982a). In addition, imaging with ^{123}I-labeled HMFG-2 has been employed to assess the response to chemotherapy and to provide evidence of recurrence of known ovarian cancer (Britton *et al.*, 1985). An important clinical requirement with this type of evaluation is that the labeled antibody is taken up by viable and not necrotic cells and the results of this study with HMFG-2 suggested that this was indeed the case. The application of gamma camera immunoscintigraphy for external radiodiagnostic imaging of ovarian tumors presents fewer problems in comparison to imaging of primary breast tumors and their local recurrences. These, being close to the central blood pool, are often obscured by the nonlocalized radioactivity present in the circulation. The use of antibody fragments here, which are cleared more rapidly from the circulation than intact immunoglobulins, represents one approach for increasing tumor/nontumor ratios of radioactivity and hence improving resolution of tumors.

Developing from studies on the localization of tumors with radiolabeled antibodies, Epenetos (1984) has employed the HMFG-2 antibody as a vehicle for delivering therapeutic doses of radiation to tumors that were exposed to high levels of radioactivity by administering ^{131}I-labeled HMFG-2 directly into the appropriate body compartment (i.e., regional antibody-guided irradiation). Subsequently, in a larger clinical study, further encouraging results were obtained using antibody-guided irradiation of malignant pleural and pericardial effusions (Pectasides *et al.*, 1986).

The development of antibodies for the targeting of toxins or cytotoxic drugs to tumors is without doubt a major objective in new therapeutic strategies (Baldwin, 1985). Clearly, antibodies showing greater specificity for tumors would increase the possibilities for eliminating or minimizing adverse side reactions.

8. CONCLUSIONS

At present, there are few studies that would confirm the existence of tumor antigens with a unique association to breast cancer. There is, however, an increasing realization that epigenetic modifications of normal tissue components, as synthesized by the malignant cell, lead to the generation of epitopes that may be exploited as tumor-associated antigens. For example, the loss or reduction in glycosyl transferases in tumors results in the accumulation of incomplete precursor molecules that appear elevated in tumors. Even epitopes related to normal breast differentiation, such as those defined by anti-milk-fat-globule antibodies, may be exploited for antibody targeting of radionuclides for diagnostic and therapeutic purposes. Since these antibodies may be reactive with a variety of tumors, this further increases the scope for developing clinical applications for individual antibodies.

The antibodies described in this chapter are directed against a class of high molecular weight glycoproteins that evidently are potent immunogens in the mice employed for their original production. Therefore, to develop more specific antibodies, or at least antibodies reactive with other putative tumor antigens, perhaps one approach would be to eliminate these molecules from the initial immunizing material or induce tolerance in the mice against these antigens. With this approach, it should be feasible to evaluate a whole range of less immunodominant molecules as candidates for breast-carcinoma-associated antigens.

ACKNOWLEDGMENTS. These studies were supported by the Cancer Research Campaign.

Sincere thanks are expressed to the following for providing samples of their monoclonal antibodies: J. Taylor-Papadimitriou and J. Burchell for HMFG-1 and HMFG-2; J. Hilgers and J. Hilkens for 115D8, 115F5, and 115G2; P. A. W. Edwards for M8, M18, and M24; M. Bramwell, W. D. Smith, and H. Harris for Ca1, Ca2, and Ca3; M. I. Colnaghi for MBr1, MBr8, MOv2, MOv8, and MOv15; and G. Rowland for 11.285.14.

The skilful technical assistance of Susan Edwards and Glenn Crocker is gratefully acknowledged. Helen Beverley-Clarke is thanked for providing purified preparations of the NCRC-11 antibody.

9. REFERENCES

Arklie, J., Taylor-Papadimitriou, J., Bodmer, W., Egan, M., and Millis, R., 1981, Differentiation antigens expressed by epithelial cells in the lactating breast are also detectable in breast cancers. *Int. J. Cancer* **28**:23–29.

Ashall, F., Bramwell, M. E., and Harris, H., 1982, A new marker for human cancer cells. 1. The Ca antigen and the Ca1 antibody. *Lancet* **ii**:1–6.

Baldwin, R. W., 1985, Monoclonal antibody targeting of anti-cancer agents: Mühlbock Memorial Lecture. *Eur. J. Cancer Clin. Oncol.* **11**:1281–1285.

Berry, N., Jones, D. B., Smallwood, J., Taylor, I., Kirkham, N., and Taylor-Papadimitriou, J., 1985, The prognostic value of the monoclonal antibodies HMFG-1 and HMFG-2 in breast cancer. *Br. J. Cancer* **51**:179–186.

Bramwell, M. E., Bhavanandan, V. P., Wiseman, G. and Harris, H., 1983, Structure and function of the Ca1 antigen. *Br. J. Cancer* **48**:177–183.

Bramwell, M. E., Ghosh, A. K., Smith, W. D., Wiseman, S., Spriggs, A., and Harris, H., 1985, Ca2 and Ca3. New monoclonal antibodies as tumor markers in serous effusions. *Cancer* **56**:105–110.

Britton, K. E., Granowska, M., and Shepherd, J., 1985, Localisation of cancer of the ovary and metastases using ^{123}I-labelled monoclonal antibody HMFG-2 compared to surgical findings. In *Monoclonal Antibodies for Cancer Detection and Therapy* (R. W. Baldwin and V. S. Byers, eds.), pp. 201–221, Academic Press, London.

Burchell, J. M., and Taylor-Papadimitriou, J., 1985, Monoclonal antibodies to breast cancer and their application. In *Monoclonal Antibodies for Cancer Detection and Therapy* (R. W. Baldwin and V. S. Byers, eds.), pp. 1–15, Academic Press, London.

Burchell, J., Durbin, H., and Taylor-Papadimitriou, J., 1983, Complexity of expression of antigenic determinants recognized by monoclonal antibodies HMFG-1 and HMFG-2, in normal and malignant human mammary epithelial cells. *J. Immunol.* **131**:508–513.

Burchell, J., Wang, D., and Taylor-Papadimitriou, J., 1984, Detection of the tumour-associated antigens recognized by the monoclonal antibodies HMFG-1 and 2 in serum from patients with breast cancer. *Int. J. Cancer* **34**:763–768.

Campbell, A. M., McCormack, M. A., Ross, C. A., and Leake, R. E., 1986, Immunological analysis of the specificity of the autologous humoral response in breast cancer patients. *Br. J. Cancer* **53**:7–11.

Canevari, S., Fossati, G., Balsari, A., Sonnino, S., and Colnaghi, M. I., 1983, Immunochemical analysis of the determinant recognized by a monoclonal antibody (MBr1) which specifically binds to human mammary epithelial cells. *Cancer Res.* **43**:1301–1305.

Ceriani, R. L., Thompson, K., Peterson, J. A., and Abraham, S., 1977, Surface differentiation antigens of human mammary epithelial cells carried on the human milk fat globule. *Proc. Natl. Acad. Sci. U.S.A.* **74**:582–586.

Cordell, J., Richardson, T. C., Pulford, K. A. F., Ghosh, A. K., Gatter, K. C., Heyderman, E., and Mason, D. Y., 1985, Production of monoclonal antibodies against human epithelial membrane antigen for use in diagnostic immunocytochemistry. *Br. J. Cancer* **52**:347–354.

Corvalan, J. R. F., Axton, C. A., Brandon, D. R., Smith, W., and Woodhouse, C., 1984, Classification of anti-CEA monoclonal antibodies. In *Protides of the Biological Fluids* (H. Peeters, ed.), Vol. 31, pp. 921–924, Pergamon Press, Oxford.

Dearnaley, D. P., Sloane, J. P., Ormerod, M. G., Steele, K., Coombes, R. C., Clink, H. Mc.D., Powles, T. J., Ford, H. T., Gazet, J.-C., and Neville, A. M., 1981, Increased detection of mammary carcinoma cells in marrow smears using antisera to epithelial membrane antigen. *Br. J. Cancer* **44**:85–90.

De Kretser, T., Thorne, H. J., Picone, D., and Jose, D. G., 1986, Biochemical characterization of the monoclonal antibody-defined ovarian carcinoma-associated antigen SGA. *Int. J. Cancer* **37**:705–712.

Edwards, D. P., Grzyb, K. T., Dressler, L. G., Mansel, R. E., Zava, D. T., Sledge, G. W., and McGuire, W. L., 1986, Monoclonal antibody identification and characterization of a M_r43,000 membrane glycoprotein associated with human breast cancer. *Cancer Res.* **46**:1306–1317.

Edwards, P. A. W., 1985, Heterogeneous expression of cell surface antigens in normal epithelia and their tumours, revealed by monoclonal antibodies. *Br. J. Cancer* **51**:149–160.

Edwards, P. A. W., and Brooks, I. M., 1984, Antigenic subsets of human breast epithelial cells distinguished by monoclonal antibodies. *J. Histochem. Cytochem.* **32**:531–537.

Ellis, I. O., Robins, R. A., Elston, C. W., Blamey, R. W., Ferry, B., and Baldwin, R. W., 1984, A monoclonal antibody, NCRC-11, raised to human breast carcinoma. I. Production and immunohistological characterization. *Histopathology* **8**:501–506.

Ellis, I. O., Hinton, C. P., MacNay, J., Robins, R. A., Elston, C. W., Owainati, A., Blamey, R. W., Baldwin, R. W., and Ferry, B., 1985, Immunocytochemical staining of breast carcinoma with the monoclonal antibody NCRC-11—A new prognostic indicator. *Br. Med. J.* **290**:881–883.

Epenetos, A. A., 1984, Antibody guided irradiation of malignant lesions: Three cases illustrating a new method of treatment—A report of the Hammersmith Oncology Group and the Imperial Cancer Research Fund. *Lancet* **i**:1441–1443.

Epenetos, A. A., Britton, K. E., Mather, S., Shepherd, J., Granowska, M., Taylor-Papadimitriou, J., Nimmon, C. C., Durbin, H., Hawkins, L. R., Malpas, J. S., and Bodmer, W., 1982a, Targeting of iodine-123-labelled tumour-associated monoclonal antibodies to ovarian, breast, and gastrointestinal tumours. *Lancet* **ii**:999–1004.

Epenetos, A. A., Canti, G., Taylor-Papadimitriou, J., Curling, M., and Bodmer, W., 1982b, Use of two epithelium specific monoclonal antibodies for diagnosis of malignancy in serous effusions, *Lancet* **ii**:1004–1006.

Foster, C. S., and Neville, A. M., 1984, Monoclonal antibodies to the human mammary gland: III. Monoclonal antibody LICR-LON-M18 identifies impaired expression and excess sialylation of the I(Ma) cell-surface antigen by primary breast carcinoma cells, *Hum. Pathol.* **15**:502–513.

Garcia, M., Capony, F., Derocq, D., Simon, D., Pau, B., and Rochefort, H., 1985, Characterization of monoclonal antibodies to the estrogen-regulated M_r52,000 glycoprotein and their use in MCF-7 cells. *Cancer Res.* **45**:709–716.

Gatter, K. C., Alcock, C., Heryet, A., Pulford, K. A., Heyderman, E., Taylor-Papadimitriou, J., Stein, H., and Mason, D., 1984, The differential diagnosis of routinely processed anaplastic tumors using monoclonal antibodies. *Am. J. Clin. Pathol.* **82**:33–43.

Heyderman, E., Steele, K., and Ormerod, M. G., 1979, A new antigen on the epithelial membrane: Its immunoperoxidase localization in normal and neoplastic tissues. *J. Clin. Pathol.* **32**:35–39.

Heyderman, E., Strudley, I., Powell, G., Richardson, T. C., and Mason, D., 1985, A new monoclonal antibody to epithelial membrane antigen (EMA)—E29. A comparison of its immunocytochemical reactivity with polyclonal antibodies and with another monoclonal antibody, HMFG-2. *Br. J. Cancer* **52**:355–361.

Hilkens, J., Buijs, F., Hilgers, J., Hagemann, Ph., Calafat, J., Sonneberg, A., and Van der Valk, M., 1984, Monoclonal antibodies against human milk-fat globule membranes detecting differentiation antigens of the mammary gland and its tumors. *Int. J. Cancer* **34**:197–206.

Hilkens, J., Kroezen, V., Buijs, F. Hilgers, J., Van Vleigt, M., De Voogd, W., Bonfrer, J., and Bruning, P. F., 1985, MAM-6, A carcinoma associated marker: Preliminary characterization and detection in sera of breast cancer patients. In *Proceedings of the International Workshop on Monoclonal Antibodies and Breast Cancer* (R. L. Ceriani, ed.), pp. 28–42, Martinus Nijhoff, Boston.

Hilkens, J., Kroezen, V., Bonfrer, J. M. G., De Jong-Bakker, M., and Bruning, P. F., 1986, MAM-6 antigen, a new serum marker for breast cancer monitoring, *Cancer Res.* **46**:2582–2587.

Janossy, G. J., Thomas, J. A., Pizzolo, G., Granger, S. M., McLaughlin, J., Stansfeld, A. G., and Sloane, J., 1980, Immuno-histological diagnosis of lymphoproliferative diseases by selected combinations of antisera and monoclonal antibodies, *Br. J. Cancer* **42**:224–242.

Karlsson, S., Swallow, D. M., Griffiths, R., Corney, G., Hopkinson, D. A., Dawney, A., and Carton, J. P., 1983, A genetic polymorphism of a human urinary mucin. *Ann. Hum. Genet.* **47**:263–269.

Köhler, G., and Milstein, C., 1975, Continuous cultures of fused cells secreting antibody of predefined specificity. *Nature* **256**:495–497.

Mastroianni, L., Winternitz, W. W., and Lowi, N. P., 1958, The in vitro metabolism of the human endosalpinx. *Fertil. Steril.* **9**:500–508.

McGee, J. O'D., Woods, J. C., Ashall, F., Bramwell, M. E., and Harris, H., 1982, A new marker for human cancer cells. 2. Immunohistochemical detection of the Ca antigen in human tissues with the Ca1 antibody. *Lancet* **ii**:7–10.

McIlhinney, R. A. J., Patel, S., and Gore, M. E., 1985, Monoclonal antibodies recognizing epitopes carried on both glycolipids and glycoproteins of the human milk fat globule membrane, *Biochem. J.* **227**:155–162.

Ménard, S., Tagliabue, E., Canevari, S., Fossati, G., and Colnaghi, M. I., 1983, Generation of monoclonal antibodies reacting with normal and cancer cells of human breast. *Cancer Res.* **43**:1295–1300.

Miotti, S., Aguanno, S., Canevari, S., Diotti, A., Orlandi, R., Sonnino, S., and Colnaghi, M. I., 1985, Biochemical analysis of human ovarian cancer-associated antigens defined by murine monoclonal antibodies. *Cancer Res.* **45**:826–832.

Nuti, M., Mottolese, M., Viora, M., Perrone Donnorso, R., Schlom, J., and Natali, P. G., 1986, Use of monoclonal antibodies to human breast tumor associated antigens in fine needle aspirate cytology. *Int. J. Cancer* **37**:493–498.

Ormerod, M. G., Steele, K., Westwood, J. H., and Mazzini, M. N., 1983, Epithelial membrane antigen: Partial purification, assay and properties. *Br. J. Cancer* **48**:533–541.

Ormerod, M. G., McIlhinney, J., Steele, K., and Shimizu, M., 1985, Glycoprotein PAS-O from the human milk fat globule membrane carries determinants for epithelial membrane antigen. *Molec. Immunol.* **22**:265–269.

Papsidero, L. D., and Johnson, E. A., 1986, Physicochemical purification and immunological characteristics of ductal carcinoma antigen. *Int. J. Cancer* **37**:697–703.

Papsidero, L. D., Croghan, G. A., O'Connel, M. J., Valenzuela, L. A., Nemoto, T., and Ming Chu, T., 1983, Monoclonal antibodies (F36/22 and M7/105) to human breast carcinoma. *Cancer Res.* **43**:1741–1747.

Pectasides, D., Stewart, S., Courtney-Luck, N., Rampling, R., Munro, A. J., Krausz, T., Dhokia, B., Snook, D., Hooker, G., Durbin, H., Taylor-Papadimitriou, J., Bodmer, W. and Epenetos, A. A., 1986, Antibody-guided irradiation of malignant pleural and pericardial effusions. *Br. J. Cancer* **53**:727–732.

Price, M. R., Brown, A., Armitage, N. C., and Baldwin, R. W., 1985a, Application of a microradioimmunoassay to the analysis of monoclonal antibody-defined epitopes on antigen preparations from human colonic cancer. *IRCS Med. Sci.* **13**:366–367.

Price, M. R., Edwards, S., Owainati, A., Bullock, J. E., Ferry, B., Robins, R. A., and Baldwin, R. W., 1985b, Multiple epitopes on a human breast carcinoma associated antigen. *Int. J. Cancer* **36**:567–574.

Price, M. R., Edwards, S., Powell, M., and Baldwin, R. W., 1986a, Epitope analysis of monoclonal antibody NCRC-11 defined antigen isolated from human ovarian and breast carcinomas. *Br. J. Cancer* **54**:393–400.

Price, M. R., Edwards, S., Robins, R. A., Hilgers, J., Hilkens, J., and Baldwin, R. W., 1986b, Epitopes with diagnostic and prognostic significance co-expressed on a human breast carcinoma-associated antigen. *Eur. J. Cancer Clin. Oncol.* **22**:115–117.

Rainsbury, R. M., Westwood, J. H., Coombes, R. C., Neville, A. M., Ott, R. J., Kalirai, T. S., McCready, V. R., and Gazet, J.-C., 1983, Location of metastatic breast carcinoma by monoclonal antibody chelate labelled with indium-111, *Lancet* **ii**:934–939.

Schlom, J., Wunderlich, D., and Teramoto, V. A., 1980, Generation of human monoclonal antibodies reactive with human mammary carcinoma cells. *Proc. Natl. Acad. Sci. U.S.A.* **77**:6841–6845.

Shimizu, M., and Yamauchi, K., 1982, Isolation and characterization of mucin-like glycoprotein in human milk fat globule membrane. *J. Biochem.* **91**:515–524.

Sloane, J. P., and Ormerod, M. G., 1981, Distribution of epithelial membrane antigen in normal and neoplastic tissues and its value in diagnostic tumour pathology. *Cancer* **47**:1786–1795.

Swallow, D. M., Griffiths, B., Bramwell, M., Wiseman, G., and Burchell, J., 1986, Detection of the urinary "PUM" polymorphism by the tumour binding monoclonal antibodies Ca1, Ca2, Ca3, HMFG-1 and HMFG-2, *Disease Markers* **4**:247–254.

Tagnon, H. J., 1986, Some changing concepts of the natural history of human mammary cancer and their effect on diagnosis and treatment. *Eur. J. Cancer Clin. Oncol.* **22**:123–128.

Wilkinson, M. J. S., Howell, A., Harris, M., Taylor-Papadimitriou, J., Swindell, R., and Sellwood, R. A., 1984, The prognostic significance of two epithelial membrane antigens expressed by human mammary carcinomas. *Int. J. Cancer* **33**:299–304.

Chapter 2

The Immunochemistry of Some Blood Group Antigens

Relation to Erythrocyte Membrane Structure and to Hemagglutination

A. H. Merry, H. C. Gooi, and D. Voak

1. INTRODUCTION

Certain antigens expressed on the erythrocyte surface of one individual may be recognized by antibodies in the serum of another individual, and some of these are commonly known as blood group antigens. In other species, some of the same antigens, for example, the ABH antigens, are only expressed on non-erythroid cells (Oriel *et al.*, 1986). Indeed, in humans several of the so-called blood group antigens are not only expressed on erythrocytes but also on a variety of other cell types (e.g., the expression of ABH and Lewis antigens on intestinal cells described in Chapter 4). Some antigens, such as those of the Ii system, may also be expressed on soluble glycoproteins (Burnie, 1973; DeBoissezon *et al.*, 1970) or glycolipids in the plasma or secretions as are those of the Lewis system (Marcus and Cass, 1969; Mollison *et al.*, 1963; Sneath and Sneath, 1955). In this

A. H. Merry Blood Group Reference Laboratory, Radcliffe Infirmary, Oxford OX2 6HE, United Kingdom. **H. C. Gooi** Regional Blood Transfusion Centre, Leeds LS15 7TW, United Kingdom. **D. Voak** Regional Transfusion and Immuno-Haematology Centre, Cambridge CB2 2PT, United Kingdom.

chapter, however, we propose to concentrate on the blood group antigens that are present on the erythrocyte membrane.

1.1. Blood Transfusion and Blood Group Antigens

The discovery of the ABO blood group system by Landsteiner at the start of this century (Landsteiner, 1901) led to the demonstration that it was possible to carry out safely the transfusion of blood from one individual to another, provided a cross-matching test for ABO compatibility between donor and recipient was performed. The presence of an antibody in the serum of the recipient to the donor erythrocytes does not always indicate that the transfusion will be incompatible as some antibodies such as low titre anti-I rarely cause destruction of the donor cells because such antibodies do not generally bind to erythrocytes at 37°C although they are detectable in serological tests carried out at lower temperatures. It is important, however, that the specificity of any antibody found in cross matching is always determined. The ABO cross-matching test can be performed quite simply since the anti-A and anti-B antibodies can generally be detected by their ability to agglutinate cells and the clumps of cells are easily recognized. This discovery heralded the introduction of blood transfusion into clinical medicine. (The reader is referred to Mollison, 1983, for a detailed account of the development of blood transfusion.) It was also demonstrated that the ABO blood groups were Mendelian inherited characteristics (von Dungern and Hirzfeld, 1910). It is now recognized that there are a large number of other blood group antigens that are also present. The discovery and occurrence of blood group antigens are detailed elsewhere (Issitt and Issitt, 1975; Mollison, 1984; Race and Sanger, 1975).

1.2. Blood Group Systems

When the inheritance of these antigens is considered, it is possible to show that some of the antigens are related to one another, and they may be considered part of the same blood group system. Apart from the ABH system other major blood group systems are Rh, Lewis, Ii, P, MNS, Lutheran, Kell, and Kidd. Sometimes the antigens have been shown to be independent of any of these systems. These systems are defined by the reactions of antibodies to the corresponding antigens, and the classifications were generally made before elucidation of the immunochemistry of the antigens. The antibodies that define the presence of antigens may be "naturally occurring," which simply means that the nature of the immunizing agent is not certain but occurs early in life. Such naturally occurring ABO antibodies are generally found in the absence of the corresponding antigens since these antigens are ubiquitously distributed in nature, for example, in several strains of bacteria and in plants (Springer *et al.*, 1962) and the antibodies are produced in response to stimulation by these antifollowing blood transfusion or the exposure to fetal blood during pregnancy, for example, anti-D. In some cases, they may also be produced by the injection of human erythrocytes or blood group substances into animals, for example, anti-M and anti-N (Landsteiner and Levine, 1927). Certain membrane components have now been shown to be associated with specific blood group activities. Some of these are summarized in Table I.

The advent of hybridoma technology for producing monoclonal antibodies with defined specificities has led to many applications both in routine blood group serology (Parsons, 1985; Voak *et al.*, 1983) and in research into the immunochemistry of blood group antigens (Anstee and Edwards, 1982; Parsons, 1985) (see Section 4.3 and Table IX).

In addition to the antigens assigned to the major blood group systems, there are also those that are found very infrequently (sometimes called "private"

Table I
Erythrocyte Membrane Components Associated with Blood Group Antigen Activity[a]

Component	Blood group system	Major antigens	Apparent molecular weight	Approximate abundance (copies per cell)	
				Component	Antigen
Glycolipid	ABH	A_1		>1,000,000	1,000,000[b]
		A_2			250,000
		B			750,000
		H			1,000,000(0 cells)
	P	P		~10,000,000	(globoside)
		P^k		~3,000,000	(trihexose ceramide)
		P^1			
	I	I,i			
	Lewis	Le^a,Le^b			4,500–7,300(Le^a)
Band 3	ABH, Ii		90,000–100,000	~1,000,000	1–500,000[c]
Band 4.5	ABH, Ii		55,000	50,000	
Alpha SGP	MN	M,N	43,000	600,000	(dimers)
Delta SGP	Ss	S,s	25,000	80–100,000	12,000(s)
Beta SGP	Gerbich		39,000	50–70,000	
Peptide(s)	Rh	D	~30,000		9,900–33,000
		-D-			88–170,000
		D^{VI}			1,700–3,200
		D^u			110–8,000
Peptide	Rh	C,c	~30,000		35,000–85,000
Peptide	Rh	E,e	~30,000		13,499–24,000
Glycopeptide	Duffy	Fy^a,Fy^b	40,000–50,000		≈15,000
Glycopeptide (DAF)	Cromer	Cr^a,Tc^a,Dr^a, Es^a,WES^b,IFC	70,000	10–15,000	10–15,000(IFC)
(Glyco)peptide	Kell	K1,K2,K7, K14,K22	93,000	3.1–5,900	3–6,000(K)
Glycopeptide	LW	LW^a,LW^b	37,000–47,000		2,500–5,000
Glycopeptide	Lutheran	Lu^a,Lu^b	78 + 85,000		750–4,000(Lu^b)

[a]Adapted with permission from Anstee (1987).
[b]Some 70% of A antigen is thought to be on glycoprotein and the remainder on glycolipid.
[c]Value for I antigen on adult cells, most of the I antigen thought to be on glycoprotein.

antigens) or those that have a very high frequency and are almost always present (sometimes called "public" antigens). Examples of low-frequency antigens are Webb (Wb) and Wright (Wr^a) and examples of high-frequency antigens are Gerbich (Ge) and Cromer. The immunochemistry of the Webb and Gerbich antigens is discussed in Section 3.2.2.2. For further details of high- and low-frequency antigens see Race and Sanger (1975).

The techniques employed in the identification of antibodies and of blood group antigens are mainly those of blood group serology as discussed in Section 4. Since the original discovery of blood group antigens there has been considerable speculation on the nature of these antigens, which has prompted research into the components of the erythrocyte membrane that carry blood group antigens. Biochemical and immunochemical techniques are now being used increasingly to characterize blood group antigen structures. A considerable body of knowledge has now been accumulated on the structure of certain components with blood group antigen activity—notably those that carry antigens of the ABH, Lewis, P, Ii, and MNS systems—and this will be discussed in detail. In comparison with most other types of cell, it is possible to obtain relatively large quantities of erythrocytes and to prepare cell plasma membranes from the mature erythrocyte, which is non-nucleated, readily lysed, and has no intracellular organelles. As a consequence of this, the erythrocyte membrane has been the subject of considerable research and its structure is now known in some detail.

2. STRUCTURE OF THE ERYTHROCYTE MEMBRANE

2.1. Components

In common with other cell membranes, the erythrocyte membrane is composed of lipid, protein, and carbohydrate. The relative proportions of these components vary in different membrane types and are related to the function of the cell or organelle. In the erythrocyte membrane, the composition is of roughly equal amounts of lipid and protein with about 10% carbohydrate. There are several reviews on the erythrocyte membrane in which views of the structure and functional relation of the various components are discussed (Schrier, 1985; Singer and Nicholson, 1972; Steck, 1974; Tanner 1983). Some aspects of the structure that are relevant to the expression and detection of blood group antigens and to hemagglutination will now be considered.

2.1.1. Lipids

The most abundant lipid component is phospholipid, which has been estimated to be from 45 to 65% of the total by weight. Phospholipid in the erythrocyte membrane is organized as a bilayer with the hydrophobic side chains forming the core. The major phospholipids present are phosphatidyl choline (PC, 28%), phosphatidyl ethanolamine (PE, 27%), sphingomyelin (SM, 26%), and phophatidyl serine (PS, 13%) (Schrier, 1985). The remainder is present as neutral lipid, which comprises mainly nonesterified cholesterol (Hui *et al.*, 1980) that is inserted between the hydrophobic fatty acid chains in the core of the bilayer and influences the rigidity of the membrane. A small proportion of the membrane lipid is glycosylated to form glycolipid. The distribution of phospholipid in the membrane has been shown to be asymmetric with respect to the inner (cytoplasmic) half of the lipid bilayer and the outer half. Approximately 70–80% of the PC and SM are in the outer layer, whereas 80–100% of the PE and PS are in the inner layer (Marinetti and Crane, 1978; Verkeij *et al.*, 1973). The phospholipids have a relatively rapid lateral mobility in the membrane of 2–4 $\times$ 10^{-9} cm^2/s at 25°C (Thompson and Axelrod, 1980) and may be visualized as flowing across the membrane surface. In contrast, there is very little movement of phospholipid between the inner and outer halves of the membrane. It would appear, however, that blood group antigens in general have little lateral mobility. The presence of cholesterol in the hydrophobic core of the bilayer (Hui *et al.*, 1980) enhances the lateral movement of phospholipid. It has been shown to cross the membrane freely, and to exchange with plasma cholesterol (Cooper, 1978).

The glycolipids are situated only on the outside of the membrane (Anstee, 1986) and several of these components are associated with blood group antigen activity. There are a series of complex lipid molecules derived from a basic ceramide structure in which the carbohydrate is linked covalently by linkage from a glucose residue to the ceramide moiety (Figure 1).

The sugar chains found on the glycolipids are of variable length and complexity (see Section 3.1.2). It is possible that some of the minor components have not yet been recognized or the structure of some more complex chains not yet fully characterized. This complexity results from the sequential action of a variety of glycosyl tranferases that each add a specific sugar residue to the chain as shown by the groups of Morgan, Watkins, and Hakomori. Many of the biosynthetic pathways leading to the structures found on the erythrocyte membrane have now been elucidated (Hakomori, 1981a; Watkins, 1980). The different sequences of sugar residues and of the linkages between them resulting from the interaction of the transferases lead to the complexity of the glycolipids found in the erythrocyte membrane (see Section 3.1.2).

2.1.2. Proteins

A large number of different protein components that are associated with the erythrocyte membrane have now been identified. These may be present on the interior or exterior surface of the membrane or may traverse the membrane as

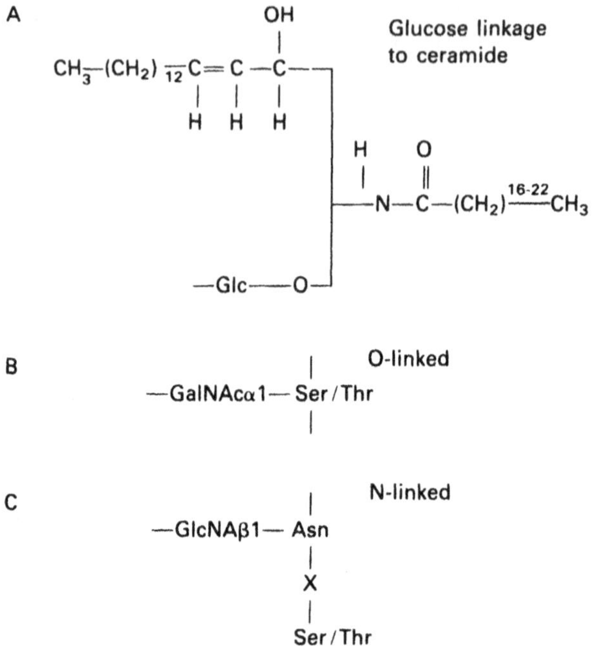

FIGURE 1. Linkage of oligosaccharides to ceramide and peptide chains. Structures of (A) linkage of glucose to ceramide in glycolipids, (B) O-linkage of GalNAc to serine/threonine in glycoproteins, and (C) N-linkage of GlcNAc to asparagine in glycoproteins.

transmembrane proteins. The proteins may be separated from the lipid matrix by solubilization of the membrane with a detergent such as sodium dodecyl sulfate (SDS). Analysis by SDS–polyacrylamide gel electrophoresis (SDS–PAGE) reveals a number of bands separated on the basis of differences in molecular weight. These were originally designated by numbers in order of decreasing apparent molecular weight as bands 1–8 (Steck, 1974). As the resolution of SDS–PAGE improved, particularly through the use of discontinuous buffer systems (Laemmli, 1970), other bands were recognized and so there may be subdivisions of some of these bands. Several of these proteins have now been purified and characterized.

Some of these proteins may be dissociated from the membrane of the isolated erythrocyte ghosts simply by manipulation of the ionic strength of the suspending medium and are presumably bound by ionic interactions. They are generally termed peripheral membrane proteins and are all thought to be located on the inner or cytosolic face of the membrane. The highest molecular weight bands (1 and 2) have been identified as alpha spectrin (240 kD) and beta spectrin (220 kD), respectively (Branton *et al.*, 1981). The spectrin molecules are phosphorylated at four sites. Band 2.1 has been named ankyrin or syndein and has a molecular weight of 215 kD (Bennet, 1985). Bands 4.1 and 4.9 are also situated on the internal surface (Schrier, 1985). Band 4.2 has a molecular weight of 43 kD and has recently been purified (Korgsgren and Cohen, 1986) and the binding site for the cytoplasmic portion of band 3 has been identified. The region known as band 4.5 is diffuse and probably comprises several polypeptides of similar molecular weights. Band 5 has been characterized as actin with a molecular weight of 42 kD (Branton *et al.*, 1981; Cohen, 1983).

Other proteins may only be dissociated from the membrane by the use of detergents. These are termed integral membrane proteins and are thought to have regions with a series of hydrophobic amino acids that interact with the hydrocarbon chains in the interior of the lipid bilayer. They also have hydrophilic regions that are often glycosylated and extend from the outer surface of the membrane. Some of the integral proteins traverse the membrane and also have a portion that extends into the cytosol and is not glycosylated. Several of these integral proteins have now been purified and characterized and some have been shown to carry blood group antigens. The anion transport protein, band 3, is the most abundant of the transmembrane proteins. It has been purified (Tanner *et al.*, 1980) and shown to be a glycoprotein with a molecular weight of 90–100 kD (Kopito and Lodish, 1985) and there is a single, complex, branched N-linked glycan (Figure 2) of molecular weight 11kD (Jarnfelt *et al.*, 1978; Steck, 1978). Band 3 has been shown to carry ABH and Ii antigens. Its functional role is believed to be involved with the exchange of HCO_3^- and Cl^- anions across the membrane.

The transmembrane protein responsible for glucose transport into the erythrocyte is one of the peptides that migrate in the band-4.5 region. The glycoproteins and in particular sialoglycoproteins do not stain well with conventional protein stains such as Coomassie blue. They are therefore difficult to identify on SDS gels unless specific staining techniques are used. Periodic Schiff's base (PAS) stain will reveal the major sialic acid containing glycoproteins and these are now well characterized (Anstee and Tanner, 1986). There are a number of different systems currently being used for the nomenclature for these sialoglycoproteins, and this may lead to some confusion. They were originally described in terms of PAS staining bands 1–4 (e.g., PAS 1).

Following the isolation of the major sialoglycoprotein, which was named glycophorin A by Marchesi *et al.* (1973), an alternative nomenclature in which the different sialoglycoproteins were referred to as glycophorin A, B, and C became widely accepted. Anstee *et al.* (1979) have proposed a nomenclature in which individual sialoglycoproteins are referred to by Greek symbols (as shown in Figure 3). This system has the advantage that the bands visualized by SDS–PAGE are now known to comprise complexes of sialoglycoproteins and are more readily described. It is also more convenient to describe some abnormal variant sialoglycoproteins in this way. This nomenclature will therefore be used in this

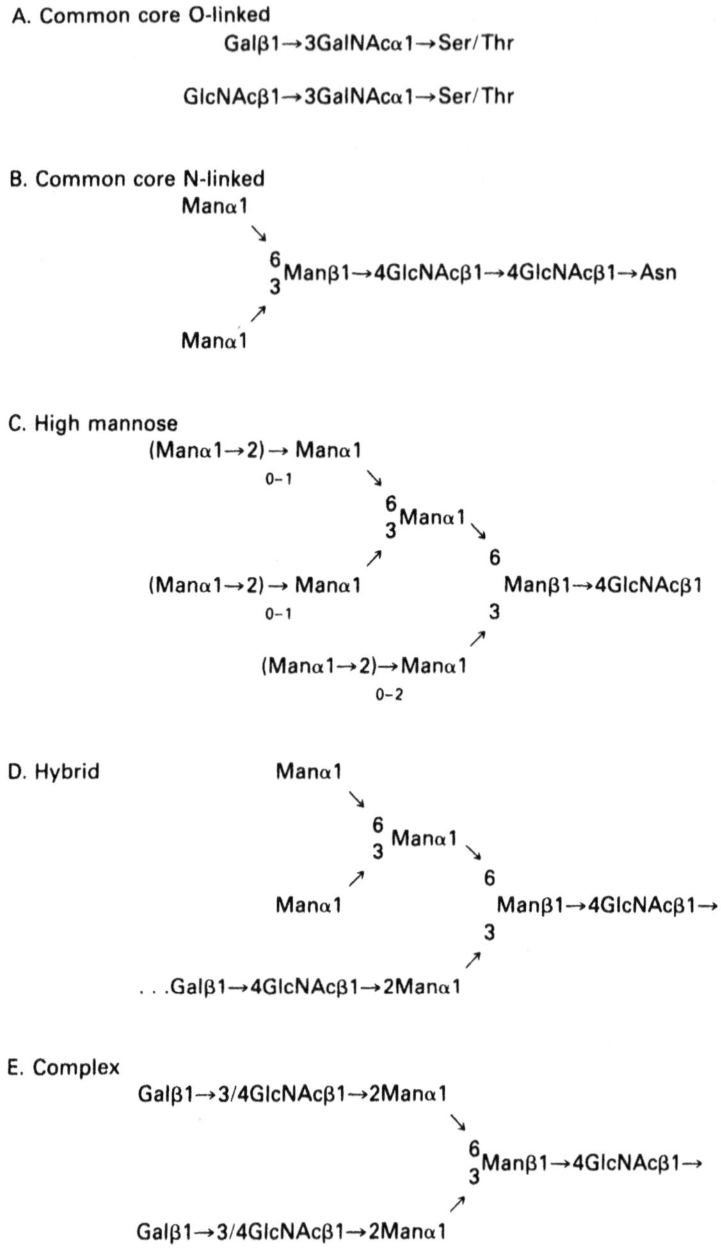

FIGURE 2. Types of O- and N-linked oligosaccharide in glycoproteins: (A) the two types of the common core sequence of O-linked glycans, (B) common core sequence of N-linked glycan, (C) high mannose type of N-linked glycan, (D) hybrid type of N-linked glycan, and (E) complex type of N-linked glycan.

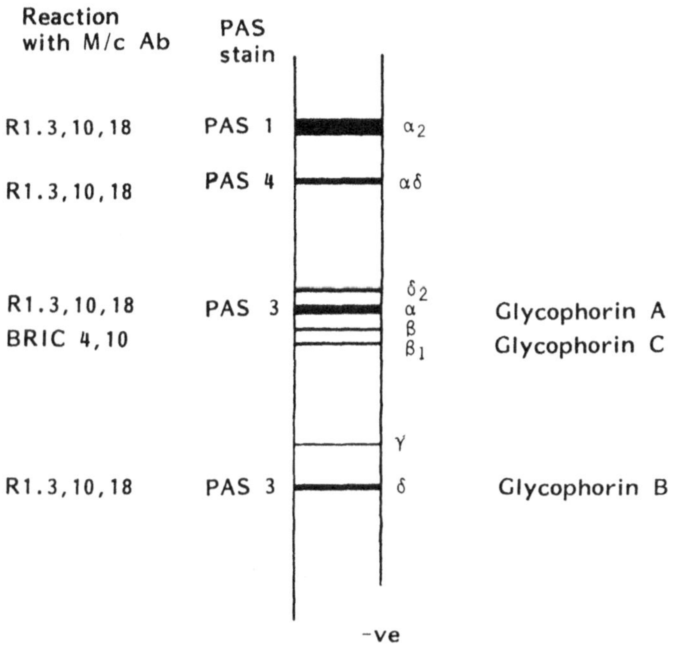

FIGURE 3. Diagrammatic representation of sialoglycoproteins separated by SDS-PAGE and stained with PAS. The nomenclature of Anstee *et al.* (1979) and of Steck (1974) are shown for comparison along with the reactions of some representative monoclonal antibodies (Anstee and Edwards, 1982) useful in immunoblotting.

chapter. The complete amino acid sequences of alpha sialoglycoprotein (glycophorin A) (Tomita *et al.*, 1977) and of delta sialoglycoprotein (glycophorin B) (Dahr *et al.*, 1981) is now known. Other sialoglycoproteins (beta, beta 1, and gamma) have also been identified (Owens *et al.*, 1971). A number of other proteins associated with blood group antigen activity have now been identified and will be described in detail in Section 3.2.2.

2.1.3. Cytoskeleton

Spectrin, actin, and ankyrin and bands 4.1 and 4.9 are associated to form the so-called cytoskeleton (for reviews see Bennet, 1985; Cohen, 1983; Marchesi, 1983; and Sheetze, 1983). It is apparent that this structure has a very important role in the maintenance of the erythrocyte shape (Lux, 1979) and gives the cells their viscoelastic properties that are of great importance in their passage through the circulation. There is considerable interest in the role of the erythrocyte cytoskeleton as there are numerous hematological disorders in which the cells have an abnormal shape and reduced deformability that may make them more susceptible to premature hemolysis (e.g., hereditary elliptocytosis). It is becoming apparent that several transmembrane proteins are attached to components of the cytoskeleton and are probably of importance in anchoring the membrane to the cytoskeleton.

The first interaction to be described was that of the anion transport protein, band 3, the cytoplasmic domain of which has been shown to bind to the cytoskeleton via ankyrin (Bennet and Stenbuck, 1979; Weaver *et al.*, 1984). The number of attachment sites has been estimated at about 100,000. Some data have suggested that sialoglycoprotein alpha is associated with band 4.1 in artificial liposomes (Anderson and Lovrien, 1984) but alpha does not seem to be associated with the cytoskeleton in the erythrocyte membrane. It has been demonstrated, however, that the sialoglycoprotein beta is associated with the cytoskeleton (Mueller and Morrison, 1981; Owens *et al.*, 1971). The peptide that carries Rh activity was shown to associate with the cytoskeleton by Ridgwell *et al.* (1984), although this interaction does not appear essential for Rh(D) antigen expression (Victoria *et al.*, 1987). It is also apparent that this peptide interacts with the membrane phospholipids, since in cells with the Rh null phenotype, which have a deficiency of this peptide, there is a change in the asymmetry of the membrane phospholipids (Kuypers *et al.*, 1985), and this suggests a role in maintaining phospholipid asymmetry (see also Chapter 3).

Recent work has shown that the protein carrying LW blood group activity (Mallinson *et al.*, 1986) and that carrying Kell activity (Parsons, 1985) are also associated with the cytoskeleton. It is difficult to assess the relative contributions of the different attachment sites but there appear to be 50,000–75,000 copies per cell of the beta SGP and around 30,000 of the Rh(D) polypeptide. These may therefore contribute significantly to the overall attachment. The Kell and LW proteins are less abundant (probably less than 5000 copies per cell for each protein); however, their attachment may play some specific role. The possible relations of those proteins associated with the cytoskeleton are shown in Figure 4.

The attachment of the component carrying blood group antigen activity to the cytoskeleton may well influence the reaction of that antigen with certain antibodies. This will in turn affect the way in which particular antigens are detected, especially when using serological techniques that rely on the agglutination of cells. The attachment of the component to the cytoskeleton will obviously restrict the lateral mobility of the component in the membrane and this restriction in mobility may well affect any steric hindrance to antibody binding. Thus, it would be expected that the characteristics of antigens that are located on glycolipids, such as the antigens of the P system, would be different from those of antigens located on proteins linked to the cytoskeleton, such as the Rh(D) antigen. For antigens in which the epitopes are present on sequences of the protein

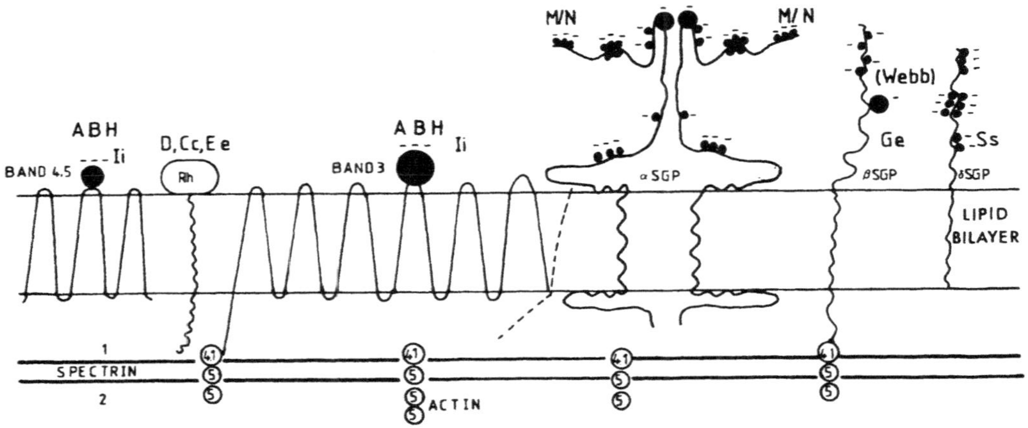

FIGURE 4. Diagrammatic representation of the organization of some proteins of the erythrocyte membrane. The structure of the anion transport protein (band 3) is taken from that proposed for the protein from murine erythrocyte membranes by Kopito and Lodish (1985) and that of alpha SGP from Welsh *et al.* (1985). The attachment sites to the cytoskeleton are from Mueller and Morrison (1981). The orientation and linkage of the Rh peptide(s) to the cytoskeleton are not yet certain (see also Chapter 3). The locations of some blood group antigens associated with the membrane proteins are also indicated.

adjacent to the position at which the peptide chain is inserted into the membrane, the phospholipid environment around the protein may be important in the expression of the antigen. It is documented that the D antigen of the Rh system requires phospholipid for its expression (Green, 1968; Hughes-Jones *et al.*, 1975) and it is found that the antigen is not detectable on the isolated peptides. Attachment to the cytoskeleton has also been reported to stabilize the structure of the Rh(D) antigen (Paradis *et al.*, 1986). It has also been reported (Rearden, 1985) that the expression of the Wr^b determinant, which is thought to be located on the alpha sialoglycoprotein between residues 60 and 70 adjacent to the transmembrane portion from residues 72–92 (Ridgwell *et al.*, 1983), is also dependent on phospholipid for its expression. The influence of the attachment of transmembrane proteins to the cytoskeleton, on the behavior and detection of blood group antigens in serological techniques, will be discussed in Section 4.1.2.

2.1.4. Effect of Transmembrane Protein–Cytoskeleton Interaction on Erythrocyte Deformability and Malarial Parasite Invasion

The attachment of certain transmembrane proteins to the cytoskeleton may explain other phenomena. One of these is the observation that the binding of certain antibodies to erythrocyte membrane proteins can affect the rigidity of the cell membrane (Chasis *et al.*, 1985). A ligand-induced interaction of alpha SGP with skeletal proteins was observed following binding of certain antibodies or monovalent Fab fragments derived from them. This was taken to indicate that conformational changes, induced by the reaction of the antibody with an epitope in the external portion of the molecule, could be transmitted through the transmembrane portion to the cytoplasmic region and could influence the association with the cytoskeleton. This is of interest in considering the way in which some antibodies, such as those with Wr^b specificity, will block the invasion of erythrocytes by the malarial parasite *Plasmodium falciparum*. It was originally thought that the effect of certain antibodies, especially those with anti-Wr^b activity, in blocking parasite invasion was due to the masking of a specific receptor for the parasite as in anti-Fy^a blocking *P. vivax* (Hadley *et al.*, 1984), as discussed by Pasvol (1984). Some of the antibodies that show this effect have a similar specificity to those that affect the membrane ridgidity (Chasis *et al.*, 1985) and it was postulated that the blocking effect may simply be a reflection of this. It has recently been shown (G. Pasvol, unpublished observations), in studies of a series of monoclonal antibodies directed against the alpha sialoglycoprotein, that the portion of the molecule to which the antibody binds can be correlated with the efficiency of blocking parasite invasion. Thus, those antibodies that bind closest to the transmembrane portion of the molecule, such as those with anti-Wr^b specificity, had the greatest effect on parasite invasion and were also the antibodies that had the most influence on erythrocyte deformability. It will be of considerable interest to see whether antibodies to determinants on other transmembrane proteins that have a similar location will also affect erythrocyte deformability. Such studies may give further insights into the organization and relations between different components in the membrane.

Our understanding of the topography of the surface of the erythrocyte membrane is very incomplete, however, and the representations of the relation of components shown in Figure 4, and those previously published, should be considered highly diagrammatic, since they may bear little resemblance to the true organization. They do serve, nevertheless, to give a general idea of the arrangement and interaction of the different proteins, the way in which they are inserted into the lipid bilayer, and the constraints that may be placed on their mobility.

3. IMMUNOCHEMISTRY

The carbohydrate antigens are secondary gene products, the primary products being the specific glycosyl transferases. The structural characterization of the antigens in the ABH, Lewis, Ii, and P blood group systems resulted from the work of the groups of Morgan, Watkins, Kabat, Feizi, Hakomori, and Marcus. Their work showed the enzymic basis for synthesis of the structures and how glycosylation steps lead to the formation of a new antigen and the masking of the precursor antigen. The shared structural organization of the sugar moieties, on both glycoproteins and glycolipids, enables the same blood group specific glycosyl transferase to synthesize the blood group antigen, irrespective of the carrier molecule. There may be variations, however, in the tissue–organ level in the expression of these antigens. In secretions, the expression of ABH and Lewis antigens is influenced by the secretor gene. The earlier work in this field is reviewed elsewhere (Hakomori and Kobata, 1974; Kabat, 1956; Watkins, 1966). On the surface of the erythrocyte and other cell membranes, and in the liquid phase, these antigens are associated with the carbohydrate side chains of macromolecules of glycoproteins and glycolipids, which are sometimes collectively referred to as glycoconjugates. These blood group antigens may also occur as free carbohydrate moieties in milk and urine.

Until recently, our knowledge of the nature of the protein antigens was limited. There have now been considerable advances in the understanding of the immunochemistry of blood group antigens, which are associated with protein components of the membrane. These advances have relied on the development of new methodology and in particular the application of biochemical and immunochemical techniques such as the use of SDS–PAGE, immunoprecipitation, protein sequencing, and immunoblotting. In the study of glycoconjugates, the techniques of gas chromatography, high-performance liquid chromatography

(HPLC), nuclear magnetic resonance (NMR), and fast atom bombardment mass spectrometry are now being used in the elucidation of structures in addition to the more classical techniques of thin layer and paper chromatography and the use of specific glycosidic enzymes. It is convenient to consider separately those antigens where the epitopes are determined only by the carbohydrate structure and those where the amino acid sequence of the peptide or protein influences the antigenic determinant.

3.1. Carbohydrate Antigens

The carbohydrate structures on glycoconjugates are organized in such a way that they may be visualized as having core, backbone, and peripheral regions. The side chains have certain organizational similarities and yet have characteristic features. (For detailed reviews see Hakomori, 1981b; Macher and Sweeley, 1978; Sadler, 1984; and Snider, 1984).

3.1.1. Glycoproteins

There are two main types of carbohydrate linkage on glycoproteins. Those that are linked by *N*-acetylgalactosamine to serine or threonine residues on the protein through an oxygen atom are known as O-glycosidic (Figure 1). N-linked chains are joined to the peptide chain through an amide nitrogen (Figure 1). O-linked chains are also known as mucin type because they occur commonly and abundantly on mucous secretions. They can be distinguished from N-linked chains by their susceptibility to mild alkali degradation. The core monosaccharide is almost always further substituted by enzymatic glycosylations. As a general rule, N-linked chains have a common branched pentasaccharide core structure. The mannosyl residues are further glycosylated, and three types of N-linked chains are recognized: high mannose, complex, and hybrid (Figure 2). Complex and high mannosyl N-linked chains may be substituted further by Galβ1→3/4GlcNAc units. While such a generalization cannot be made for O-linked structures, the disaccharides Galβ1→3GalNAcα1→ and GlcNAcβ1→3GalNAcα1→ are found as a core sequence in many O-linked glycoproteins. The O-linked carbohydrate structures that carry carbohydrate blood group antigens are elongated by Galβ1→3/4GlcNAc repeating units. The disaccharides Galβ1→3GlcNAc and Galβ1→4GlcNAc are isomers designated type 1 and type 2 chains (Rege *et al.*, 1963), respectively. They form the backbone regions of glycoproteins and may serve as precursor structures for blood group ABH and Lewis antigens. Galβ1→3GalNAc has been designated a type 3 chain (Feizi, 1985a) but this is discussed later in relation to A_1 and A_2 antigens. Repeating type 2 linear and branched sequences constitute i/I antigens (see discussion in Section 3.1.4).

Megalosaccharides of these linear and branched repeating units were originally described on O-linked glycoproteins. Such macromolecules with up to 50–60 monosaccharide (polyglycosyl peptides or polylactosylaminoglycans), which are stable at high pH (N-linked), have been isolated from erythrocytes (Krusius *et al.*, 1978) and granulocytes (Spooncer *et al.*, 1984). Certain substitutions of these backbone regions result in a number of well-characterized antigens. It is not clear why further substitutions do not occur in the peripheral structures.

3.1.2. Glycolipids

Carbohydrate blood group antigens occur on a number of glycolipid carrier molecules. These are known collectively as glycosphingolipids, and blood group activity is associated with some of these, the neutral glycosphingolipid. Like the carbohydrate side chains of glycoproteins, there are certain distinctive structural features of these moieties on glycosphingolipids. Except for the Gal series, all are linked by a glucose residue to the sphingosine base of the ceramide moiety (Figure 1). The ceramide moieties of blood group active glycosphingolipids consist of predominantly C-22 and C-24 fatty acids. The core monosaccharide is further glycosylated to form repeating type 1 and type 2 sequences in linear and branched forms. Certain of these sequences (lacto, neolacto, and globo series) are identical to those that have been described on O-linked carbohydrate chains of glycoproteins. Others in the lacto and neolacto series are identical to N-linked chains. The carbohydrate side chains of glycolipids (like glycoproteins) can be visualized as having core, backbone, and peripheral regions and blood group antigenic activities reside mostly in the backbone and peripheral regions. Megalosaccharide side chains of 30–60 sugar residues with blood group activity (polyglycosyl ceramides) have been isolated from erythrocyte glycolipids that cofractionate with glycoproteins in aqueous butanol (Koscielak *et al.*, 1976). There is a great similarity in the organization of the carbohydrate side chains of glycoproteins and glycolipids and this would account for both glycoconjugates having the same blood group antigenicity.

3.1.3. The ABH Blood Group System

The ABH blood group system is clinically the most important, since hemolysins occur naturally in those who lack the corresponding antigen, and incompatible transfusion will result in intravascular hemolysis. The ABO antigens are among the most widely used of the polymorphic markers.

It was originally proposed that the ABO locus was triallelic (Bernstein, 1924); A and B are functional alleles and the O allele nonfunctional. This has been modified to accommodate the observation that two major subgroups of A are serologically well defined and the system is best explained by the alleles A_1,

A_2, B, and O. Other subgroups of A, for example, A_x, A_3, and A_4, have been described but are generally rare. Subgroups of B, for example, B_3, B_x, and B_n, are even rarer (Table II).

The formation of A and B antigens is dependent on the presence of the H antigen. In turn, H antigen is formed from type 1 or type 2 precursor chains as a result of the enzymatic activity of α-2-fucosyl transferase. Recent data indicate that there are at least two different α-2-fucosyl transferases (Betteridge and Watkins, 1986; Kumazaki and Yoshida, 1984; LePendu *et al.*, 1985): one acting preferentially on type 1 chains and the other on type 2 chains. Both enzymes occur in secretions (Betteridge and Watkins, 1986), but the enzyme in erythrocytes preferentially acts on type 2 chains (Oriol *et al.*, 1986) and this is consistent with the observation that erythrocyte ABH antigens are associated almost exclusively with type 2 precursor chains. The presence of at least two α-2-fucosyl transferases with different tissue distribution is, according to Oriol and co-workers, the result of two genes, the *H* gene and the *Se* gene that code for the enzyme on erythrocytes and secretory glands, respectively. The classical model for the control of expression of ABH antigens on erythrocytes and in secretions envisages one structural gene *H* and two regulatory genes *Z* and *Se* controlling the expression of these antigens on erythrocytes and secretions, respectively. The *A* gene codes for *N*-acetylogalactosaminyl transferase, and the *B* gene codes for galactosyl transferase, which, respectively, transfer GalNAc and Gal to the H substance. Substitution at the C-2 position of the nonreducing terminal galactose

Table II
Antigens of the ABH Blood Group System

Gene	Gene frequency	Gene product	Genotype	Phenotype	Antibodies in serum	Phenotype frequency
A	A_1 0.2090 A_2 0.0696	NAcGalNAc transferase	A_1A_1 0.0437 A_1A_2 0.0291 A_1O 0.2760	A_1	Anti-B	0.3488
			A_2A_2 0.0048 A_2O 0.0919	A_2	Anti-B >1% Anti-A_1 in addition	0.0967
B	0.0612	Gal transferase	BB 0.0037 BO 0.0808	B	Anti-A	0.0845
A and B			A_1B 0.0256	A_1B	None	0.0256
			A_2B 0.0085	A_2B	25% Possess anti-A	0.0085
O	0.6602		OO 0.4359		Anti-A Anti-B	0.4359

by fucose is essential for A and B transferase activity. The biosynthetic pathway for the synthesis of these antigens is shown in Figure 1 in Chapter 4 (see also Watkins, 1966). For further discussion on these enzymes see Section 4.1.2.

3.1.3.1. A_1 and A_2 Subgroups. There is much interest in the structural basis of the A_1 and A_2 subgroups. It has been suggested that there are differences between A_1 and A_2 antigens and that the A_1 determinant is based on type 1 chains and A_2 on type 2 chains (Kisailus and Kabat, 1978; Moreno *et al.*, 1971). Differences in the inhibition curves of *Dolichos biflorus,* a lectin with anti-A_1 activity (Bird, 1952) and anti-A with A_1 and A_2 substances were noted. However, only type 2 based A have been found on A active glycoproteins and glycolipids isolated from A_1 and A_2 erythrocytes and this would be inconsistent with the view that A_1 and A_2 antigens are based on different precursor chains.

Recent work suggests that the basis of A_1 and A_2 subgroups is more likely a result of greater epitope density resulting from more A sites and repetitive A structures. A_1 erythrocytes have been shown to have a greater number of A sites than A_2 cells (Economidou *et al.*, 1967; Greenbury *et al.*, 1963; Williams and Voak, 1972). Polyglycosyl ceramides from A_1 eythrocytes carry more A antigens than A_2 polyglycosyl ceramides (Koscielak *et al.*, 1976). A_2 erythrocytes have a greater proportion of incomplete A^c or A^d glycolipids (Hakamori *et al.*, 1977). Recently, a repetitive A determinant linked to type 2 precursor chain and designated type 3 chain A has been isolated (Clausen *et al.*, 1985).

GalNAcα1→ 3Galβ1 →3GalNAcα1→ 3Galβ1 →4GlcNAcβ1→3Galβ1→4Glcβ1→1Cer
(Fuc α1,2 → 3Galβ1 ; Fuc α1,2 → 3Galβ1)

Type 3 chain A

Readers should not confuse the type 3 (precursor) chain as used by Feizi, Oriol, and others with the type 3 designation here. Donald (1981) originally used a type 3 chain to describe the A active tetramer:

GalNAcα1→3Galβ1→3GalNAcα1→
(Fuc α1,2 → Galβ1)

Type 3 chain A is found predominantly on A_1 erythrocytes and on linear and branched glycolipids but not on glycoproteins. A_2 erythrocytes contain trace amounts of this structure but the precursor chain, type 3H, is present in greater quantities.

The A_1 and A_2 subgroups are usually defined serologically by reactivity with the lectin from Dolichos biflorus. The lectin recognizes A-linked GalNAc residues and when used under defined conditions will only react with higher-density A antigen sites present on A_1 erythrocytes. A number of monoclonal A hybridoma antibodies have been raised (Gooi *et al.*, 1985; Voak *et al.*, 1980) that show a preferential but not exclusive reaction with A_1 erythrocytes. In hapten inhibition studies, ten anti-A hybridoma antibodies were inhibited to varying degrees by the A trisaccharide, the A trisaccharide on different carrier molecules, and even by A-like structures (Gooi *et al.*, 1985). To date, only one monoclonal antibody (TH-1) has been shown to react specifically with A_1 erythrocytes and not with A_2 and O erythrocytes (Clausen *et al.*, 1985). In direct binding assays, antibody TH-1 recognizes the repetitive type 3 chain A determinant. It does bind, however, to a small extent to globo A and type H_2 glycolipids.

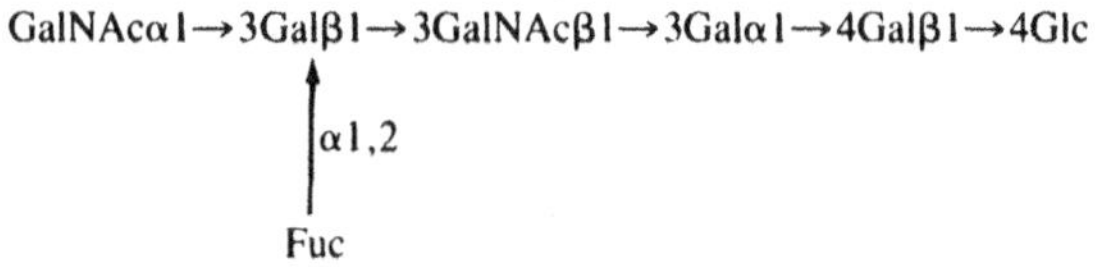

Globo A

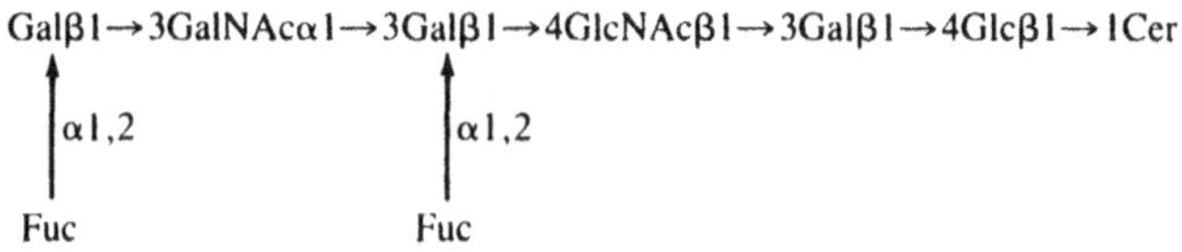

Type 3 H

It would appear from its reactivity with the three structures above and its nonreactivity with a number of type 1 and type 2 A and H and A Lewis structures that chain length of the carrier molecule of the A determinant is an important requirement for reactivity with TH-1. While a repetitive A determinant is the optimal hapten, it will be interesting to determine if the internal 3GalNAcα1 is an essential structural prerequisite for TH-1 reactivity.

Work with anti-i and anti-I cold agglutinins has shown that repetitive linear or branched poly-*N*-acetyllactosamine chains are the structural basis of the I/i antigens. Although a certain minimal structure can have antigenic activity, it is the repeating structures that are optimal antigens (Gooi *et al.*, 1984). It has also been observed that in hapten inhibition assays the carbohydrate hapten on a carrier molecule (and hence multivalent) has inhibitory activities up to 1000 times that of the free hapten (S. C. Lui, H. C. Gooi, and T. Feizi, unpublished observation). It is well known that when carbohydrate antigens are isolated from glycoconjugates, there is a drastic loss of antigen activity. Work by Tang and coworkers showed that it is possible to restore the inhibitory activity to its original level by incorporating the isolated hapten into a carrier molecule. A scheme (see Figure 5) has been proposed where an antigen may not be detectable because it is

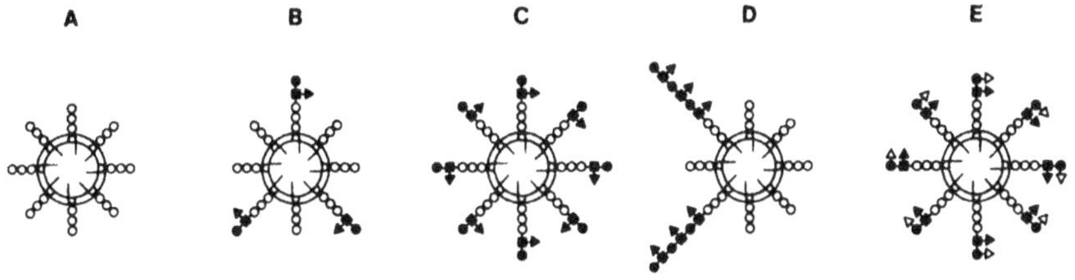

FIGURE 5. Scheme for the expression of 3-fucosyl-*N*-acetyllactosamine determinant on cells. The trisaccharide sequence, consisting of galactose (●), *N*-acetylglucosamine (■), and fucose (▲) may occur as a terminal structure on the carbohydrate chains of glycoproteins or glycolipids. Here is shown part of a membrane-associated glycolipid consisting of an oligosaccharide (○) joined to a ceramide moiety. The antigenic determinant may be lacking (as in A) or too sparse to be detectable serologically (as in B). In C and D the determinant is represented as abundantly expressed on different (C) or the same (D) oligosaccharide chains. In E, although the determinant is abundantly synthesized, its antigenicity is masked by an additional monosaccharide (△). Reproduced with permission from Gooi *et al.* (1983b).

not present or present at such an epitope density as not to be detectable by serological or immunochemical methods (Gooi *et al.*, 1983b). Antigenicity is demonstrated when the determinant occurs on numerous carrier chains or when it is a repetitive determinant on the same chain. Such a scheme would provide an immunochemical explanation of A_1 and A_2 subgroups and the selective reactivity of the Dolichos lectin when used at an appropriate dilution (Voak and Lodge, 1968). However, it is the A transferase that would ultimately determine the basis for such structural differences in A epitope density and respective A determinants. There is good evidence that A_1 transferase is the more efficient of the two enzymes (Schachter *et al.*, 1973).

3.1.4. Blood Group i and I

The I and i antigens were originally defined by cold agglutinins (Marsh and Jenkins, 1960; Weiner *et al.*, 1956). These are cold reactive monoclonal and oligoclonal antibodies that occur in chronic cold hemagglutinin disease and may complicate lymphoproliferative disorders, infectious mononucleosis, and *Mycoplasma pneumoniae* infections. Hybridoma antibodies have also been raised with cold agglutinin activity (Gooi *et al.*, 1983a). Anti-I reacts preferentially with almost all adult erythrocytes (Weiner *et al.*, 1956) and anti-i with cord erythrocytes and rare adult cells (Marsh and Jenkins, 1960). There is a reciprocal relation between the expression of I and i antigens on erythrocytes. The i antigenic activity, which is strongest during the neonatal period, diminishes and I antigenicity increases during infancy so that by 18–24 months there is almost no i activity.

The carbohydrate nature of I/i antigens was suggested by the susceptibility of the I antigen to hexosaminidase from *Clostridium tertium* (Marcus *et al.*, 1963). Subsequent work with oligosaccharides isolated from ovarian cyst glycoproteins (Feizi *et al.*, 1971) showed that the cold agglutinin, anti-I Ma, recognized the following structure:

Galβ1→4GlcNAcβ1→6Gal

Another cold agglutinin, anti-i McC, was best inhibited by the tetrasaccharide, lacto-*N*-neotetraose (Tsai *et al.*, 1976), which has the following structure:

Galβ1→4GlcNAcβ1→3Galβ1→4Glc

Since these antigens occur on erythrocytes, attempts were made to isolate them from this source. Although the yield of I/i antigens from erythrocytes proved to be poor, it suggested that I/i activity resided in gangliosides and antigenicity was enhanced by desialylation (Feizi *et al.*, 1978). It was work with bovine erythrocyte gangliosides, however, that proved to be a good source of I/i antigens and enabled the structural basis of i and I antigenicity to be established. Five out of six anti-i were inhibited by the following structure (Niemann *et al.*, 1978):

Galβ1→4GlcNAcβ1→3Galβ1→4GlcNAcβ1→3Galβ1→4Glcβ1→Cer

lacto-*N*-norhexaosyl ceramide

The minimum structural requirement for i specificity is two repeating type 2 chains, the tetrasaccharide, lacto-*N*-neotetraose:

Galβ1→4GlcNAcβ1→3Galβ1→4GlcNAc

Three anti-I were inhibited to varying degrees by lacto-*N*-norhexaosyl ceramide. The branched glycospingolipid, lacto-*N*-iso-octaosyl ceramide,

```
Galβ1→4GlcNAcβ1
               ↘
               6
                Galβ1→4GlcNAcβ1→3Galβ1→4Glcβ1→Cer
               3
               ↗
Galβ1→4GlcNAcβ1
```

inhibited ten out of 11 anti-I and two out of six anti-i tested (Feizi *et al.*, 1979; Watanabe *et al.*, 1979). A number of antibodies with anti-I specificity could be identified and classified into different groups. The first group (e.g., anti-I Ma) recognizes the Galβ1→4GlcNAcβ1→6Gal sequence of branched poly-*N*-lactosamine structures. The second group (e.g., anti-I Step and anti-I Low) reacts with the Galβ1→4GlcNAcβ1→3Gal domains. Some antibodies in this group may have additional requirement for all or part of the Galβ1→4GlcNAcβ1→6Gal domains.

The i activity is lost in the presence of the GlcNAcβ1→6Gal branch. I and i antigens are therefore associated with type 2 chains. Subsequent immunochemical studies (Gooi *et al.*, 1984) show that the optimal determinant for anti-i antibodies are at least hexasaccharides of the neolacto series and linear tetra- and hexa-neolacto saccharides can react with group 2 anti-I antibodies. The above-mentioned studies showed the structural basis of I and i antigens and their relation. The i antigen is converted to I antigen by the activity of glycosyl transferases that result in GlcNAcβ1→6Gal branching structure on the linear poly-*N*-acetyl lactosamine sequence. This is consistent with the observation that fetal erythrocyte glycosphingolipids are predominantly linear structures and that those of adult erythrocytes are branched and more complexed (Watanabe and Hakomori, 1976). I and i are therefore carbohydrate developmental antigens, rather than polymorphic markers, and the order of development is from linear to branched structures. In the mouse embryo, however, the order of expression of these antigens is reversed, that is, I before i. It has been suggested that this unexpected observation is due to a maternal branching enzyme (Feizi, 1985a).

It was noted that anti-i cold agglutinins vary in their reactivity with a given panel of adult and cord blood erythrocytes (Race and Sanger, 1975). The differences in the reactivities of these anti-I and anti-i antibodies with I and i active glycoproteins were also individually characteristic (Feizi and Kabat, 1972). The immunochemical studies with well-characterized oligosaccharides mentioned above provide further evidence of differences in the fine specificities of these monoclonal antibodies and provide an explanation for their differing reaction patterns with glycoconjugates and erythrocytes where the carbohydrate moieties are likely to be more complex.

3.1.5. Lewis System

The antigens of the Lewis blood group system were initially defined by antibodies described by Mourant (1946) and Andresen (1948). These antibodies recognize Le^a and Le^b antigens, respectively. Lewis determinants are soluble antigens found in secretions; on erythrocytes they occur on glycosphingolipids acquired from plasma (Sneath and Sneath, 1955).

Lewis antigens are based on type 1 precursor chains. The carbohydrate nature of these antigens were noted by Watkins and Morgan (1957). Their structures,

Galβ1→3GlcNAc and Galβ1→3GlcNAc
↑ α1,4 ↑ α1,2 ↑ α1,4
Fuc Fuc Fuc

Le^{a} **Le^{b}**

were established by oligosaccharides isolated from Le^{a} active (Rege *et al.*, 1964) and Le^{b} active (Marr *et al.*, 1967) ovarian cyst glycoproteins.

The expression of Lewis antigens on erythrocytes and in secretions results from the Lewis and secretor gene status of the individual. The *Le* gene codes for Le^{a} antigen and is expressed when present. The other allele at this locus, the *le* gene, is nonfunctional and does not give rise to Le^{b} determinants. Le^{b} is formed from Le^{a} by the *H*-gene-specified a-fucosyltransferase (Le^{a-b+} may be formed from type 1-based A and B substances). Karlsson (1978) isolated a glycosphingolipid from human erythrocytes with A,Le^{b} sequence. Antisera with anti-A, Le^{b} have previously been found (Crookston *et al.*, 1970; Seaman *et al.*, 1968). As Lewis antigens are secreted antigens, the Le^{b} status on erythrocytes will be dependent on the secretor status. In the individual with *Le, H,* and secretor genes, the red cells will be Le^{a-b+} and this antigen will be found in the saliva. Le^{a-b+} erythrocytes therefore occur in secretor persons. However, the converse does not always apply as some secretors of ABH substances may be Lewis gene negative and hence their erythrocytes will be Le^{a-b-}. Similarly, individuals with Le^{a+b-} erythrocytes are nonsecretors of ABH substances but not all nonsecretors have Le^{a+b-} erythrocytes. The interaction of the Lewis and secretor (and/or *H*) genes is shown in Table III.

The question of whether the erythrocytes of Lewis gene negative secretors and nonsecretors bear any "Lewis" antigens has provoked much debate. Goat antisera raised against saliva from a OLe^{a-b+} individual, agglutinated erythrocytes of all ABH secretors (Potapov, 1970). Two specificities were identified. One was anti-Le^{a-b+}. The other defined an antigen on Le^{a-b-} erythrocytes from secretors and was designated anti-Le^{d}. An antibody, ARM, described by Gunson and Latham (1972) reacted with Le^{a-b-} erythrocytes from nonsecretors and was called anti-Le^{c}. The structures of Le^{c} and Le^{d} are not yet known but a few have been proposed (Table III). Le^{c} and Le^{d} could be type 2 precursor chain-based isomers of Le^{a} and Le^{b}. This was suggested by Watkins but the inappropriateness of the nomenclature was also pointed out (see Watkins, in Gunson and Latham, 1972). Anti-Le^{a} ARM was found to be inhibited by

Galβ1→4Glc
↑ α1,3
Fuc

3-Fucosylactose

but lacto-*N*-fucopentanol III,

Galβ1→4GlcNAcβ1→3Galβ1→4Sorbitol
↑α1,3
Fuc

was a better inhibitor (Ginsburg *et al.*, 1980).

Work by Graham *et al.* (1978) with goat antisera raised against human saliva suggested that Le^{c} and Le^{d} are type 1 precursor chain-based structures and anti-Le^{d} (Graham *et al.*, 1978) was inhibited by lacto-*N*-fucopentaose 3,

Galβ1→3GlcNAcβ1→3Galβ1→4Glc
↑α1,2
Fuc

Table III
Blood Group Lewis System in *H*-Gene Positive Individuals

Genotype		Phenotype			
Lewis	Secretor	Erythrocyte	Saliva	Antigen	
Le	Se	Le^{b}	Le^{b} (and some Le^{a})	Galβ1→3GlcNAc ↑α1,2 ↑α1,4 Fuc Fuc Le^{a-b+}	
Le	se	Le^{a}	Le^{a}	Galβ1→3GlcNAc ↑α1,4 Fuc Le^{a+b-}	
le	Se	Le^{a-b-} (Le^{d})	Le^{a-b-}	Galβ1→3GlcNAc[a] ↑α1,2 Fuc	Galβ1→4GlcNAc[a] ↑α1,2 ↑α1,3 Fuc Fuc Y-hapten/Le^{y}
				Le^{d}	
le	Se	Le^{a-b-} (Le^{c})	Le^{a-b-}	Galβ1→3GlcNAc[b]	Galβ1→4GlcNAc[b] ↑α1,3 Fuc X-hapten/Le^{x}
				Le^{c}	

[a]Suggested Le^{d} antigen.
[b]Suggested Le^{c} antigen.

Further evidence that type 1 structures are involved in these specificities is provided by Lemieux *et al.* (1981) and Le Pendu *et al.* (1982). The possible antigenic structures of Le^c and Le^d are listed in Table III. A trisaccharide, 3-fucosyl-*N*-acetyllactosamine, isolated from adenocarcinoma of the colon, has been called X-hapten (Yang and Hakomori, 1971). It was subsequently designated Le^x (Kannagi *et al.*, 1982). This structure in the mouse embryo defines a stage-specific embryonic antigen, SSEA-1 (Gooi *et al.*, 1981). Le^x had previously been characterized by an antibody reacting with almost all cord erythrocytes and adult Le^{a+b-} and Le^{a-b+} erythrocytes (Andresen and Jordal, 1949; Sturgeon and Archilla, 1970). This Le^x structure has not yet been characterized but is not identical to X-hapten or SSEA-1 (Francois *et al.*, 1986).

The difucosylated type 2 structure

$$\begin{array}{ll} \text{Gal}\beta 1 & \rightarrow \text{GlcNAc} \\ \uparrow \alpha 1,2 & \uparrow \alpha 1,3 \\ \text{Fuc} & \text{Fuc} \end{array}$$

has been called Y-hapten (Hakomori and Kobata, 1974) or Le^Y (Kannagi *et al.*, 1982). It is apparent that the nomenclature is confusing and it is hoped that monoclonal antibodies will become available to help sort out the existence, structural, and genetic basis of these antigens.

3.1.6. Blood Group *P* System

Blood group P was initially defined by rabbit antisera raised against human erythrocytes (Landsteiner and Levine, 1927) and red cells thus classified were either P+ or P−. A human antibody with similar specificity was subsequently identified (Landsteiner and Levine, 1930). This specificity would be equivalent to anti-P_1 in the present but larger blood group P system. Three antigens and five phenotypes are recognized (Table IV).

The antigens of the P system on erythrocytes are generally thought to be carried on glycosphingolipids (Naiki and Marcus, 1974; Naiki *et al.*, 1975), although it has recently been reported that they may also be associated with erythrocyte glycoproteins (Haselberger and Schenkel-Brunner, 1982). The carbohydrate nature of these antigens was suggested by hemagglutination inhibition experiments using simple sugars (Watkins and Morgan, 1964). Cameron and Staveley (1957) found that hydratid cyst glycoproteins were a rich source of P_1 and P^k antigens. Proteins extracted from salmon and trout roe have also been shown to have agglutinating activity for P, P_1, and P^k cells, which could be inhibited by the addition of galactose (Voak *et al.*, 1974a). A trisaccharide and a

Table IV
Antigens of the P Blood Group System

Phenotype	Antigen on red cells	Antibodies in serum	Phenotype frequency
P_1	P_1,P	None	75%
P_2	P	Anti-P_1	25%
P_1^k	P_1,P^k	Anti-P	Very rare
P_2^k	P^k	Anti-P	Very rare
p	None	Anti-P,P_1,P^k	Very rare

disaccharide isolated from hydatid cyst fluid were found to inhibit anti-P_1 and anti-P^k, respectively (Cory *et al.*, 1974). It had been shown that P_1 antigenicity was associated with erythrocyte glycosphingolipid (Marcus, 1971) and further studies with this erythrocyte fraction established the structures of P_1, P, and P^k (Naiki and Marcus, 1974; Naiki *et al.*, 1975), as shown in Figure 6. The possible biosynthetic pathways are also shown in Figure 6. Consideration of the structures of these antigens would indicate that P^k determinant is the precursor substrate for P and the genes coding for the galactosyl and β-*N*-acetylgalactosaminyl transferases are nonallelic. The P_1 determinant is synthesized by a different pathway and the precursor substance is paragloboside. It is not clear whether the same or different genes are involved in the synthesis of P^k and P_1. Three possibilities are suggested: (1) the P^k and P_1 genes are nonallelic; (2) the P_1 gene is regulatory; and (3) the P^k and P_1 genes are allelic (Marcus *et al.*, 1981).

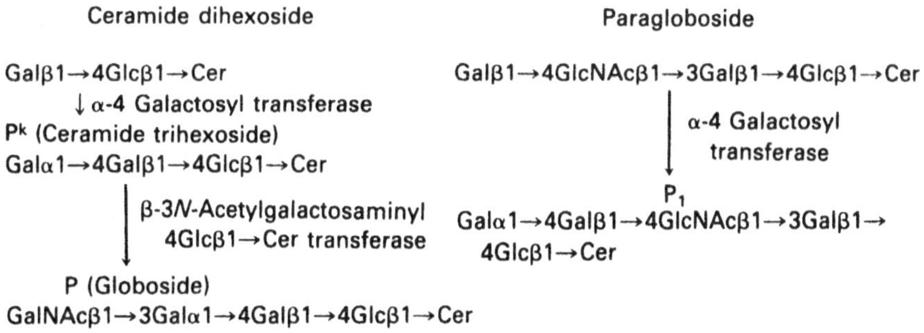

FIGURE 6. Structure and biosynthesis of P, P_1, and P^k antigens. It is not clear whether one or more α-4 galactosyl transferase enzymes are involved in the synthesis of P_1 and P^k.

3.2. Protein Antigens

In comparison with the detailed knowledge of the chemistry of the structures of carbohydrate chains with blood group antigen activity, detailed studies of protein antigens have only appeared fairly recently. By isolation of the purified protein, it has proved possible to identify the amino acid sequences that determine the MN (Blumenfeld and Adamany, 1978; Dahr *et al.*, 1976; Furthmayr, 1978) and Ss (Dahr *et al.*, 1980) antigens. Where it is only possible to obtain a partial amino acid sequence of the isolated protein, the complete amino acid sequence of the protein may be deduced by recombinant DNA techniques. From the partial sequence obtained, the complementary base codes for DNA can be derived and synthetic oligonucleotide probes constructed. These may then be used to probe libraries for the gene and to obtain the full DNA coding sequence for the protein. DNA sequences for α-sialoglycoprotein (Siebert and Fukuda, 1986a) and δ-sialoglycoprotein (Siebert and Fukuda, 1986b) have recently been published. The relative ease with which DNA may be sequenced make this an attractive alternative for proteins that are difficult to isolate and this technique has been used to derive the full sequence for one of the beta sialoglycoproteins (Colin *et al.*, 1986).

3.2.1. Techniques

Progress in this area has been restricted by limitations of the techniques for the identification and separation of the peptides or proteins that carry blood group antigens. They are often present in only small amounts in the membrane. Additionally, the isolation of the protein from its environment in the membrane in which it may be associated with skeletal or other protein components or phospholipids may affect its antigenic properties. To monitor the separation and purification of the component, it must still be able to react with antibodies that define the antigen in question. Suitable immunological techniques for demonstrating this reaction are therefore required.

3.2.1.1. SDS–PAGE. The standard analytical technique for resolving the protein components of the membrane has been SDS–PAGE, which will separate them by differences in molecular weight. Erythrocyte ghosts may readily be prepared by hypertonic lysis (Dodge *et al.*, 1963) and washed to remove free hemoglobin and cytosol proteins. The absence of intracellular organelles in the mature erythrocyte means that such a preparation consists only of the plasma membranes. The membranes may be solubilized in SDS and the membrane proteins analyzed by the discontinuous buffer technique of SDS–PAGE described by Laemmli (1970). Suitable precautions should be taken to limit proteolysis after the addition of SDS because the unfolded proteins are often more susceptible to digestion by proteolytic enzymes derived from the cell membrane.

Something on the order of 50 minor bands can be detected by Coomassie blue staining in this system in addition to those bands originally numbered 1–8 using SDS–PAGE systems with more limited resolution. The PAS Schiff's base staining technique will additionally identify the sialoglycoproteins (Figure 3) mentioned previously, but it is probable that there are other minor glycoproteins that are not visualized by this technique. To identify other proteins with blood group antigen activity, more sensitive and specific detection techniques may be required.

An increase in sensitivity may be obtained by incorporating a radioactive label into the protein. There are a number of ways in which this may be done. Surface proteins containing accessible tyrosine residues may be iodinated by means of lactoperoxidase labeling of the intact cells (Gahmberg, 1982; Moore *et al.*, 1982). Distinct protein patterns are present in autoradiographs of the labeled proteins analyzed by SDS–PAGE. Alternatively, a label may be introduced into sugar residues by the periodate/$NaB[^3H]_4$ (Gahmberg and Andersson, 1977) or neuraminidase + galactose oxidase/$NaB[^3H]_4$ (Gahmberg, 1976) techniques. The former technique labels surface proteins to varying degrees depending on the tyrosine composition and their accessibility to the enzyme. The latter techniques have the advantage of preferentially labeling the sialoglycoproteins. By using either of these techniques, it is possible to label minor components not readily detected by protein or carbohydrate staining techniques.

3.2.1.2. Immunoprecipitation. Once the component has been labeled, it may be precipitated selectively by reaction with an appropriate antibody with blood group antigen activity in an immune precipitation procedure. There are, however, a number of problems with immunoprecipitation techniques. Nonspecific coprecipitation of other components may occur, or the antigenic determinants may be destroyed by solubilization of the component from the membrane. Various ways in which these may be overcome have been described. Moore *et al.* (1982) and Gahmberg (1982) described an elegant modification of the technique in which antibody was bound to the labeled peptide component in the intact cell and the immune complex was subsequently isolated with Protein A and dissociated. The labeled protein may then be analyzed by SDS–PAGE. In this way, a sensitive and specific analysis of peptide components associated with Rh and Fy^a blood group activity has been obtained. The Rh blood group antigens are fully discussed in Chapter 3. It is also possible to purify components on a column of immobilized antibody. After washing to remove unbound material, the component may then be eluted by dissociation of the complex. The selection of antibodies with suitable specificity and affinity is critical for such procedures and if monoclonal antibodies are available or can be produced they may prove useful (Anstee and Edwards, 1982).

3.2.1.3. Immunoblotting. Another promising technique that is increasingly being used is that of immunoblotting. This technique, which has also been described as Western blotting, was developed from one first applied to the detection of nucleic acids (Southern blotting). In immuno the proteins are separated by SDS–PAGE and then electrophoretically transferred to a nitrocellulose membrane. The transferred proteins become immobilized on the nitrocellulose membrane and may then be probed for reaction with an antibody (Towbin and Gordon, 1984). Staining of the antibody, which may be carried out by incubation with a labeled second antibody, will then enable the detection of the proteins carrying a specific antigen. This technique has considerable potential in the study of protein components with blood group antigen activity (Mallinson *et al.*, 1986; Parsons *et al.*, 1987). The sensitivity is such that antigens, which may only be present in small quantities, may readily be detected by the use of suitable antibodies (Huisman, 1986).

It has recently been shown that human antisera with blood group antigen activity may perform well in immunoblotting systems even if they have weak activity in serological tests (Spring *et al.*, 1987). Problems with nonspecific reactions when using such sera may be reduced if the antibody is adsorbed onto, and then eluted from, cells that are positive for the appropriate antigen (Spring *et al.*, 1987; Tanner *et al.*, 1987). Since antisera are available with specificities for many of the blood group antigens whose immunochemistry is as yet unknown, this may provide a means for identifying the components involved.

There are certain problems with the technique. The antibodies that are reactive must be able to recognize epitopes present on the denatured protein bound to the nitrocellulose membrane. These may not be the same as those epitopes recognized by antibodies of blood group specificity on protein in situ in the cell membrane. Screening of antibodies that are reactive with the protein bound to nitrocellulose may be accomplished by the dot blot technique (Herbrink *et al.*, 1982), in which the whole extract of solubilized membrane proteins are linked to the nitrocellulose membrane without separation and then probed with the antibody.

A more serious problem is that of nonspecific binding. This may be due to either the binding of the blood group antibody to the nitrocellulose or the nonspecific binding of antibody to other membrane proteins on the blot. These may generally be reduced by blocking the surplus binding sites on the nitrocellulose with nonimmune serum or with inert proteins such as albumin, gelatin, or caseine. A nonionic detergent, such as Tween (Batteiger *et al.*, 1982), is also usually included in the washing and antibody solutions. These measures should reduce nonspecific binding of antibody due to protein–protein interactions, but some generally remains and experimental conditions may need to be adjusted to minimize this. It should be mentioned that the binding to an apparently nonrelated protein may not in fact be due to a nonspecific interaction (Richards *et al.*, 1975) as some monoclonal antibodies may react with related epitopes on quite distinct proteins although they do so with different affinities (Ghosh and Campbell, 1986). Such interactions may be detected in a sensitive technique such as immunoblotting and are enhanced when there is a high epitope density as may be found in the resolved protein bands. The denaturing conditions employed in the technique may also contribute to nonspecific interactions and the conformational changes induced after SDS–PAGE and subsequent transfer to nitrocellulose are not well defined.

Interpretation of the results obtained by immunoblotting is not always easy. In addition to the problems of nonspecific binding, the protein band detected may be quite diffuse if it is a glycoprotein since there is generally incomplete glycosylation giving rise to microheterogeneity of the protein. Indeed, there may be distinct bands of the same protein with different glycosylation states. Artifactual bands may arise because of proteolysis by endogenous membrane proteolytic enzymes, which can be accelerated when the proteins are unfolded by SDS. The proteins may also be found in complexes, as has been demonstrated for the sialoglycoproteins (Anstee *et al.*, 1979; Dahr *et al.*, 1976). Thus, the same protein component may be present on several different bands on the gel. These complexes may reflect associations between proteins that are present in the intact membrane or they may be the result of aggregation during the separation procedure and thus artifactual. Whatever the reason, it is important to distinguish the individual components that carry the blood group antigen being studied. Interpretation of the results of immunoblotting experiments should always be carried out with caution.

3.2.1.4. Selective Glycosidic Enzymes. Another technique that is proving to be very helpful is the use of selective glycosidic enzymes (Mallinson *et al.*, 1986; Parsons *et al.*, 1987; Tanner *et al.*, 1987). The majority of the proteins on the exterior surface of the erythrocyte membrane are thought to be glycoproteins. The enzyme neuraminidase, which specifically cleaves *N*-acetyl neuraminic acid attached to the end of the oligosaccharide chains, has been available for some time and has been used widely in the serology of blood group antigens (Bird and Wingham, 1970; Springer and Ansell, 1958). Purified glycosidases that will cleave carbohydrate chains present as N-linked glycans attached to asparagine residues are now available (Elder *et al.*, 1982; Plummer *et al.*, 1984). Endo-β-galactosidase will cleave the carbohydrate chain at galactose residues in polylactosaminyl type chains in the outer branched structure of such N-linked glycans. The enzyme endo-*N*-acetyl glucosamidase-F and peptidyl *N*-glycanase, which are both present in cultures from *Flavobacterium meningosepticum* (Plummer *et al.*, 1984), will selectively cleave the chain at *N*-acetyl glucosamine or asparagine, respectively. A reduction in the apparent molecular weight of a band following treatment with these enzymes indicates the presence of the appropriate sugar chain. Thus, the presence of sialic acid in the O-linked glycan present on the sialoglycoproteins can be demonstrated by an increase in their mobility on SDS–PAGE following treatment with neuraminidase (Hadley *et al.*, 1984). The presence of N-linked glycans may be inferred if a similar effect is noted after treatment with endo-F/endo-*N*-glycanase (Tanner *et al.*, 1987). Treatment with endo-β-galactosidase may cause a sharpening of the bands owing to removal of the terminal portion of the sugar chains of N-linked glycans, which are often glycosylated to differing degrees and give rise to microheterogeneity. The use of these enzymes is playing an increasingly important part in determining the nature of blood group active membrane glycoproteins (Tanner *et al.*, 1987).

3.2.2. Examples of Blood Group Antigens That Have Been Shown to Be Associated with Specific Membrane Proteins

A number of blood group antigens have now been shown to be associated with specific erythrocyte membrane proteins (see also Table I). The antigens of the Rh system are an example of these and this is discussed in Chapter 3. The sialoglycoproteins have been shown to carry a number of blood group antigens as discussed in Sections 3.2.2.1 and 3.2.2.2. Application of the techniques described in Section 3.2.1 is now beginning to give information on the location of several other blood group antigens on membrane proteins. Two examples of these are the Kell and Lutheran antigens and these will be discussed briefly. For a more detailed account of these and of other blood group antigens carried on membrane proteins, the reader is referred to the recent review of Anstee (1987).

3.2.2.1. Sialoglycoproteins Associated with MNSs, Wr^b, and En^a Blood Group Activity. The best characterized of the protein components associated with blood group antigen activity are those glycoproteins known as sialoglycoproteins (SGPs), so called because they contain a high proportion of sialic acid residues, mainly in the form of O-linked tetrasaccharide (Thomas and Winzler, 1969). These have been the subject of several comprehensive reviews (Anstee, 1981; Anstee and Tanner, 1986; Anstee *et al.*, 1982; Dahr, 1983) and only the main points are summarized here. As previously explained, the erythrocyte membrane SGP may readily be identified on SDS–PAGE of SDS membrane extracts as they are selectively stained by the PAS staining technique (Steck, 1974) or by immunostaining techniques.

The entire sequence of the major SGP, alpha (glycophorin A), has been determined and comprises some 131 amino acids (Tomita *et al.*, 1978). It has been shown to have three distinct regions with a highly glycosylated N-terminal portion, a transmembrane portion from residues 73 to 92 that contains hydrophobic amino acids and a cytoplasmic C-terminal portion that is not glycosylated. A model of the molecular organization of alpha has been proposed by Welsh *et al.* (1985), based on circular dichroism studies and confirmational prediction from amino acid sequence. The disposition across the membrane is also indicated (Figure 7). When it was first isolated, it was recognized that at

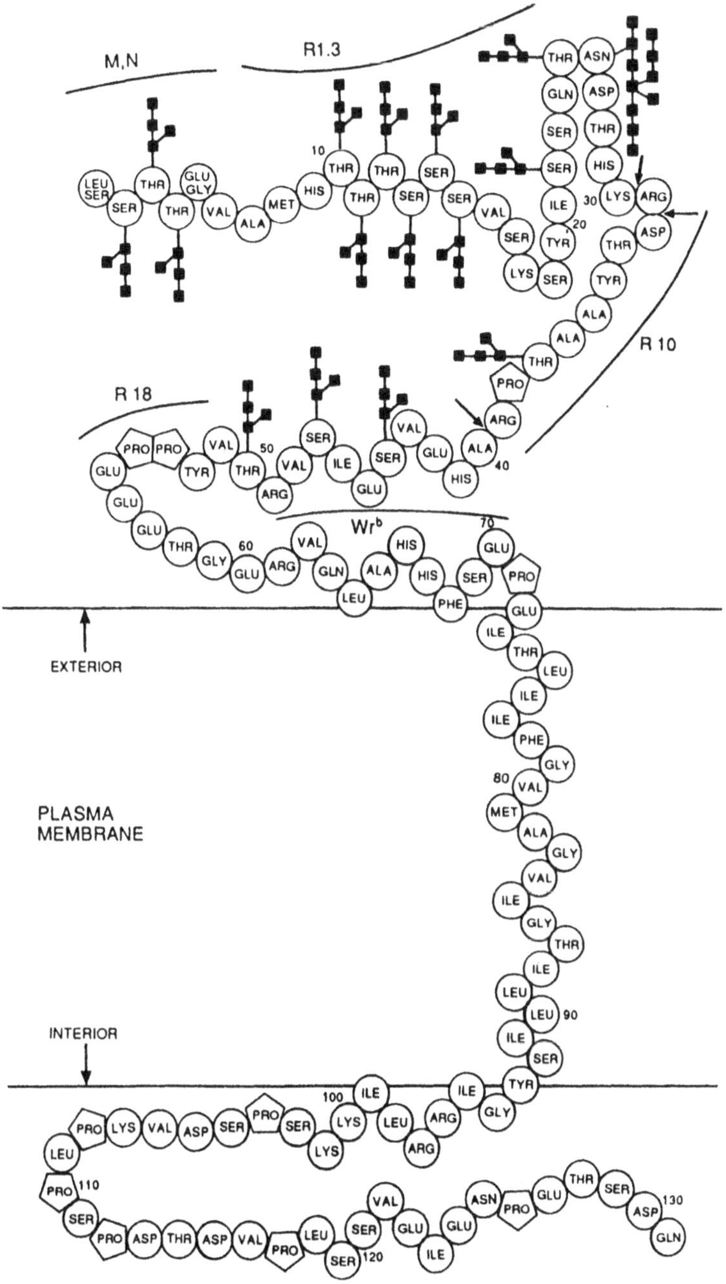

FIGURE 7. Molecular organization of alpha SGP. Reproduced with permission from Welsh *et al.* (1985). The location of the epitope recognized by some monoclonal antibodies as deduced by immunochemical (Mawby *et al.*, 1981) and by epitope mapping (B. Gardner *et al.*, unpublished observations) are illustrated. Arrows denote cleavage sites for trypsin for cell-bound SGP.

positions 1 and 5 in the N-terminal portion there was an amino acid polymorphism that was determined by the MN blood group of the cells used to prepare the SGP (Tomita *et al.*, 1978). The remainder of the sequence was identical for either M- or N-derived SGPs. Thus, the alpha SGP with leucine at position 1 and glycine at position 5 has N activity, whereas that with serine at position 1 and glutamine at position 5 has M activity.

The delta SGP is approximately 96 amino acid residues in length (Dahr *et al.*, 1985b) and the sequence of the first 26 residues is identical to that of N-type alpha SGP. Thus, the N-terminal sequence of delta SGP is always leucine at position 1 and glycine at position 5 in normal cells irrespective of the MN type (Dahr *et al.*, 1980, 1985; Furthmayr, 1978). Thus, all normal cells will have some N antigen that is recognized by certain anti-N antisera and generally termed N. Unlike the N antigen on alpha, this activity is not removed by trypsin treatment of the cells, although the antigen is sensitive to chymotrypsin (Judson and Anstee, 1977). There are a large number of variant cells in which there are deficiencies or abnormalities in these two SGPs (Moulds, 1983) and in some of these the structure of the variant produced has been determined (Blanchard *et al.*, 1985; Dahr *et al.*, 1985b). These are outlined below but for a more complete discussion of this subject see Anstee (1983) and Anstee and Tanner (1986).

In general, the abnormalities may be classified into certain types as follows:

1. Those in which the SGP is completely absent. Examples of these are En(Fin) in which alpha is not present (Anstee *et al.*, 1977), $S^-s^-U^-$ in which delta is absent (Dahr *et al.*, 1978), M^k in which both alpha and delta are absent (Anstee and Tanner, 1978; Dahr *et al.*, 1977; Furthmayr, 1978; Tokunuaga *et al.*, 1979), and the Leach phenotype in which beta and gamma SGPs are absent (Anstee *et al.*, 1984a).
2. Those in which there is an amino acid substitution as in Miltenberger class I or class II cells (Blanchard *et al.*, 1983; Dahr *et al.*, 1984).
3. Those in which an abnormal SGP is produced that is a hybrid molecule, as in Miltenberger class V, or cells carrying the St^a or Dantu antigens.

It has now been established (Anstee *et al.*, 1982) that there are several different types of molecule postulated to have been formed by hybridization of the genes for alpha and delta SGPs. Those hybrids in which the N terminal of the protein is derived from delta SGP with a C terminal derived from alpha have been termed the Lepore type of hybrid (Anstee *et al.*, 1982) by analogy with the similar type of unequal crossover event giving rise to hemoglobin Lepore (Weatherall and Clegg, 1979). An example of this type is the hybrid in St^a cells. Other hybrids have been described (Anstee *et al.*, 1979) in which the N terminal is derived from alpha SGP and the C terminal from delta SGP and these are termed anti-Lepore. An example with this type of hybrid are Miltenberger class V cells.

These crossover events have been postulated to arise by chromosome misalignment as discussed by Anstee *et al.* (1982) and Dahr *et al.* (1980). There is considerable sequence homology between these two proteins as first pointed out by Dahr *et al.* (1980). Apart from the homology of the first 26 residues, there are also regions of homology with 27–35 of delta with overlapping sequences at 56–64 and 59–67 on alpha (Dahr *et al.*, 1980) and also regions of homology in the transmembrane regions of delta and alpha (Dahr *et al.*, 1985b). Thus, there is the possibility of misalignment of the homologous sequences of the chromosomes carrying the genes for these proteins and of such crossover events occurring.

Once the biochemical basis of the MNS system was understood, it was possible to explain the relation of some of the other antigens that had been detected by serological tests. Thus, the antigen Hil, which is associated with Miltenberger class V, is probably an antigen resulting from novel sequence created by the alpha–delta crossover. In a similar way, the St^a and Dantu antigens have been shown to be located at the approximate positions 25–30 and 26–42 of the hybrid molecules (Blanchard *et al.*, 1985; Dahr *et al.*, 1985b) by sequence analysis. The sequences in these regions of the two hybrid molecules are a result of the delta–alpha crossovers. Recent work indicates that in the complex Miltenberger system, the same serological determinants that define the different classes within this system may arise by either amino acid substitution or crossover.

3.2.2.2. Sialoglycoproteins Associated with Gerbich and Webb Blood Group Antigen Activity. The antigens of the Gerbich and Webb blood group systems have recently been shown to be associated with other sialoglycoproteins—namely, beta, beta 1, and gamma. These sialoglycoproteins are less abundant than the alpha sialoglycoprotein but do have some structural similarity. The gene coding for beta has been isolated and its sequence determined Colin *et al.* (1986). The deduced amino acid sequence suggests that it is a transmembrane protein with a glycosylated N-terminal portion and a C-terminal portion that extends from the cytoplasmic side of the membrane.

Association of the Gerbich blood group system with these SGPs was suggested when it was noticed that the normal forms of these SGPs were absent from cells that had the Gerbich negative phenotype (Anstee *et al.*, 1984a). Three different phenotypes were investigated. In one (the Leach phenotype), there was an apparent absence of all three SGPs, whereas in two others abnormal components were detected that are now thought to be hybrids of the beta and gamma SGPs (Anstee *et al.*, 1984b), which have arisen by unequal crossing over in an analogous fashion to that producing the hybrid SGPs in St^a and Mi^V cells. As mentioned above, there is evidence that the beta component is associated with the cytoskeleton via band 4.1 (Anstee *et al.*, 1984a; Mueller and Morrison, 1981) and it is of considerable interest that a proportion of the cells with the Gerbich negative phenotype have an abnormal shape. Some of the cells with the Leach phenotype are elliptocytic (Anstee *et al.*, 1984a) and in a family study it was noted that heterozygous cells may have a higher proportion of elliptocytic cells than normal (Daniels *et al.*, 1986), although fewer than in the homozygous cells. These studies provide evidence that the beta, beta 1, and possibly gamma SGPs play a role in the maintenance of the shape of normal erythrocytes.

The blood group antigen Webb (Wb) is a very low-frequency antigen that is independent of all the major blood group systems. Studies on cells positive for this phenotype have shown that there is an abnormal SGP beta component in these cells as reported by Reid *et al.* (1985) and subsequently by Macdonald and Gerns (1986). The abnormal component has a decreased apparent molecular weight of some 2700 less than the normal beta component. Since removal of the N-linked sugar from normal beta by *endo*-glycosidase F results in a similar decrease in molecular weight (Reid *et al.*, 1985) and this enzyme has no effect on the abnormal component in Webb positive cells, it has been postulated that this component lacks the N-linked oligosaccharide. Presumably, the Webb antigen is located in the region to which the N-linked oligosaccharide is normally attached but is only exposed when this sugar chain is absent.

3.2.2.3. Kell Blood Group System. The membrane components that carry the antigens of the Kell blood group system have recently been shown to be proteins. An indication that the antigens are located on a protein component came from the demonstration that treatment of cells with disulfide bond reducing agents such as dithiothreitol and aminoethylthioproprionate inactivated antigens of the Kell system (Advani *et al.*, 1982), presumably because of the unfolding of the peptide chain (Branch *et al.*, 1983). The isolation of a membrane protein with Kell activity has now been reported by two groups. Redman *et al.* (1984) were able to isolate a protein of 93 kD by immunoprecipitation with human anti-Kell antiserum or with a murine monoclonal anti-Kell. In similar experiments, Wallas *et al.* (1986) isolated a component of approximately 90 kD by elution of the protein from purified anti-Kell immobilized on protein A agarose. The component isolated under nonreducing conditions was capable of neutralizing anti-Kell antisera. When analyzed on SDS–PAGE, the unreduced material with Kell activity gave bands of molecular weight 200, 170, 115, and 85 kD (Redman *et al.*, 1984). The reduced material gave bands of apparent molecular weight 93 and 85 kD and it was assumed that the 93 kD band was an unfolded form of the 85 kD band.

3.2.2.4. Protein Components Associated with Lutheran Blood Group Antigens. Recent evidence has suggested that the Lu^b antigen of the Lutheran blood group system is associated with glycoprotein(s). A monoclonal antibody described by Parsons *et al.* (1987), which has been shown to have Lu^b specificity, was found to react with two membrane proteins of M_r 78,000 and 85,000 on immunoblotting. The proteins were shown to be N-glycosylated as there was a reduction in their molecular weight after treatment with endoglycosidase F (Parsons *et al.*, 1987). This antibody did not react with cells treated with disulfide bond reducing reagents, suggesting that intrachain disulfide bonds are important for maintaining the configuration of this protein in which the Lu^b epitope is expressed.

Cells that are negative for antigens of the Lutheran system may arise from the inheritance of either recessive (Darnborough *et al.*, 1963,) X-linked (Norman *et al.*, 1986) or dominant (Tippett, 1971) genes. The monoclonal antibody described by Parsons *et al.* (1987) does not bind to cells with the recessive or X-linked phenotypes (Merry *et al.*, 1987) but other monoclonal antibodies such as H86 described by Knowles *et al.* (1982), A3D8 by Telen *et al.* (1983), and that described by Udden *et al.* (1987) react with these cells but will not react with cells of the dominant Lu(a−b−) type. A3D8 recognizes epitopes on glycoproteins of M_r 80,000 and 170,000. The dominant type of Lu(a−b−) cells have suppressed expression of antigens in several other blood group system. It has been postulated that the deficiency in these cells is due to abnormal glycosylation in a carbohydrate sequence common to some glycoproteins such as that expressing Lu^b epitopes and to some glycolipids (Udden *et al.*, 1987).

3.2.3. Relative Amounts of Proteins with Blood Group Antigen Activity

Quantitative determination of the number of antigen sites for blood group antigens has shown a wide variability in the level of expression of different antigens on the erythrocyte surface. There have been estimated to be over 1,000,000 copies of the A antigen on certain cells (Economidu *et al.*, 1967) and as few as 850 copies of the Lu^b antigen (Merry *et al.*, 1987) on some normal erythrocytes (see also Table I).

It has been estimated that there are several million molecules of glycosphingolipid on the erythrocyte membrane surface. Blood group A and B determinants have been shown to be located on the oligosaccharides associated with both glycosphingolipids and also the band-3 and -4.5 glycoproteins (see above). The number of A antigen sites has been estimated to be from 140,000 on A_2 to 1,170,000 on A_1 cells (Economidu *et al.*, 1967). On rare cells giving very weak serological reactions with anti-A as few as 700 sites per cell (Cartron, 1976) were detected. Group B cells have on the order of 750,000 B antigen sites (Economidu *et al.*, 1967). The polyglycosyl carbohydrate chains attached to the anion transporter, band 3, and band 4.5 have been estimated to carry some 70% of the blood group A antigens as estimated by experiments in which the A specific GalNAc transferase was used to transfer ^{14}C-labeled GalNAc to H cells (Schenkel-Brunner, 1980; Wilczynska *et al.*, 1980). It is possible that this means of estimation of the distribution of A antigen between glycoprotein and glycolipid may not completely reflect the distribution following the in vivo action of the transferase.

Expression of the Rh(D) antigen of the Rh system is dependent on the Rh phenotype of the cells. Rochna and Hughes-Jones (1967) found there to be from 9900–14,600 sites on cells with the phenotype R_1r to 15,800–33,000 sites on cells with the R_2R_2 phenotype. Certain cells have an abnormally weak or strong expression of Rh(D) antigen. Some low-incidence cells called D^u have a much reduced expression of the Rh(D) antigen (Bush *et al.*, 1974) with some examples having less than 500 D antigen sites per cell. On the other hand, rare cells with the -D- phenotype were estimated to have from 110,000 to 202,000 sites per cell (Hughes-Jones *et al.*, 1971) and those with the .D. phenotype 56,000 (Contreras *et al.*, 1979). Some rare examples of cells with the R_2R_2 phenotype give strong serological reactions (Renton and Hancock, 1956) and quantitative tests suggest there may sometimes be a much higher level of expression of the Rh(D) antigen, which may approach that found on -D- cells (N. C. Hughes-Jones, personal communication; Merry *et al.*, in press).

The number of sites for other antigens of the Rh system was estimated at 35,000–85,000 for c and 13,400–24,000 for e by Hughes-Jones *et al.* (1971). Quantitative assay of the number of binding sites for the monoclonal antibody R6a, which has Rh-related specificity since it fails to react with Rh_{null} cells (Anstee and Edwards, 1982), showed there to be 46,000–104,000 per cell. Experiments with different monoclonal antibodies with D specificity indicate that some antibodies only react with certain categories of D variant cells (Leader *et al.*, submitted for publication).

The number of MN antigen sites can be inferred from the number of alpha SGP sites present. This has been estimated at around 600,000 in normal cells (Merry *et al.*, 1986) although this probably represents the number of dimer molecules (B. Gardner *et al.*, unpublished observations). A value of 1,000,000 for the number of alpha SGP molecules was estimated by Gahmberg *et al.* (1979) using radioimmunoassay inhibition techniques. The number of potential Ss antigen sites inferred from the number of delta SGP sites would be on the order of 80,000–100,000, although these may not all be accesible since a figure of 12,000 sites per cell for s antigen was reported by Masouredis *et al.* (1980).

The number of I sites on adult cells has been reported as 500,000 by Evans *et al.* (1965) and as 100,000 by Doinel *et al.* (1976), who suggested that there there may sometimes be a much higher level of expression of the Rh(D) antigen, which may approach that found on -D- cells (N. C. Hughes-Jones, personal communication; Merry *et al.*, in press).

The number of Kell antigen sites estimated by labeled anti-K was 3500 on cells with the Kk phenotype and 6100 on cells with the KK phenotype (Hughes-Jones and Gardner, 1971). The number of k sites has been estimated at 2000–

5000 (Masouredis *et al.*, 1980). Similar values of 2500–5900 were obtained with a monoclonal antibody, BRIC-18, which recognizes a Kell-related protein (Merry *et al.*, 1984).

The combined number of beta and beta-1 SGPs has been estimated by two monoclonal antibodies BRIC-4 and BRIC-10 to be around 50,000 (Anstee *et al.*, 1984a) and recent work suggests a figure of 50,000–75,000 which would indicate that this is expressed at a similar level to the delta SGP (Merry *et al.*, 1986).

The LW antigen is phenotypically related to the Rh(D) antigen since both are lacking from Rh_{null} cells. The number of LW sites is, however, considerably less than that of Rh(D) and only around 4400 sites were detected on Rh(D) positive cells and about 2800 on Rh(D) negative cells (Mallinson *et al.*, 1986).

The number of Lu^b antigen sites has been found to be low and a variable level of expression was indicated in the cells tested. Values of around 1600–4000 for homozygous cells and 850 to around 1800 for heterozygous cells were reported (Merry *et al.*, 1987).

The number of antigen sites for Fy^a, Fy^b, Jk^a, U, and Di^b have all been reported to be similar and in the range of 14,000–23,000 by Masouredis *et al.* (1980) using quantitative immunoferritin microscopy. Because of the limitations of this technique, however, the authors suggested that these values for ghost membranes may not be comparable to those for intact erythrocytes.

4. SEROLOGICAL ASPECTS OF BLOOD GROUP ANTIGENS

4.1. Hemagglutination Techniques: Practical Aspects

Despite developments of alternative techniques such as enzyme-linked immunoassay (Greenwalt *et al.*, 1975; Leikola and Perkins, 1980), immunoradiometric assay (Jenkins *et al.*, 1977; Masouredis, 1962; Merry *et al.*, 1982; Rochna and Hughes-Jones, 1965), fluorometric assay (Schreiber *et al.*, 1981), and solid-phase techniques (Moore, 1983; Sinor *et al.*, 1985), which are used in some specialist laboratories and may have wider application in the future, the standard technique of blood group serology remains hemagglutination. It has the advantage of technical simplicity and, when performed correctly, has proved over the years to be sufficiently sensitive and reliable for antibody detection. There are many refinements to the technique and it may be performed by manual or by automated methods. The standard technique relies on the ability of the antibody to agglutinate the cells into a visible clump. This may be performed in a variety of ways but is now generally carried out by examination of the agglutination, after centrifugation at low speed, in tubes (spin-tube technique), or wells in microtiter plates. The centrifugation is sometimes omitted and the cells allowed to sediment under gravity. The ability of IgM antibodies to agglutinate cells in saline suspension is much greater than that of IgG antibodies and is related to the difference in distance between the antibody combining sites of some 300 Å for IgM as opposed to around 160 Å for IgG (see Section 4.2).

4.1.1. Enhancement of Hemagglutination Tests

Modifications to the hemagglutination technique, which have been applied to enhance or to potentiate the reaction, may be broadly divided as follows:

1. Addition of various water-soluble polymers such as bovine serum albumin or polybrene.
2. Modification of the cell surface by proteolytic enzymes such as trypsin, papain, or bromelin.
3. The addition of a second antibody to the immunoglobulin and or bound complement components in the antiglobulin technique of Coombs *et al.* (1945), which greatly increases the sensitivity of hemagglutination.

All these techniques are in current use in blood group serology.

4.1.2. Sensitivity of Different Methods for Detecting IgG Antibodies

The relative sensitivity of the various methods used in blood group serology is demonstrated in Table V, using IgG anti-D. The two-stage papain test, described by Goldsmith (1955) (in which the erythrocytes are pretreated with the enzyme, washed, and then treated with the antibody) and antiglobulin tests (Coombs *et al.*, 1945) are of similar high sensitivity. The albumin displacement test described by Case (1959) (in which an albumin solution is layered over a centrifuged suspension of erythrocytes) was found to be 7–10 times less sensitive

Table V
Relative Sensitivity of Enhancement Tests for IgG Anti-D

Test method (15 minute spin-tube)	Minimum level of anti-D detected[a] macroscopically with D positive (R_1r) cells (μg/mL)
Albumin suspension	0.9
Albumin displacement (+20 min saline)	0.05
Papain (two-stage)	0.007
Antiglobulin (spin-tube)	0.007

[a]The minimum concentration of anti-D detectable by serological tests varies with the avidity of the anti-D. The results presented are for one anti-D serum.

than the antiglobulin test. The albumin suspension test procedure may be used as an intermediate test phase before proceeding to an antiglobulin test. The antiglobulin test is the most important antibody detection method used in antibody screening and compatibility tests for IgG antibodies, because it should detect any clinically significant antibody of importance in blood transfusion. The other enhancement tests are supportive tests, which vary in their degree of usefulness with the various antibody specificities involved. Thus, pretreatment of the erythrocytes with proteolytic enzymes, such as papain, provides an excellent and simple rapid test for Rh antibodies, provided that the reagent and technique are standardized (Scott *et al.*, 1987). Albumin displacement tests are useful for the detection of antibodies to enzyme-sensitive antigens and also anti-K antibodies. Albumin tests are not as sensitive, however, as antiglobulin tests and cannot be relied on to detect weak antibodies missed through faulty antiglobulin test procedure.

4.1.3. Overlapping Specificities of the A and B Glycosyltransferases

Until 1979, it was believed that these enzymes had an absolute specificity, in that the *A*-gene specified transferase utilized only UDP–*N*-acetylgalactosamine and *B*-gene-specified transferase only UDP–galactose. Greenwell *et al.* (1979) demonstrated, however, by immunochemical procedures that the B transferase could, in vitro, transfer *N*-acetylgalactosamine from UDP–*N*-acetylgalactosamine to make an A-active structure. A series of immunochemical studies followed; Watkins *et al.* (1981), Yates *et al.* (1982, 1983, 1984), Greenwell *et al.* (1986).

The relevance of these studies to blood group serology was not recognized at that time. It was recently found, however, that a monoclonal anti-A (MHO4), which gave strong reactions in serological tests with all group A cells, also reacted with certain group B red cells. Serological investigations by Voak following communication with Watkins and Yates provided evidence for the in vivo development of some A antigens by the B-specific transferase. This interesting phenomenon in which group B red cells have a very small amount of A antigen on their surface, and have been termed B(A) cells, was studied in depth (Voak, Cambridge, U.K., Davies *et al.*, Ortho, Newark, U.S.A., and Beck *et al.*, Kansas City, U.S.A.). The research was discussed at an Ortho Symposium in 1986 and is reviewed by Voak *et al.* (1987).

4.1.4. The B(A) Phenomenon

A blend of two monoclonal anti-As (3D3 and MHO4) has been prepared by Ortho Diagnostics. This anti-A blend, called Biclone, is an exceptional anti-A in that it reacts with weak subgroups of A that are not detected by conventional anti-

A (group B) sera and most other monoclonal anti-A reagents. This Biclone anti-A was found to give very weak agglutination with some samples of group B red cells.

Investigations with the individual monoclonal antibodies (see Table VI) revealed that MHO4 was the anti-A component causing these reactions. This murine anti-A (MHO4) was prepared accidentally in attempts to make platelet antibodies; in serological tests it is the strongest anti-A ever examined by one of us (D. Voak) and agglutinated 20 out of 24 samples of the weak A subgroup called A_x.

This weak agglutination with certain group B red cells, B(A) cells, can be eliminated by dilution of the MHO4 anti-A, although there is a careful balance between the potency of the anti-A to detect A_x and the ability to detect the A in reactive group B red cells (Voak *et al.*, 1987). The incidence of fairly strong B(A) reactions with Biclone anti-A was generally very low (10 in tests on over 1 million ABO typing tests). The incidence of very weak B(A) bloods was 0.1–0.3% in some centers; only Beck reported an incidence of 0.9%, mainly in black people, and others found none (for a review see the Ortho Symposium publication, 1987).

4.1.5. Method Sensitivity of B(A) Reactions

The agglutination produced by MHO4 anti-A, used at a 30 times higher concentration than in Biclone, is shown by spin-tube and sedimentation tech-

Table VI
Saline Titers of Tissue Culture Supernatant Monoclonal Anti-A Reagents[a]

	Anti-A reagent	
Cells	3D3	MH04
A_1	1024	1024
A_2	512	512
A_1B	512	512
A_2B	64	256
A_2B (weak)	4	256
A_3B	0	256
A_3	4	256
A_3	0	64
B(A) "strong"[b]	0	32

[a]Saline agglutination tests negative with group B and O red cells.
[b]3+ very fragile agglutination, easily dispersed.

Table VII
Agglutination Reactions of Monoclonal Anti-A MH04 with B(A) Red Cells

	Group B donor cells					B(A) cells	
Test	1	2	3	4	5	23509	21865
Spin-tube[a]							
Saline	0	0	0	0	0	+	+++
Papain	+	+	+	++	+++	++	+++
Sedimentation							
Saline	0	0	0	0	0	0	w
Papain	w	0	w	0	0	w	++

[a]The strength of agglutination in a spin-tube test (15 s at 500*g*; 0 to +++ = grades of agglutination) using MHO4 monoclonal anti-A.

niques using papain-treated and untreated cells (see Table VII). Spin-tube tests read by a very careful gentle agitation (or ''tip and roll'') technique give loose agglutination with B(A) red cells. This agglutination is missed by most workers because their standard procedure for ABO grouping destroys the fragile agglutinates. In the United Kingdom, the use of a reading technique involving a pipette to transfer the cell button to the microscope slide also destroys this type of agglutination. This latter technique is a very reproducible and reliable procedure for routine blood grouping work. The treatment of erythrocytes with the proteolytic enzymes papain, bromelin, or ficin enhances the B(A) agglutination reactions and all adult group B cells can be shown to have some A antigen by this technique at high concentrations of MHO4 anti-A.

It was demonstrated that the B(A) reaction is due to A and is not a cross-reaction with B, since the anti-A reaction of MHO4 with B(A) cells is specifically inhibited by the A substance of A secretor saliva. The weak agglutination imposes severe limitations in studies of the B(A) phenomenon as the anti-A is so loosely bound to the A of B(A) red cells that it is removed by washing the red cells and so absorption and elution studies give very unreliable results.

Yates *et al.* (1984) presented in vitro evidence that the B transferases can make a normal blood group A-active tetrasaccharide. However, the in vivo production of the A on B(A) red cells results in an antigen that only reacts very weakly with MHO4 and is nonreactive with normal anti-A of group B sera. We suspect that the A structure produced by B transferases is only present in very small amounts or a different structure to that produced by in vitro studies. Further proof that the B(A) reaction is due to A reaction was provided by showing that the anti-A of high titer anti-A,B (group O) serum, absorbed by B cells to remove anti-B, can also agglutinate B(A) and A_x red cells (Table VIII). Also note the reactions are enhanced by proteolytic enzyme treatment of the red cells.

Table VIII
Reactions with B(A) and Ax Cells of High Titer Group O Anti-A Serum Following Absorption with B Cells

Cell	Test	Dilution of anti-A (Lavil)[a] Neat	2	4	8	16	32	64
B(A) 21865	Saline	+	+	w	0	0	0	0
B(A) 21865	Papain	+++	+++	++	++	+	+	+
B	Saline	0	0	0	0	0	0	0
B	Papain	++	+	w	w	w	0	0
A_x	Saline	++	++	+	+	+	w	w

[a]All negative at further dilutions.

4.1.6. Aberrant Blood Group Expression B(A) and A(B) Occurs in Malignant Tissues

The reverse A(B) phenomenon has not been seen on erythrocytes, but Yates and Watkins (1982) have shown that A transferase can make a small amount of B, but the rate of B antigen formation by A transferases is much lower than that for A antigen by B transferases. However, both A and B aberrant antigens have been found in malignant tissues, for example, in localized foci of gastric and colonic tumor tissue (Breimer, 1980; Denk *et al.*, 1974; Hakkinen, 1970; Kapadia *et al.*, 1981).

The aberrant production of A and/or B depends on the pH and/or relative concentrations of nucleotide sugars, and Yates and Watkins (1982) postulated that under suitable conditions the *A*- or *B*- gene-specified enzyme will transfer the "wrong" sugar, giving rise to A or B antigens that are apparently incompatible with the blood group of the individual. We would point out, however, that while this theory may explain the occurrence of A antigen in group B individuals, and the occurrence of B antigen in group A individuals, it does not explain the occurrence of A antigen in group O individuals, which was observed by Hakkinen (1970).

4.2. Hemagglutination Techniques: Theory and Relation to Immunochemistry

Although hemagglutination is such a widely used technique, our knowledge of its theoretical basis is still incomplete. In order for erythrocytes to be agglutinated, the repulsive forces that normally keep the cells apart must be overcome by the formation of sufficient antibody molecule bridges to support agglutination. When the cells are in the circulation, it is obviously important that separation between them is maintained. The mechanisms that have evolved to ensure that this happens are undoubtedly sophisticated and efficient and represent an important function of the erythrocyte membrane. Thus, in agglutination reactions, the conditions must be such that this fundamental property of erythrocytes can be counteracted.

The hemagglutination reaction may be considered as taking place in two stages. In the first stage, the antigen binding site on an antibody molecule combines with the blood group antigen on one cell. This binding reaction has been shown to be governed by thermodynamic considerations, which apply to any antibody–antigen reaction and follow the law of mass action (Hughes-Jones, 1975; Hughes-Jones *et al.*, 1963, 1964; van Oss *et al.*, 1978). It has been shown that the binding is related to the association constant of the antibody, the concentrations of antibody and antigen, temperature, and the ionic strength of the medium; the effect of all these parameters may be determined experimentally (Hughes-Jones *et al.*, 1964; Masouredis, 1962).

The second stage of agglutination is the reaction of the second binding site of antibody molecules with antigens on another cell in sufficient number to overcome repulsive forces. Agglutination may be brought about by bivalent attachment of IgG antibodies to antigens on adjacent cells. IgM antibodies are more effective because they have a greater span between the binding sites. With IgM antibodies there is also the theoretical possibility of multivalent attachment of the antibody molecule to more than one antigen site on the same cell. The bivalent attachment will be affected by the same parameters as the binding of antigen at the first site, but in addition the cells must be able to approach each other sufficiently closely for this bridging to take place. The strength of binding must be greater than that of the repulsive force between the cells.

Investigation of the repulsive forces between artificial phosphatidyl choline bilayers has shown that the force is related exponentially to the separation distance between the lipid bilayers (LeNeuveu *et al.*, 1976). In the presence of excess water, there was an equilibrium separation in which the repulsive and attractive forces are balanced at 27.5 Å. Addition of dextran solution will remove water associated with phospholipid by osmotic pressure and the separation between the bilayers was found to decrease in proportion to the dextran concentration. Thus, the water of hydration attracted to the phospholipid molecules is an important factor in the repulsive force between the bilayers in this artificial membrane system, and it seems probable that it is of similar importance in the repulsive force between erythrocytes. The water of hydration also has an effect on the first stage of agglutination as measurements of the binding energies in antigen antibody interactions (van Oss and Neumann, 1977) have shown that the free-energy change in the reaction taking place in the presence of water is less than when water is absent. They stressed the importance of the displacement of water in the formation of antigen–antibody bonds and that dehydration will allow the closest fit and will strengthen the bond. The forces keeping the cells apart are of great practical importance in hemagglutination tests, since any factor that will affect the strength of the repulsive force will determine the sensitivity of the test.

The nature of the repulsive force is complex and is influenced by the hydration of molecules at the cell surface and by the charge on the cells (van Oss *et al.*, 1978). There are numerous theories on the relative importance of these repulsive forces that normally keep the cells apart (see discussion by Steane and Greenwalt, 1980). The exterior of the cell membrane possesses a negative charge of which more than 80% is due to the presence of a charged carboxyl ion on the sialic acid residues present in some sugar chains. The majority of the sialic acid residues are found in the tetrasaccharide attached to the sialoglycoproteins and it has been postulated that this charge may have considerable importance in maintaining cell separation. The electrostatic repulsion as formulated by Derjaguin and Landau (1948) is influenced by the ionic strength of the medium, pH, dielectric constant, and viscosity. This was investigated by Pollack *et al.* (1964) who proposed that the zeta potential was the major factor keeping the cells apart. This potential exists between the negative charge of the cell surface and the positively charged ionic cloud that surrounds the cell. The thickness of this diffuse ionic cloud, and hence the zeta potential, is affected by the dielectric constant of the medium. They claimed that this could be changed by the addition of macromolecules such as bovine serum albumin or polyvinyl pyrolidone, which enhance agglutination. Such a theory had previously been postulated for the agglutination of bacteria. Stratton *et al.* (1973) and Voak *et al.* (1974b) pointed out that since neuraminidase treatment of the cells, which causes a marked reduction in zeta potential, does not greatly enhance agglutination with IgG anti-D, factors other than zeta potential must be of greater importance. Subsequent work has shown that technical errors caused by electrode polarization (Brooks, 1971) probably invalidate Pollack's measurements of the dielectric constant after the addition of macromolecules.

A reduction in the net negative charge of 60–70% may be brought about by the removal of sialic acid residues by treatment with neuraminidase. In the case of the O-linked glycans on sialoglycoproteins, which constitute 80% of the membrane sialic acid, this will leave the disaccharide structure GalNAc→Gal, which is the T antigen readily recognized by peanut lectin (Bird, 1964). If, however, the cells are treated with a proteolytic enzyme such as trypsin or papain, there is a considerable increase in sensitivity for an IgG anti-D. There has been some debate on the reason for this (Steane and Greenwalt, 1980). Stratton *et al.* (1973) and Voak *et al.* (1974) hold the view that neuraminidase still leaves peptide chains of the sialoglycoprotein intact, whereas the proteolytic enzymes will cleave some peptide chains, thus causing a reduction in the steric hindrance to binding of the anti-D. It has also been argued (Voak, 1986) that removal of the hydrophilic peptide chains will decrease the water of hydration associated with these structures as described by Stean and Greenwalt (1980). Masouredis *et al.* (1977) concluded that no single factor could account for proteolytically produced changes in enhancing hemagglutination but that a fundamental change in the biophysical properties of the membrane was responsible.

The validity of the zeta potential theory was also questioned by van Oss *et al.* (1978) after investigation of parameters such as cell shape, extracellular osmotic pressure, and the degree of hydration. It was pointed out that some macromolecules that enhance agglutination either increase or do not affect the zeta potential. For some polymers such as dextran, the ability of the polymer to form bridges between cells was considered important. Osmotic effects on cell shape caused by macromolecules, which may induce a change from the normal biconcave disc form, were considered to allow a greater part of the surfaces of the cells to approach each other. It seems probable that there are a variety of forces that are responsible for the repulsion between erythrocytes (Steane and Greenwalt, 1980). The relative magnitude of these will vary with the distance between the cells.

Since the secondary structures of the divalent IgG molecules and the IgM molecule, which probably has a valency of 5 for cell-bound antigens, are known in considerable detail (Beale and Feinstein, 1976), it is possible to calculate the maximum distance between two antibody combining sites in each of the molecules (Kehoe and Capra, 1971; Wu and Kabat, 1970). Calculations show that a maximum distance across the IgM pentamer is some 300 Å (Beal and Feinstein, 1976) and some 160 Å for the IgG molecule (Bunting and Cathou, 1973). This would account for the differences in behavior of antibodies of these classes in hemagglutination tests because the cells do not need to approach each other so closely for them to be cross linked by the IgM molecule. The multivalent attachment of IgM antibodies to antigens on each cell has been observed for anti-A, which has a high site density (Williams and Voak, 1972) and this would increase the strength of binding. The ability of IgM antibodies to do this is limited to those antigens that are less than 300 Å apart on the cell membrane. For many blood group antigens, including the Rh antigens, the abundance and distribution on the surface of the membrane probably prevent this multivalent attachment, although on cells that have the rare phenotype -D- and that have 3–5 times as many Rh(D) antigen sites as cells with the R_2R_2 phenotype (Hughes-Jones *et al.*, 1971), such multivalent attachment has been observed with anti-D.

There are other observations that indicate that the distance between cells over which the cross linking has to take place before the cells are agglutinated is critical. The high level of sensitivity of the antiglobulin test, which may detect the addition of as few as 10–20 molecules of IgG per cell (Merry *et al.*, 1986) is due to the bridging between IgG molecules on adjacent cells by a second antibody operating at intracellular distances out of the range of repulsive forces. IgG antibodies, which are not capable of directly agglutinating cells, may be able to do so after the reduction and alkylation of interchain disulfide bonds in the molecule, which allows unfolding of the molecule (Romans *et al.*, 1977). It has been calculated that this may increase the span between the antigen combining sites by as much as 140 angstrom units. It has been shown that this modification can be applied to some anti-D (Romans *et al.*, 1977) and anti-Kell antisera. It appears, however, that only antibodies of higher affinity may be converted into directly agglutinating antibodies by this procedure.

The linking of cells by lectin or antibody molecules may be visualized by electron microscopy with ferritin-labeled lectin (Williams and Voak, 1972) or antibody (Voak and Williams, 1971; Voak *et al.*, 1974b) or by scanning electron microscopy of agglutinated cells (Masouredis *et al.*, 1977; Salsbury and Clark, 1967; van Oss and Mohn, 1970). It appears that an intracellular distance of 180–200 Å is critical for agglutination and that the repulsive forces acting on the cells over this distance are such that they cannot be overcome by the strength of the antigen–antibody interaction. Cells of the Rh phenotype R_2R_2 when maximally coated with nonagglutinating IgG antibodies to give 35,000 molecules per cell do not agglutinate even when subjected to centrifugal forces of 1000*g* for 220 s, indicating the strength of the repulsive forces. When cells with the -D- phenotype—in which the Rh(D) antigen site density is greater—are coated with 42,000 IgG molecules, weak agglutination can be observed in a standard 90 min sedimentation or spin-tube test and strong agglutination after centrifugation at 1000*g* for 20 s. Williams and Voak (1972) and Voak *et al.* (1974b) observed the distribution of A and Rh(D) antigens and agglutination of erythrocytes by a ferritin-labeled specific anti-A lectin and anti-D antibody using electron microscopy. They concluded that the ability to form multiple bonds between the cells and to form large areas of cell-to-cell contact were important in bringing about strong hemagglutination. The formation of multiple bonds will be influenced by the multiple factors of antigen site density, the antibody span, antibody affinity/avidity, and the antibody valency. Measurement taken from the electron micrographs of IgG anti-D agglutinated cells after albumin or protease enhancement indicated a maximum separation between the unit membrane of 110 ± 10 Å (Voak *et al.*, 1974b). The flexibility of the cell membrane was also shown to be an important factor in the agglutinability of erythrocytes by allowing large areas of cell-to-cell contact (Voak *et al.*, 1974b). The effect of flexibility of the cell membrane is also discussed by Steane and Greenwalt (1980).

The effects of proteolytic enzymes on proteins in the intact cell membrane are uncertain since steric effects may restrict the accessibility of the enzyme to potential cleavage sites. As an example, the number of theoretical cleavage sites for trypsin, which has a specificity for the peptide bond linking basic amino acids such as lysine and arginine, can be calculated from the amino acid sequence of alpha SGP. There are six such sites in the extracellular portion of the chain, but in the intact membrane some of these sites are inaccessible and others are cleaved at only a slow rate. Cleavage does take place, however, at positions 31–32, 32–33, and 39–40 and this results in the loss of a major fragment. The effects of other proteolytic enzymes are even more difficult to predict since they are often used as impure preparations and the specificity is not so well defined. The St^a and Dantu antigens are carried on hybrid delta–alpha SGPs that contain sequences known to be cleaved by trypsin and chymotrypsin (Blanchard *et al.*, 1985; Dahr *et al.*, 1985b), and there is cleavage of these bonds in normal alpha and delta components (Judson and Anstee, 1977). In the hybrids, however these cleavage sites appear to be more inaccessible to the enzymes as the antigens have increased resistance to proteolytic digestion (Blanchard *et al.*, 1985, Dahr *et al.*, 1985). It is also possible that cleavage of bonds between amino acid residues by proteolytic enzymes may not remove peptides from all membrane proteins. There may be noncovalent interactions by which the cleaved peptide remains associated with the protein or, as in the case of the anion transport protein band 3, the peptide may still remain attached at one end of the membrane if the protein traverses the membrane more than once.

The reactions of IgG anti-A, anti-B, or Rh(D) antibodies with their respective antigens are often discussed in relation to the relative accessibility of the two antigens. There are problems in making this comparison, as there are differences in the relative number of antigen sites. It has been estimated that there are something in the region of 1 million A sites on A_1 cells and around 250,000 on A_2 cells, whereas the number of D sites varies from approximately 10,000 to 30,000 per cell depending on the Rh phenotype (see Table I). The density and mobility will also affect agglutination since with more closely packed sites multiple linkages between the cells may be formed at the point of contact. IgG anti-D normally does not agglutinate red cells suspended in saline, but experimental evidence from the reaction of anti-D with cells of the rare -D- phenotype on which there may be up to 180,000–200,000 Rh(D) antigen sites per cell (Hughes-Jones *et al.*, 1971) has shown that these cells may be agglutinated when 40,000 antibody molecules are bound (D. Voak and A. H. Merry, unpublished observations).

Another major difference between these antigens is that the D antigen is apparently only carried on a defined peptide, whereas the GalNAc-containing structure recognized as the A antigen is apparently carried on a variety of glycolipid and glycoprotein molecules. The distribution of the A antigen has variously been estimated to be from 50 to 70% on glycoprotein (Finne *et al.*, 1980). Although the precise location of the Rh(D) antigen on peptide or peptides is not yet known, evidence points to it being in a region near the insertion into the membrane [see Chapter 3 for details of immunochemistry of the Rh(D) polypeptide]. The A antigen may be present in a variety of different environments from that of the membrane glycolipids to that of the N-linked glycan of bands 3 and 4.2 and possibly to a lesser extent on other glycoproteins. It is therefore not possible to define its location in the membrane with certainty and the component carrying the antigenic determinant recognized by anti-A antibodies may change under different experimental conditions. The behavior of the A antigen on glycoprotein and glycolipid with respect to agglutination would not be expected to be the same. Clustering of glycolipid may occur quite readily, whereas a large proportion of the band 3 molecules are anchored to the cytoskeleton.

In view of this, the question is how the effect of the location of the blood group antigen on its behavior in agglutination tests can best be studied. In light of the developments now taking place in our understanding of the immunochemistry of blood group antigens, it should be possible to compare hemagglutination reactions of antibodies of different specificities, where the nature of the antigenic determinants and the component with which it is associated are well defined. In addition, the location and orientation of the components should be known and the antigens under study are of similar abundance. If the studies are conducted with monoclonal antibodies, it is also possible to define the antigenic determinants recognized and the physicochemical properties of the antibody. We are now in a position to perform such experiments and the results should contribute greatly to our knowledge of the basis of hemagglutination reactions.

4.3. Monoclonal Antibodies with Blood Group Antigen Activity

Following the production of monoclonal antibodies with A specificity (Barnstable *et al.*, 1978) a large number of murine monoclonal antibodies with blood group antigen activity have now been produced and a few of these are listed in Table IX. They have proved invaluable in research into the structure of blood group antigens and several are now in current use as blood grouping reagents (Parsons, 1985; Voak, 1983, 1984, 1986).

There are also a number of human monoclonal antibodies of anti-D specificity available that have been produced by either EB virus transformation of lymphoblastoid cell lines (Crawford *et al.*, 1983), or cell fusion of human–mouse cells (Thomson *et al.*, 1986), or cell fusion of human–human cells (Lowe *et al.*, 1986).

It would appear that there is considerable potential for monoclonal antibodies for blood grouping reagents or for therapeutic use in the prevention of Rh hemolytic disease of the newborn, although the mechanism of Rh prophylaxis is not understood and requires further investigation. The change from polyclonal antisera to monoclonal antibodies should be carried out with caution, however,

Table IX
Some Representative Examples of Monoclonal Antibodies with Blood Group Antigen Activity

Antibody specificity	Representative examples
A	Very many examples including 3D3 (Lowe *et al.*, 1984), MH04 (Voak *et al.*, 1987), BRIC 66 (Parsons, 1985)
A_1	TH-1 (Clausen *et al.*, 1985), several examples of anti-A may react preferentially, but not exclusively with A_1
B	Very many examples including NB1, 5A5, 3B4 (Voak *et al.*, 1973), BRIC 30 (Parsons, 1985)
A,B	Moore *et al.* (1984)
H	Doniel *et al.* (1983)
I	Feizi (1985[b])
Le^a	Young *et al.* (1983), Fraser *et al.* (1985)
Le^b	Messeter *et al.* (1984)
P	Borne *et al.* (1986), Kannagi *et al.* (1982)
LKE	Tippet *et al.* (1986)
Lu^b	S. F. Parsons *et al.* (1987)
M	Fraser *et al.* (1985), Nichols *et al.* (1985)
N	Fraser *et al.* (1982), Fletcher and Harbour (1984)
Wr^b	Anstee and Edwards (1982)
Kell	Parsons *et al.* (1982)
k (Cellano)	Sonneborn *et al.* (1983)
LW^a	Sonneborn *et al.* (1984)
T	Rahman and Longnecker (1982)
Tn	S. F. Parsons (unpublished observations)
MER2	Daniels *et al.* (1987)
D	Crawford *et al.* (1983), Doyle *et al.* (1985), Thompson *et al.* (1987)
e	Bourel *et al.* (1987), see also Voak and Tills (1983)

as the very precise specificity of monoclonal antibodies or the high concentrations in which they may be produced may give rise to unexpected reactions in serological tests as in the B(A) phenomenon. This underlines the need for very thorough research on the characteristics of monoclonal antibodies for use in either blood group serology or therapeutically.

5. CLOSING REMARKS AND FUTURE DEVELOPMENTS

We have outlined the present state of our knowledge of the membrane components associated with blood group antigen activity; however, there are still a large number of antigens that have not been assigned to definite components. As the sensitivity and selectivity of immunochemical and biochemical techniques increase, there is no doubt that in a few years time many of these will have been identified. In particular, the isolation and characterization of the peptides and oligosaccharides of membrane glycoproteins can be expected to yield much useful information.

There is also likely to be greater understanding of the pathways by which the membrane glycoproteins are synthesized. It is to be hoped that we will also be able to learn more of the organization of the different components on the cell surface and of the relations between them. This may enable us to understand better the function of such molecules as the sialoglycoproteins.

The fact that many of the blood group antigens are not confined to erythrocytes and that changes in malignancy detected by alterations in erythrocyte membrane proteins may reflect a change in the same component in a tissue that is connected with the pathophysiology of the disease is also a very interesting area.

Finally, the field of recombinant DNA technology has yet to make its impact on the study of blood group antigens. The genes for the alpha, delta, and beta SGPs have now, been isolated and sequenced (Siebert and Fukuda, 1986a,b). Work on the isolation of genes coding for other proteins with blood group antigen activity and for glycosyltransferases determining specific blood groups is now actively underway. The next few years should see considerable progress in this area and will greatly add to our knowledge of the nature and genetics of blood group antigens.

6. REFERENCES

Advani, H., Zamor, W., Judd, W. J., Johnson, C. L., and Marsh, W. L., 1982, Inactivation of Kell blood group antigens by 2-aminoethylisothiouronium bromide. *Br. J. Haematol.* **51**:107–115.

Anderson, R. A., and Lovrien, R. E., 1984, Glycophorin is linked by band 4.1 protein to the human erythrocyte membrane skeleton. *Nature* **307**:655–658.

Andresen, P. H., 1948, The blood group system L. A new blood group L_2. A case of epistasy within the blood groups. *Acta Pathol. Microbiol. Scand.* **25**:728–732.

Andresen, P. H., and Jordal, K., 1949, An incomplete agglutinin related to the L-(Lewis) system. *Acta Pathol. Microbiol. Scand.* **26**:636–654.

Anstee, D. J., 1981, The blood group MNSs-active sialoglycoproteins. *Semin. Haematol.* **18**:13–31.

Anstee, D. J., 1983, Contribution of monoclonal antibodies to the study of red cell membrane sialoglycoproteins. In *Red Cell Membrane Glycoconjugates and Related Genetic Markers* (J.-P. Cartron, P. Rouger, and Ch. Salmon, eds.), pp. 37–42, Librairie Arnette, Paris.

Anstee, D. J., 1985, Blood group MNSs-active sialoglycoproteins of the human erythrocyte membrane. In *Immunobiology of the Erythrocyte* (S. G. Sandler, J. Nusbacher, and M. S. Schanfield, eds.), pp. 55–67, A. R. Liss, New York.

Anstee, D. J., 1986, Red cell membrane antigens. In *Red Cell Antigens and Antibodies* (G. Garratty, ed.), pp. 1–16, American Association of Blood Banks, Arlington, VA.

Anstee, D. J., and Edwards, P. A. W., 1982, Monoclonal antibodies to human erythrocytes. *Eur. J. Immunol.* **12**:228–232.

Anstee, D. J., and Tanner, M. J. A., 1986, Red cell membrane sialoglycoproteins. *Br. J. Haematol.* **64:**211–217.

Anstee, D. J., Mawby, W. J., and Tanner, M. J. A., 1979, Abnormal blood group Ss active sialoglycoproteins in the membrane of Miltenberger class III,IV and V human erythrocytes. *Biochem. J.* **183:**193–204.

Anstee, D. J., Mawby, W. J., and Tanner M. J. A., 1982, Structural variation in human erythrocyte sialoglycoproteins, In *Membranes and Transport* (A. Martinosi, ed.), Vol. 2. pp. 427–433, Plenum Press, New York.

Anstee, D. J., Barker, D. M., Judson, P. A., and Tanner, M. J. A., 1977, Inherited sialoglycoprotein deficiencies in human erythrocytes of type En(a−). *Br. H. Haematol.* **35:**309–320.

Anstee, D. J., Parsons, S. F., Ridgwell, K., Tanner, M. J. A., Merry, A. H., Thomson, E. E., Judson, P. A., Johnson, P., Bates, S., and Fraser, I. D., 1984a, Two individuals with elliptocytic red cells apparently lack three minor erythrocyte membrane sialoglycoproteins. *Biochem. J.* **218:**615–619.

Anstee, D. J., Ridgwell, K., Tanner, M. J. A., Daniels, G. L., and Parsons, S. F., 1984b, Individuals lacking the Gerbich blood group antigen have alterations in the human erythrocyte membrane sialoglycoproteins beta and gamma. *Biochem. J.* **221:**97–104.

Barnstable, C. J., Bodmer, W. F., and Brown, G., 1978, Production of monoclonal antibodies to group A erythrocytes, HLA and other human cell surface antigens—New tools for genetic research. *Cell* **14:**9–20.

Batteiger, B., Newhall, W. J., and Jones, R. B., 1982, The use of Tween-20 as a blocking agent in the detection of proteins transferred to nitrocellulose membranes. *J. Immunol. Methods* **55:**297–307.

Beal, P., and Feinstein, A., 1975, Structure and function of the constant region of immunoglobulins. *Q. Rev. Biophys.* **9:**135–180.

Beale, D., and Feinstein, A., 1976, Structure and function of the constant region of immunoglobulins. *Q. Rev. Biophys.* **9:**135–180.

Bennet, V., 1985, The membrane skeleton of human erythrocytes and its implication for more complex cells. *Ann. Rev. Biochem.* **54:**273–304.

Bennet, V., and Stenbuck, P. J., 1979, Association between ankyrin and the cytoplasmic domain of band 3 isolated from the human erythrocyte membrane. *J. Biol. Chem.* **255:**2540–2548.

Bernstein, F., 1924, Ergebnisse einer biostatischen zusammenfassenden Bertrachtung uber die ehrlichen Blutstrukturen des Menschen. *Klin. Wochenschr.* **3:**1495–1497.

Betteridge, A., and Watkins, W. M., 1986, Acceptor substrate specificities of human α-1-L-fucosyltransferases from different tissues. *Biochem. Soc. Trans.* **13:**1126–1127.

Bird, G. W. G., 1952, Relationship of the blood sub-groups A_1, A_2, and A_2B, A_2B to haemagglutinins present in the seeds of *Dolichos biflorus. Nature* **170:**674.

Bird, G. W. G., 1964, Anti-T in peanuts. *Vox Sang.* **9:**748–749.

Bird, G. W. G., and Wingham, J., 1970, *N*-Acetylneuraminic (sialic) acid and human blood group antigen structure. *Vox Sang.* **18:**240–252.

Blanchard, D., Asseraf, A., Prignet, M.-J., and Cartron, J.-P., 1983, Miltenberger class I and II erythrocytes carry a variant of glycophorin A. *Biochem. J.* **213:**399–404.

Blanchard, D., Dahr, W., and Cartron, J.-P., 1985, Structural analysis of a blood group active hybrid sialoglycoprotein from human red cell membranes. In *Proceedings of the 8th International Symposium on Glycoconjugates* (A. Davidson, *et al.*, eds.), pp 40–42, Praeger Scientific, New York.

Blumenfeld, O. O., and Adamany, A. M., 1978, Structural polymorphism within the amino terminal regions of MM, NN and MN glycoproteins (glycophorins) of the human erythrocyte membrane. *Proc. Natl. Acad. Sci. U.S.A.* **75:**2727–2731.

Borne, von dem, A. E. G., Bos, M. J. E., Joustra-Mass, N., Tromp, J. F., van Wijngaarden-dubois, R., and Tetteroo, P. A. T., 1986, A murine monoclonal IgM antibody specific for blood group P antigen (globoside). *Br. J. Haematol.* **63:**35–46.

Branch, D. R., Muensch, H. A., Sy Siok Hian, A. L., and Petz, L. D., 1983, Disulphide bonds are a requirement for Kell and Cartwright (Yt[a]) blood group antigen integrity. *Br. J. Haematol.* **54:**573–578.

Bourel, D., Lecointre, N., Genenet, M., Gueguen-Duchesne, M., and Genetet, B., 1987, Murine monoclonal antibody suitable for use as an Rh reagent, anti-e. *Vox Sang.* **52:**85–88.

Branton, D., Cohen, C. M., and Tyler, J., 1981, Interaction of cytoskeletal proteins of the human erythrocyte membrane. *Cell* **24:**24–32.

Breimer, M. E., 1980. Adaptation of mass spectrometry for the analysis of tumour antigens as applied to blood group glycolipids of a human gastric carcinoma. *Cancer Res.* **40:**897–908.

Brooks, D. E., 1971, Theoretical and experimental studies on the interaction of neutral polymers with cell surfaces, Ph.D. thesis, University of Oregon, pp. 127–137.

Bunting, J. R., and Cathou, R. E., 1973, Energy transfer distance measurements in immunoglobulins II. Localization of the hapten binding sites and the interheavy chain disulfide bond in rabbit antibody. *J. Mol. Biol.* **77:**223–235.

Burnie, K. L., 1973, Ii antigens and antibodies. *Can. J. Med. Technol.* **35:**5–15.

Bush, M., Sabo, B., Stroop, M., and Masouredis, S. P., 1974, Red cell D antigen sites and titration scores in a family with weak and normal D[u] phenotypes inherited from a homozygous D[u] mother. *Transfusion* **14:**433–439.

Cameron, G. L., and Staveley, J. M., 1957, Blood group P substance in hydratid cyst fluid. *Nature* **179:**147–148.

Cartron, J.-P., 1976, Etude quantative et thermodynamique des phenotypes erythrocytaires "A faible." *Rev. Fr. Trans. Immuno-hematol.* **19:**35–54.

Case, J., 1959, The albumin layering method for D typing. *Vox Sang.* **4:**403–410.

Chasis, J. A., Mohandas, N., and Shohet, S. B., 1985, Erythrocyte membrane rigidity induced by glycophorin A–ligand interaction: Evidence for a ligand-induced association between glycophorin A and skeletal proteins. *J. Clin. Invest.* **75:**1919–1926.

Clausen, H., Levery, S. B., Nudelman, E., Tsuchiya, S., and Hakamori, S., 1985, Repetitive A epitope (type 3 chain A) defined by blood group A_1-specific monoclonal antibody TH-1: Chemical basis of qualitative A_1 and A_2 distinction. *Proc. Natl. Acad. Sci. U.S.A.* **82:**1199–1203.

Cohen, C. M., 1983, The molecular organisation of the red cell membrane skeleton. *Semin. Haematol.* **20:**141–158.

Colin, Y., Rakael, C., London, J. Romeo, P. H., d'Auriol, L., Galibert, F., and Cartron, J. P., 1986, Isolation of cDNA clones and complete amino acid sequence of human erythrocyte glycophorin C. *J. Biol. Chem.* **261:**229–233.

Contreras, M., Armitage, S., Daniels, G. L., and Tippett, P., 1979, Homozygous .D. *Vox Sang.* **35:**81–84.

Coombs, R. R. A., Mourant, A. E., and Race, R. R., 1945, A new test for the detection of weak and "incomplete" Rh agglutinins. *Br. J. Exp. Pathol.* **26:**255–266.

Cooper, R. A., 1978, Influence of increased membrane cholesterol on membrane fluidity and cell function in human red blood cells. *J. Supramol. Struct.* **8:**413–430.

Crawford, D. H., Barlow, M. J., Harrison, J. F., Winger, L., and Huens, E. R., 1983, Production of human monoclonal antibody to Rhesus D antigen. *Lancet* **i:**386–388.

Cory, H. T., Yates, A. D., Donald, A. S. R., Watkins, W. M., and Morgan, W. T. J., 1974, The nature of the human blood group P_1 determinant. *Biochem. Biophys. Res. Commun.* **61:**1289–1296.

Crookston, M. C., Tilley, C. A., and Crookston, J. H., 1970, Human blood chimaera with seeming breakdown of immune tolerance. *Lancet* **ii:**1110–1112.

Dahr, W., 1983, Biochemical studies of erythrocyte membrane glycoprotein variants. In *Red Cell Membrane Glycoconjugates and Related Genetic Markers* (J.-P. Cartron, P. Rouger, and Ch. Salmon, eds.), pp. 27–36, Librairie Arnette, Paris.

Dahr, W., Uhlenbruck, G., and Knott, H., 1977, The deficiency of Mk erythrocytes as revealed by sodium-dodecyl sulphate polyacrylamide gel electrophoresis. *J. Immunogenet.* **4:**191–200.

Dahr, W., Uhenbruck, G., Janssen, E., and Schmalisch, R., 1977, Heterogeneity of human red cell membrane sialoglycoproteins. *Blut* **32:**171–184.

Dahr, W., Uhlenbruck, G., Leikola, J., and Wagstaff, W., 1978, Studies on the membrane glycoprotein defect of En(a−) erythrocytes: III N-terminal amino acids of sialoglycoproteins from normal and En(a−) erythrocytes. *J. Immunogenet.* **5:**117–127.

Dahr, W., Beyreuther, K., Steinbach, H., Gielen, W., and Kruger, J., 1980, Structure of the Ss blood group antigens II. A methionine/threonine polymorphism within the N-terminal sequence of the Ss glycoprotein. *Hoppe Seylers Z. Physiol. Chem.* **361:**895–906.

Dahr, W., Beyreuther, K., Kordowicz, M., and Kruger, J., 1981, N-terminal amino acid sequence of sialoglycoprotein D (glycophorin C) from human erythrocyte membranes. *Eur. J. Biochem.* **125:**57–62.

Dahr, W., Newman, R. A., Contreras, M., Kordowicz, M., Teesdale, P., Beyreuther, K., and Kruger, J., 1984, Structures of Miltenberger class I and II specific major human erythrocyte membrane sialoglycoproteins. *Eur. J. Biochem.* **138:**259–265.

Dahr, W., Muller, T., Moulds, J., Baumeister, G., Issit, P., Wilkinson, S., and Garratty, G., 1985a, High frequency antigen of the human erythrocyte membrane sialoglycoproteins: 1. En^a receptors in the glycosylated domain of the Mn sialoglycoprotein. *Biol. Chem. Hoppe Seyler* **366:**41–51.

Dahr, W., Blanchard, D., Hummel, M., Cartron, J.-P., and Beyreuther, K., 1985b, Preparative purification of the Ss and D sialoglycoproteins and amino acid sequence analysis of their intramembranous domains. In *Proceedings of the 8th International Symposium on Glycoconjugates* (A. Davidson *et al.*, eds.), pp. 21–22, Praeger Scientific, New York.

Daniels, G. L., Shaw, M.-A., Judson, P. A., Reid, M. E., Anstee, D. J., Colpitts, P., Cornwall, S., Moore, B. P. L., and Lee, S., 1986, A family demonstrating inheritance of the Leach phenotype: A Gerbich-negative phenotype associated with elliptocytosis. *Vox Sang.* **50:**117–121.

Daniels, G. L., Tippett, P., Palmer, D. K., Miller, Y. E., Geyer, D., and Jones, C., 1987, MER2, a red cell polymorphism defined by monoclonal antibodies. *Vox Sang.* **52:**107–111.

Darnborough, J., Firth, R., Giles, C. M., Goldsmith, K. L. G., and Crawford, M. N., 1963, A "new" antibody anti-Lu^aLu^b and two further examples of the phenotype Lu(a−b−). *Nature* **198:**796–798.

Deboissezon, J. F., Marty, Y., Ducos, J., and Abbal, M., 1970, Presence constante d'une substance inhibitrice de l'anticorps anti-i dans le serum humain normal. *C.R. Acad. Sci. (Paris)* **271:**1448–1460.

Denk, H., Tappeiner, G., Davidowitz, A., Eckerstorfer, R., and Holzner, J. H., 1974, Carcinoembryonic antigens and blood group substances in carcinomas of the stomach and colon. *J. Natl. Cancer Inst.* **53:**933–942.

Dodge, J. T., Mitchell, C., and Hanahan, D. J., 1963, Preparation and chemical characteristics of haemoglobin-free ghosts of human erythrocytes. *Arch. Biochem. Biophys.* **100:**119–130.

Doinel, C., Ropars, C., and Salmon, C., 1976, Quantitative and thermodynamic measurements on I and i antigens of human red blood cells. *Immunology* **30:**289–295.

Donald, A. S. R., 1981, A-active trisaccharides isolated from A_1 and A_2 blood group specific glycoproteins. *Eur. J. Biochem.* **120:**243–249.

Doyle, A., Jones, T. J., Bidwell, J. L., and Bradley, B. A., 1985, In vitro development of human monoclonal-secreting plasmacytomas *Hum. Immunol.* **13:**199–209.

Dungern, E. von, and Hirzfeld, L., 1910, Uber verebung gruppenpezifischer des blutes. Struckturen des Blutes. *Z. Immunitaets forsch.* **6:**284–292.

Economidou, J., Hughes-Jones, N. C., and Gardner, B., 1967, Quantitative measurements concerning A and B antigenic sites. *Vox Sang.* **12:**321–328.

Elder, J. F., and Alexander, S., 1982, Endo-β-*N*-acetylgalactosamidase from *Flavobacterium menigosepticum* cleaves both high mannose and complex glycoproteins. *Proc. Natl. Acad. Sci. U.S.A.* **79:**4540–5544.

Feizi, T., 1985a, Carbohydrate antigens in human cancers. In *Cancer Surveys* (W. F. Bodmer, ed.), pp. 245–269, Oxford University Press, Oxford.

Feizi, T., 1985b, Demonstration by monoclonal antibodies that carbohydrate structures of glycoproteins and glycolipids are oncodevelopmental antigens. *Nature* **314:**53–57.

Feizi, T., and Kabat, E. A., 1972, Immunochemical studies on blood groups. LIV. Classification of anti-I and anti-i sera into groups based on reactivity patterns with various antigens related to the blood group A, B, H, Le^a, and Le^b precursor substances. *J. Exp. Med.* **135:**1247–1258.

Feizi, T., Kabat, E. A., Vicari, G., Anderson, B., and Marsh, W. L., 1971, Immunochemical studies on blood groups. XLIX. The I antigens complex: Specificity differences amongst anti-I sera revealed by quantitative precipitation studies; partial structure of the I determinant specific for one anti-I serum. *J. Immunol.* **117:**717–721.

Feizi, T., Childs, R. A., Hakomori, S., and Powell, M. E., 1978, Blood group Ii active gangliosides of human erythrocyte membranes. *Biochem. J.* **173:**245–254.

Feizi, T., Childs, R. A., Watanabe, K., and Hakomori, S., 1979, Three types of blood group I specificity among monoclonal anti-I autoantibody revealed by analysis of a branched erythrocyte glycolipid. *J. Exp. Med.* **149:**975–980.

Finne, J., 1980, Identification of the blood group ABH-active glycoprotein components of the human erythrocyte membranes. *Eur. J. Biochem.* **104:**181–189.

Fletcher, A., and Harbour, C., 1984, An interesting monoclonal anti-N produced following immunisation with human group O, NN erythrocytes. *J. Immunogenet.* **11:**121–126.

Francois, A., Sansonetti, N., Mollicone, R., LePendu, J., Galton, J., Jankness, B., and Oriel, R., 1986, Heterogeneity of Lewis antibodies. A comparison of reaction of human and animal reagents with synthetic oligosaccharides. *Vox Sang.* **50:**227–234.

Fraser, R. H., Munroe, A. C., Williamson, A. R., Barrie, E. K., Hamilton, E. A., and Mitchell, R., 1982, Mouse monoclonal anti-N. I. Production and serological characterisation. *J. Immunogenet.* **9:**295–301.

Fraser, R. H., Inglis, G., Mackie, A., Munroe, A. C., Allan, E. K., Mitchell, R., Sonneborn, H. H., and Uthemann, H., 1985, Mouse monoclonal antibodies reacting with M blood group-related antigens. *Transfusion* **25:**261–266.

Furthmayr, H., 1978, Structural composition of glycophorins and immunochemical analysis of genetic varaints. *Nature* **271:**519–524.

Gahmberg, C. G., 1976, External labeling of human erythrocyte glycoproteins. Studies with galactose oxidase and fluorography. *J. Biol. Chem.* **251:**510–515.

Gahmberg, C. G., 1982, Molecular identification of the human Rh_o(D) antigen. *FEBS Lett.* **140:**93–97.

Gahmberg, C. G., and Andersson, L. C., 1977, Selective radioactive labelling of cell-surface sialoglycoproteins by periodate/tritiated borohydride. *J. Biol. Chem.* **252:**5888–5894.

Gahmberg, C. G., and Karhi, K. K., 1984, Association of Rh_o(D) polypeptide with the membrane in Rh_o(D)-positive human red cells. *J. Immunol.* **133:**334–337.

Gahmberg, C. G., Jokinnen, M., and Andersson, C. C., 1979, Expression of the major red cell sialoglycoprotein, glycophorin A, in the human leukemic cell line K562. *J. Biol. Chem.* **254**:7442–7448.

Ghosh, S., and Campbell, A. M., 1986, Multispecific monoclonal antibodies. *Immunol. Today* **7**:217–222.

Ginsburg, V., McGinniss, M. H., and Zopf, D. A., 1980, Biochemical basis for some blood groups. *Prog. Clin. Biol. Res.* **43**:45–53.

Goldsmith, K., 1955, Papain-treated red cells in the detection of incomplete antibodies. *Lancet* **i**:76–80.

Gooi, H. C., Feizi, T., Kapechia, A., Knowles, R. B., Salter, D., and Evans, M. J., 1981, Shape specific embryonic antigen involves 1–3 fucosylated type 2 blood group chains. *Nature* **292**:156–158.

Gooi, H. C., Kumura, K.-I., Edwards, P. A. W., Foster, C. S., Pickering, N., and Feizi, T., 1983a, Two mouse hybridoma antibodies against human milk fat globules recognise the I(Ma) determinant β-D-Galp(1–4)β-D-GlcpNAc(1–6). *Carbohydr. Res.* **120**:293–307.

Gooi, H. C., Thorpe, S. J., Hounsell, E. F., Rumpold, H., Kraft, D., Forster, O., and Feizi, T., 1983b, Marker of peripheral blood granulocytes and monocytes of man recognised by two monoclonal antibodies, VEP8 and VEP9, involves the trisaccharide 3 fucosyl-*N*-acetyllactosaime. *Eur. J. Immunol.* **13**:306–312.

Gooi, H. C., Veyrieres, A., Alais, J., Scudder, P., Hounsell, E. F., and Feizi, T., 1984, Further studies of the specificities of monoclonal anti-i and anti-I antibodies using chemically synthesised linear oligosaccharides of the poly-*N*-acetyllactosamine series. *Mol. Immunol.* **21**:1099–1104.

Gooi, H. C., Hounsell, E. F., Picard, J. K., Lowe, A. D., Voak, D., Lennox, E. S., and Feizi, T., 1985, Differing reactions of monoclonal anti-A antibodies with oligosaccharides related to blood group A. *J. Biol. Chem.* **260**:13218–13224.

Graham, H. A., Sinay, P., Hirsch, H. F., and Jacquinet, J. C., 1978, Inhibition of anti-Le^{dH} and Ulex anti-H lectin with oligosaccharides. In *XV Congress of the International Society of Blood Transfusion*, p. 552, Karger, Basel (Abstract).

Green, F. A., 1968, Phospholipid requirement of Rh antigenic activity. *J. Biol. Chem.* **243**:5519–5525.

Greenbury, C. L., Moore, D. H., and Nunn, L. A. C., 1963, Reaction of 7S and 19S components of immune rabbit antisera with human A and AB red cells. *Immunology* **6**:421–433.

Greenwalt, T. J., Swierk, E. Mc.F., and Steane, E. A., 1975, Use of horseradish peroxidase-labelled antiglobulin for the colorimetric determination of erythrocyte antibodies. *J. Immunol. Methods* **8**:351–362.

Greenwell, P., Yates, A. D., and Watkins, W. M., 1979, Blood group A synthesising activity of the blood-group *B* gene specified -3-galactosyl transferase. In *Glycoconjugates* (R. Schaur, E. Boer, M. F. Buddecke, J. Kramer, C. F. Vliegenthart, and H. Wiegendt, eds.), pp. 268–269, Thieme, Stuttgart.

Greenwell, P., Yates, A. D., and Watkins, W. M., 1986, UDP-*N*-acetyl-D-galactosamine as a donor substrate for the glycosyltransferase encoded by the *B* gene at the human blood group ABO locus. *Carbohydr. Res.* **149**:149–170.

Gunson, H. H., and Latham, V., 1972, An agglutinin in human serum reacting with cells from Le(a−b−) non-secretor individuals. *Vox Sang.* **22**:344–353.

Hadley, T. J., David, P. H., McGinniss, M. H., and Miller, L. H., 1984, Identification of an eythrocyte component carrying the Duffy blood group Fy^a. *Science* **223**:597–599.

Hakkinen, I., 1970, A-like blood-group antigen in gastric cancer of patients in blood groups O or B. *J. Natl. Cancer Inst.* **44**:1183–1193.

Hakomori, S., 1981a, Blood groups ABH and I: Antigens of human erythrocytes; Chemistry, polymorphism and their developmental change. *Semin. Haematol.* **18**:39–62.

Hakomori, S., 1981b, Glycosphingolipids in cellular interaction, differentiation, and oncogenesis. *Annu. Rev. Biochem.* **50**:733–764.

Hakomori, S., and Kobata, K., 1974, Blood group antigens. In *The Antigens* (M. Sela, ed.), pp. 79–140, Academic Press, New York.

Hakomori, S., Watanabe, K., and Laine, R. A., 1977, Glycosphingolipids with blood group A, H, and I activity: Their status in group A_1 and A_2 erythrocytes and their changes associated with onogeny and oncogeny. In *Human Blood Groups* (J. F. Mohn, R. W. Plunkett, R. K. Cunningham, and R. M. Lambert, eds.). pp. 150–163, Karger, Basel.

Haselberger, C. G., and Schenkel-Brunner, H., 1982, Evidence for erythrocyte membrane glycoproteins being carriers of blood group P_1 determinants. *FEBS Lett.* **149**:126–128.

Herbrink, P., Bussel, van, F. J., Warnaar, S. O., 1982, The antigen spot test (AST): A highly sensitive assay for the detection of antibodies. *J. Immunol. Methods* **48**:293–298.

Hughes-Jones, N. C. 1975, Red cell antigens, antibodies and their interaction. *Clin. Haematol.* **4**:29–40.

Hughes-Jones, N. C., and Gardner, B., 1971, The Kell system: Studies with radioactively labelled anti-K. *Vox Sang.* **21**:154–158.

Hughes-Jones, N. C., Gardner, B., and Telford, R., 1963, Studies on the reaction between the blood group antibody anti-D and erythrocytes. *Biochem. J.* **88**:435–440.

Hughes-Jones, N. C., Gardner, B., and Telford, R., 1964, The effect of pH and ionic strength on the reaction between anti-D and erythrocytes. *Immunology* **7**:72–81.

Hughes-Jones, N. C., Gardner, B., and Lincoln, P., 1971, Observations on the number of available c,D,e and E antigen sites on red cells. *Vox Sang.* **21**:210–216.

Hughes-Jones, N. C., Green, E. J., and Hunt, V. M., 1975, Loss of Rh antigen activity following the action of phospholipase A_2 on red cell stroma. *Vox Sang.* **29**:184–195.

Hui, S. W., Stewart, C. M., Carpenter, M. P., and Stewart, T. P., 1980, Effects of cholesterol on lipid organisation in human erythrocyte membrane. *J. Cell Biol.* **85**:283–291.

Huisman, J. G., 1986, Immunoblotting: An emerging technique in immunohaematology. *Vox Sang.* **50**:129–136.

Issitt, P. D., and Issitt, C. H., 1975, *Applied Blood Group Serology*, Spectra Biologicals, Oxnard, CA.

Jarnfelt, J., Rush, J., Yu-Teh, L., and Laine, R. A., 1978, Erythroglycan, a high molecular weight glycopeptide with the repeating structure [galactosyl(1-4)-2-deoxy-2-acetamido-glucosyl(1-3)] comprising more than one third of the protein bound carbohydrate of human erythrocyte stroma. *J. Biol. Chem.* **253**:8006–8009.

Jenkins, D. E., Moore, W. H., and Hawiger, A., 1977, A method for radioactive antiglobulin testing with ^{125}I labelled anti-IgG. *Transfusion* **17**:16–25.

Judson, P. A., and Anstee, D. J., 1977, Comparative effects of trypsin and chymotrypsin on blood group antigens. *Med. Lab. Sci.* **34**:1–6.

Kabat, E. A., 1956, *Blood Group Substances: Their Chemistry and Immunochemistry*, Academic Press, New York.

Kannagi, R., Nudelmann, E., Levery, R. B., and Hakamori, S., 1982, A series of human erythrocyte glycosphingolipids reacting to the monoclonal antibody directed to a developmentally related antigen, SSEA-1. *J. Biol. Chem.* **257**:14865–14874.

Kapadia, A., Feizi, T., Jewell, D., Keeling, J., and Slavin, G., 1981, Immunocytochemical studies of blood group A, H.I and i antigens in gastric mucosae of infants with normal gastric histology and of patients with gastric carcinoma and chronic benign peptic ulcerations. *J. Clin. Pathol.* **34**:320–337.

Karlsson, K. A., 1978, Glycolipid patterns of blood and other tissues. In *Proceedings of the XVth Congress of the International Society of Blood Transfusion*, p. 641, Karger, Basel.

Kehoe, J. M., and Capra, J. D., 1971, Localisation of two additional hypervariable regions in immunoglobulin heavy chains. *Proc. Natl. Acad. Sci. U.S.A.* **68**:2019–2021.

Kisailus, E. C., and Kabat, E. A., 1978, Immunochemical studies on blood groups. LXVI. Competitive binding assays of A_1 and A_2 substances with insolubilised anti-A serum and insolubilised agglutinin from *Dolichos biflorus*. *J. Exp. Med.* **147**:830–843.

Knowles, R. W., Bai, Y., Lomas, C., and Tippett, P., 1982, Two monoclonal antibodies detecting high frequency antigens absent from red cells of the dominant type of Lu(a−b−) Lu:-3. *J. Immunogenet.* **9**:353–357.

Kopito, R. R., and Lodish, H. F., 1985, Primary structure and transmembrane orientation of the murine anion exchange protein. *Nature* **316**:234–238.

Korgsgren, C., and Cohen, C. M., 1986, Purification and properties of human erythrocyte band 4.2; association with the cytoplasmic domain of band 3. *J. Biol. Chem.* **25**:5536–5543.

Koscielak, J., Miller-Podraza, H., Krause, R., and Piasek, A., 1976, Isolation and characterisation of poly(glycosyl)ceramides (megaloglycolipids) with A.H. and I blood group activities. *Eur. J. Biochem.* **71**:9–18.

Krusius, T., Finne, J., and Rauvala, H., 1978, The poly(glycosyl) chains of glycoproteins: Characterisation of a novel type of glycoprotein saccharide from human erythrocyte membranes. *Eur. J. Biochem.* **92**:289–300.

Kumazaki, T., and Yoshida, A., 1984, Biochemical evidence that secretor gene *Se* is a structural gene encoding a specific fucosyltransferase. *Proc. Natl. Acad. Sci. U.S.A.* **81**:4193–4197.

Kuypers, F., Linde-Sibelius-Trip, M., and Roelofsen, B., 1985, The phospholipid organisation in the membrane of McLeod and Leach phenotype erythrocytes. *FEBS Lett.* **184**:20–24.

Laemmli, U. K., 1970, Cleavage of structural proteins during the assembly of the head of bacteriophage T4. *Nature* **227**:680–681.

Landsteiner, K., 1901, Uber agglutinationserscheinungen normalen menschlichen Blutes. *Klin. Wochenschr.* **14**:1132–1134.

Landsteiner, K., and Levine, P., 1927, Further observations on individual differences of human blood. *Proc. Soc. Exp. Biol. Med.* **24**:941–942.

Landsteiner, K., and Levine, P., 1930, On the inheritance and racial distribution of agglutinable properties of human blood. *J. Immunol.* **18**:87–100.

Leader, K. A., Kuntel, B. M., Toole, G. D., Kirkwood, J. T., Merry, A. H., and Bradley, B. A., 1988, Human monoclonal anti-Rh(D) antibodies III. Antibodies reacted with category D_{IV} red cells and their use in blood grouping estimation of D_{IV} antigen density and determination of the population incidence of D_{IV}. *Vox Sang.* (submitted).

Leikola, J., and Perkins, H. A., 1980, Enzyme-linked antiglobulin test: An accurate and simple method to quantify red cell antibodies. *Transfusion* **20**:138–144.

Lemieux, R. H., Baker, D. A., Weinstein, W. M., and Sweitzer, P. M., 1981, Artificial antigens: Antibody preparations for the localisation of Lewis determinants in tissues. *Biochemistry* **20**:199–205.

LeNeveu, D. M., Rand, R. P., and Parsegian, V. A., 1976, Measurement of forces between lecithin bilayers. *Nature* **259**:601–603.

LePendu, J., Lemieux, R. U., and Oriol, R., 1982, Purification of anti-Le^c antibodies with specificity for β-D-Gal(1–3)–β-D-GlcNAc using a synthetic immunoabsorbent. *Vox Sang.* **43**:188–195.

LePendu, J., Cartron, J.-P., Lemeieux, R. U., and Oriol, R., 1985, The presence of at least two different H-blood group related β-D-1-fucosyl transferases in human serum and the genetics of blood group H substances. *Am. J. Hum. Genet.* **37**:749–760.

Lowe, A. D., Lennox, E., and Voak, D., 1984, A new monoclonal anti-A: Culture supernatants with the performance of hyperimmune human reagents. *Vox Sang.* **40**:99–104.

Lowe, A. D., Green, S. M., Voak, D., Gibson, T., and Lennox, E. S., 1986, A human–human monoclonal anti-D by direct fusion with a lymphoblastoid line. *Vox Sang.* **51**:212–216.

Lux, S. E., 1979, The spectrin–actin membrane skeleton of normal and abnormal red blood cells. *Semin. Haematol.* **16**:21–33.

Macdonald, E. B., and Gerns, L. M., 1986, An unusual sialoglycoprotein associated with the Webb-positive phenotype. *Vox Sang.* **50:**112–116.

Macher, B. A., and Sweeley, C. C., 1978, Glycosphingolipids: Structure, biological source and properties. *Methods Enzymol.* **50:**236–251.

Mallinson, G., Martin, P. G., Anstee, D. J., Tanner, M. J. A., Merry, A. H., Tills, D., and Sonneborn, H. H., 1986, Identification and partial characterisation of the human erythrocyte membrane component(s) that express the antigens of the LW blood group system. *Biochem. J.* **234:**649–652.

Marchesi, V. T., 1983, The red cell membrane cytoskeleton: Recent progress. *Blood* **61:**1–11.

Marchesi, V. T., Jackson, R. L., Segrest, J. P., and Kahane, I., 1973, Molecular features of the major glycoprotein of the human erythrocyte membrane. *Fed. Proc.* **32:**1833–1837.

Marcus, D. M., 1971, Isolation of a substance with blood group P_1 activity from human erythrocyte stroma. *Transfusion* **11:**16–18.

Marcus, D. M., and Cass, L. E., 1969, Glycosphingolipids with Lewis blood group activity: Uptake by human erythrocytes. *Science* **164:**553–555.

Marcus, D. M., Kabat, E. A., and Rosenfield, R. E., 1963, The action of enzymes from *Clostridium tertium* on the I determinant of human erythrocytes. *J. Exp. Med.* **118:**175–194.

Marcus, D. M., Kundu, S. K., and Suziki, A., 1981, The P blood group system: Recent progress in immunochemistry and immunogenetics. *Semin. Haematol.* **18:**63–71.

Marinetti, G. V., and Crain, R. C., 1978, Topology of amino-phospholipids in the human red cell membrane. *J. Supramol. Struct.* **8:**191–213.

Marr, A. M. S., Donald, A. S. R., Watkins, W. M., and Morgan, W. T. J., 1967, Molecular and genetic aspects of human blood group Le^b specificity. *Nature* **215:**1345–1349.

Marsh, W. L., and Jenkins, W. J., 1960, Anti-i: A new cold antibody. *Nature* **188:**200–209.

Masouredis, S. P., 1962, Reaction of ^{131}I anti-Rh_0(D) with enzyme treated red cells. *Transfusion* **2:**363–367.

Masouredis, S. P., Sudora, E. J., and Victoria, E. J., 1977, Immunological and electron microscopic analysis of IgG anti-D saline haemagglutination of neuraminidase- and proteinase-modified red cells. *J. Clin. Lab. Med.* **90:**929–948.

Masouredis, S. P., Sudora, E., Mahan, L., and Victoria, E. L., 1980, Quantitative immunoferritin microassay of Fy^a, Fy^b, Jk^a, U and Di^b antigen site numbers on human red cells. *Blood* **56:**969–977.

Mawby, W. J., Anstee, D. J., and Tanner, M. J. A., 1981, Immunochemical evidence for hybrid sialoglycoproteins of human erythrocytes. *Nature* **291:**161–162.

Merry, A. H., and Thomson, E. E., 1985, The role of quantitative antiglobulin tests in immunohaematology; determination of the significance of erythrocyte-bound IgG; investigation of some parameters affecting the sensitivity of the antiglobulin test. *Biotest Bull.* **2:**137–144.

Merry, A. H., Thomson, E. E., Anstee, D. J., and Stratton, F., 1984, The quantification of erythrocyte antigen sites with monoclonal antibodies. *Immunology* **51:**793–800.

Merry, A. H., Hodson, C., Thomson, E. E., Mallinson, G., and Anstee, D. J., 1986, The use of monoclonal antibodies to quantify the levels of sialoglycoproteins alpha and delta and variant sialoglycoproteins in human erythrocyte membranes. *Biochem. J.* **238:**93–98.

Merry, A. H., Gardner, B., Parsons, S. F., and Anstee, D. J., 1987, Estimation of the number of binding sites for a murine monoclonal anti-Lu^b on human erythrocytes. *Vox Sang.* **53:**57–60.

Merry, A. H., Hodson, C., and Moore, S., 1988, Variation in D antigen expression. *Transfusion* (in press).

Messeter, L., Brodin, T., and Chester, M. A., 1984, Mouse monoclonal antibodies. *Vox Sang.* **46:**185–194.

Mollison, P. L., 1983, *Blood Transfusion in Clinical Medicine*, Blackwell Scientific Publications, Oxford.

Mollison, P. L., Polley, M. J., and Crome, P., 1963, Temporary suppression of Lewis blood-group antibodies to permit incompatible transfusion. *Lancet* **i**:909.

Moore, H. H., 1983, Automated reading of red cell antibody identification tests by a solid phase antiglobulin technique. *Transfusion* **24**:218–221.

Moore, S., Woodrow, C. F., and McClelland, D. B. L., 1982, Isolation of membrane components associated with human red cell antigens Rh(D), (C), (E), and Fy^a. *Nature* **295**:529–531.

Moore, S., Chirnside, A., Micklem, L. R., and McClelland, B., 1984, A mouse monoclonal antibody with anti-A,(B) specificity which agglutinates A_x cells. *Vox Sang.* **47**:427–434.

Moreno, C., Lunblad, A., and Kabat, E. A., 1971, Immunochemical studies on blood groups. LI. A comparative study of the reaction of A_1 and A_2 group glycoproteins with human anti-A. *J. Exp. Med.* **134**:349–357.

Moulds, J., 1983, Immunochemistry of the MNSs blood group system. In *Red Cell Membrane Glycoconjugates and Related Genetic Markers* (J.-P. Cartron, P. Rouger, and Ch. Salmon, eds.), pp. 17–26, Librairie Arnette, Paris.

Mourant, A. E., 1946, A "new" human blood group antigen of frequent occurrence. *Nature* **158**:237–238.

Mueller, T. J., and Morrison, M., 1981, Glycoconnectin (PAS2), a membrane attachment site for the human erythrocyte cytoskeleton. *Proc. Clin. Biol. Res.* **56**:95–116.

Naiki, M., and Marcus, D. M., 1974, Human erythrocyte P and P^k blood group antigens: Identification as glycosphinolipids. *Biochem. Biophys. Res. Commun.* **60**:1105–1111.

Naiki, M., Fong, J., Ledeen, R., and Marcus, D. M., 1975, Structure of human erythrocyte blood group P_1 glycosphingolipid. *Biochemistry* **14**:4831–4837.

Nichols, M. E., Rosenfield, R. E., and Rubenstein, P., 1985, Two blood group M epitopes disclosed by monoclonal antibodies. *Vox Sang.* **49**:138–148.

Niemann, H., Watanabe, K., Hakomori, S., Childs, R. A., and Feizi, T., 1978, Blood group i and I activities of lacto-*N*-norhexaosylceramide and its analogues: The structural requirements for i specificities. *Biochem. Biophys. Res. Commun.* **81**:1286–1293.

Norman, P. C., Tippett, P., and Beal, R., 1986, A Lu(a−b−) phenotype caused by an X-linked recessive gene. *Vox Sang.* **51**:49–52.

Oriel, R., LePendu, J., and Mollicone, R., 1986, Genetics of ABO, H. Lewis, X and related antigens, *Vox Sang.* **51**:161–171.

Ortho Symposium, 1987, A Scientific Forum on Blood Grouping Serum Anti-A (Murine Monoclonal Blend Biclone), Ortho Diagnostic Systems, Raritan, NJ.

Owens, J. W., Mueller, T. J., and Morrison, M., 1971, A minor sialoglycoprotein of the human erythrocyte membrane. *Arch. Biochem. Biophys.* **204**:247–254.

Paradis, G., Bazin, R., and Lemieux, R., 1986, Protective effect of the membrane skeleton on the immunologic reactivity of the human red cell Rh_o(D) antigen. *J. Immunol.* **137**:240–244.

Parsons, S. F., 1985, Monoclonal antibodies in blood group serology. *Med. Lab. Sci.* **42**:361–366.

Parsons, S. F., Judson, P. A., and Anstee, D. J., 1982, BRIC-18: A monoclonal antibody with a specificity related to the Kell blood group system. *J. Immunogenet.* **9**:377–382.

Parsons, S. F., Mallinson, G., Judson, P. A., Anstee, D. J., Tanner, M. J. A., and Daniels, G. L., 1987, Evidence that the antigens of the Lutheran blood group system are located on erythrocyte membrane glycoproteins of M_r 85,000 and 78,000. *Transfusion* **27**:61–63.

Pasvol, G., 1984, Receptors on red cells for *Plasmodium falciparum* and their interaction with merozoites. *Philos. Trans. R. Soc. Lond.* [*Biol.*] **307**:189–200.

Plummer, T. H., Elder, J. H., Alexander, S., Phelan, A. W., and Trentino, A. L., 1984, Demonstration of peptide-*N*-glycosidase F activity in endo-β-*N*-acetylglucosamidase F preparations. *J. Biol. Chem.* **259**:10700–10704.

Pollack, W., Hager, H. J., Reckel, R., Toren, D. A., and Singher, H. O., 1964, A study of forces involved in the second stage of haemagglutination. *Transfusion* **5**:158–183.

Potapov, M. I., 1970, Detection of the antigen of the Lewis system, characteristic of the erythrocytes of group Le(a−b−) secretor. *Prog. Haematol.* **11**:45–49.

Race, R. R., and Sanger, R., 1975, *Blood Groups in Man,* 6th ed., Blackwell Scientific Publications, Oxford.

Rahman, R. A. F., and Longnecker, B. M., 1982, A monoclonal antibody specific for the Tomsen-Freidenreich cryptic T antigen. *J. Biol. Chem.* **129**:2021–2024.

Rearden, A., 1985, Phospholipid dependence of Wr^b antigen expression in human erythrocyte membranes. *Vox Sang.* **49**:346–353.

Redman, C. M., and Marsh, W. L., 1985, Kell blood group glycoproteins in normal and Ko cells. *Transfusion* **24**:470.

Redman, C. M., Marsh, W. L., Mueller, G. P., Avellino, G. P., and Johnson, C. L., 1984, Isolation of Kell-active protein from the red cell membrane. *Transfusion* **24**:176–178.

Rege, V. P., Painter, T. J., Watkins, W. M., and Morgan, W. T. J., 1963, Three new trisaccharides obtained from human blood group A,B,H, and Le^a substances: Possible sugar sequences in the carbohydrate chains. *Nature* **200**:532–34.

Rege, V. P., Painter, T. J., Watkins, W. M., and Morgan, W. T. J., 1964, Isolation of a serologicaly active fucose containing trisaccharide from human blood group Le^a substance. *Nature* **204**:740–742.

Reid, M. E., Shaw, M.-A., Rowe, G., Anstee, D. J., and Tanner, M. J. A., 1985, Abnormal minor human erythrocyte sialoglycoprotein (beta) in association with the rare blood group antigen Webb (Wb). *Biochem. J.* **232**:289–291.

Renton, P. H., and Hancock, J. A., 1956, Variability of the Rh antigen D. *Br. J. Haematol.* **4**:89–95.

Richards, F. F., Konigsberg, W. H., and Rosenstern, R. W., 1975, On the specificity of antibodies. *Science* **187**:130–137.

Ridgwell, K., Tanner, M. J. A., and Anstee, D. J., 1983, The Wr^b antigen, a receptor for *Plasmodium falciparum* malaria, is located on a helical region of the major membrane sialoglycoprotein of human red blood cells. *Biochem. J.* **209**:273–276.

Ridgwell, K., Tanner, M. J. A., and Anstee, D. J., 1984, The rhesus (D) polypeptide is linked to the human erythrocyte cytoskeleton. *FEBS Lett.* **174**:7–10.

Rochna, E., and Hughes-Jones, N. C., 1965, The use of purified ^{125}I-labelled anti-gamm globulin in the determination of the number of D antigens sites on red cells of different phenotypes. *Vox Sang.* **10**:675–686.

Romans, D. G., Tilley, C. A., Crookston, M. C., Falk, R. E., and Dorrington, K. J., 1977, Conversion of incomplete antibodies to direct agglutinins by mild reduction: Evidence for segmental flexibility within the Fc fragment of immunoglobulin G. *Proc. Natl. Acad. Sci. U.S.A.* **74**:2531–2535.

Sachs, S., and Lennox, E., 1981, Monoclonal anti-B a new blood typing reagent, *Vox Sang.* **40**:99–104.

Sadler, J. E., 1984, Biosynthesis of glycoproteins: Formation of O-linked oligosaccharides. In *Biology of Carbohydrates,* Vol. 2 (V. Ginesbug and P. W. Robbins, eds.), pp. 199–228, Wiley, New York.

Salsbury, A. J., and Clarke, J. A., 1967, Surface changes in red blood cells undergoing agglutination. *Rev. Fr. d'Etudes Clin. Biol.* **12**:345–352.

Schachter, H., Michaels, M. A., Tilley, C. A., Crookston, M. C., and Crookston, J., 1973, Qualitative differences in the *N*-acetyl-D-galactosaminyltranferases produced by human A_1 and A_2 genes. *Proc. Natl. Acad. Sci. U.S.A.* **70**:220–224.

Schenkel-Brunner, H., 1980, Blood group ABH antigens of human erythrocytes. Quantitative studies on the distribution of H antigenic sites among different classes of membrane constituents. *Vox Sang.* **38**:310–314.

Schrier, S. S., 1985, Red cell membrane biology—Introduction. *Clin. Haematol.* **14**:1–12.

Scott, M., Voak, D., and Downie, D. M., 1988, Evaluation of optimum papain enzyme technique: An explanation for the poor performance of the one stage method. *Med. Lab. Sci.* **45:**1–28.
Seaman, M. J., Chalmers, D. G., and Franks, D., 1968, Siedler: An antibody which reacts with A_1Le^{a-b+} red cells. *Vox Sang.* **15:**25–30.
Sinor, L. T., Rachel, J. M., Beck, M. L., Bayer, W. L., Coenen, W. M., and Plapp, F. V., 1985, Solid-phase ABO grouping and Rh typing. *Transfusion* **25:**21–23.
Sheetz, M. P., 1983, Membrane skeletal dynamics; role in modulation of red cell deformability, mobility of transmembrane proteins, and shape. *Semin. Haematol.* **20:**175–188.
Siebert, P. D., and Fukuda, M., 1986a, Isolation and sequence of a human glycophorin A-specific cDNA clone. *Proc. Natl. Acad. Sci. U.S.A.* **83:**1665–1669.
Siebert, P. D., and Fukuda, M., 1986b, Human glycophorin A and B are encoded by separate, single copy genes coordinately regulated by a tumor-promoting phorbol ester. *J. Biol. Chem.* **261:**12433–12436.
Singer, S. J., and Nicholson, G. L., 1972, The fluid mosaic model of the structure of cell membranes. *Science* **175:**720–731.
Sneath, J. S., and Sneath, P. H. A., 1955, Transformation of the Lewis groups of human red cells. *Nature* **176:**172–187.
Snider, M. D., 1984, Biosynthesis of glycoproteins: Formation of N-linked oligosaccharides. In *Biology of Carbohydrates* (V. Ginesbug and P. W. Robbins, eds.), Vol. 2, pp. 163–198, Wiley, New York.
Sonneborne, H. H., Utheman, H., and Pfeffer, A., 1983, Monoclonal antibody specific for human blood group k (Cellano). *Biotest Bull.* **1:**328–330.
Sonneborn, H. H., Utheman, H., Tills, D., Lomas, C. G., Shawe, M. A., and Tippett, P., 1984, Monoclonal anti-LW^{ab}. *Biotest Bull.* **2:**145–148.
Spooncer, E., Fukuda, M., Klock, J. C., Oates, J. E., and Dell, A., 1984, Isolation and characterisation of a polyfucosylated lactosaminoglycan from human granulocytes. *J. Biol. Chem.* **259:**4792–4801.
Spring, F. A., Judson, P. A., Daniels, G. L., Parsons, S. F., Mallinson, G., and Anstee, D. J., 1987, A human cell surface glycoprotein which carries Cromer-related blood group antigens on erythrocytes and is also expressed on leucocytes and platelets. *Immunology* **62:**307–313.
Springer, G. F., and Ansell, N. J., 1958, Inactivation of human erythrocyte agglutinogens M and N by influenza virus and receptor-destroying enzyme. *Proc. Natl. Acad. Sci. U.S.A.* **44:**182–190.
Springer, G. F., Horton, R. E., and Forbes, M., 1959, Origin of anti-human blood group B agglutinins in white leghorn chicks. *J. Exp. Med.* **110:**221–238.
Springer, G. F., Williamson, P., and Readler, B. L., 1962, Blood group active Gram-negative bacteria and higher plants. *Ann. N.Y. Acad. Sci.* **97:**104–115.
Steane, E. A., and Greenwalt, T. J., 1980, Erythrocyte agglutination. In *Immunobiology of the Erythrocyte* (S. G. Sandler, J. Nusbacher, and M. S. Schanfield, eds.), pp. 171–188, A.R. Liss, New York.
Steck, T. L., 1974, The organisation of proteins in the human red blood cell membrane. *J. Cell Biol.* **62:**1–19.
Steck, T. L., 1978, The band 3 protein of the human red cell membrane: A review. *J. Supramol. Struct.* **8:**311–324.
Stratton, F., Rawlinson, V. I., Gunson, H. H., and Phillips, P. K., 1973, The role of zeta potential in Rh agglutination. *Vox Sang.* **24:**273–279.
Sturgeon, P. H., and Achilla, M. B., 1970, Studies on the secretion of blood group substances. I. Observations on red cell phenotype Le(a+b+x+). *Vox Sang.* **18:**301–310.
Tang, P. W., Gooi, H. C., Hardy, M., Lee, Y. C., and Feizi, T., 1985, Novel approach to the study of the antigenicities and receptor functions of carbohydrate chains of glycoproteins. *Biochem. Biophys. Res. Commun.* **132:**474–480.

Tanner, M. J. A., 1983, Erythrocyte membrane structure and function. In *Malaria and the Red Cell*, pp. 3–23, Ciba Foundation Symposium, Pitman, London.

Tanner, M. J. A., Williams, D. G., and Jenkins, R. E., 1980, Structure of the erythrocyte anion transport protein. *Ann. N.Y. Acad. Sci.* **341**:455–464.

Tanner, M. J. A., Anstee, D. J., Mallinson, G., Ridgwell, K., Martin, P. G., Avent, N. D., and Parsons, S. F., 1987, Effect of endoglycosidase F/peptidyl *N*-glycosidase F preparations on the surface components of the human erythrocyte. *Carbohydr. Res.* (in press).

Telen, M. J., Eisenbarth, G. S., and Haynes, B. F., 1983, Regulation of expression of a novel erythrocyte surface antigen by the inhibitor Lutheran In(Lu) gene. *J. Clin. Invest.* **71**:1878–1886.

Thomas, D. B., and Winzler, R. J., 1969, Structural studies on human erythrocyte sialoglycoproteins: Alkali-labile oligosaccharides. *J. Biol. Chem.* **255**:5943–5946.

Thomson, K. M., Melamed, M. D., Eagel, K., Gorick, B. D., Gibson, T., Holburn, A. M., and Hughes-Jones, N. C., 1986, Production of human monoclonal IgG and IgM antibodies with anti-D(Rhesus) specificity using heterohybridomas. *Immunology* **28**:157–160.

Thompson, N. L., and Axelrod, D., 1980, Reduced lateral mobility of a fluorescent lipid probe in cholesterol depleted erythrocyte membranes. *Biochim. Biophys. Acta* **597**:155–165.

Tippett, P., Andrews, P. W., Knowles, B. B., Solter, D., and Goodfellow, P. N., 1986, Red cell antigens P(Globoside) and Luke: Identification by monoclonal antibodies defining the murine stage specific embryonic antigens -3 and -4 (SSEA-3 and SSEA-4). *Vox Sang.* **51**:53–56.

Tokunuaga, E., Sasakawa, S., Tamaka, K., Kawamata, H., Giles, C. M., Ikin, E. W., Poole, J., Anstee, D. J., Mawby, W. J., and Tanner, M. J. A., 1979, Two apparently healthy Japanese individuals of type M^kM^k have erythrocytes which lack both the blood group MN and Ss-active sialoglycoproteins. *J. Immunogenet.* **6**:383–390.

Tomita, M., Furthmayer, H., and Marchesi, V. T., 1977, Primary structure of human erythrocyte glycophorin A. Isolation and characterisation of peptides and complete amino acid sequence. *Biochemistry* **17**:4756–4770.

Towbin, H., and Gordon, J., 1984, Immunoblotting and dot immunoblotting—current states and outlook. *J. Immunol. Methods* **72**:313–340.

Tsai, C. M., Zopf, D. A., Wistar, T. R., and Ginsburg, V., 1976, A human cold agglutinin which binds lacto-*N*-neotetraose. *J. Immunol.* **117**:717–721.

Udden, M. M., Umeda, M., Hirano, Y., and Marcus, D. M., 1987, New abnormalities in the morphology, cell surface receptors, and electrolyte metabolism of In(Lu) erythrocytes. *Blood* **69**:52–57.

van Oss, C. J., and Mohn, J. F., 1970, Scanning electron microscopy of red cell agglutination. *Vox Sang.* **19**:432–443.

van Oss, C. J., and Neuman, A. W., 1977, Comparison between antibody binding energies and interfacial free energies. *Immunol. Commun.* **6**:341–354.

van Oss, C. J., Mohn, J. F., and Cunningham, R. K., 1978, Influence of various physicochemical factors on haemagglutination. *Vox Sang.* **34**:351–361.

Vengelen-Tyler, V., Anstee, D. J., Issit, P. D., Pavone, B. G., Ferguson, S. J., Mawby, W. J., Tanner, M. J. A., Blajchman, M. A., and Lorque, P., 1981, Studies on the blood of an Mi^V homozygote. *Transfusion* **21**:1–14.

Verkeij, A. J., Zwall, R. F. A., and Roelofsen, B., 1973, The asymmetric distribution of phospholipids in the human red cell membrane. A combined study using phopholiposomes and freeze-etch electron microscopy. *Biochem. Biophys. Acta* **323**:178–193.

Victoria, E. J., Branks, M. J., and Masouredis, S. P., 1987, Immunoreactivity of the Rh_o(D) antigen in cytoskeleton-free vesicles. *Transfusion* **27**:32–35.

Voak, D., 1984, Monoclonal antibodies in immuno-haematology. In *Proceedings of the 7th Biotest Symposium Biotest Bull.* 279–389.

Voak, D., 1986, Monoclonal antibodies: Applications to blood group serology. *Labmedica*, August/September 1986.

Voak, D., and Lodge, T. W., 1968, The role of H in the development of A. *Vox Sang.* **15:**345–252.

Voak, D., and Williams, M. A., 1971, An explanation of the failure of the direct antiglobulin test to detect erythrocyte sensitisation in ABO haemolytic disease of the newborn and observations on the pinocytosis of IgG anti-A antibodies by infant (cord) red cells. *Br. J. Haematol.* **20:**9–20.

Voak, D., Todd, G. M., and Pardoe, G. I., 1974a, A study of the serological behaviour and nature of the anti-B/P/P^k activity of *Salmonidae* roe protectins. *Vox Sang.* **27:**176–188.

Voak, D., Cawley, J. C., Emmines, J. P. and Barker, C. R., 1974b, The role of enzymes and albumen in haemagglutination reactions. *Vox Sang.* **27:**156–170.

Voak, D., Sachs, S., and Alderson, T., 1980, Monoclonal anti-A from a hybrid myeloma: Evaluation as a blood grouping reagent. *Vox Sang.* **14:**9–20.

Voak, D., Lowe, A. D., and Lennox, E., 1983, Monoclonal antibodies: ABO serology. *Biotest Bull.* **4:**291–299.

Voak, D., Davies, D., Beck, M., and Yates, A. D., 1988, The B(A) phenomenon: A monoclonal anti-A (MHO4) demonstrates traces of A on normal group B red cells. *Vox Sang.* (submitted).

Wallas, C., Simon, R., Sharpe, M. A., and Byler, C., 1986, Isolation of a Kell-reactive protein from red cell membranes. *Transfusion* **26:**173–176.

Watanabe, K., and Hakomori, S., 1976, Status of blood group carbohydrate chains in ontogenesis and oncogenesis. *J. Exp. Med.* **144:**644–653.

Watanabe, K., Hakomori, S., Childs, R. A., and Feizi, T., 1979, Characterisation of a blood group I-active ganglioside. *J. Biol. Chem.* **254:**3221–3228.

Watkins, W. M., 1966, Blood group substances. *Science* **152:**172–181.

Watkins, W. M., 1980, Biochemistry and genetics of the ABO, Lewis and P blood group systems. In *Advances in Human Genetics* (H. Harris and K. Hirschhorn, eds.), Vol. 10, pp. 1–136, Plenum Press, New York.

Watkins, W. M., and Morgan, W. T. J., 1957, Specific inhibition studies relating to the Lewis blood group system. *Nature* **180:**1038–1040.

Watkins, W. M., and Morgan, W. T. J., 1964, Blood groups P_1 substance II. Immunological properties. In *Proceedings of the 9th Congress of the International Society of Blood Transfusion*, pp. 230–234, Karger, Basel.

Watkins, W. M., Greenwell, P., and Yates, A. D., 1981, The genetic and enzymic regulation of the synthesis of the A and B determinants in the ABO blood group system. *Immunol. Commun.* **10**(2):83–100.

Weatherall, D. J., and Clegg, J. B., 1979, Recent developments in the molecular genetics of human haemoglobins. *Cell* **16:**467–479.

Weaver, D. C., Pasternack, G. R., and Marchesi, V. T., 1984, The structural basis of ankyrin function II. Identification of two functional domains. *J. Biol. Chem.* **259:**6170–6175.

Weiner, A. S., Unger, L. J., Cohen, L., and Feldman, J., 1956, Type-specific cold antibodies as a cause of acquired haemolytic anaemia and haemolytic transfusion reactions; biologic test with bovine red cells. *Ann. Intern. Med.* **44:**221–240.

Welsh, E. J., Thom, D., Morris, E. R., and Rees, D. A., 1985, Molecular organization of glycophorin A: Implications for membrane interactions. *Biopolymers* **24:**2301–2332.

Wilczynska, Z., Miller-Podraza, H., and Kosielac, I., 1980, The contribution of different glycoconjugates to the total ABH blood group activity of human erythrocytes. *FEBS Lett.* **112:**277–279.

Williams, M. A., and Voak, D., 1972, Studies with ferritin-labelled *Dolicos biflorus* lectin on the numbers and distribution of A sites on A_1 and A_2 erythrocytes, and on the nature of its specificity and enhancement by enzymes. *Br. J. Haematol.* **23:**427–441.

Wu, T. T., and Kabat, E. A., 1970, An analysis of the sequence of the variable regions of Bence–

Jones proteins and myeloma light chains and their implications for antibody complementarity. *J. Exp. Med.* **132:**21–50.

Yates, A. D., and Watkins, W. M., 1982. The biosynthesis of blood group B determinants by the blood group A gene-specified -3-*N*-acetyl-D-galactose-aminyl transferase. *Biochem. Biophys. Res. Commun.* **109**(23):958–965.

Yates, A. D., Greenwell, P., and Watkins, W. M., 1983, Overlapping specificities of the glycosyltransferase specified by the blood-group A and B genes: A possible explanation for aberrant blood-grouping expression in malignant tissues. *Biochem. Soc. Trans.*, pp. 300–301 (602nd meeting, London).

Yates, A. D., Feeney, J., Donald, A. S. R., and Watkins, W. M., 1984, Characterisation of A blood group A-active tetrasaccharide synthesised by a blood-group B gene-specified glycosyltransferase. *Carbohydr. Res.* **130:**251–260.

Young, H., and Hakomori, S., 1971, A sphingolipid having a novel type of ceramide, *N*-lacto-*N*-fucopentose III. *J. Biochem.* **246:**1192–1200.

Young, W. W., Johnson, H. S., and Tamura, Y., 1983, Characterisation of monoclonal antibodies directed against the human Le^b blood group antigen. *J. Biol. Chem.* **256:**13223–13225.

Chapter 3

Molecular Characteristics of the Blood Group Rh_0(D) Molecule

Carl G. Gahmberg

1. INTRODUCTION

1.1. Discovery of the Rh_0(D) Antigen

Already in the seventeenth century jaundice of newborn infants had been described. The condition was first thought to be similar to that of adults, which often was due to occlusion of the bile ducts. The fact, however, that there often were several cases in the same family argued against this view (Clarke, 1982).

An important breakthrough occurred in 1938 when Ruth Darrow gave a thorough clinical description of what was then known as *Icterus gravis neonatorum* (Darrow, 1938). She also offered an explanation for the development of the disease: an antigen–antibody reaction between the fetus and the mother. She thought that fetal hemoglobin, known already at that time to be different from that of adults, had passed the placenta resulting in immunization of the mother. The antibodies from the circulation of the mother would then pass the placenta to the fetus and destroy its red cells. The principle proved to be right but she missed by a hair's breadth the actual mechanism.

One year later, Levine and Stetson (1939) described a case where intrauterine fetal death was associated with severe anemia. They postulated that this

Carl G. Gahmberg Department of Biochemistry, University of Helsinki, SF-00170 Helsinki, Finland.

was due to fetal red cell destruction caused by maternal antibodies, which had passed through the placenta and reacted with the membrane of the fetal cells. Soon thereafter, Landsteiner and Wiener (1940) reported experiments where they had immunized guinea pigs and rabbits with red cells from rhesus monkeys. The resulting antisera agglutinated the red cells from about 85% of a random human population. Wiener and Peters (1940) then showed that the transfusion reactions often observed were due to anti-Rh antibodies. In 1941, Levine and co-workers (Levine *et al.*, 1941) gave a detailed description of the hemolytic disease of the newborn and also proposed that the antisera produced against rhesus monkey red cells had a similar specificity.

They showed that 90% of the mothers with affected children were Rh-negative, in contrast to 15% in the general population. All husbands and affected children were Rh-positive.

These studies resulted in the well-known concept of the importance of Rh isoimmunization in the hemolytic disease of the newborn. During the second or later pregnancies of a $Rh_0(D)$-negative mother carrying a Rh(D)-positive fetus, clinically important levels of anti-$Rh_0(D)$ antibodies may arise.

Whereas one marriage out of seven is between a Rh-negative woman and a Rh-positive father, the incidence of Rh isoimmunization is much more rare. Levine (1943) first observed that ABO incompatibility was associated with a lower incidence of hemolytic disease of the newborn. However, Race and Sanger (1950) first put forward the view that ABO incompatibility actually could prevent the Rh isoimmunization by destruction of the incompatible fetal cells, and these studies were further extended by Nevanlinna and Vainio (1956). Support for this came from other groups (Clarke *et al.*, 1958).

The hemolytic disease was first successfully treated with transfusion of Rh-positive blood to the fetus. Whereas this treatment often had a positive outcome there remained problems, mainly because of the fact that antibodies remained in the fetal circulation for several days, resulting in hemolysis of the transfused cells. Better results were obtained by extensive exchange transfusion of the blood.

Today, most cases of Rh isoimmunization can be prevented. This is one of the major breakthroughs in modern medicine and has had wide general applications. Ronald Finn first pointed out that the protection affected by ABO incompatibility could be mimicked in ABO-compatible mothers and fetuses using other antibodies to destroy the Rh-positive fetal cells (Finn, 1960; Finn *et al.*, 1961). Stern and Berger (1960) showed that Rh-negative volunteers failed to make anti-Rh antibodies if first injected with Rh-positive cells coated with anti-$Rh_0(D)$ antibodies. A breakthrough came when Clarke and co-workers showed that most Rh isoimmunizations could be prevented by injecting pregnant mothers with plasma containing anti-$Rh_0(D)$ antibodies (Clarke *et al.*, 1963).

Freda and co-workers (Freda *et al.*, 1964) introduced the use of anti-D gamma globulin in the prevention, which proved to be much more convenient than the previously used plasma.

Among the Rh antigens, the Rh_o(D) antigen is the most important clinically and has been studied most intensively. It seems to be specific for the erythroid lineage. The Rh complex, however, is serologically complicated and is composed of several antigens. The Cc, Dd, and Ee are the best known, but several other antigens have been defined serologically. However, practically nothing is known about these other molecules, and therefore they fall out of the scope of this chapter. Importantly, -D-/-D- (super D) cells with increased amounts of Rh_o(D) antigen have been described, as have Rh_{null} cells lacking all Rh antigens. These will be dealt with in more detail in Section 3.3.

Because of its theoretical and clinical importance, there has existed an urgent need to isolate and characterize the Rh_o(D) antigen in detail. This has proved difficult, however, and although several reports have appeared in the literature, only recently has some agreement been reached on the molecular nature of the antigen. To be able to understand the problems involved and to appreciate the work done on the molecular characterization of the Rh_o(D) molecule, it is important to have a solid background knowledge on the structure of the human red cell membrane. I will therefore first describe this membrane in some detail.

2. STRUCTURE OF THE HUMAN RED CELL MEMBRANE

The human red cell membrane is no doubt the best-known mammalian cell plasma membrane. There are several reasons for this. First, it is easy to isolate in pure form; second, blood banks can often provide large amounts of fresh or outdated blood for research; and third, there are several diseases resulting from structural abnormalities of the red cell membrane (Parker and Berkowitz, 1986) and they give useful information about functional properties of membrane components. Several reviews on the erythrocyte membrane have appeared, and they give detailed information about different aspects on this topic (Gahmberg *et al.*, 1986; Marchesi, 1983; Marchesi *et al.*, 1976; Steck, 1974).

2.1. The Red Cell Membrane Lipids

All membranes are formed by lipids and proteins. The major red cell membrane lipids are cholesterol, the phospholipids phosphatidylcholine, phosphatidylethanolamine, phosphatidylserine, sphingomyelin, and phosphatidylinositol, and the glycolipids globoside, ceramide trihexoside, and ceramide dihexoside. Cholesterol may have a role in stabilizing the membrane and is evidently distributed in both halves of the bilayer. The phospholipids have an asymmetric distribution with most of the phosphatidylcholine in the outer half, 80% of the phosphatidylethanolamine in the inner half, and all phosphatidylserine at the cytoplasmic aspect of the membrane. Most of the sphingomyelin is in the outer half of the bilayer (Bretscher, 1972; Verkleij *et al.*, 1973).

The glycolipids may function as receptors for microbes, lectins, and various antibodies (Hakomori, 1981). Some of them constitute important blood group antigens like the ABH and P blood group systems (Hakomori and Kobata, 1975). The glycolipids are most probably completely asymmetrically distributed in the two halves of the lipid bilayers. No glycolipids have been found at the cytoplasmic aspect of the membrane (Gahmberg and Hakomori, 1973; Steck and Dawson, 1974). Using galactose oxidase/$NaBD_4$ labeling, we have determined that 30–60% of the major glycolipid globoside is accessible at the cell surface (Lampio *et al.*, 1984). It is possible that the rest is located externally but tightly associated with glycoproteins, because neuraminidase treatment of intact cells, which primarily removes sialic acids from the glycophorins, resulted in increased glycolipid exposure (Lampio *et al.*, 1986).

The lipids can move in the horizontal plane of the bilayer and around their own axis. However, transbilayer movement of phospholipids and glycolipids is extremely rare in normal erythrocytes.

The reason for the asymmetric distribution of the phospholipids between the two halves of the bilayer is not known, but evidently proteins like spectrin and the $Rh_o(D)$ antigen are involved (see Section 3.3).

2.2. The Red Cell Integral Membrane Proteins

The integral membrane proteins are directly associated with the membrane lipids. This is possible because these proteins contain hydrophobic amino acid sequences facing the lipid milieu. They are by definition not soluble in normal aqueous buffers but need detergents or lipid solvents for solubilization. The major and best studied proteins in this group are glycophorin A,B, and C. the anion transport protein, band 3, and the proteins in the band 4.5 region (see Figure 1). These are all synthesized in erythroid precursor cells, evidently at defined stages of differentiation (Gahmberg *et al.*, 1986).

Glycophorin A was the first integral membrane protein that was sequenced (Tomita and Marchesi, 1975) and among the first integral membrane proteins for which the biosynthesis was studied in detail (Gahmberg *et al.*, 1980; Jokinen *et al.*, 1979, 1981, 1985). There are about 10^6 glycophorin A polypeptides per cell (Gahmberg *et al.*, 1979). Glycophorin cDNA has recently been made from K562 cells (Siebert and Fukuda, 1986), and small differences were observed between the sequence obtained and the published sequence obtained by Tomita and Marchesi (1975). As suggested before (Jokinen *et al.*, 1981), the in vitro synthesized protein contains a signal sequence, and this was found to be 19 amino acids long (Siebert and Fukuda, 1986). Glycophorin A is composed of 131 amino

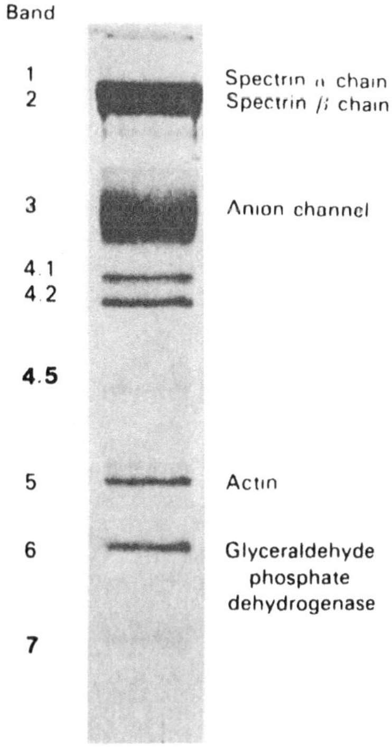

FIGURE 1. Coomassie blue-stained polyacrylamide gel of erythrocyte membranes run in presence of sodium dodecyl sulfate. The major stained proteins are indicated. The major bands are indicated to the left, and the common names of some of them to the right.

acids forming three separate domains (Figure 2). The NH_2-terminal region at the external aspect of the cell membrane contains all the carbohydrate. In the mature red cell, there is one asparagine-linked complex-type oligosaccharide located at asparagine-26, and 15 serine- and threonine-linked O-glycosidic oligosaccharides (Tomita and Marchesi, 1975). The protein carries the MN blood group activities. The amino acid sequence in the NH_2-terminal part is essential for distinguishing M and N blood groups. Thus, M-type glycophorin A contains serine in position 1 and glycine in position 5, whereas N-type glycophorin A has leucine and glutamic acid in those positions, respectively (Dahr *et al.*, 1977; Furthmayr, 1978; Wasniowska *et al.*, 1977). The O-glycosidic carbohydrate is also essential and neuraminidase treatment abolishes MN activity (Mäkelä and Cantell, 1958; Springer and Ansell, 1958). The protein spans the membrane and the intrabilayer part has an α-helix hydrophobic sequence. The cytoplasmic end is hydrophilic and interacts with other proteins like bands 3 (Nigg *et al.*, 1980) and 4.1, a membrane skeletal protein (Anderson and Lovrien, 1984). Part of the glycophorin A molecules are phosphorylated at serine 102 (Gahmberg *et al.*, 1983; Shapiro and Marchesi, 1977).

Glycophorin A shows developmental structural changes. Thus, glycophorin A from immature K562 erythroleukemia cells did not show much MN activity when tested with M- and N-specific monoclonal antibodies (Gahmberg *et al.*,

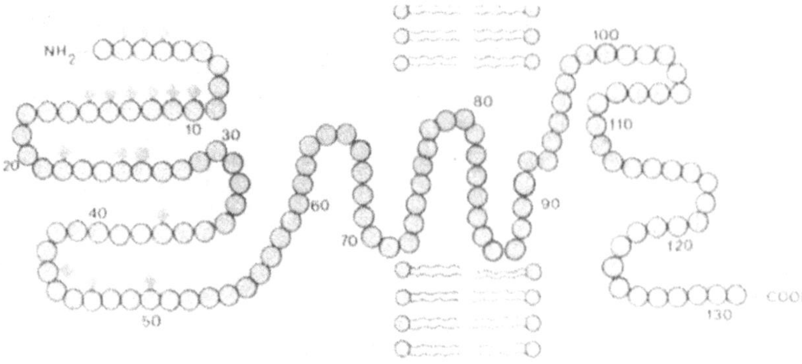

FIGURE 2. Schematic structure of the major red cell sialoglycoprotein, glycophorin A. The NH_2-terminal portion is outside the lipid bilayer and contains the carbohydrates. ♦ = O-glycosidic oligosaccharide; ● = N-glycosidic oligosaccharide. The polypeptide segment within the lipid bilayer is hydrophobic. The COOH-terminal portion protrudes into the cytoplasm and interacts with other integral and membrane-skeleton-associated proteins.

1984). The same held true for glycophorin A in immature erythroid cells from normal bone marrow and from malignant cells from patients with erythroleukemia (Andersson *et al.*, 1979; Greaves *et al.*, 1983). Whereas rabbit anti-glycophorin A antiserum readily reacted with the immature cells, and a glycophorin A molecule could be immune precipitated, monoclonal antibodies reacted poorly (Ekblom *et al.*, 1985). This discrepancy was solved when it was shown that glycophorin A from immature red cells was less O-glycosylated, which apparently resulted in conformational changes leading to decreased immunoreactivity (Gahmberg *et al.*, 1984).

Glycophorin B has an identical NH_2-terminal sequence as glycophorin A (Furthmayr, 1978); also, glycophorin C, which is 127 amino acids long, has a very similar structure (Colin *et al.*, 1986). Glycophorin B is always blood group N-active, whereas the Gerbich blood group is associated with glycophorin C. Glycophorin C carries a single N-glycosidic oligosaccharide in the NH_2-terminal part of the molecule (Colin *et al.*, 1986).

Band 3 is a large membrane glycoprotein that, like glycophorin A, probably exists as a dimer in the membrane. It functions as an anion transporter (Cabantchik and Rothstein, 1974). However, it has a completely different structure from glycophorin A. Its NH_2 terminal is in the cytoplasm and it contains the binding sites for several other proteins like ankyrin (Bennett and Stenbuck, 1980), hemoglobin (Shaklai *et al.*, 1977), and glycolytic enzymes (Strapazon and Steck, 1976; Yu and Steck, 1975). The primary structure of the murine protein has been determined (Kopito and Lodish, 1985), and it shows extensive homology with those parts of the human protein that have been sequenced (Brock *et al.*, 1983; Mawby and Findlay, 1983). The protein traverses the membrane 12 times. Five of the membrane spanning sequences are hydrophobic, resembling that of glycophorin A. Seven are less hydrophobic, and their structure is compatible with the hypothesis that they form membrane-spanning regions with hydrophilic cores. Evidently, together they form the anion channel through the membrane. The protein contains one exceptionally large oligosaccharide (Gahmberg *et al.*, 1976), which has a polylactosamine structure (Järnefelt *et al.*, 1978; Krusius *et al.*, 1978). It is probably located at residue 660 in the murine protein and evidently in the same or a closely similar location in the human protein (Jay, 1986). Band 3 does not contain an NH_2-terminal signal sequence and it can be post-translationally incorporated into the membrane.

Band 3 carries ABO and Ii blood group antigens (Finne, 1980; Karhi and Gahmberg, 1980), which are formed by its carbohydrate. Thus, the blood group A determinant is nonreducing α-*N*-acetylgalactosamine, the B determinant is α-galactose, and the backbone structure D-galactose (fucose α1→2) β1–3 (4)-*N*-acetylglucosamine forms the H antigen. The i antigen is built up by the repeating disaccharide D-galactose β1→4 *N*-acetyl-D-glucosamine β1→3, whereas the I antigen contains the branched structure D-galactose β1→6 (D-galactose β1→4) *N*-acetyl-D-glucosamine β1→3 (Fukuda *et al.*, 1979).

Like band 3, the band 4.5 protein shows a diffuse staining when studied by polyacrylamide gel electrophoresis in the presence of sodium dodecyl sulfate. This is certainly due to heterogeneity in its single N-glycosidic oligosaccharide. This oligosaccharide is probably similar to that of band 3, and it is ABH and Ii blood group active (Karhi and Gahmberg, 1980). This protein is responsible for glucose transport through the membrane. The primary structure has been determined (Mueckler *et al.*, 1985). It contains 492 amino acids and the N-glycosidic oligosaccharide is located at asparagine-45. Both the NH_2-terminal and COOH-terminal ends seem to reside in the cytoplasm. Like band 3, this protein traverses the membrane 12 times and part of the membrane-spanning sequence has amphipathic α-helical structures (Mueckler *et al.*, 1985).

The glucose transport protein is also in that sense similar to band 3 because it can insert post-translationally into microsomes (Mueckler and Lodish, 1986). It contains no cleavable NH_2-terminal signal peptide, and therefore it is likely that it uses internal hydrophobic peptide sequences for translocation.

The glycoprotein carrying the Kell blood group activity is similar in size to band 3, but peptide maps show very different primary structures (Redman *et al.*, 1986). Only part of the protein molecules seem to be available at the cell surface.

Not much is known about the Duffy blood group antigen. Moore *et al.* (1982) immune precipitated a major molecule with an apparent molecular weight of 39,500 from surface-labeled cells, but other bands were also obtained. Hadley *et al.* (1984) found that the protein had an apparent molecular weight of 35,000–43,000. It was glycosylated because neuraminidase treatment changed its apparent molecular weight. The desialylated protein still bound anti-Duffy antibodies.

2.3. The Red Cell Membrane Skeleton

The erythrocyte membrane skeleton is the insoluble residue remaining after extraction with nonionic detergents like Triton X-100 (Steck, 1974). It is located at the cytoplasmic aspect of membrane and is formed by several different polypeptides, which extensively interact with each other and with the cytoplasmic portions of membrane-spanning integral membrane proteins (Marchesi, 1983).

The best-known proteins in this meshwork are the α- and β-spectrin polypeptides, ankyrin, bands 4.1 and 4.2, actin (band 5), and glyceraldehyde 3-phosphate dehydrogenase (band 6). The spectrin molecules (see Figures 2 and 3) have apparent molecular weights of 250,000 and 225,000. They are both composed of distinct domains and form extended rodlike heterodimer structures. The spectrin molecules easily form higher oligomers in vitro and are no doubt essential for the formation of the red cell skeleton and for the biconcave shape of the erythrocyte. The β-chain is phosphorylated, and the phosphate is located near the end of the chain, which interacts with the α-chain during oligomer formation. Whether the phosphate is important for the function of spectrin is not known.

When red cell membranes are incubated in buffers of low ionic strength and EDTA, spectrin molecules dissociate from the membrane. Spectrin readily binds to spectrin-depleted membranes, and the binding protein has been identified. It is called ankyrin (band 2.1) and has an apparent molecular weight of 215,000

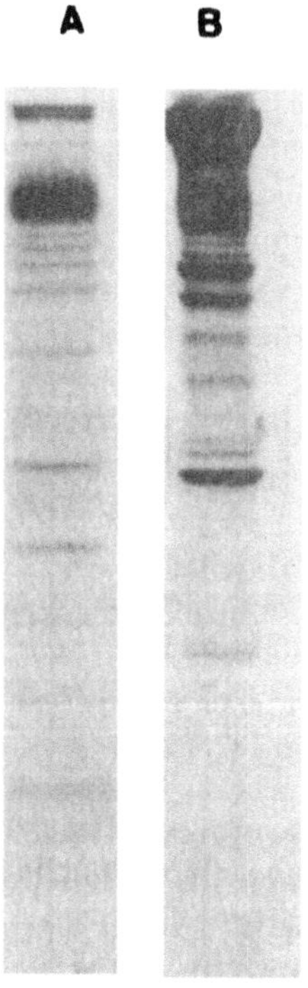

FIGURE 3. Solubilization of the red cell membrane by nonionic detergents. Red cell membranes were treated with 1% Triton X-100 in PBS and centrifuged. The solubilized proteins (A) and membrane-skeleton-associated proteins (B) were studied by polyacrylamide gel electrophoresis in the presence of sodium dodecyl sulfate. A major component in (A) is band 3, whereas (B) is enriched in spectrin, bands 4.1, 4.2, and 5.

(Bennett and Stenbuck, 1980). Ankyrin is tightly associated with the membrane, but it is not an integral membrane protein; therefore, treatment with 1-M salt releases it (Bennett and Stenbuck, 1980). The binding site for ankyrin in the membrane is located in the NH_2-terminal portion of band 3. Thus, a complicated interactive system is responsible for the membrane–spectrin binding.

Actin is a major component of the red cell membrane and binds to spectrin. Interestingly, this association is promoted by band 4.1, and a fragment of 4.1 has been isolated, which is responsible for this activity (Correas *et al.*, 1986; Marchesi, 1983). The importance of 4.1 for the integrity of the red cell membrane was recently directly shown. Red cells deficient in 4.1 are fragile and the affected individuals have severe hemolytic anemia (Tchernia *et al.*, 1981). Addition of 4.1 to 4.1-deficient membranes resulted in normal stable membranes (Takakuwa *et al.*, 1986). In addition, 4.1 binds to glycophorin A (Anderson and Lovrien, 1984), and this binding seems to be regulated by phosphatidylinositol 4,5-bisphosphate (Anderson and Marchesi, 1985). In the presence of the lipid, the binding between glycophorin A and 4.1 substantially increased. This finding may be important because of the fact that inositol 1,4,5-triphosphate and 1,2-diacylglycerol, formed from phosphatidylinositol 4,5-bisphosphate, both act as second messengers in response to cellular stimulators. But the phospholipid breakdown may also affect the interaction between integral membrane proteins and the membrane skeleton. This may of course happen not only in erythroid cells but also in other, nucleated cells.

Band 6 or glyceraldehyde 3-phosphate dehydrogenase is bound to the membrane through the NH_2-terminal portion of band 3 as discussed above. It is unclear whether this protein is important for the stability of the membrane. It has been shown that its enzyme activity can be affected by binding to the membrane.

3. CHARACTERIZATION OF THE Rh_0(D) MOLECULE

3.1. Early Work

The description above of the red cell membrane shows that even a cell membrane considered relatively simple is extremely complex at the molecular level. Only some of the major proteins have been studied in any detail, and we know almost nothing about their locations in the membrane with respect to neighboring molecules. However, progress in this field is evident and important advances are being made, albeit slowly.

In the 1970s, much was learned about dealing with membrane molecules. The introduction of different types of detergent that substituted for the membrane lipids especially resulted in a breakthrough in membrane molecular research. For a review on useful detergents for membrane research see Helenius and Simons (1975).

Isolation and characterization of membrane components per se is much more difficult when working with molecules easily solubilized in aqueous buffers. This especially holds true when the molecule in question only represents a small portion of the starting material, when it is labile, and when it is difficult to assay. All these facts are unfortunately applicable to the $Rh_0(D)$ antigen. Some aspects on the current knowledge of $Rh_0(D)$ biochemistry have recently been reviewed (Gahmberg and Karhi, 1986; Issitt, 1986).

In several blood group systems, carbohydrate is essential. The ABO and MN blood groups have been mentioned above in connection with well-characterized membrane glycoproteins. Further blood group systems in this category include at least the Lewis and P blood glycolipids. Therefore, it first seemed reasonable to assume that the $Rh_0(D)$ antigen could also be formed by carbohydrate. However, no support for this hypothesis has been obtained. Instead, several years ago it was shown that proteases acting on isolated membranes or $Rh_0(D)$ preparations (but not on the intact cell) destroyed the $Rh_0(D)$ activity. This does not necessarily mean that protein is responsible for the antigen activity but it certainly strengthens this possibility.

Important results were obtained by Green and co-workers. They first showed that lipid was essential for $Rh_0(D)$ activity (Green, 1972), By extracting membranes with butanol, the antigen activity was lost, and it could partially be restored by adding phospholipids. This indicated that the $Rh_0(D)$ antigen needed a membranelike environment for activity. The same group then showed that the antigen could be solubilized by deoxycholate, and that activity was regained after removal of the detergent (Lorusso and Green, 1975). They further showed that the $Rh_0(D)$ activity could partially be resolved from blood group A activity.

Litten *et al.* (1978) partially characterized the $Rh_0(D)$ activity by enzyme treatments. They reported solubilization of the antigen by treatment of isolated membranes with EDTA and 2-mercaptoethanol. This treatment is similar to that used for removal of spectrin from membranes and could well have resulted in partial solubilization of the membrane-skeleton-associated antigen (see Section 3.3). No effect was obtained by using phospholipases A and C, by neuraminidase, leucine aminopeptidase, bovine pancreatic carboxypeptidase A, or pig pancreatic carboxypeptidase B. However, the proteases pronase, trypsin, and papain destroyed the antigen activity.

There are several other early reports in the literature on the characterization of the $Rh_0(D)$ antigen but evidently they have not appreciably advanced our knowledge in the field, and these are therefore left out.

3.2. Later Studies

The major membrane components band 3 and glycophorin A are both present in about 10^6 copies per mature red cell. The number of $Rh_0(D)$ antigens on $Rh_0(D)$-positive cells has been estimated using radioactively labeled anti-D antibodies. Such an approach can be criticized because of a number of reasons. In fact, the estimated values vary a lot. Cunningham *et al.* (1985) found about 4000 $Rh_o(D)$ sites on $Rh_o(D)$ positive cells. Others have found 20,000–40,000 antigen sites per cell (Masouredis *et al.*, 1976; Rearden and Chiu, 1983). Anyway, the results clearly show that the $Rh_o(D)$ antigen is a relatively minor membrane component. Furthermore, Rh_{null} cells lacking all Rh antigens (Vos *et al.*, 1961) did not show any observable protein abnormality on protein-stained polyacrylamide gels (Ballas *et al.*, 1984).

One approach to identify the Rh antigens could be to study their expression during red cell differentiation. The appearances of glycophorin A (Ekblom *et al.*, 1985; Gahmberg *et al.*, 1978; Robinson *et al.*, 1981; Yurchenko and Furthmayr, 1980), blood group A antigens (Karhi *et al.*, 1981; Sieff *et al.*, 1982), and band 3 (Rearden and Chiu, 1983; Sieff *et al.*, 1982) have been studied using bone marrow cells and immunological techniques. Glycophorin A appeared already on proerythroblasts, whereas band 3 and blood group A antigen appeared at the basophilic normoblast stage of differentiation.

Rearden and Chiu (1983) used human anti-D antibodies and found that the $Rh_o(D)$ antigen was present on pronormoblasts in relatively low amounts. They obtained a number of 6000–9000 sites per cell. A progressive increase in antigen site number was observed on more differentiated cells reaching 25,000–35,000 on erythrocytes. Protease treatment of precursor cells exposed, however, more $Rh_o(D)$ antigen sites (Rearden and Masouredis, 1981). Rearden and Chiu could not find any $Rh_o(D)$ antigens on BFU-E or CFU-E precursor cells. It is possible, however, that their antibodies did not bind complement, and this could explain why no inhibition of burst- or colony-forming units was observed. Falkenburg *et al.*, (1985), in contrast, using fluorescent antisera and fluorescence-activated cell sorting, obtained evidence that the CFU-E cells contained $Rh_o(D)$ antigens, but not the BFU-E cells.

The appearance of $Rh_o(D)$ antigen activity during erythroid cell differentiation correlated with the presence of band 3, which implied that band 3 carried $Rh_o(D)$ activity, but obviously firm conclusions cannot be drawn from this. In fact, hemoglobin levels would probably have correlated as well!

Plapp *et al.* (1979) first tried affinity chromatography to purify the $Rh_o(D)$ antigen. Anti-$Rh_o(D)$IgG was coupled to Sepharose, and after solubilization in deoxycholate, bound molecules were eluted with dimethylformamide. Protein molecules with apparent molecular weights of around 7000 were specifically eluted. The low molecular weight peptides could be fragments released by the strong conditions or could be the result of proteolysis of larger molecules.

Using a different approach, Victoria *et al.* (1981) observed binding of band 3-like molecules to anti-$Rh_o(D)$ antibodies.

Evans *et al.* (1982) separated membrane proteins from $Rh_o(D)$-positive and -negative erythrocytes and determined antigen activity from polyacrylamide gel slices by an enzyme-linked immunoassay. Surprisingly, a positive peak was obtained from both Rh_0(D)-positive and -negative membranes with an apparent molecular weight between 13,000 and 30,000. The same group has claimed that in Rh_0(D)-negative cells antigen activity is found on the cytoplasmic aspect of the Rh_0(D)-negative membranes (Evans *et al.*, 1983; Plapp *et al.*, 1981), but this finding could not be confirmed by others (Gahmberg and Karhi, 1984; Kleeman *et al.*, 1982).

Two attempts have been made to determine the size of the Rh_0(D) antigen by radiation inactivation of isolated membranes. Folkerd and co-workers obtained a value of 174,000 (Folkerd *et al.*, 1977), and Green and co-workers obtained a value of 60,000 (Green *et al.*, 1983). No one yet knows how large the Rh_0(D) complex actually is in the membrane.

3.3. Approaching a Consensus on the Nature of the Rh_0(D) Molecule

The results described above aiming at determining the molecular size of the Rh_0(D) antigen, although giving much valuable information, made the picture confusing and emphasized the difficulties associated with the characterization of the antigen. However, the most recent work has, in my opinion, completely changed the picture.

In 1982, Gahmberg and Moore and co-workers (Gahmberg, 1982; Moore *et al.*, 1982) independently showed that, by using ^{125}I-lactoperoxidase surface-labeled erythrocytes and immune isolation, a major Rh_0(D) active molecule with an apparent molecular weight of 28,000–33,000 was obtained. This protein was only observed in Rh_0(D)-positive cells. Rh_0(D)-negative membrane gave no labeled band. Furthermore, Moore and co-workers showed that the c and E antigens had similar molecular weights.

A more detailed investigation of the molecule was made by Gahmberg (1983). First, it was shown that ^{125}I-lactoperoxidase-labeled -D-/-D- membranes contained a strong band in the 30,000 molecular weight region on polyacrylamide gel electrophoresis in the presence of sodium dodecyl sulfate. Normal Rh_0(D)-positive membranes showed a more weakly labeled band, and Rh_{null} membrane lacked all labeled proteins in this region of the gel. This strongly indicated that this component was associated with Rh_0(D) activity. Furthermore, the 30,000 molecular weight polypeptide was immune precipitated by anti-Rh_0(D) antiserum from -D-/-D- membranes (Figure 4C). When intact cells were treated with pronase, the 30,000 molecular weight band was not cleaved whereas many other surface proteins including glycophorin A and band 3 were degraded (Gahmberg, 1983). This was expected, because previous work had shown that protease treatment of intact cells did not destroy the Rh_0(D) antigen activity.

Interestingly, the Rh_0(D) molecule does not seem to contain carbohydrate. In mammalian cells, surface proteins are generally *glycoproteins* (Gahmberg, 1976), and other exceptions to this rule are to the best of my knowledge not

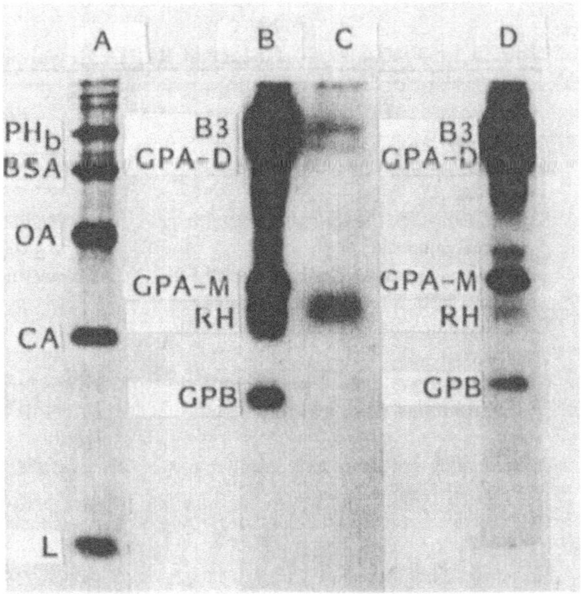

FIGURE 4. Polyacrylamide slab gel electrophoresis of ^{125}I-labeled red cell membranes and immune precipitate obtained with anti-Rh_0(D) antiserum. (A) Pattern of molecular weight standard proteins: PH_b = phosphorylase b; BSA = bovine serum albumin; OA = ovalbumin; CA = carbonic anhydrase; L = lysozyme. (B) Pattern of ^{125}I-lactoperoxidase-labeled -D-/-D- membranes: B3 = band 3; GPA-D = glycophorin A dimer; GPA-M = glycophorin A monomer; RH = Rh_0(D) molecule; GPB = glycophorin B. (C) Immune precipitate obtained with anti-Rh_0(D) antiserum from ^{125}I-labeled -D-/-D- membranes. (D) Pattern of normal Rh_0(D)-positive ^{125}I-labeled red cell membranes. Note that the Rh_0(D) molecule is relatively more weakly labeled in D.

known. The Rh_o(D) protein did not bind to any of several lectins studied, and endo-β-D-*N*-acetylglucosaminidase H, endo-β-galactosidase, or mild alkali (to liberate O-glycosidic carbohydrate) did not have any effect on the molecule. Furthermore, it was not labeled by the galactose oxidase/NaB^3H_4 or periodate/NaB^3H_4 surface-labeling techniques specific for galactose/*N*-acetylgalactosamine and sialic acid residues, respectively. In addition, the polypeptide did not seem to be phosphorylated (Gahmberg, 1983).

Further support for the association of the 28,000–33,000 molecular weight component with Rh_o(D) activity came when Ridgwell *et al.* (1983) showed that Rh_{null} cells lack two surface proteins with apparent molecular weights of 32,000 and 34,000. They used a SH-reactive radioactive cell surface-specific probe to label the proteins because previous work had shown that SH groups are essential for Rh_o(D) activity (Abbott and Schachter, 1976; Green, 1965).

Using acrylamide gels of different concentrations, Gahmberg postulated that the Rh_o(D) antigen is relatively hydrophobic (Gahmberg, 1982). Others have provided further evidence for this. After extraction with chloroform–meth-

Table I
Important Events in the History of $Rh_o(D)$ Research

Year	Findings	Reference
1938	Description of the hemolytic disease of the newborn	Darrow (1938)
1939	Suggestion that fetal red cells are destroyed by maternal antibodies	Levine and Stetson (1939)
1940	Antisera made in rhesus monkeys reacted with red cells from part of the human population	Landsteiner and Wiener (1940)
1941	Full description of the hemolytic disease of the newborn	Levine *et al.* (1941)
1960–1963	Introduction of preventive measures against isoimmunization using anti-$Rh_o(D)$ antisera	Stern and Berger (1960); Finn *et al.* (1961); Clarke *et al.* (1963)
1972	Demonstration that lipids are needed for $Rh_o(D)$ activity	Green (1972)
1975	Solubilization and reconstitution of $Rh_o(D)$ activity	Lorusso and Green (1975)
1982	Molecular identification of the $Rh_o(D)$ antigen	Gahmberg (1982); Moore *et al.* (1982)
1983	More detailed characterization of $Rh_o(D)$ antigen	Gahmberg (1983)
1984	Demonstration of an association of the $Rh_o(D)$ molecule with the membrane skeleton	Ridgwell *et al.* (1984); Gahmberg and Karhi (1984)

anol, a 35,000 molecular weight protein was obtained in the chloroform phase that showed $Rh_o(D)$ activity (Brown *et al.*, 1983; Sinor *et al.*, 1984). The preparation, however, was not pure.

More recent work has shown that the $Rh_o(D)$ antigen partially is attached to the membrane skeleton. Using different immunological and surface-labeling techniques, scientists found that most of the antigen remained bound to the insoluble residue after extraction of membranes with Triton X-100 (Gahmberg and Karhi, 1984; Ridgwell *et al.*, 1984). The simplest explanation for its membrane association is that the $Rh_o(D)$ polypeptide spans the membrane and interacts with presently unknown cytoskeletal components. The importance of the cytoskeleton was recently further emphasized when Paradis *et al.* (1986) found that the skeleton is essential for preserving $Rh_o(D)$ activity.

Table I shows some of the major events in elucidating the properties of the $Rh_o(D)$ antigen.

4. GENERAL DISCUSSION

Presently, it seems reasonably certain that the 28,000–33,000 molecular weight surface-exposed polypeptide enriched in -D-/-D- cells and absent from

Rh_{null} cells is the Rh_o(D) antigen. Most probably, the other Rh antigens are located on similar polypeptides. Furthermore, it seems possible that Rh_o(D)-negative cells contain a similar protein, which lacks the D-antigen site. No positive evidence for this is available, however.

Recently, some monoclonal antibodies against the D antigen have been described (Crawford *et al.*, 1983; Ridgwell *et al.*, 1983). By using these or conventional human anti-D antisera and biochemical techniques, it should become possible to purify the protein in milligram amounts. This would be important because of at least two reasons. First, by amino acid sequencing a partial structure could be obtained. With this information one could make corresponding synthetic peptides and strong antipolypeptide antisera using the peptides for immunization. The antisera could then be used for large-scale purification. Furthermore, one could make synthetic oligonucleotides and use these to clone the Rh_o(D)-antigen genes. In this way, the primary structure could be obtained and perhaps the D site identified after comparison with Rh genes from Rh_o(D)-negative cells.

Second, it would be enormously useful to have a pure Rh_o(D) antigen for production of clinically useful anti-Rh_o(D) antisera. These could be used for blood typing and especially for prevention of the hemolytic disease of the newborn.

The availability of the Rh_o(D) antigen on the erythrocyte surface to lactoperoxidase-catalyzed iodination and its reactivity with anti-Rh_o(D) antisera obviously mean that part of the molecule is exposed to the external milieu. On the other hand, it is also bound to the membrane skeleton. This does not prove, but makes it likely, that the protein spans the membrane. One possibility is of course that there is a connecting polypeptide between the Rh_o(D) molecule and the membrane skeleton. The evidently hydrophobic nature of the Rh_o(D) polypeptide strengthens the possibility that it itself spans the membrane. In Figure 5 is shown an idealized version of the Rh_o(D) molecule in the red cell membrane.

The absence of Rh antigens in Rh_{null} cells results in functional and structural abnormalities. Rh_{null} individuals develop hemolytic anemia (Levine *et al.*, 1973; Sturgeon, 1970) and have a decreased content of Na^+ and K^+ (Ballas *et al.*, 1984). Their Na^+–K^+-activated ATPase activity was increased about two times over normal values (Ballas *et al.*, 1984; Lauf and Joiner, 1976). Furthermore, the phospholipids of Rh_{null} membranes showed an abnormal distribution. All phosphatidylcholine molecules in Rh_{null} cells could be exchanged with a phospholipid exchange protein, whereas only 75 in normal red cells were exchangeable (Kuypers *et al.*, 1984). Also, phosphatidylethanolamine was more accessible in Rh_{null} cells. These facts indicate that the Rh antigens are important in maintaining normal lipid bilayer–membrane skeleton interactions.

The apparent absence of sugars in the Rh_o(D) antigen is intriguing. One can only speculate why this is so. One obvious possibility is of course that there is no N-glycosylation consensus sequence (–asparagine–X–threonine/serine) in the

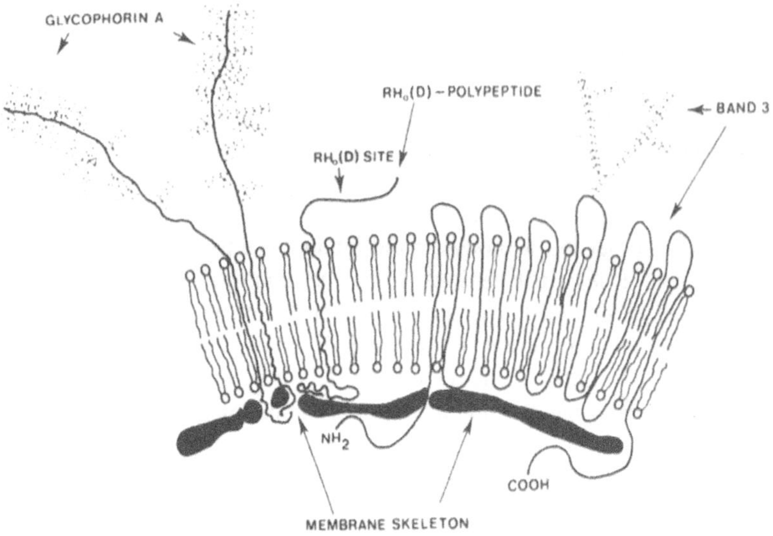

FIGURE 5. An idealized version of the $Rh_0(D)$-positive red cell membrane. The $Rh_0(D)$ molecule is thought to span the membrane and interact with the membrane skeleton. It probably lacks carbohydrate and is relatively deeply embedded in the membrane surrounded by lipids and major glycoproteins like glycophorin A and band 3. The $Rh_0(D)$-antigen-active site must be located externally, but nothing is known in detail about it. Glycophorin A is heavily glycosylated and probably exists as a dimer. Band 3 contains only one oligosaccharide, which has a complex structure containing Ii and ABH blood group activities. It traverses the membrane 12 times, and both its NH_2 and COOH terminals are probably in the cytoplasm.

protein. The polypeptide requirements for O-glycosylation are not known. A further possibility could be that the protein is synthesized free in the cytoplasm and post-translationally attached to the cytoskeleton. Because of its hydrophobic character, it would then diffuse into the lipid bilayer. Thus, it would never enter the intraluminal glycosylation systems in the endoplasmic reticulum and the Golgi apparatus. A further, but more remote, possibility is that it is cotransported with another protein (which probably should be a glycoprotein) to the cell surface. This is the case, for example, for the nonglycopolypeptide β_2-microglobulin in class I transplantation antigens (Lee *et al.*, 1980; Owen *et al.*, 1980; Ploegh *et al.*, 1979; Sege *et al.*, 1981).

It has long been known that the LW (Landsteiner–Wiener) blood group system somehow is associated with the Rh system. Rh_{null} erythrocytes also lack the LW antigens, and $Rh_o(D)$-positive cells give a stronger reaction with anti-LW reagents than Rh(D)-negative cells.

Two individuals, however, are known who have normal Rh antigens but lack the LW antigens (Mallinson *et al.*, 1986). Moore and co-workers (1983) and Mallingon *et al.* (1986) have clearly shown that the LW antigens are located on a diffusely migrating band of 37,000–47,000 apparent molecular weight on polyacrylamide gel electrophoresis in the presence of sodium dodecyl sulfate. The LW antigens clearly contained carbohydrate; neuraminidase treatment sharpened the band on electrophoresis and endo-*N*-acetylglycosaminidase F treatment to remove N-glycosidic oligosaccharides lowered their apparent molecular weights. Furthermore, the monoclonal anti-LW antibodies reacted weakly after carbohydrate removal (Mallinson *et al.*, 1986). The remote possibility still remains, however, that the LW antigens are glycosylated forms of the Rh_0(D) antigen.

5. CONCLUDING REMARKS

The Rh_0(D) antigen seems to be an erythroid cell-specific nonglycosylated, membrane-skeleton-bound polypeptide with an apparent molecular weight of 28,000–33,000. Its absence results in hemolytic anemia. It is clinically and biochemically one of the most important and interesting blood group antigens, and therefore there exists an urgent need to purify larger amounts of the protein and determine its primary structure. Probably, the difference between the Rh_0(D) molecule and other Rh molecules is the result of sequence differences in the 28,000–33,000 molecular weight polypeptides.

ACKNOWLEDGMENTS. The original work reported from the author's laboratory was supported by the Academy of Finland, the Sigrid Juselius Foundation, and the National Cancer Institute Grant 26294. I thank Yvonne Heinilä for expert secretarial assistance.

REFERENCES

Abbott, R. E., and Schachter, D., 1976, Inactivation of the Rh_0(D) antigen of intact human erythrocytes by glutathione maleimide. *J. Biol. Chem.* **251**:7176–7183.

Anderson, R. A., and Lovrien, R. E., 1984, Glycophorin is linked by band 4.1 protein to the human erythrocyte skeleton. *Nature* **307**:655–658.

Anderson, R. A., and Marchesi, V. T., 1985, Regulation of the association of membrane skeletal protein 4.1 with glycophorin by a polyphosphoinositide. *Nature* **318**:295–298.

Andersson, L. C., Gahmberg, C. G., Teerenhovi, L., and Vuopio, P., 1979, Glycophorin A as a marker for early erythroid differentiation in acute leukemia. *Int. J. Cancer* **24**:717–720.

Ballas, S. K., Clark, M. R., Mohandas, N., Colfer, H. F., Caswell, M. S., Bergren, M. O., Perkins, H. A., and Shohet, S. B., 1984, Red cell membrane and cation deficiency in Rh null syndrome. *Blood* **63**:1046–1055.

Bennett, V., and Stenbuck, P. J., 1980, Human erythrocyte ankyrin. *J. Biol. Chem.* **255**:2540–2548.

Bretscher, M. S., 1972, Asymmetrical lipid bilayer structure for biological membranes. *Nature New Biol.* **236**:11–12.

Brock, C. J., Tanner, M. J. A., and Kempf, C., 1983, The human erythrocyte anoin-transport protein. *Biochem. J.* **213**:577–586.

Brown, P. J., Evans, J. P., Sinor, L. T., Tilzer, L. T., and Plapp, F. V., 1983, The rhesus D antigen. A dicyclohexyl-carbodiimide-binding proteolipid. *Am. J. Pathol.* **110**:127–134.

Cabantchik, Z. I., and Rothstein, A., 1974, Membrane protein related to anion permeability of human red blood cells. I. Localization of disulfonic stilbene binding sites in proteins involved in permeation. *J. Membr. Biol.* **15**:207–226.

Clarke, C. A., 1982, Rhesus haemolytic disease of the newborn and its prevention. Historical annotation. *Br. J. Haematol.* **52**:525–535.

Clarke, C. A., Finn, R., McConnell, R. B., and Sheppard, P. M., 1958, The protection afforded by ABO incompatibility against erythroblastosis due to rhesus anti-D. *Int. Arch. Allergy* **13**:380.

Clarke, C. A., Donohoe, W. T. A., McConnell, R. B., Woodrow, J. C., Finn, R., Krevans, J. R., Kulke, W., Lehane, D., and Sheppard, P. M., 1963, Further experimental studies on the prevention of Rh haemolytic disease. *Br. Med. J.* **i**:979–984.

Colin, Y., Rahuel, C., London, J., Roméo, P.-H., d'Auriol, L., Galibert, F., and Cartron, J.-P., 1986, Isolation of cDNA clones and complete amino acid sequence of human erythorocyte C. *J. Biol. Chem.* **261**:229–233.

Correas, I., Leto, T. L., Speicher, D. W., and Marchesi, V. T., 1986, Identification of the functional site of erythrocyte protein 4.1 involved in spectrin–actin associations. *J. Biol. Chem.* **261**:3310–3315.

Crawford, D. H., Barlow, M. J., Harrison, J. F., Winger, L., and Huehns, E. R., 1983, Production of human monoclonal antibody to rhesus D antigen. *Lancet* **i**:386–388.

Cunningham, N. A., Zola, A. P., Hui, H. L., Taylor, L. M., and Green, R. A., 1985, Binding characteristics of anti-Rh_0(D) antibodies to Rh_0(D)-positive and D^u red cells. *Blood* **66**:765–768.

Dahr, W., Uhlenbruck, G., Janssen, E., and Schmalisch, R., 1977, Different N-terminal amino acids in the MN glycoproteins from MM and NN erythrocytes. *Hum. Genet.* **35**:335–343.

Darrow, R., 1938, *Icterus gravis* (erythroblastosis) *neonatorum*. *Arch. Pathol.* **25**:378–417.

Ekblom, M., Gahmberg, C. G., and Andersson, L. C., 1985, Late expression of M and N antigens on glycophorin A during erythroid differentiation. *Blood* **66**:233–236.

Evans, J. P., Sinor, L. T., Brown, P. J., Tilzer, L. L., and Plapp, F. V., 1982, Identification of Rh_0(D) antigen in polyacrylamide gels by an enzyme-linked immunoassay. *Mol. Immunol.* **19**:671–675.

Evans, J. P., Brown, P. J., Sinor, L. T., Tilzer, L. L., Beck, M., and Plapp, F. V., 1983. Detection of an antigen on the inner surface of Rh negative erythrocytes which binds anti-D IgG. *Mol. Immunol.* **20**:529–536.

Falkenburg, J. H. F., Fibbe, W. E., van der Vaart-Duinkerken, N., Nichols, M. E., Rubinstein, P., and Jansen, J., 1985, Human erythroid progenitor cells express rhesus antigens. *Blood* **66**:660–663.

Finn, R., 1960, Erythroblastosis. *Lancet* **i**:526.

Finn, R., Clarke, C. A., Donohoe, W. T. A., McConnell, R. B., Sheppard, P. M., Lehane, D., and Kulke, W., 1961, Experimental studies on the prevention of Rh haemolytic disease. *Br. Med. J.* **i**:1486–1490.

Finne, J., 1980, Identification of the blood-group ABH-active glycoprotein components of human erythrocyte membrane. *Eur. J. Biochem.* **104**:181–189.

Folkerd, E. J., Ellory, J. C., and Hughes-Jones, N. C., 1977, A molecular size determination of Rh(D) antigen by radiation inactivation. *Immunochemistry* **14**:529–531.

Freda, V. J., Gorman, J. G., and Pollack, W., 1964, Successful prevention of experimental Rh sensitisation in man with an anti-Rh gammaglobulin antibody preparation. *Transfusion* **4**:26–32.

Fukuda, M., Fukuda, M. N., and Hakomori, S., 1979, Developmental change and genetic defect in the carbohydrate structure of band 3 glycoprotein of human erythrocyte membrane. *J. Biol. Chem.* **254**:5458–5465.

Furthmayr, H., 1978, Structural comparison of glycophorins and immunochemical analysis of genetic variants. *Nature* **271**:519–524.

Gahmberg, C. G., 1976, External labeling of erythrocyte glycoproteins. Studies with galactose oxidase and fluorography. *J. Biol. Chem.* **251**:510–515.

Gahmberg, C. G., 1982, Molecular identification of the human Rh_0(D) antigen. *FEBS Lett.* **140**:93–97.

Gahmberg, C. G., 1983, Molecular characterization of the human red cell Rh_0(D) antigen. *EMBO J.* **2**:223–227.

Gahmberg, C. G., and Hakomori, S., 1973, External labeling of cell surface galactose and galactosamine in glycolipid and glycoprotein of human erythrocytes. *J. Biol. Chem.* **248**:4211–4317.

Gahmberg, C. G., and Karhi, K. K., 1984, Association of Rh_0(D) polypeptides with the membrane skeleton in Rh_0(D)-positive human red cells. *J. Immunol.* **133**:334–337.

Gahmberg, C. G., and Karhi, K. K., 1986, Chemistry of ABH/Ii, MN/Ss, and Rh_0(D) blood group-active proteins of the human red-cell membrane. In *Receptors in Cellular Recognition and Developmental Processes* (R. M. Gorczynski, ed.), pp. 251–266, Academic Press, New York.

Gahmberg, C. G., Myllylä, G., Leikola, J., Pirkola, A., and Nordling, S., 1976, Absence of the major sialoglycoprotein in the membrane of human En(a-) erythrocytes and increased glycosylation of band 3. *J. Biol. Chem.* **251**:6108–6116.

Gahmberg, C. G., Jokinen, M., and Andersson, L. C., 1978, Expression of the major sialoglycoprotein (glycophorin) on erythroid cells in human bone marrow. *Blood* **52**:379–387.

Gahmberg, C. G., Jokinen, M., and Andersson, L. C., 1979, Expression of the major red cell sialoglycoprotein, glycophorin A, in the human leukemic cell line K562. *J. Biol. Chem.* **254**:7442–7448.

Gahmberg, C. G., Jokinen, M., Karhi, K. K., and Andersson, L. C., 1980. Effect of tunicamycin on the biosynthesis of the major human red cell sialoglycoprotein, glycophorin A, in the leukemia cell line K562. *J. Biol. Chem.* **255**:2169–2175.

Gahmberg, C. G., Jokinen, M., Karhi, K. K., Kämpe, O., Peterson, P. A., and Andersson, L. C., 1983, Glycophorin A, in vitro biogenesis and processing. *Methods Enzymol.* **96**:281–298.

Gahmberg, C. G., Ekblom, M., and Andersson, L. C., 1984, Differentiation of human erythroid cells is associated with an increased O-glycosylation of the major sialoglycoprotein, glycophorin A. *Proc. Natl. Acad. Sci. U.S.A.* **81**:6752–6756.

Gahmberg, C. G., Ekblom, M., and Andersson, L. C., 1986, Markers of normal and malignant erythroid differentiation: Role in clinical practice. In *Tumour Markers: Concepts and Applications* (A. S. Daar, ed.), pp. 379–429, Blackwell Scientific, Oxford.

Greaves, M. F., Sieff, C., and Edwards, P. A. W., 1983, Monoclonal antiglycophorin as a probe for erythroleukemias. *Blood* **61**:645–651.

Green, F. A., 1965, Studies on the Rh(D) antigen. *Vox Sang.* **10**:32–53.

Green, F. A., 1972, Erythrocyte membrane lipids and Rh antigen activity. *J. Biol. Chem.* **247**:881–887.

Green, F. A., Owens, N. A., Hui, J. L., Jung, C. Y., and Cuppoletti, J., 1983, Molecular size of the Rh_0(D) antigen of the human erythrocyte in situ by radiation inactivation. *Mol. Immunol.* **20**:361–365.

Hadley, T. J., David, P. H., McGinniss, M. H., and Miller, L. H., 1984, Identification of an erythrocyte component carrying the Duffy blood group Fy^a antigen. *Science* **223**:597–599.

Hakomori, S., 1981, Glycosphingolipids in cellular interaction. differentiation, and oncogenesis. *Annu. Rev. Biochem.* **50:**733–764.

Hakomori, S., and Kobata, A., 1975, Blood group antigens. In *The Antigens* (M. Sela, ed.), Vol. 2, p. 80, Academic Press, New York.

Helenius, A., and Simons, K., 1975, Solubilization of membranes by detergents. *Biochim. Biophys. Acta* **415:**29–79.

Issitt, P. D., 1986, Biochemistry of the Rh blood group system. In *Recent Advances in Blood Group Biochemistry* (V. Vengelen-Tyler and W. J. Judd, eds.), pp. 105–138, American Association of Blood Banks, Arlington, VA.

Jay, D. G., 1986, Glycosylation site of band 3, the human erythrocyte anion-exchange protein. *Biochemistry* **25:**554–556.

Järnefelt, J., Rush, J., Li, Y.-T., and Laine, R. A., 1978, Erythroglycan, a high molecular weight glycopeptide with the repeating structure (galactosyl(1–4)-2-deoxy-2-acetamidoglucosyl(1–3)) comprising more than one-third of the protein-bound carbohydrate of human erythrocyte stroma. *J. Biol. Chem.* **253:**8006–8009.

Jokinen, M., Andersson, L. C., and Gahmberg, C. G., 1985, Biosynthesis of the major human red cell sialoglycoprotein, glycophorin A. O-glycosylation. *J. Biol. Chem.* **260:**11314–11321.

Jokinen, M., Gahmberg, C. G., and Andersson, L. C., 1979, Biosynthesis of the major human red cell sialoglycoprotein, glycophorin A, in a continuous cell line. *Nature* **279:**604–607.

Jokinen, M., Ulmanen, I., Andersson, L. C., Kääriäinen, L., and Gahmberg, C. G., 1981, Cell-free synthesis and glycosylation of the major human red cell sialoglycoprotein, glycophorin A. *Eur. J. Biochem.* **114:**393–397.

Karhi, K. K., and Gahmberg, C. G., 1980, Identification of blood group A-active glycoproteins in the human erythrocyte membrane. *Biochim. Biophys. Acta* **622:**344–354.

Karhi, K. K., Andersson, L. C., Vuopio, P., and Gahmberg, C. G., 1981, Expression of blood group A antigens in human bone marrow cells. *Blood* **57:**147–151.

Kleeman, J. E., Masouredis, S. P., and Victoria, E. J., 1982, Exposure of the Rh_0(D) antigen on the surface and cytoplasmic domains of the red cell membrane. *Immunology* **45:**27–30.

Kopito, R. R., and Lodish, H. F., 1985, Primary structure and transmembrane orientation of the murine anion exchange protein. *Nature* **316:**234–238.

Krusius, T., Finne, J., and Rauvala, H., 1978, The poly(glycosyl) chains of glycoproteins. Characterization of a novel type of glycoprotein saccharides from human erythrocyte membranes. *Eur. J. Biochem.* **92:**289–300.

Kuypers, F., van Linde-Sibenius-Trip, M., Roelofsen, B., Tanner, M. J. A., Anstee, D. J., and Op den Kamp, J. A. F., 1984, Rh_{null} human erythrocytes have an abnormal membrane phospholipid organization. *Biochem. J.* **221:**931–934.

Lampio, A., Finne, J., Homer, D., and Gahmberg, C. G., 1984, Exposure of the major human red cell glycolipid, globoside, to galactose oxidase. *Eur. J. Biochem.* **145:**77–82.

Lampio, A., Rauvala, H., and Gahmberg, C. G., 1986, Exposure of major neutral glycolipids in red cells to galactose oxidase. Effect of neuraminidase. *Eur. J. Biochem.* **157:**611–616.

Landsteiner, K., and Wiener, A. S., 1940, An agglutinable factor in human blood recognised by immune sera for rhesus blood. *Proc. Soc. Exp. Biol. Med.* **43:**223.

Lauf, P. K., and Joiner, C. H., 1976, Increased potassium transport and ouabain binding in human Rh_{null} red blood cells. *Blood* **48:**457–468.

Lee, J. S., Trowsdale, J., and Bodener, W. F., 1980, Synthesis of HLA antigens from membrane-associated messenger RNA. *J. Exp. Med.* **152:**3S–10S.

Levine, P., 1943, Serological factors as possible causes in spontaneous abortions. *J. Hered.* **34:**71–80.

Levine, P., and Stetson, R. E., 1939, An unusual case of intragroup agglutination. *J. Am. Med. Assoc.* **113:**126–127.

Levine, P., Burnham, L., Katzin, W. M., and Vogel, P., 1941, The role of iso-immunization in the pathogenesis of erythroblastosis fetalis. *Am. J. Obstet. Gynecol.* **42**:925–937.

Levine, P., Tripodi, D., Struck, J., Jr., Zmijewski, C. M., and Pollack, W., 1973, Hemolytic anemia associated with Rh_{null} but not with Bombay blood. *Vox Sang.* **24**:417–424.

Litten, J., Culpepper, R., and Bakerman, S., 1978, Studies on the characterization of the Rh_o(D) antigen. *Biochem. Biophys. Acta* **543**:226–234.

Lorusso, D. J., and Green, F. A., 1975, Reconstitution of Rh(D) antigen activity from human erythrocyte membranes solubilized by deoxycholate. *Science* **188**:66–67.

Mäkelä, O., and Cantell, K., 1958, Destruction of M and N blood group receptor of human red cells by some influenza viruses. *Ann. Med. Exp. Biol. Fenn.* **36**:366–374.

Mallinson, G., Martin, P. G., Anstee, D. J., Tanner, M. J. A., Merry, A. H., Tills, D., and Sonneborn, H. H., 1986, Identification and partial characterization of the human erythrocyte membrane component(s) that express the antigens of the LW blood-group system. *Biochem. J.* **234**:649–652.

Marchesi, V. T., 1983, The red cell membrane skeleton: Recent progress. *Blood* **61**:1–11.

Marchesi, V. T., Furthmayr, H., and Tomita, M., 1976, The red cell membrane. *Annu. Rev. Biochem.* **45**:667–698.

Masouredis, S. P., Sudora, E. J., Mahan, L., and Victoria, E. J., 1976, Antigen site densities and ultrastructural distribution patterns of red cell Rh antigens. *Transfusion* **16**:94–106.

Mawby, W. J., and Findlay, J. B. C., 1983, Characterization and partial sequences of di-iodosulphophenyl siothiocyanate-binding peptide from human erythrocyte anion-transport protein. *Biochem. J.* **205**:465–475.

Moore, S., 1983, Identification of red cell membrane components associated with rhesus blood group antigen expression. In *Red Cell Membrane Glycoconjugates and Related Genetic Markers* (J.-P. Cartron, R. Rouger, and C. Salmon, eds.), pp. 97–106. Librarire Arnette, Paris.

Moore, S., Woodrow, C. F., and McClelland, D. B. L., 1982, Isolation of membrane components associated with human red cell antigens Rh(D), (c), (E) and Fy^a. *Nature* **295**:529–531.

Mueckler, M., and Lodish, H. F., 1986, The human glucose transporter can insert posttranslationally into microsomes. *Cell* **44**:629–637.

Mueckler, M., Caruso, C., Baldwin, S. A., Panico, M., Blench, I., Morris, H. R., Allard, W. J., Lienhard, G. E., and Lodish, H. F., 1985, Sequence and structure of a human glucose transporter. *Science* **229**:941–945.

Nevanlinna, H. R., and Vainio. T., 1956, The influence of mother–child ABO incompatibility on Rh immunisation. *Vox Sang.* **1**:26–36.

Nigg, E. A., Bron, C., Girardet, M., and Cherry, R. J., 1980, Band 3-glycophorin A association in erythrocyte membranes demonstrated by combining protein diffusion measurements with antibody-induced cross-linking. *Biochemistry* **19**:1887–1892.

Owen, M. J., Kissonerghis, A. M., and Lodish, H. F., 1980, Biosynthesis of HLA-A and HLA-B antigen *in vivo*. *J. Biol. Chem.* **255**:9678–9684.

Paradis, G., Bazin, R., and Lemieux, R., 1986, Protective effect of the membrane skeleton on the immunologic reactivity of the human red cell Rh_o(D) antigen. *J. Immunol.* **137**:240–244.

Parker, J. C., and Berkowitz, L. R., 1986, Genetic variants affecting the structure and function of the human red cell membrane. In *Physiology of Membrane Disorders* (T. E. Andreoli, J. F. Hoffman, J. C. Parker, L. R. Berkowitz, D. D. Fanestil, and S. G. Schultz, eds.), pp. 785–814, Plenum Press, New York.

Plapp, F. V., Evans, J. P., and Tilzer, L. L., 1981, Detection of Rh_o(D) antigen on the inner surface of Rh negative erythrocyte membranes. *Fed. Proc.* **40**:208.

Plapp, F. V., Kowalski, M. M., Tilzer, L., Brown, P. J., Evans, J., and Chiga, M., 1979, Partial purification of Rh_o(D) antigen from Rh positive and negative erythrocytes. *Proc. Natl. Acad. Sci. U.S.A.* **76**:2964–2968.

Ploegh, H. L., Cannon, L. E., and Strominger, J. L., 1979, Cell-free translation of the mRNAs for the heavy and light chains of HLA-A and HLA-B antigens. *Proc Natl. Acad. Sci. U.S.A.* **76**:2273–2277.

Race, R. R., and Sanger, R., 1950, *Blood Groups in Man,* pp. 234–236, Blackwell Scientific, Oxford.

Rearden, A., and Chiu, P., 1983, Lack of rhesus antigen expression by human committed erythroid progenitors. *Blood* **61**:525–529.

Rearden, A., and Masouredis, S. P., 1981, Protease modification enhances anti-D binding to nucleated red blood cell precursors. *Vox Sang.* **41**:160–164.

Redman, C. M., Avellino, G., Pfeffer, S. R., Mukherjee, T. K., Nichols, M., Rubinstein, P., and Marsh, W. L., 1986, Kell blood group antigens are part of a 93,000-dalton red cell membrane protein. *J. Biol. Chem.* **261**:9521–9525.

Ridgwell, K., Roberts, S. J., Tanner, M. J. A., and Anstee, D. J., 1983, Absence of two membrane proteins containing extracellular thiol groups in Rh_{null} human erythrocytes. *Biochem. J.* **213**:267–269.

Ridgwell, K., Tanner, M. J. A., and Anstee, D. J., 1984, The rhesus (D) polypeptide is linked to the human erythrocyte cytoskeleton. *FEBS Lett.* **174**:7–10.

Robinson, J., Sieff, C., Delia, D., Edwards, P. A. W., and Greaves, M., 1981, Expression of cell-surface HLA-DR, HLA-ABC and glycophorin during erythroid differentiation. *Nature* **289**:68–71.

Sege, K., Rask, L., and Peterson, P. A., 1981, Role of β_2-microglobulin in the intracellular processing of HLA antigens. *Biochemistry* **20**:4523–4530.

Shaklai, N., Yguerabide, J., and Ranney, H. M., 1977, Classification and localization of hemoglobin binding sites on the red blood cell membrane. *Biochemistry* **16**:5593–5597.

Shapiro, D. L., and Marchesi, V. T., 1977, Phosphorylation in membranes of intact human erythrocytes. *J. Biol. Chem.* **252**:508–517.

Siebert, P. D., and Fukuda, M., 1986, Isolation and characterization of human glycophorin A cDNA clones by a synthetic oligonucleotide approach: Nucleotide sequence and mRNA structure. *Proc. Natl. Acad. Sci. U.S.A.* **83**:1665–1669.

Sieff, C., Bicknell, D., Caine, G., Robinson, J., Lam, G., and Greaves, M. F., 1982, Changes in cell surface antigen expression during hemopoietic differentiation. *Blood* **60**:703–713.

Sinor, L. T., Brown, P. J., Evans, J. P., and Plapp, F. V., 1984, The Rh antigen specificity of erythrocyte proteolipid. *Transfusion* **24**:179–180.

Springer, G. F., and Ansell, N. J., 1958, Inactivation of human erythrocyte agglutinogens M and N by influenza viruses and receptor-destroying enzyme. *Proc. Natl. Acad. Sci. U.S.A.* **44**:182–189.

Steck, T. L., 1974, The organization of proteins in the human red blood cell membrane. A review. *J. Cell Biol.* **62**:1–19.

Steck, T. L., and Dawson, G., 1974, Topographical distribution of complex carbohydrates in the erythrocyte membrane. *J. Biol. Chem.* **249**:2135–2142.

Stern, K., and Berger, M., 1960, Experimental isosensitization to hemoantigens in man. Paper given at the 13th Annual Meeting of the American Association of Blood Banks, San Francisco.

Strapazon, E., and Steck, T. L., 1976, Binding of rabbit muscle aldolase to band 3, the predominant polypeptide of the human erythrocyte membrane. *Biochemistry* **15**:1421–1424.

Sturgeon, P., 1970, Hematological observations in the anemia associated with blood type Rh_{null}. *Blood* **36**:310–320.

Takakuwa, Y., Tchernia, G., Rossi, M., Benabadji, M., and Mohandas, N., 1986, Restoration of normal membrane stability to unstable protein 4.1-deficient erythrocyte membranes by incorporation of purified protein. 4.1. *J. Clin. Invest.* **78**:80–85.

Tchernia, G., Mohandas, N., and Shohet, B., 1981, Deficiency of skeletal membrane protein band

4.1 in homozygous hereditary elliptocytosis: Implications for erythrocyte membrane stability. *J. Clin. Invest.* **68**:454–460.

Tomita, M., and Marchesi, V. T., 1975, Amino acid sequence and oligosaccharide attachment sites of human erythrocyte glycophorin. *Proc. Natl. Acad. Sci. U.S.A.* **72**:2964–2968.

Verkleij, A. J., Zwaal, R. F. A., Roelofsen, B., Confucius, P., Kastelijn, D., and van Deenen, L. L. M., 1973, The asymmetric distribution of phospholipids in the human red cell membrane. *Biochim. Biophys. Acta* **323**:178–193.

Victoria, E. J., Mahan, L. C., and Masouredis, S. P., 1981, Anti-Rh_o(D) binds to band 3 glycoprotein of the human erythrocyte membrane. *Proc. Natl. Acad. Sci. U.S.A.* **78**:2898–2902.

Vos, G. H., Vos, D., Kirk, R. L., and Sanger, R., 1961, A sample of blood with no detectable Rh antigens. *Lancet* **i**:14–15.

Wasniowska, K., Drzeniek, Z., and Lisowska, E., 1977, The amino acids of M and N blood group glycopeptides are different. *Biochem. Biophys. Res. Commun.* **76**:385–390.

Wiener, A. S., and Peters, H. R., 1940, Hemolytic reactions following transfusions of blood of the homologous group, with three cases in which the same agglutinogen was responsible. *Ann. Intern. Med.* **13**:2306–2322.

Yu, J., and Steck, T. L., 1975, Associations of band 3, the predominant polypeptide of the human erythrocyte membrane. *J. Biol. Chem.* **250**:9176–9184.

Yurchenko, P. D., and Furthmayr, H., 1980, Expression of red cell membrane proteins in erythroid precursor cells. *J. Supramol. Struct.* **13**:255–269.

Chapter 4

Expression of the ABH, Lewis, and Related Antigens on the Glycoproteins of the Human Jejunal Brush Border

F. R. Green, P. Greenwell, L. Dickson, B. Griffiths, J. Noades, and D. M. Swallow

1. INTRODUCTION

1.1. The Genetics, Structure, and Molecular Distribution of the ABH and Lewis Antigens

The ABH and Lewis antigens are carbohydrate structures that occur on the surface of certain hemopoietic and epithelial cells and in many body fluids. The role of the *ABO*, *Se*, *H*, and *Le* gene loci and their glycosyltransferase products in determining the structures and the genetic polymorphisms of these terminal carbohydrate antigens has been reviewed in detail elsewhere (Oriol *et al.*, 1986; Watkins, 1980). The biochemical pathways are shown in summary form in Figure 1.

It is known that in body fluids the antigenic determinants occur on soluble mucin-type glycoproteins (Watkins, 1972) and also on oligosaccharides (Kobata,

F. R. Green, L. Dickson, B. Griffiths, J. Noades, and D. M. Swallow MRC Human Biochemical Genetics Unit, University College, London NW1 2HE, United Kingdom. **P. Greenwell** Division of Immunochemical Genetics, MRC Clinical Research Centre, Harrow HA1 3UJ, United Kingdom.

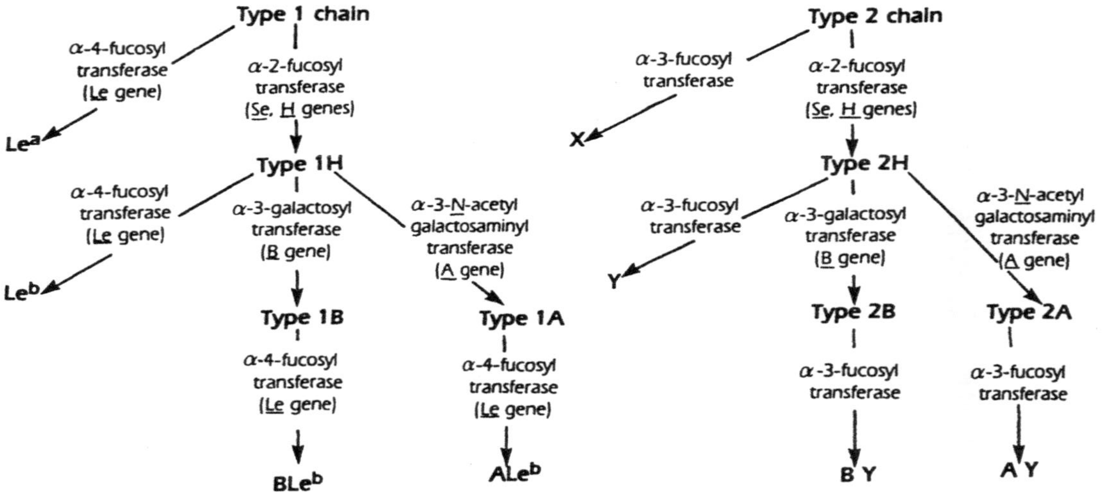

FIGURE 1. Biochemical pathways for the formation of the H, Le^a, Le^b, ALe^b, BLe^b, X, Y, AY, and BY determinants. Genes involved in coding for the transferases are shown in brackets. The α-3-*N*-acetylgalactosaminyltransferase (A) and α-3-galactosyltransferase (B) are thought to be allelic products of the *ABO* gene locus (Greenwell *et al.*, 1986). Structure: type 1 precursor, Galβ1→3GlcNAc-R; type 2 precursor, Galβ1→4GlcNAc-R; type 1H, Galβ1→3GlcNAc-R (Galβ1 ↑ 1,2 Fucα); type 2H, Galβ1→4GlcNAc-R (Galβ1 ↑ 1,2 Fucα); type 1A, GalNAcα1→3Galβ1→3GlcNAc-R (Galβ1 ↑ 1,2 Fucα); type 2B, Galα1→3Galβ1→4GlcNAc-R (Galβ1 ↑ 1,2 Fucα); X, Galβ1→4GlcNAc-R (GlcNAc ↑ 1,3 Fucα); Y, Galβ1→4GlcNAc-R (Galβ1 ↑ 1,2 Fucα; GlcNAc ↑ 1,3 Fucα); Le^a, Galβ1→3GlcNAc-R (GlcNAc ↑ 1,4 Fucα); Le^b, Galβ1→3GlcNAc-R (Galβ1 ↑ 1,2 Fucα; GlcNAc ↑ 1,4 Fucα).

1972; Strecker *et al.*, 1976), while on cell surfaces the A,B,H, and I (the precursor structure) antigens occur on both integral membrane glycolipids and glycoproteins, as has been shown in the case of red cells (Childs *et al.*, 1978; Finne *et al.*, 1978; Hakomori and Kobata, 1974; Jarnefelt *et al.*, 1978; Karhi and Gahmberg, 1980; Krusius *et al.*, 1978; Takasaki *et al.*, 1978). Several different glycoproteins of the red cell membrane have been shown to carry the ABH antigens. These include band 3 (the anion transport protein), glycophorin A, and band 4.5 (Finne, 1980). This group of antigens has also been shown to occur on well-characterized membrane glycoproteins of other cells, for example, on the epidermal growth factor receptor of A431 carcinoma cells, as well as on the surface glycolipids of these cells (Le Pendu *et al.*, 1985b; Parker *et al.*, 1984). The ABH structures can occur both on O-linked and N-linked side chains of glycoproteins (Finne, 1980; Wilczynska *et al.*, 1980).

The occurrence of each of the ABH and Lewis carbohydrate structures on the glycan units of glycoproteins and glycolipids depends on the provision of the correct structures as acceptor substrates and nucleotide sugars as donor substrates for each of the glycosyltransferases, as well as the expression of the whole series of transferases themselves. The enzymes appear not to discriminate between whether the structures occur on glycolipid or glycoprotein, possibly because the precursor disaccharide [Galβ1→3GlcNAc-type 1; Galβ1→4GlcNAc-type 2; Galβ1→3GalNAc-type 3] is well separated from the lipid or protein moiety.

The expression of the A and B antigens is dependent on the presence of fucose linked α1→2 to the terminal galactose of the precursor disaccharide, since fucose in this position is required as part of the acceptor substrate for the allelic A and B transferases. On the red cells the presence of this fucose is dependent on the genotype of the individual with respect to the gene designated *H*, which probably represents a structural gene locus coding for α-2-fucosyltransferase. Mutations at this gene locus leading to the rare deficiency of α-2-fucosyltransferase in plasma give rise to the Bombay phenotype (see Watkins, 1980), namely, the absence of ABH antigen on the red cells and in secretions. In body fluids the expression of α-2-fucosyltransferase activity, and thus the occurrence of H antigen, is under the control of a second gene *Se*. This gene, which was for a long time considered to be a regulatory gene (Watkins, 1959; Watkins and Morgan, 1959), determines the level of α-2-fucosyltransferase activity in the submaxillary glands and probably many other epithelial cells. The α-2-fucosyltransferase activity found in such tissues has some different physicochemical properties from that found in plasma, in particular showing differences in substrate preference (Betteridge and Watkins, 1985; Kumazaki and Yoshida, 1984; Le Pendu, 1983). It is now widely believed that *Se* is a structural gene locus coding for an α-2-fucosyltransferase distinct from that coded for by the *H* gene (Kumazaki and Yoshida, 1984; Le Pendu, 1983; Le Pendu *et al.*, 1985a; Oriol *et al.*, 1981), and that the reported segregation pattern in families

(Gerard *et al.*, 1982) is due to the close linkage of these two genes. However, it cannot at present be formally excluded (Betteridge and Watkins, 1985) that *Se* is not linked to *H* and codes for some kind of modifier subunit that alters the substrate specificity as occurs with α-lactalbumin and lactose synthetase (Hill and Brew, 1975), or with activator proteins that allow acid hydrolases to bind to lipid substrates (Conzelmann *et al.*, 1982; Li *et al.*, 1979; Stevens *et al.*, 1981).

Another factor that restricts the structures synthesized is that the type 1 precursor structure is required for the synthesis of the Lewis Le^a and Le^b antigens (Figure 1). The absence of this precursor on the integral membrane glycolipids and glycoproteins of the red cells accounts for the absence of Le^a and Le^b on these molecules. The Le^a and Le^b antigens present on red cells are absorbed from the plasma (Crookston and Tilley, 1977; Marcus and Cass, 1969; Sneath and Sneath, 1955).

1.2. The ABH and Lewis Antigens in the Gastrointestinal Tract

It has long been known that the A, B, and H antigens occur in the human gastrointestinal tract. The early evidence of this came from the finding that mucosal scrapings were very effective inhibitors of red cell agglutination by anti-A and anti-B (Hartmann, 1941). The quantity of the antigen present in these scrapings was greatest in the stomach and decreased along the length of the intestine (Hartman, 1941), being lowest in the rectosigmoid colon (Wiley *et al.*, 1981). Immunohistological studies have shown the A,B, and H determinants within many of the cells of the surface epithelium and glands of the gastrointestinal tract (Brown *et al.*, 1984; Glynn *et al.*, 1957; Mollicone *et al.*, 1985, 1986; Szulman, 1960, 1962). In the small intestine they occur in the goblet cells (Szulman, 1960, 1962) and also the absorptive cells (Mollicone *et al.*, 1986; Roth *et al.*, 1986, 1987).

The expression of the ABH antigens in the gastrointestinal tract appears to be largely under the control of the *Se* gene locus, particularly in the surface epithelium (Glynn and Holborow, 1959; Glynn *et al.*, 1957; Hartmann, 1941; Szulman, 1960, 1962) but varying numbers of cells in the deeper part of the crypts of the small intestine, the Brunner's glands of the duodenum, and the pyloric glands of the stomach appeared to carry the antigens in nonsecretors (Glynn and Holborow, 1959; Glynn *et al.*, 1957; Szulman, 1960, 1962). The distribution of the antigens in duodenum and stomach in individuals of known secretor status and Lewis type have recently been more thoroughly investigated with a range of monoclonal antibodies of defined specificity (Mollicone *et al.*, 1986). The relation of the expression of the antigens to the secretor status is rather complex. In the stomach, the rabbit polyclonal and mouse monoclonal anti-A reagents with wider binding specificity, and some of the anti-B reagents, stain the surface epithelium and pyloric glands of both secretor and nonsecretor individuals of A or B blood group, respectively. In contrast, certain other monoclonal anti-A and anti-B reagents with narrower specificity discriminate between secretors and nonsecretors, staining the surface epithelium (but not the pyloric glands) of secretors only. In the duodenum, the expression of A and B is dependent on the secretor status in the villi and crypts of Lieberkühn but independent in the Brunner's glands, which like the pyloric glands also show heterogeneity of expression with clusters of positive and negative glands, and occasional discrepant cells.

The expression of Le^a and Le^b antigens in the human intestine is also well documented (Lemieux *et al.*, 1981; Mollicone *et al.*, 1985). The Le^a and Le^b antigens are expressed in the surface pyloric epithelium and in the duodenal villi and crypts of Lieberkühn where the expression depends on the secretor status of the individual. Lewis antigens are not expressed in the deep pyloric glands and Brunner's glands (Mollicone *et al.*, 1985).

Unfortunately, detailed immunofluorescence studies of the kind described by Mollicone *et al.* (1985, 1986) have not yet been reported on the human jejunum.

2. PREVIOUSLY PUBLISHED EVIDENCE FOR THE OCCURRENCE OF ABH DETERMINANTS ON SPECIFIC SMALL-INTESTINAL GLYCOPROTEINS

Various lines of evidence have indicated that many of the hydrolytic enzymes that are membrane glycoproteins of the brush border of human small-intestinal absorptive cells carry ABH determinants as terminal carbohydrate structures on their glycan side chains. The first evidence for this came from sucrase–isomaltase, which when purified is a very effective inhibitor of red cell agglutination (Kelly and Alpers, 1973). Enzyme isolated from three blood group A individuals inhibited agglutination of type A erythrocytes by anti-A serum, whereas enzyme from three B individuals inhibited agglutination of B red cells by anti-B. Evidence that this antigenicity was not due to contaminating glycolipid, oligosaccharide, or mucin-type blood group substance came from the observations that the binding by anti-A or anti-B antibodies caused a shift in position of the appropriate enzyme activities on sucrose gradients. Also, bound antibody altered the mobility of the enzyme on acrylamide gel electrophoresis, and enzyme activity was selectively retained by blood group specific antibody bound to Sepharose. The latter approach was subsequently used to demonstrate A determinants on intestinal alkaline phosphatase purified from blood group A individuals (Komoda and Sakagishi, 1978).

In rabbits, where a polymorphism exists for the presence or absence of human blood group A-like determinants in the digestive tract (Zweibaum and Bouhou, 1973), the group of Suzanne Maroux has given clear evidence of the presence of A antigen on many brush border glycoproteins, in particular on the aminopeptidase of A+ rabbits (Feracci *et al.*, 1982; Gorvel *et al.*, 1982; Massey and Maroux, 1985). This was shown by using crossed immunoelectrophoresis, immunoprecipitation, and, more recently, immunoblotting, that is, SDS gel electrophoresis and electrophoretic transfer onto nitrocellulose followed by immunological detection of A antigen with anti-A serum (Massey and Maroux, 1985).

Evidence that the ABH determinants occur on a wide range of human small-intestinal glycoproteins came from the work of Triadou *et al.* (1983), who used immunoblotting after SDS gel electrophoresis of membrane extracts of ileal epithelium and detection of the ABH antigens using specific antisera and lectins. This approach showed rather clearly that the ABH determinants were linked covalently to a series of glycoproteins, many of which could be identified as specific hydrolases from their characteristic electrophoretic mobility. Furthermore, Triadou *et al.* (1983) were able to show that expression of the antigens was influenced by the secretor status of the individual.

3. OUR EXPERIMENTAL EVIDENCE FOR THE OCCURRENCE OF ABH AND LEWIS DETERMINANTS ON SPECIFIC JEJUNAL GLYCOPROTEINS AND THE DEPENDENCE ON BLOOD GROUP AND SECRETOR STATUS*

We have used the experimental approach adopted by Triadou *et al.* (1983) outlined in the previous section, except that we have extended the studies of these workers by making use of monoclonal antibodies (Table I) to isolate the individual jejunal brush border hydrolases before further analysis. We have also investigated the occurrence of the Lewis Le^a and Le^b as well as an Le^a-related determinant detected by a monoclonal antibody mbbm-4 (Dickson *et al.*, 1987) on these glycoproteins.

3.1. Methods

3.1.1. Determination of the ABO and Lewis Blood Group and Secretor Status

Since postmortem material was used for these experiments, the ABO and Lewis blood groups and ABH secretor status were determined or deduced from hemagglutination tests on red cells from the spleen and hemagglutination inhibi-

*A preliminary report of this work has been made (Green *et al.*, 1987b).

Table I
Monoclonal Antibodies Used for Isolation of Lactase, Sucrase–Isomaltase, and Aminopeptidase

Name of antibody	Enzyme polypeptides precipitated	Source/reference
HL1	Lactase	Kind gift of Dr. H.-P. Hauri, Hauri *et al.* (1985)
mlac-1 (4E1)	Lactase	Swallow *et al.* (1985)
mlac-2	Lactase	Previously unpublished. Isolated from a fusion between P3-NSI/1-Ag4-1 and mouse spleen cells after immunization with jejunal epithelial membranes (fusion 3)
mlac-5	Lactase	
mlac-8	Lactase	
mlac-9	Lactase	
mglu-1	Sucrase–isomaltase	
mglu-2	Sucrase–isomaltase	
2/614/88	Sucrase–isomaltase	Kind gift of Dr. H.-P. Hauri, Hauri *et al.* (1985)
mami-1	Aminopeptidase (L-alanyl *p*-nitroanilide hydrolase)	Previously unpublished. Isolated from a fusion between P3-NSI/1-Ag4-1 and mouse spleen cells after immunization with detergent-extracted high molecular weight jejunal glycoproteins (fusion 2)

tion assays on stomach mucosal scrapings. These scrapings, which contain gastric mucins and extracted cellular antigens, were used since it is generally accepted that the concentration of antigens present in such material is a reflection of the salivary ABH secretor status of the individual. Hartmann (1941) had demonstrated a close correlation between the amount of A and/or B antigen in stomach mucosal scrapings and the amount detected in submaxillary glands and had shown that the discrimination between secretors and nonsecretors was greater in the stomach than in other parts of the intestine. Picard *et al.* (1978), who compared the A,B,H,Le^a, Le^b, and I(Ma) activities in the gastric mucins of 60 apparently healthy people of known secretor phenotype and 13 mucosal extracts of patients without gastric cancer, similarly showed a difference between secretors and nonsecretors though there were some exceptional individuals.

To obtain additional information about the relation of the expression of the antigens on the jejunal glycoproteins to the primary products of the *ABO*, *Le*, *Se*, and *H* genes, namely, the glycosyltransferases, we have measured these enzyme activities in most of the jejunal samples used in this study, including some for which stomach scrapings were not available.

3.1.1.1. Serology. The ABO blood group of the individuals from whom jejunal samples had been obtained was determined by spleen red cell agglutination using human anti-A and anti-B and the lectin from *Ulex europeaus* as an anti-H reagent. Extracts of stomach scrapings were prepared by homogenizing 0.1 g in 0.4 mL 0.15 M sodium chloride solution and taking the supernatant after low speed (1,500 *g*) centrifugation. Doubling dilutions were used to quantify the inhibition of red cell agglutination by previously titrated anti-A, anti-B, anti-H, anti-Le^a, and anti-Le^b. The preparation of the anti-Le^a and anti-Le^b sera is shown in Table II. Individuals whose stomach scrapings showed greater inhibition of agglutination of OLe(a−b+) cells by anti-Le^b than inhibition of agglutination of OLe(a+b−) cells by anti-Le^a were typed as Le(a−b+) and assumed to be secretors, while those showing marked inhibition of OLe(a+b−) agglutination and little or none of OLe(a−b+) were typed as Le(a+b−) and as nonsecretors. Assessment of the amount of ABH in the stomach scrapings was used to provide supporting evidence of the interpretations, particularly in the case of Lewis negative individuals.

3.1.1.2. Glycosyltransferase Assays. One gram of jejunal mucosa was finely minced and homogenized using an Ultraturrax homogenizer for 3 × 30 s, at 4°C in 2 mL 0.15 M sodium chloride, and the homogenate centrifuged for 30 min at 10,000 *g* at 4°C. The supernatant to which Triton X-100 was added to a final concentration of 5 mg/mL was used for the glycosyltransferase assays, which were done as described previously (Watkins *et al.*, 1981). The α-3-fucosyltransferase (E.C.2.4.1.152) was assayed by the same procedure as the α-4-fucosyltransferase except that *N*-acetyllactosamine replaced lacto-*N*-biose I as acceptor substrate.

3.1.2. Isolation of Lactase, Sucrase–Isomaltase, and Aminopeptidase

3.1.2.1. Electrophoresis. The membrane extracts of whole jejunum used for electrophoresis were prepared as described previously (Potter *et al.*, 1985). Sodium dodecyl sulfate (SDS) 5–15% gradient polyacrylamide gel electrophoresis was done as described by Karlsson *et al.* (1983).

3.1.2.2. Monoclonal Antibodies Used. The monoclonal antibodies used for immunoprecipitation of lactase, sucrase–isomaltase. and aminopeptidase are shown in Table I. All these antibodies bind to the native and enzymically active proteins as shown by enzyme immunobinding assay carried out by an adaptation of the method described in Swallow *et al.* (1985), in which rabbit antimouse immunoglobulins (RAM) were used to bind the monoclonal antibodies to the Microtiter plate. The RAM (Dakopatts) was applied to the plate for 2 h or more at a concentration of 0.5 mg/mL in phosphate buffered saline [0.01 M NaH_2PO_4/Na_2HPO_4, 0.15 M NaCl(pH 7.2)–PBS], prior to blocking the plate with bovine serum albumin and application of the monoclonal antibodies.

3.1.2.3. Immunoprecipitation. Immunoprecipitations of lactase, sucrase–isomaltase, and aminopeptidase were done using a single monoclonal antibody for each immunoprecipitation. Unless otherwise specified, this was

Table II
Antibodies and Lectins Used for Detection of the ABH and Lewis Antigenic Determinants

Name and specificity of the antibody	Antigen used for immunization	Species immunized	Purification	Source
Anti-Le^a (for serology)	Le^a glycoprotein[a]	Rabbit	Serum absorbed with OLe(a−b+), A_1Le(a−b+), and BLe(a−b+) red cells	Clinical Research Centre
Anti-Le^b (for serology)	Le^b glycoprotein[a]	Rabbit	Serum absorbed with OLe(a+b−), A_1Le(a+b−), and BLe(a+b−) red cells	Clinical Research Centre
Anti-Le^a (for blots)	Le^a glycoprotein[a]	Rabbit	Synsorb Le^a	Clinical Research Centre
Anti-Le^b (for blots)	Le^b glycoprotein[a]	Rabbit	Synsorb Le^b	Clinical Research Centre
Anti-A	A glycoprotein[a]	Rabbit	Synsorb A	Clinical Research Centre
Anti-B	B glycoprotein[a]	Rabbit	Synsorb B	Clinical Research Centre
Anti-H	H glycoprotein[a]	Rabbit	Synsorb H type 1, H type 2	Clinical Research Centre
mbbm-1 anti-A	High molecular weight jejunal glycoproteins	Mouse	NIL (monoclonal)	MRC HBGU
mbbm-4 anti-Le^a-like	Jejunal epithelial membranes	Mouse	NIL (monoclonal)	MRC HBGU

Full name of lectin	Abbreviation	Specificity	Method of labeling	Source
Helix pomatia lectin	HPA	Anti-A, αGalNAc	^{125}I	Boehringer
Ulex europeaus lectin 1	UEA-1	Anti-H	^{125}I	Vector Laboratories
Bandeiraea simplicifolia[b] lectin 1	BS-1	Anti-B (contains some anti-A)	Biotin	Sigma[c]
Concanavalin A	CON-A	α-Mannosyl residues	^{125}I	Vector Laboratories
			Biotin	Sigma[c]

[a]Glycoproteins isolated from ovarian cyst fluid.
[b]Also known as *Griffonia simplicifolia*.
[c]Purchased biotin-labeled.

mlac-1 for lactase, 2/614/88 for sucrase–isomaltase, and mami-1 for aminopeptidase. The general procedure is described in Swallow *et al.* (1985) except that rabbit antimouse immunoglobulins (Dakopatts), conjugated to CNBr Sepharose (Pharmacia) according to the manufacturers instructions, was used as second antibody in place of *Staphylococcus* protein A adsorbant.

3.1.2.4. Immunoadsorbant Chromatography. To purify lactase, immunoadsorbant chromatography was done using mlac-1 conjugated to CNBr–Sepharose as described previously (Potter *et al.*, 1985).

3.1.3. Immunological and Lectin Affinity Detection of the ABH and Lewis Antigens

3.1.3.1. Antibodies and Lectins Used. The blood-group-specific antibodies and lectins used to detect the jejunal glycoproteins after electrophoresis are shown in Table II. The rabbit antisera (raised at the Clinical Research Centre) were absorbed and eluted from the appropriate blood group specific affinity columns (Synsorbs obtained from Chembiomed Ltd., Alberta, Canada). Binding of the antibodies to the Synsorb packed in a small column was achieved by recirculation of the antiserum diluted in PBS, at room temperature for 1–2 h and elution was effected by rapid application of 1% NH_4OH in 0.15 M NaCl, followed by immediate neutralization and dialysis against PBS.

3.1.3.2. Detection Procedures. Detection of the glycoproteins after electrophoresis was carried out in most cases after electrophoretic transfer of the separated proteins onto nitrocellulose. Antibody detection was done using a two-stage procedure involving a peroxidase-conjugated second antibody (rabbit antimouse immunoglobulins or swine antirabbit immunoglobulins as appropriate) (Swallow *et al.*, 1986). The biotinylated BS-1 was also used as described previously (Swallow *et al.*, 1986). The other lectins were iodinated with ^{125}I and used to probe the gels directly after nonspecific protein staining with Coomassie brilliant blue (Karlsson *et al.*, 1983).

3.2. Lectin and Monoclonal Antibody Binding Analysis of the Jejunal Glycoproteins of Individuals of Known Blood Group and Secretor Status

3.2.1. Determination of the ABO and Lewis Blood Groups and Secretor Status and Relation to the Glycosyltransferase Activities within the Jejunal Cells

The ABO and Lewis blood groups and secretor status of the individuals from whom jejunal extracts were selected for use in the experiments described in Section 2.2 are shown in Table III. The α-3-*N*-acetylgalactosaminyltransferase

Table III
Blood Group, Secretor Status, Jejunal Glycosyltransferase Activities, and Lectin and Antibody Detection of Blood Group Antigens on Jejunal Glycoproteins

	Assessed by serology[a]			Glycosyltransferase activity (pmol/hr per g wet wt.)					Detection of ABH antigens on jejunal glycoproteins separated by electrophoresis					
									Lectins			Antisera		
Sample number	ABO blood group	Lewis blood group	Secretor status	A[b]	B[c]	H[d]	H%[e]	Le[f]	HPA	BS-1	UEA-1	A	B	H
1	A	·[g]	·	61,000	0	2,200	40	3,500	++	−	+	·	·	·
2	A	Le(a−b+)	Secretor	63,000	1,300	1,500	17	7,900	++	−	−	++	−	++
3	A	Le(a−b+)	Secretor	·	·	·	·	·	++	−	−	++	·	·
4	A	Le(a−b+)	Secretor	2,700	0	80	4	1,200	++	±	−	·	·	·
5	A	Le(a−b+)	Secretor	170,000	0	8,800	68	5,000	++	±	+	++	−	·
6	A	·	·	92,000	0	130	·	·	−	−	−	−	−	·
7	A	Le(a−b−)	Nonsecretor	92,000	360	50	0.5	80	++	±	−	++	·	++
8	A	Le(a+b−)	Nonsecretor	13,000	800	840	5.6	24,800	−	−	·	−	−	±
9	A	Le(a+b−)	Nonsecretor	44,000	0	80	0.7	4,600	−	−	·	−	−	−
10	B	·	·	1,700	17,000	3,500	·	·	−	++	±	·	·	·
11	B	·	·	0	2,400	2,000	24	3,700	−	++	−	·	·	·
12	B	·	·	·	·	·	·	·	−	++	−	·	·	·
13	B	·	·	0	1,600	1,600	17	3,800	−	++	−	·	·	·
14	B	·	·	0	14,000	14,000	116	8,500	−	++	−	−	++	·
15	B	Le(a−b+)	Secretor	0	15,000	2,100	17	9,100	−	++	−	·	·	·
16	B	Le(a−b+)	Secretor	0	11,000	3,800	26	6,900	−	++	−	−	++	++
17	O	·	·	0	0	3,200	44	300	−	−	+	·	·	+
18	O	Le(a−b+)	Secretor	0	0	7,100	148	5,500	−	−	+	·	·	++
19	O	Le(a−b+)	Secretor	0	0	140	3	5,700	−	−	±	·	·	·
20	O	Le(a−b+)	Secretor	0	0	4,100	30	7,600	−	−	+	·	·	++
21	O	·	·	0	0	0	0	6,600	−	−	−	−	−	−
22	O	·	·	2,100	0	80	·	·	−	−	−	−	−	++
23	O	Le(a+b−)	Nonsecretor	0	0	120	1	8,600	−	−	−	−	·	−
24	AB	Le(a−b+)	Secretor	33,100	9,800	7,700	23	24,700	·	·	·	·	·	·

[a]The samples are listed in groups according to their blood group and assumed secretor status, assessed from stomach scrapings or deduced from enzyme assays.
[b]A = α-3-*N*-acetylgalactosaminyl transferase.
[c]B = α-3-galactosyltransferase.
[d]H = α-2-fucosyltransferase measured with phenyl-β-galactoside as acceptor.
[e]H% = α-2-fucosyltransferase calculated as a percentage of α-3-fucosyltransferase.
[f]Le = α-4-fucosyltransferase.
[g]·, not tested.

(A, E.C.2.4.1.40), α-3-galactosyltransferase (B, E.C.2.4.1.37), α-2-fucosyltransferase (H, E.C.2.4.1.69), and α-4-fucosyltransferase (Le, E.C.2.4.1.65) activities measured directly in the jejunal extracts are also shown. The activity of the A transferase was very high in all the blood group A individuals, except individual 4, who showed a somewhat lower level, and the B transferase was high in all blood group B individuals. Both were active in the one AB individual tested. The level of H transferase activity correlated fairly well with the secretor status determined from stomach mucus, particularly when measured as a percentage of the nonpolymorphic α-3-fucosyltransferase activity. There were, however, two cases (numbers 4 and 19) where the α-2-fucosyltransferase activities were very low in jejunum of individuals grouped as secretors. In sample 4, this was almost certainly due to the condition of the specimen since all the enzyme activities were low. In contrast, the low activity in sample 19 is less easily dismissed. The Lewis transferase was very low in two individuals, 7 and 17, one of whom had shown no Le^a or Le^b in the stomach mucosal scrapings, and the other from whom no stomach tissue was available. It seems likely that both these individuals are Lewis negative, Le(a-b-).

3.2.2. Lectin and Antibody Analysis

Figure 2 shows the blood-group-specific lectin binding of glycoproteins in a typical series of jejunal samples. All the samples showed similar profiles when the gels were probed with the non-blood-group-specific lectin Concanavalin A, or with the protein stain Coomassie brilliant blue (not shown). Binding by the A-specific lectin HPA to glycoproteins, in the apparent molecular weight range 300,000 and less, was seen only in individuals of blood group A, while the B-specific lectin BS-1 bound the most prominently to glycoproteins of B individuals (Figure 2, Table III). With one exception (individual 7), all the A individuals whose glycoproteins showed strong binding by HPA had been classified as secretors. Glycoproteins of apparent molecular weight 300,000 or less were not detected at all by HPA in jejunal extracts of three of the four nonsecretors. Similar results were obtained using the A-specific antibodies, strong binding being shown by all the secretors and also individual number 7. Weak binding of the glycoproteins was also seen in two of the other A nonsecretor individuals with the very high titer monoclonal anti-A, mbbm-1 (not shown). UEA-1 showed strongest binding to the glycoproteins of the O-secretors and to one or two of the A and B secretors (Figure 2). Fewer samples were tested with the rabbit anti-H but it was evident that the pattern of reaction was rather different and the distinction between secretors and nonsecretors was less pronounced; in particular, one apparent nonsecretor (individual 22) showed strong binding (Table III).

It is noteworthy that the jejunal glycoproteins of individual 7 stained strong-

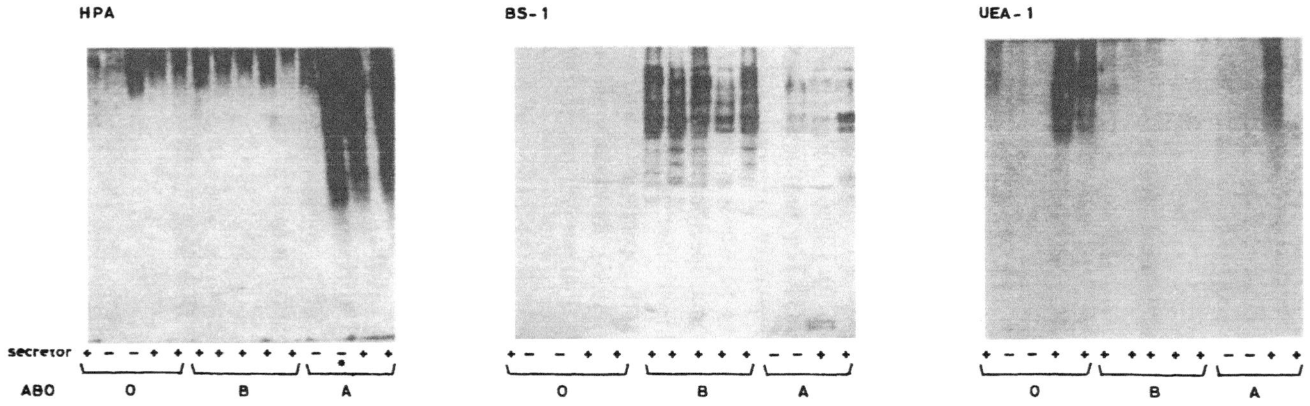

FIGURE 2. Photographs of SDS polyacrylamide gel electrophoresis of crude jejunal membrane extracts from a number of individuals of different ABO and secretor phenotype (secretor +, nonsecretor −). The gels were stained with ^{125}I-*Helix pomatia* lectin (HPA), ^{125}I-*Ulex europaeus* lectin (UEA-1), or electroblotted and probed with biotinylated *Bandieraea simplicifolia* lectin (BS-1). *The A nonsecretor jejunal extract, which shows HPA binding, is the Lewis negative individual 7 in Table III. All the other individuals are Lewis positive.

ly with anti-H, as well as the anti-A reagents, and thus were indistinguishable from those of A secretors. It is therefore perhaps not surprising that extract of jejunal mucosal scrapings, in contrast to those of the stomach from this individual, inhibited agglutination of A red cells by anti-A and O red cells by anti-H.

3.2.3. Immunoprecipitation of Jejunal Hydrolases Using Monoclonal Antibodies

Figure 3 illustrates the specificity of the immunoprecipitation method used in these experiments. Lactase, aminopeptidase, and sucrase–isomaltase are all abundant proteins and can be detected with Coomassie brilliant blue. The mature aminopeptidase and lactase polypeptides comigrate on the electrophoretic system used with an apparent molecular weight of 140,000–150,000. However, they can be distinguished since under the conditions we use aminopeptidase is not bound by the lectin CON-A, unlike both lactase and sucrase–isomaltase, which show a strong signal (Figures 3 and 4). Lactase and sucrase–isomaltase differ in apparent molecular weight but can also be distinguished and shown not to cross contaminate each other in the immunoprecipitations by immunoblotting, using mlac-1 to identify lactase, 2/614/88 which is sucrase subunit specific, and 3/705/60 which is isomaltase subunit specific (Hauri *et al.*, 1985) (not shown). Although the mature sucrase and isomaltase subunits are dissociated on SDS gel electrophoresis and show a different apparent molecular weight on some electrophoretic systems (Hauri *et al.*, 1985), they comigrate with an apparent molecular weight of 130,000–140,000 on our gels (see legend to Figure 5).

3.2.4. Detection of A, B, and H Determinants on Jejunal Lactase, Sucrase–Isomaltase, and Aminopeptidase with Specific Lectins and Antibodies

To provide direct evidence as to whether lactase, aminopeptidase, and sucrase–isomaltase each carry the ABH determinants, we selected samples from two A secretors (2 and 5), all the A nonsecreters including the anomalous individual number 7, two B secretors (14 and 16), two O secretors (17 and 18), and two O nonsecretors (21 and 22), and the one AB secretor (24) and carried out immunoprecipitations with antilactase, anti-sucrase–isomaltase, and anti-aminopeptidase. (Unfortunately, there were no B nonsecretors in our series.) Some typical results are shown in Figures 4–7.

3.2.4.1. Sucrase–Isomaltase. An experiment using immunoprecipitated sucrase–isomaltase (SI) is shown in Figure 4. A determinants were detected on the SI of the A secretors, the AB secretor, and individual 7 using either HPA, rabbit anti-A (Figure 4), or monoclonal anti-A (not shown). B determinants were

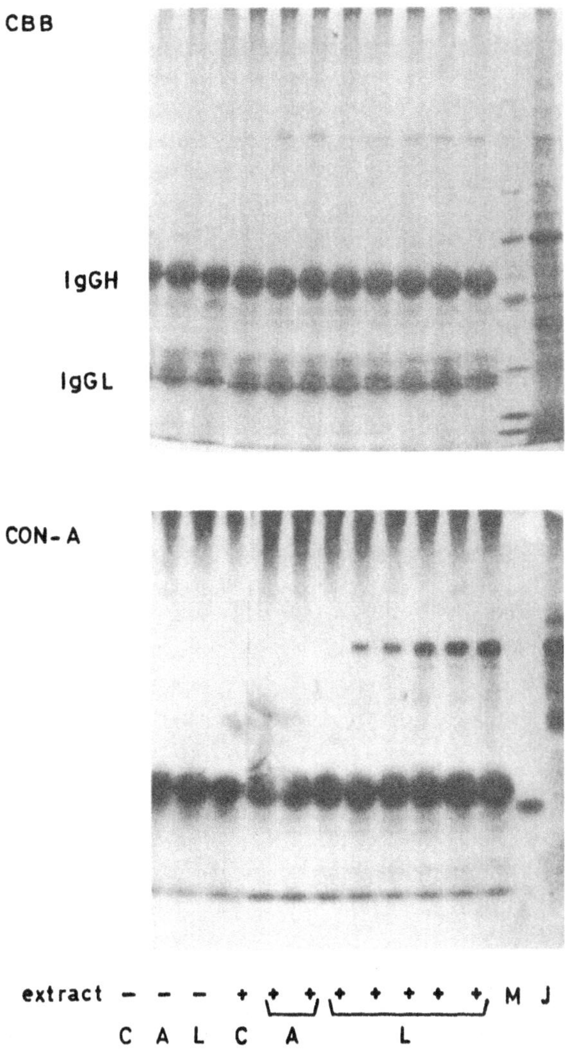

FIGURE 3. Electrophoretic analysis of immunoprecipitations of lactase (L) by three different antilactase monoclonal antibodies (left to right—mlac-2, mlac-2, HL1, HL1, mlac-1) and aminopeptidase (A) by mami-1. The controls are immunoprecipitations by an irrelevant antibody (C) and immunoprecipitations done in the absence of tissue extract (–). The gel was first stained with Coomassie brilliant blue (CBB) and then with ^{125}I CON-A. Both CBB and CON-A bind to the immunoglobulin heavy and light chains (IgGH and IgGL). The sizes of the molecular weight markers (M) are shown in Figure 2. The track J contains the jejunal extract used for the immunoprecipitations.

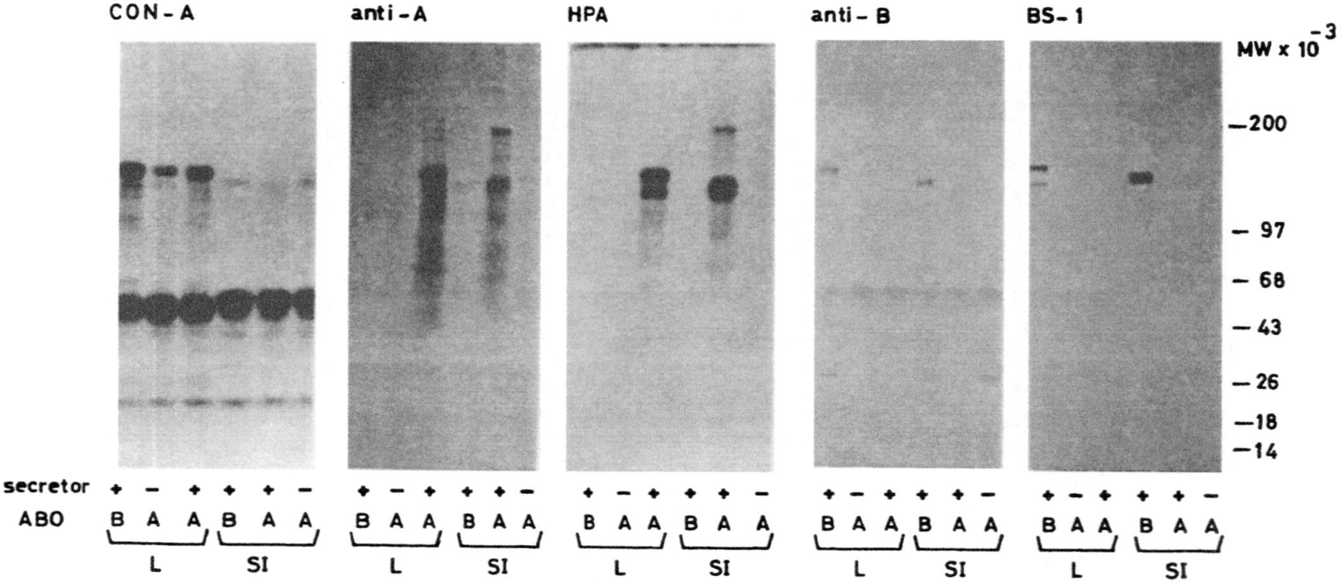

FIGURE 4. Electrophoretic analysis of immunoprecipitations of lactase (L) and sucrase–isomaltase (SI) from three Lewis-positive individuals of known ABO and secretor phenotype (secretor +, nonsecretor −). The gels were stained with ^{125}I CON-A, ^{125}I HPA, or electroblotted and probed with rabbit anti-A, rabbit anti-B, or biotinylated BS-1. In addition to the major lactase (MW ≃ 155,000) and sucrase–isomaltase polypeptides (MW ≃ 145,000), precursors (MW ≃ 200,000) and smaller components, which are probable degradation products, can be seen. IgGH and IgGL, which are stained by CON-A, are also stained faintly on the anti-B filter.

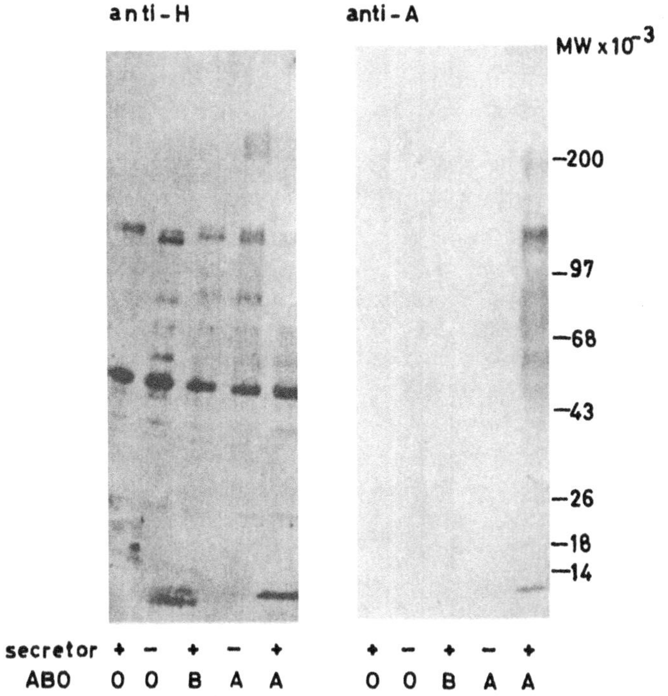

FIGURE 5. Electrophoretic analysis of immunoprecipitations of sucrase-isomaltase from individuals of known ABO blood group and secretor phenotype (secretor +, nonsecretor −). The O secretor shown is individual 17, who was probably Lewis negative. The gels were electroblotted and stained with rabbit anti-H and rabbit anti-A. IgGH and IgGL can also be seen on the anti-H filter. Although two SI bands can be detected in some of the samples, these are probably not the individual sucrase and isomaltase subunits, since they can both be detected with the subunit specific monoclonal antibodies. The slight differences in size probably result from limited proteolysis.

detected on SI of B secretors and the AB secretor using BS-1 or rabbit anti-B. Rabbit anti-H, in contrast, surprisingly bound to SI from all the individuals (Figure 5), both secretors and nonsecretors.

3.2.4.2. Lactase. Similar findings were obtained for lactase using this same approach (Figure 4) and also using lactase purified by immunoadsorbant chromatography (Figure 6).

3.2.4.3. Aminopeptidase. Aminopeptidase also gave similar results (Figure 7) to those found for sucrase–isomaltase and lactase. Thus, aminopeptidase shares some common features of glycosylation despite the fact that it differs from the disaccharidases with respect to CON-A binding.

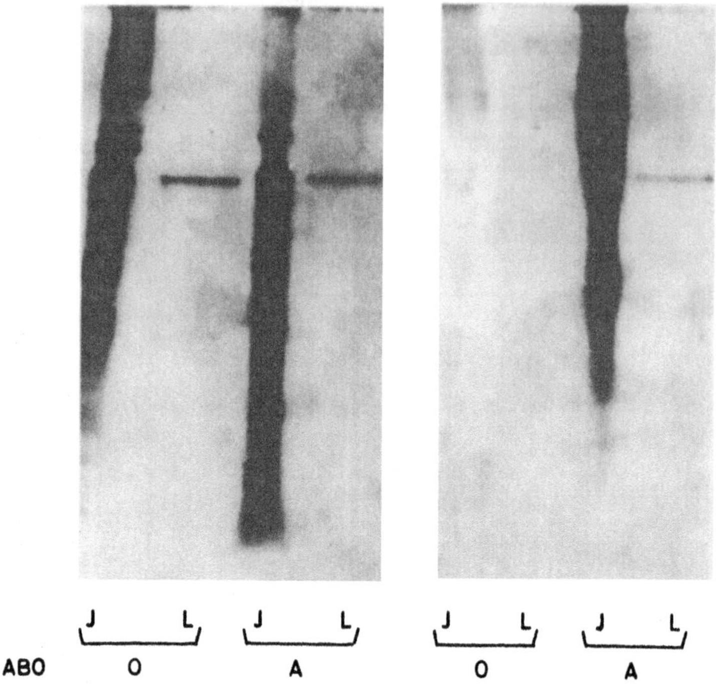

FIGURE 6. Lectin binding analysis of purified lactase (L) from two individuals of different blood group—O secretor and A secretor. The pieces of gel were stained with CON-A (left) and HPA (right). The tracks labeled J contain the original extracts from which the lactase was purified.

3.2.5. Absence of ABH on Kidney Aminopeptidase

An aminopeptidase present on the human kidney brush border membrane, which is closely related to the one in jejunum (Kenny and Maroux, 1982), can also be immunoprecipitated by mami-1 and seems likely to be a product of the same gene as that which codes for jejunal aminopeptidase. We have therefore examined kidney aminopeptidase from three of the same individuals of known blood group from whom jejunal samples were also available (Figure 7). The A, B, and H antigens were weak or absent from all the high molecular weight glycoproteins in the kidney membrane extracts and could not be detected on isolated aminopeptidase. Measurement of the glycosyltransferase activities in the three kidney samples used for the immunoprecipitation experiment revealed very low α-2-fucosyltransferase activities, in contrast to high levels of the A and B transferases. The H transferase activities were, in picomoles product formed per hour per gram wet weight, 160 (B secretor individual 16), 14 (A secretor indi-

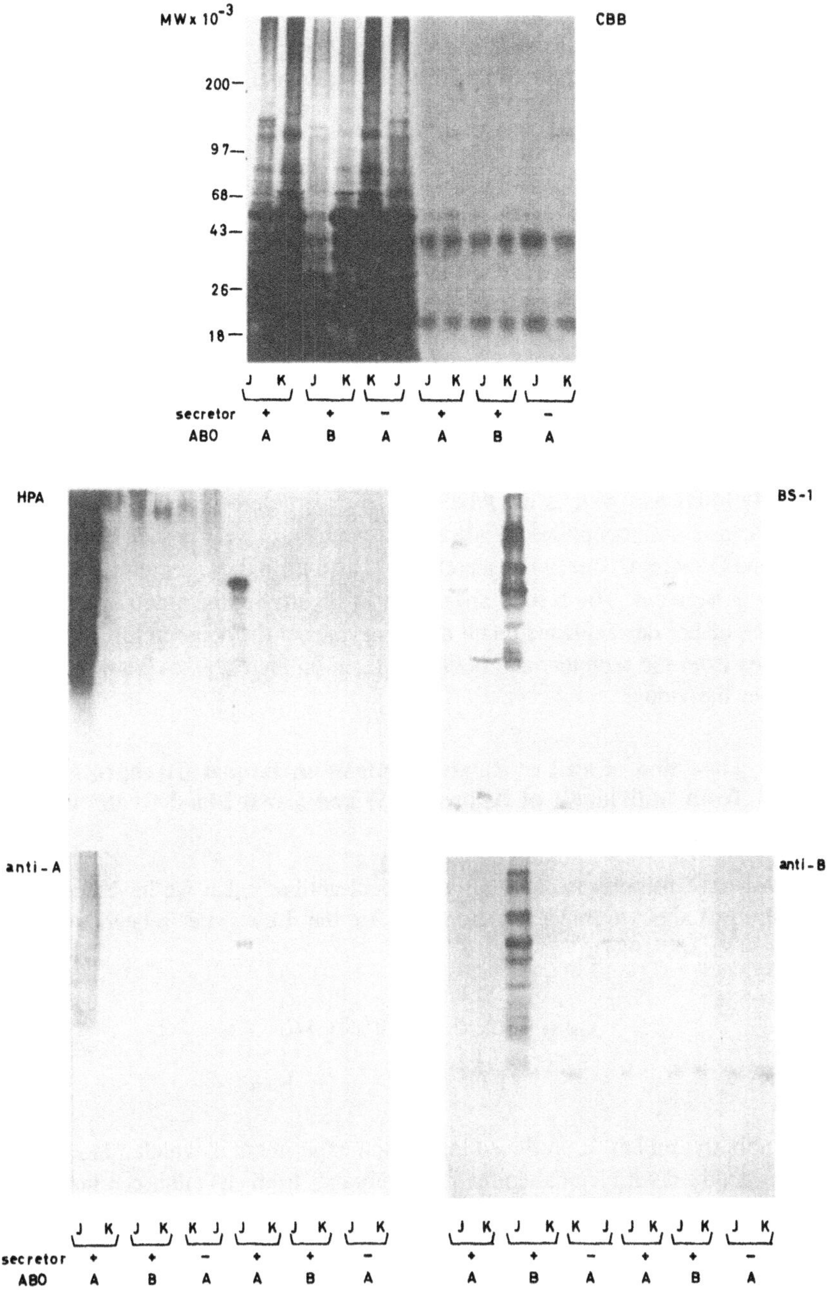

FIGURE 7. Electrophoretic analysis of immunoprecipitations of aminopeptidase from jejunum (J) and kidney (K) from three different individuals of known ABO blood group and secretor phenotype (secretor +, nonsecretor −). The gels were either stained directly with Coomassie brilliant blue followed by ^{125}I-labeled HPA or electroblotted and stained with rabbit anti-A, anti-B, and biotinylated BS-1. The first six tracks contain the original extracts and tracks 7–12 contain the immunoprecipitates.

vidual 5), and 80 (A nonsecretor individual 8). (See Table III for the activities of this enzyme in the jejunal extracts of these individuals.)

3.2.6. Detection of Le^a and Le^b on Jejunal Lactase, Sucrase–Isomaltase, and Aminopeptidase with Anti-Le^a and Anti-Le^b

Jejunal membrane glycoproteins of 13 individuals of known blood group and secretor status were examined by immunoblotting with monospecific anti-Le^a and anti-Le^b. Le^a determinants were detected strongly on the glycoproteins of the nonsecretors but not on those of the secretors. Le^b determinants were prominent on the glycoproteins of the Lewis positive O secretors and weakly detectable on the glycoproteins of the A and B secretors and the O nonsecretors (Figure 8). The low binding of anti-Le^b to the glycoproteins of A and B secretors is probably a reflection of the specificity of the purified antiserum, which presumably binds less well to the ALe^b and BLe^b structures than Le^b.

Lactase, aminopeptidase, and sucrase–isomaltase were immunoprecipitated from one O secretor, one O nonsecretor, and also from the O secretor believed to be Lewis negative. The results are shown in Figure 8, which demonstrates the presence of Le^a determinants on all three enzymes of the nonsecretor, Le^b on the enzymes from the secretor, and neither antigen on the enzymes from the Lewis negative individual.

3.2.7. Detection of an Le^a Related Antigen on Jejunal Glycoproteins from Individuals of Known ABO and Lewis Blood Group and Secretor Status

We have recently isolated a monoclonal antibody that we have shown by inhibition studies to bind preferentially to the Lewis Le^a-related structure (Dickson *et al.*, 1987):

$$\begin{array}{l} \text{Gal}\beta 1\rightarrow 3\text{GlcNAc}\beta 1\rightarrow 3\text{Gal}\beta 1\rightarrow 4\text{Glc} \\ \qquad\qquad\uparrow 1{,}4 \qquad\qquad\qquad\qquad \uparrow 1{,}3 \\ \qquad\qquad \text{Fuc}\alpha \qquad\qquad\qquad\qquad\ \text{Fuc}\alpha \end{array}$$

This antibody, mbbm-4, produced in a fusion experiment in which a mouse had been immunized with jejunal epithelial membranes from an O nonsecretor, binds strongly to a large series of jejunal glycoproteins in nonsecretors (with the exception of individual 7), but binds only weakly to a few components in secretors (Figure 9) irrespective of the ABO blood group of the individual. These observations suggest that an Le^a-related structure, similar or identical to the structure shown above, is present and abundant on jejunal glycoproteins. In secretors, the presence of fucose linked α1→2 to the terminal galactose gener-

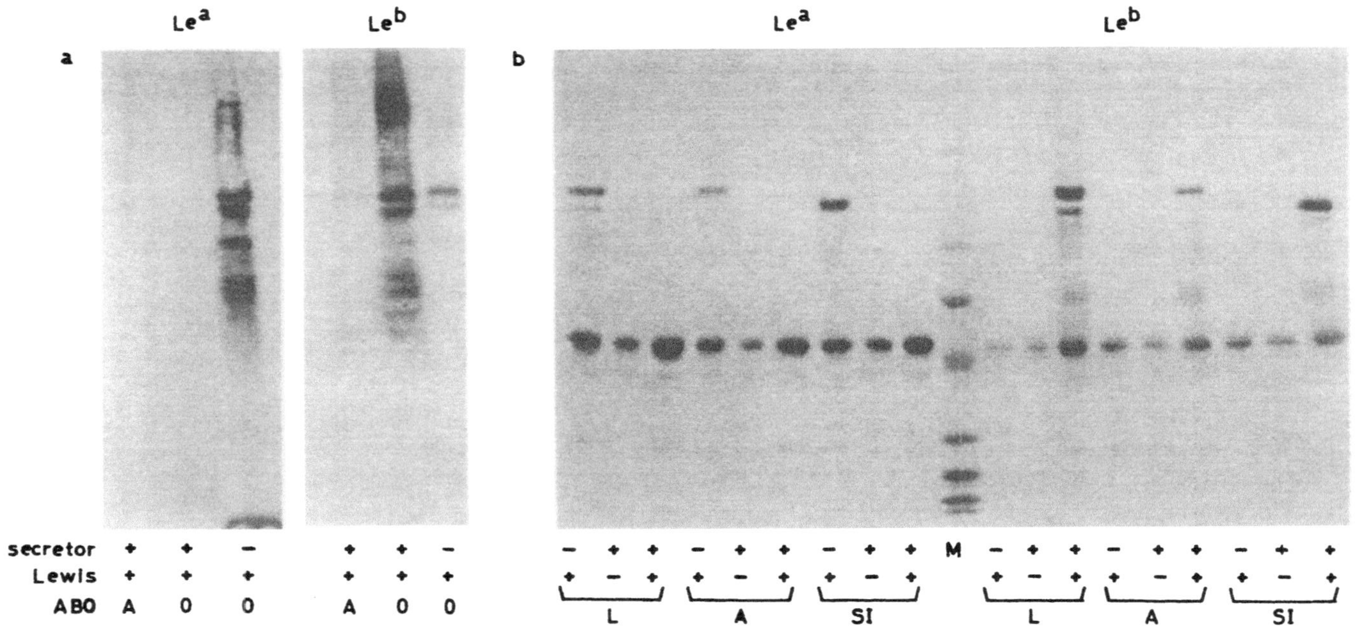

FIGURE 8. Electrophoretic and antibody binding analysis of jejunal glycoproteins from individuals of known ABO blood group, Lewis, and secretor status (+, −) using anti-Le^a and anti-Le^b. The first two panels (a) show jejunal extracts from individuals of known ABO, Lewis, and secretor phenotype. The two right-hand panels (b) show immunoprecipitations of lactase (L, using mlac-9), aminopeptidase (A, using mami-1), and sucrase–isomaltase (SI, using mglu-2) from jejunal extracts of three blood group O individuals of different Lewis and secretor status (+, −). The sizes of the molecular weight markers (M) are shown in Figure 2.

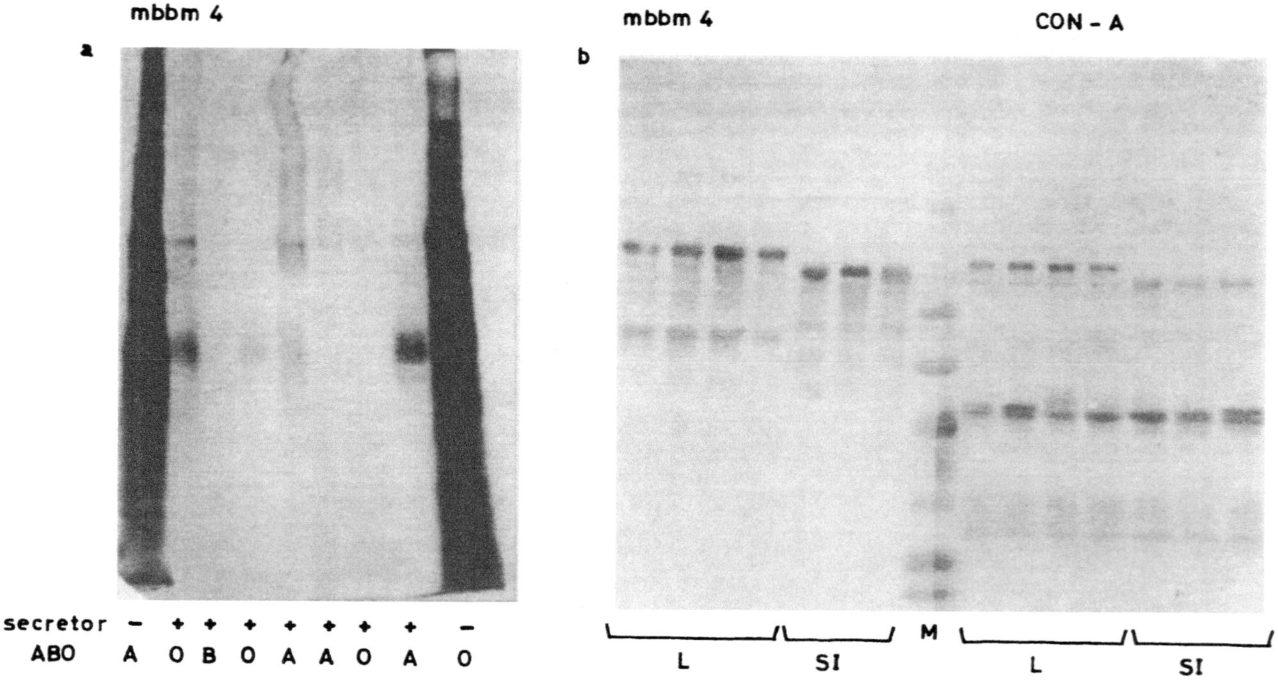

FIGURE 9. (a) Electrophoretic and antibody analysis of glycoproteins in jejunal extracts from individuals of known blood group and secretor status (+, −) using mbbm-4 to detect the Le^a-related antigen. (b) Immunoprecipitations of lactase (L), using left to right mlac-8, mlac-5, mlac-9, and mlac-1, and of sucrase–isomaltase using left to right mglu-1, mglu-2, and 2/614/88, from a jejunal extract of an O nonsecretor. Pieces of gel were immunoblotted and probed with mbbm-4 or biotinylated CON-A. With CON-A IgGH and IgGL can also be seen.

ates the Le^b-related epitope, which is not bound by mbbm-4. The Le^a-related epitope recognized by mbbm-4 is carried on lactase, aminopeptidase, and sucrase–isomaltase as shown by immunoprecipitation of these enzymes from the jejunal extract of a nonsecretor and immunoblotting with mbbm-4 (Figure 9, aminopeptidase not shown). The enzymes from either a Lewis positive O secretor or the Lewis negative secretor do not carry this determinant (not shown).

3.3. Role of the Secretor Gene (*Se*) in Determining the α-2-Fucosyltransferase Activity within the Jejunal Epithelial Cells

In the majority of samples tested, the level of expression of the ABH and Lewis antigens on the jejunal glycoproteins appeared to depend on the secretor status of the individual, as assessed from stomach mucin. There was also a reasonably good correlation of the level of the α-2-fucosyltransferase activity within the jejunal epithelial cells and the secretor status deduced serologically (Table III). Since there is evidence that the α-2-fucosyltransferase activities that are under the control of the *H* and *Se* genes differ in their substrate specificity, we have examined the relative activities of jejunal samples of secretors and nonsecretors toward lacto-*N*-biose I (Galβ1→3GlcNAc) and *N*-acetyllactosamine (Galβ1→4GlcNAc) as type i and type 2 acceptors, respectively. In some individuals, we have also examined the distribution of activities toward these substrates along the length of the intestine from the stomach to colon.

3.3.1. Substrate Specificity of Jejunal α-2-Fucosyltransferase

The distribution of the ratios of the activities toward lacto-*N*-biose I and *N*-acetyllactosamine are shown in Figure 10. Lacto-*N*-biose I tends to be the pre-

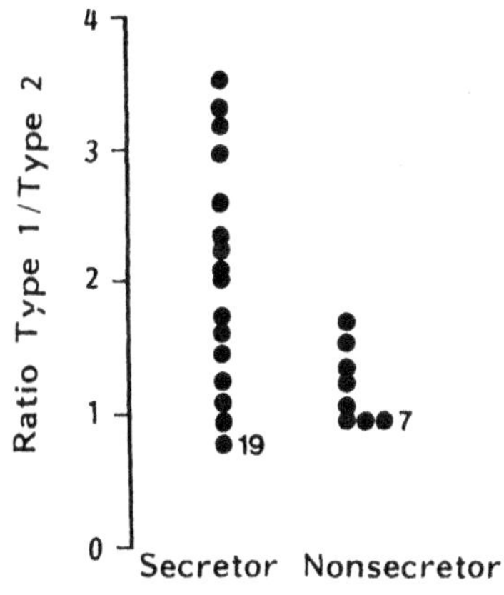

FIGURE 10. α-2-Fucosyltransferase in human jejunal samples from 24 different individuals: Ratio of activity with type 1 (lacto-*N*-biose I) and type 2 (*N*-acetyllactosamine) acceptor substrates in secretors and nonsecretors. The points representing results from individuals 7 and 19 (see Table III) are shown.

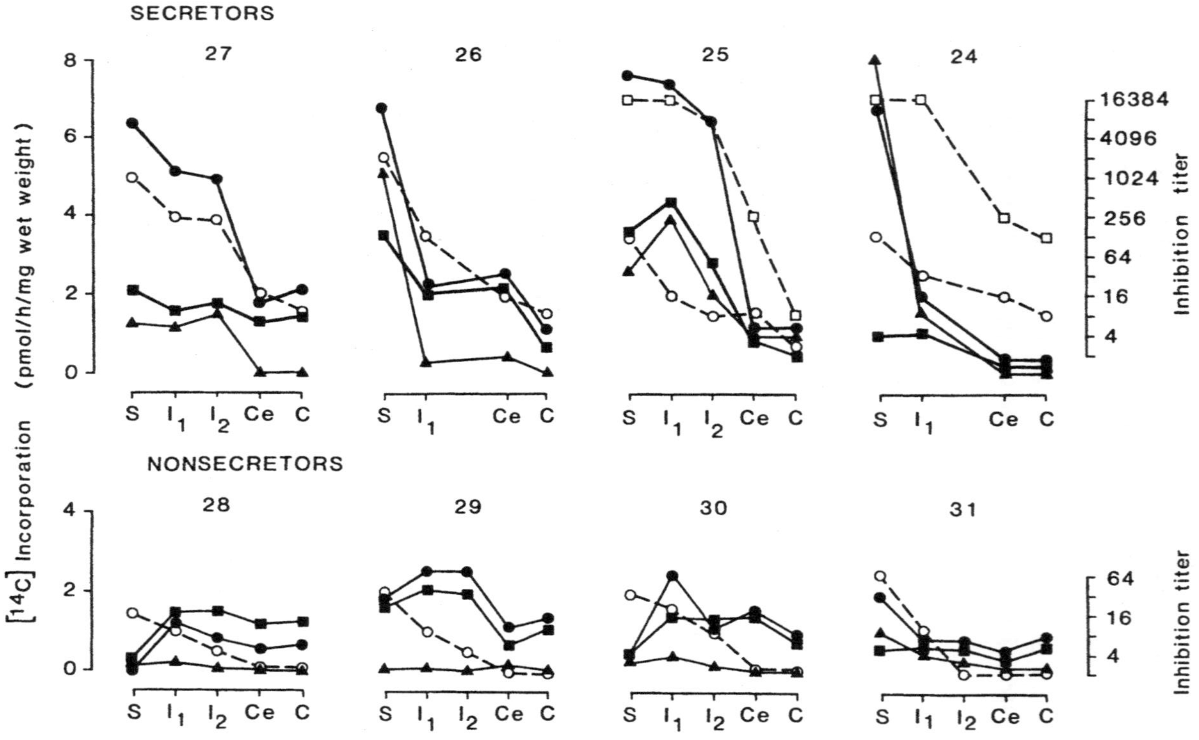

FIGURE 11. α-2-Fucosyltransferase activity measured along the length of the human intestine from eight different individuals (four secretors, four nonsecretors) using three different acceptor substrates—lacto-*N*-biose I -●-, *N*-acetyllactosamine -■-, and phenyl β-galactoside -▲-—and distribution of H antigen -○- and A antigen -□-: S—stomach, I_1—duodenum and jejunum, I_2—ileum, Ce—cecum, and C—colon. Individuals 25–31 were not shown in Table III but the secretor status was determined by the same methods. Individual 24 was blood group AB, and 25, group A.

ferred acceptor in all the individuals but in the secretors there is a much larger range of values, with about half the group showing rather higher values than in the nonsecretors.

3.3.2. Distribution and Substrate Specificity of α-2-Fucosyltransferase along the Intestine

The intestine was divided into five sections—stomach (S), duodenum and jejunum (I_1), ileum (I_2), cecum (Ce), and colon (C)—and the complete mucosal layer of each of these sections was scraped away from the muscle and used for aqueous extractions for measurement of α-2-fucosyltransferase activity with three acceptors (Figure 11) that were done in parallel on the same extracts. Although the overall difference in levels of activity between the secretors and nonsecretors is clear, particularly in the stomach, there is a marked overlap in the values. The activities measured in the jejunum of the secretor individual 26, for example, are indistinguishable from those seen in the nonsecretors. There is also no very clear-cut pattern of relative substrate preference characteristic of either the organ or the individual. It is not possible to explain the variations in activity simply by the quality of the postmortem material unless the substrate specificity of the enzymes alters as they denature and degrade.

4. SUMMARY AND DISCUSSION OF THE EVIDENCE

It is clear from this study and those published previously in the literature that the ABH determinants are present on many small-intestinal glycoproteins (Table IV). They are also expressed on cell-surface glycoproteins of the colon carcinoma cell lines, which show enterocytic differentiation (Chantret *et al.*, 1987). We and others (Kelly and Alpers, 1973; Triadou *et al.*, 1983) have shown that the expression of the determinants depends on the phenotype of the individual with respect to the ABO blood group polymorphism. Our findings are essentially in agreement with those of Triadou *et al.* (1983) that presence of the determinants is (at least in part) dependent on the secretor status of the individual. The discovery of an apparent A nonsecretor, number 7, with high level of expression of A antigen in the jejunum could simply reflect the unreliability of stomach scrapings for determining secretor status, but the α-2-fucosyltransferase activity in the jejunal extract of this individual was very low with all three substrates, in keeping with, but not proof of, the nonsecretor phenotype (Table III, Figure 11). It is intriguing, but possibly coincidental, that this exceptional individual appeared also to be Lewis negative.

Our results also show evidence overall of the relation of the expression of the antigens to the level of activity of the appropriate glycosyltransferases within

Table IV
Intestinal Glycoproteins Expressing A,B, H, and Le Antigens[a]

Protein	Evidence	Reference
Sucrase–isomaltase (E.C.3.2.1.4810)	Inhibition of hemagglutination; shift in position on sucrose gradients; shift in electrophoretic mobility; retention by anti-blood-group affinity column	Kelly and Alpers (1973)
	Immunoblotting-identified by electrophoretic mobility	Triadou *et al.* (1983)
	Immunoblotting of isolated sucrase–isomaltase	This chapter
Alkaline phosphatase (E.C.3.1.3.1)	Retention by anti-blood-group affinity column	Komoda and Sakagishi (1978); Komoda *et al.* (1981)
Enterokinase (E.C.3.4.21.9)	Deduced from production of anti-A sera after immunization with pure enzyme	Hermon-Taylor *et al.* (1977)
Lactase (E.C.3.2.1.23)	Immunoblotting-identified by electrophoretic mobility	Triadou *et al.* (1983)
	Immunoblotting of isolated lactase	This chapter
Aminopeptidases A and N (E.C.3.4.11.2/7)	Immunoblotting-identified by electrophoretic mobility	Triadou *et al.* (1983)
Aminopeptidase N (E.C.3.4.11.2.?)	Immunoblotting of isolated aminopeptidase	This chapter
Maltase-glucoamylase (E.C.3.2.1.20)	Immunoblotting-identified by electrophoretic mobility	Triadou *et al.* (1983)
Dipeptidyl peptidase IV (E.C.3.4.14.5)	Immunoblotting-identified by electrophoretic mobility	Triadou *et al.* (1983)

[a]See section 2 for details of evidence.

the jejunal cells. However, the correlation of the level of α-2-fucosyltransferase activity and the expression of the ABH antigens was not perfect and the relation of the level and specificity of the α-2-fucosyltransferase to the secretor status of the individual was complex. The data cannot readily be explained by the expression of two distinct enzymes differing in their preference for type 1 and type 2 acceptors, one of which, the type 1 preferring enzyme, is deficient in nonsecretors. Even in the stomach where the activity appears to be the most clearly under the control of the *Se* gene, consistent proportions of activity toward the three substrates were not seen (Figure 11).

These observations suggest that additional genetic factors may be involved in determining the level and specificity of α-2-fucosyltransferase in the human intestine. Some suspicion of the involvement of additional genes had already come from the immunohistological work of Mollicone *et al.* (1985, 1986), which hinted at a complex control of the expression of the H antigen in human intestinal cells.

Our results also suggest the presence of more than one kind of H antigen on jejunal glycoproteins. The Le^a, Le^b, and Le^a-related structure recognized by mbbm-4 must be built on type 1 precursor and results with the rabbit anti-Le^b serum would suggest that the $\alpha 1 \rightarrow 3$-linked *N*-acetylgalactosamine occupies the end of most of these side chains in ALe(a−b+) individuals and $\alpha 1 \rightarrow 3$-linked galactose in BLe(a−b+) individuals. The results using *Ulex* lectin (UEA-1), which is said to be specific for type 2H (Hindsgaul *et al.*, 1985), suggest that type 2H is also present on jejunal glycoproteins; if so, this appears to be under the control of the *Se* gene (Figure 2). However, in our experiments, the signal with UEA-1 was not strong and we were not able to show more than trace binding to the immunoprecipitated hydrolases. Type 3 structures (Gal$\beta 1 \rightarrow 3$GalNAc) may also occur on these glycoproteins but at present there is no experimental evidence available as to whether or not this is the case.

An additional complexity was the evidence, using rabbit anti-H, of H determinants on the hydrolases of nonsecretors (Figure 5, lactase and aminopeptidase not shown), which are not converted to A determinants, suggesting perhaps that there are some cells in which the A and B enzymes are not expressed, which at the same time express an H transferase that is not under the control of the *Se* gene. However, it is possible that this observation is simply a reflection of the specificity of the purified antibody, which may conceivably cross-react with other structures.

5. GENERAL DISCUSSION

5.1. Other Candidate Glycoproteins That May Carry ABH and Lewis Determinants

Most of the small-intestinal glycoproteins shown so far to carry the ABH determinants are known to occur on the cell surface and more particularly they are anchored into the membrane of the brush border (Fransen *et al.*, 1985; Hauri *et al.*, 1985; Semenza, 1986). The disaccharidases and peptidases probably all share common structural features such as anchorage into the membrane by the N-terminal end of the molecules (Kenny and Maroux, 1982; Semenza, 1986). The bulk of each molecule projects into the lumen of the gut and probably carries both O-linked and N-linked carbohydrate side chains. Other intestinal proteins that share these properties undoubtedly also carry the ABH determinants. These include γ-glutamyltransferase (E.C.2.3.2.2. Laperche *et al.*, 1986) and possibly also two other peptidases identified on the enterocyte microvillar membrane, angiotensin I-converting enzyme and *p*-aminobenzoic acid peptide hydrolase (Hauri *et al.*, 1985).

Other candidate proteins known to occur on the enterocyte membrane are the peptide and polypeptide hormone and growth factor receptors (Bergeron *et al.*, 1980; Binder *et al.*, 1980; Broyart *et al.*, 1981; Forgue-Lafitte *et al.*, 1979, 1980). It is noteworthy that the epidermal growth factor receptor on the surface of A431 epidermoid carcinoma cells carries A and Y (Le Pendu *et al.*, 1985b; Parker *et al.*, 1984). The apolipoproteins A-I, A-II, and B-48, synthesized within the absorptive cells of the jejunum, may also be glycosylated (Herbert *et al.*, 1983; Vauhkonen, 1986; Vauhkonen *et al.*, 1985) and carry these determinants.

Blood-group-related carbohydrate structures have been shown on β-galactosyltransferase isolated from human milk (Childs *et al.*, 1986). β-Galactosyltransferase is present in small-intestinal enterocytes and there is evidence, though disputed (Boyle *et al.*, 1986), of its occurrence in the brush border (Roth *et al.*, 1985). It therefore seems more than likely that intestinal β-galactosyltransferase also carries the ABH and Lewis determinants.

The expression of the ABH determinants on β-galactosyltransferase, which is involved in the synthesis of the "precursor" structures, raises the intriguing question as to whether they are also present on the A, B, H, and Le transferases themselves. So far the answer to this is not known but immunization of A- rabbits with human A transferase does not produce a significant elevation in the anti-A titer as was found after immunization with enterokinase (Hermon-Taylor *et al.*, 1977) from A secretor individuals and aminopeptidase from A+ rabbits (Gorvel *et al.*, 1982).

5.2. Possible Functional Significance

Although the widespread occurrence of the ABH and Lewis antigens on the glycoproteins of the jejunal brush border is clear, little is known about the number of carbohydrate side chains that carry these determinants or whether they are N-linked to asparagine or O-linked to serine or threonine. There is evidence that both pig and rabbit aminopeptidase have both N- and O-linked side chains (Danielsen *et al.*, 1983; Massey and Maroux, 1985) and that the A determinants may occur on the O-linked side chains of aminopeptidase in A+ rabbits (Massey and Maroux, 1985). In sucrase–isomaltase, there is a serine- and threonine-rich region in the hypothetical "stalk" of the molecule (Hunziker *et al.*, 1986) that links the membrane anchor and the bulk of the molecule, in both rabbit and human (Green *et al.*, 1987a; Hunziker *et al.*, 1986), and it has been suggested that glycosylation in this area has a role in protection of the stalk from the action of pancreatic proteases (Hunziker *et al.*, 1986). Indeed, it seems plausible that, in general, glycosylation of these molecules protects against proteolysis. However, it is more questionable whether the genetically determined polymorphisms of the carbohydrate structures have any functional significance.

Although there are examples of associations of ABO blood group and secretor phenotype with disease of the gastrointestinal tract, such as duodenal ulcer and gastric cancer (e.g., Clarke *et al.*, 1955; Doll *et al.* 1961), these associations are more likely to be explained by variations in terminal carbohydrate structure on the mucins secreted by the intestinal goblet cells rather than of the carbohydrates attached to the membrane proteins, since these mucins form a thick layer over the epithelial surface. However, it does appear that the variation in structure of the terminal carbohydrates on these molecules has some physiological consequences. This is shown in the case of intestinal alkaline phosphatase. The serum level of this enzyme, in addition to being influenced by the lipid intake in the diet, is highly correlated with the ABO blood group and secretor status of the individual (Bamford *et al.*, 1965; Langman *et al.*, 1966). The highest levels are found in O and B secretors while much lower levels are found in O and B nonsecretors and in all A individuals. One obvious explanation of these observations, though not the only one, is that the alkaline phosphatases may differ in half-life in the serum according to their "blood group," owing to different rates of clearance by carbohydrate binding cell-surface receptors.

Whether or not the presence of varying levels of intestinal alkaline phosphatase in the serum is likely to be of any clinical or biological relevance is hard to say since the in vivo role of intestinal alkaline phosphatase is not known. However, it will be of interest to know whether any other enterocytic proteins with a better defined biological role vary in serum level in individuals of different blood group and secretor status. In this context, it is intriguing to know whether the jejunal apolipoproteins carry these blood group antigens, and whether this may have any relevance to the reported associations between serum cholesterol level and ABO blood group and secretor status (Beckman *et al.*, 1970; Flatz, 1970; Lamba *et al.*, 1974; Oliver *et al.*, 1969) and between incidence of coronary heart disease and myocardial infarction and ABO blood group (Goldbourt and Neufeld, 1986).

6. CONCLUDING REMARKS

The ABH and Lewis antigenic determinants are present on a very wide range of small-intestinal glycoproteins. This is a reflection of the expression, within the epithelial cells, of the series of glycosyltransferases needed to catalyze the transfer of the sugars comprising these structures. The person-to-person differences in the levels of the enzymes and of the structures present result, at least in part, from the common allelic variation at three gene loci *ABO*, *Se*, and *Le*, but additional genetic variation may be involved. Presumably, in other cell types of the body where the full complement of transferase enzymes is expressed, a similar situation pertains with many glycoproteins also carrying the ABH and Lewis determinants. However, the absence of any one enzyme required in the pathway would prevent the completion of the structures. This appears to happen in the kidney and colon where there is very little α-2-fucosyltransferase activity and where the A and B antigens are very poorly expressed. In these organs, it is probable that the Le^a structure is carried on a range of glycoproteins.

ACKNOWLEDGMENTS. We would like to thank all the colleagues who have helped in many invaluable ways in the production of this chapter, in particular, Dr. Winifred Watkins, for all her help and advice throughout; Drs. D. A. Hopkinson, Rafael Oriol, Patricia Tippett, and Alain Zweibaum for advice with the manuscript; Mrs. Doreen Trinder for secretarial assistance; Mrs. Nona Parry-Jones and Mr. Alan Yates for preparing the figures; and Dr. Hans-Peter Hauri for providing us with some of the antibodies. P.G. is in receipt of an I.C.R.F. Fellowship.

REFERENCES

Bamford, K. F., Harris, H., Luffman, J. E., Robson, E. B., and Cleghorn, T. E., 1965, Serum-alkaline–phosphatase and the ABO blood-groups. *Lancet* **i**:530–531.

Beckman, L., Olivecrona, T., and Hernell, O., 1970, Serum lipids and their relation to blood groups and serum alkaline phosphatase isozymes. *Hum. Hered.* **20**:569–579.

Bergeron, J. J. M., Rachubinski, R., Searle, N., Borts, D., Sikstrom, R., and Posner, B. I., 1980, Polypeptide hormone receptors in vivo: Demonstration of insulin binding to adrenal gland and gastrointestinal epithelium quantitative radioautography. *J. Histochem. Cytochem.* **8**:824–835.

Betteridge, A., and Watkins, W. M., 1985, Variant forms of α-2-L-fucosyltransferase in human submaxillary glands from blood group ABH "secretor" and "non-secretor" individuals. *Glycoconjugate J.* **2**:61–78.

Binder, H. J., Lemp, G. F., and Gardner, J. D., 1980, Receptors for vasoactive intestinal peptide and secretin on small intestinal epithelial cells. *Am. J. Physiol.* **238**:G190–G196.

Boyle, F., Snape, S., Duane, P., Cook, N., and Peters, T., 1986, Galactosyltransferase activity is not localized to the brush border membrane of human small intestine. *Biosci. Rep.* **6**:171–175.

Brown, A., Ellis, I. O., Embleton, M. J., Baldwin, R. W., Turner, D. R., and Hardcastle, J. D., 1984, Immunohistochemical localization of Y hapten and the structurally related H type-2 blood-group antigen on large-bowel tumors and normal adult tissues. *Int. J. Cancer* **33**:727–736.

Broyart, J.-P., Dupont, C., Laburthe, M., and Rosselin, G., 1981, Characterization of vasoactive intestinal peptide receptors in human colonic epithelial cells. *J. Clin Endocrinol. Metab.* **52**:715–721.

Chantret, I., Chevalier, G., Dussaulx, E., and Zweibaum, A., 1987, A and H blood group antigens as markers of sucrase–isomaltase from enterocyte-like differentiated human colon carcinoma cell lines HT-29 and Caco-2. *Cancer Res.* **47**:1426–1433.

Childs, R. A., Feizi, T., Fukuda, M., and Hakomori, S.-I., 1978, Blood-group-I activity associated with band 3, the major intrinsic membrane protein of human erythrocytes. *Biochem. J.* **173**:333–336.

Childs, R. A., Berger, E. G., Thorpe, S. J., Aegerter, E., and Feizi, T., 1986, Blood-group-related carbohydrate antigens are expressed on human milk galactosyltransferase and are immunogenic in rabbits. *Biochem. J.* **238**:605–611.

Clarke, C. A., Cowan, W. K., Edwards, J. W., Howel-Evans, A. W., McConnell, R. B., Woodrow, J. C., and Sheppard, P. M., 1955, The relationship of the ABO blood groups to duodenal and gastric ulceration. *Br. Med. J.* **2**:643–646.

Conzelmann, E., Burg, J., Stephan G., and Sandhoff, K., 1982, Complexing of glycolipids and their transfer between membranes by the activator protein for degradation of lysosomal ganglioside G_{M2}. *Eur. J. Biochem.* **123**:455–464.

Crookston, M. C., and Tilley, C. A., 1977, A and B and Lewis antigens in plasma. In *Human Blood Groups* (J. F. Mohn, R. W. Plunkett, R. K. Cunningham, and R. M. Lambert, eds.), pp. 246–256, Karger, Basel.

Danielsen, E. M., Cowell, G. M., Noren, O., Sjöström, H., and Dorling, P. R., 1983, Biosynthesis of intestinal microvillar proteins: The effect of swainsonine on post-translational processing of aminopeptidase N. *Biochem. J.* **216**:325–331.

Dickson, L., Swallow, D. M., and Donald, A. S. R., 1987, A new monoclonal antibody recognising a Lewis-related structure. *Biochem. Soc. Trans.* **15**:400–401.

Doll, R., Drane, H., and Newell, A. C., 1961, Secretion of blood group substances in duodenal, gastric and stomal ulcer, gastric carcinoma, and diabetes mellitus. *Gut* **2**:352–359.

Feracci, H., Bernadac, A., Gorvel, J. P., and Maroux, S., 1982, Localization by immunofluorescence and histochemical labeling of aminopeptidase N in relation to its biosynthesis in rabbit and pig enterocytes. *Gastroenterology* **82**:317–324.

Finne, J., 1980, Identification of the blood-group ABH-active glycoprotein components of human erythrocyte membrane. *Eur. J. Biochem.* **104**:181–189.

Finne, J., Krusius, T., Rauvala, H., Kekomäki, R., and Myllylä, G., 1978, Alkali-stable blood group A- and B-active poly(glycosyl)-peptides from human erythrocyte membrane. *FEBS Lett.* **89**:111–115.

Flatz, G., 1970, Serum-cholesterin, ABO-Blutgruppen und Hämoglobintyp: Beitrag zur Frage der genetischen Beeinflussung der Serum-Cholesterinkonzentration. *Humangenetik* **10**:318–328.

Forgue-Lafitte, M.-E., Horvat, A., and Rosselin, G., 1979, Insulin binding by a cell line (HT 29) derived from human colonic cancer. *Mol. Cell Endocrinol.* **14**:123–130.

Forgue-Lafitte, M.-E., Laburthe, M., Chamblier, M.-C., Moody, A. J., and Rosselin, G., 1980, Demonstration of specific receptors for EGF-urogastrone in isolated rat intestinal epithelial cells. *FEBS Lett.* **114**:243–246.

Fransen, J. A. M., Ginsel, L. A., Hauri, H.-P., Sterchi, E., and Blok, J., 1985, Immuno-electronmicroscopical localization of a microvillus membrane disaccharidase in the human small-intestinal epithelium with monoclonal antibodies. *Eur. J. Cell Biol.* **38**:6–15.

Gerard, G., Vitrac, D., Le Pendu, J., Muller, A., and Oriol, R., 1982. H-deficient blood groups (Bombay) of Reunion Island. *Am. J. Hum. Genet.* **34**:937–947.

Glynn, L. E., Holborow, E. J., 1959, Distribution of blood-group substances in human tissues. *Br. Med. Bull.* **15**:150–153.

Glynn, L. E., Holborow, E. J., and Johnson, G. D., 1957, The distribution of blood-group substances in human gastric and duodenal mucosa. *Lancet* **ii**:1083–1088.

Goldbourt, U., and Neufeld, H. N., 1986, Genetic aspects of arteriosclerosis. *Arteriosclerosis* **6**:357–377.

Gorvel, J. P., Wisner-Provost, A., and Maroux, S., 1982, Identification of glycoproteins bearing human blood group A determinants in rabbit enterocyte plasma membranes. *FEBS Lett.* **143**:17–20.

Green, F., Edwards, Y., Hauri, H.-P., Povey, S., Ho, M.-W., Pinto, M., and Swallow, D., 1987a, Isolation of a cDNA probe for a human jejunal brush-border hydrolase, sucrase–isomaltase, and assignment of the gene locus to chromosome 3. *Gene* **57:**101–110.

Green, F., Greenwell, P., Griffiths, B., Dickson, L., and Swallow, D., 1987b, The ABH and Lewis antigens on the human intestinal hydrolases. *Biochem. Soc. Trans.* **15:**401–402.

Greenwell, P., Yates, A. D., and Watkins, W. M., 1986, UDP-*N*-acetyl-D-galactosamine as a donor substrate for the glycosyltransferase encoded by the *B* gene at the human blood group ABO locus. *Carbohyd. Res.* **149:**149–170.

Hakomori, S.-I., and Kobata, A., 1974, Blood group antigens. In *The Antigens* (M. Sela, ed.), Vol. 2, pp. 79–140, Academic Press, New York.

Hartmann, G., 1941, *Group Antigens in Human Organs.* Munksgaard, Copenhagen.

Hauri, H.-P., Sterchi, E. E., Bienz, D., Fransen, J. A. M., and Marxer, A., 1985, Expression and intracellular transport of microvillus membrane hydrolases in human intestinal epithelial cells. *J. Cell. Biol.* **101:**838–851.

Herbert, P. N., Assmann, G., Gotto, A. M., Jr., and Fredrickson, D. S., 1983, Familial lipoprotein deficiency: A betalipoproteinemia, hypobetalipoproteinemia, and Tangier disease. In *The Metabolic Basis of Inherited Disease* (J. B. Stanbury, J. B. Wyngaarden, D. S. Fredrickson, J. L. Goldstein, and M. S. Brown, eds.), pp. 589–622, McGraw–Hill, New York.

Hermon-Taylor, J., Perrin, J., Grant, D. A. W., Appleyard, A., Bubel, M., and Magee, A. I., 1977, Immunofluorescent localisation of enterokinase in human small intestine. *Gut* **18:**259–265.

Hill, R. L., and Brew, K., 1975, Lactose synthetase. *Adv. Enzymol.* **43:**411–490.

Hindsgaul, O., Khare, D. P., Bach, M., and Lemieux, R. U., 1985, Molecular recognition. III. The binding of the H-type 2 human blood group determinant by the lectin of *Ulex europaeus. Can. J. Chem.* **63:**2653–2658.

Hunziker, W., Spiess, M., Semenza, G., and Lodish, H. F., 1986, The sucrase–isomaltase complex: Primary structure, membrane-orientation, and evolution of a stalked, intrinsic brush border protein. *Cell* **46:**227–234.

Jarnefelt, J., Rush, J., Li, Y.-T., and Laine, R. A., 1978, Erythroglycan, a high molecular weight glycopeptide with the repeating structure [galactosyl-(1–4)-2-deoxy-2-acetamido-glucosyl (1–3)] comprising more than one-third of the protein-bound carbohydrate of human erythrocyte stroma. *J. Biol. Chem.* **253:**8006–8009.

Karhi, K. K., and Gahmberg, C. G., 1980, Identification of blood group A-active glycoproteins in the human erythrocyte membrane. *Biochim. Biophys. Acta* **622:**344–354.

Karlsson, S., Swallow, D. M., Griffiths, B., Corney, G., Hopkinson, D. A., Dawnay, A., and Cartron, J. P., 1983, A genetic polymorphism of a human urinary mucin. *Ann. Hum. Genet.* **47:**263–269.

Kelly, J. J., and Alpers, D. H., 1973, Blood group antigenicity of purified human intestinal disaccharidases. *J. Biol. Chem.* **218:**8216–8221.

Kenny, A. J., and Maroux, S., 1982, Topology of microvillar membrane hydrolases of kidney and intestine. *Physiol. Rev.* **62:**91–127.

Kobata, A., 1972, Isolation of oligosaccharides from human milk. *Methods Enzymol.* **28:**262–271.

Komoda, T., and Sakagishi, Y., 1978, The function of carbohydrate moiety and alteration of carbohydrate composition in human alkaline phosphatase isoenzymes. *Biochim. Biophys. Acta* **523:**395–406.

Komoda, T., Sakagishi, Y., and Sekine, T., 1981, Multiple forms of human intestinal alkaline phosphatases: Chemical and enzymatic properties, and circulating clearances of the fast- and slow-moving enzymes. *Clin. Chim. Acta* **117:**167–187.

Krusius, T., Finne, J., and Rauvala, H., 1978, The poly(glycosyl) chains of glycoproteins: Characterisation of a novel type of glycoprotein saccharides from human erythrocyte membrane. *Eur. J. Biochem.* **92:**289–300.

Kumazaki, T., and Yoshida, A., 1984, Biochemical evidence that secretor gene, *Se*, is a structural gene encoding a specific fucosyltransferase. *Proc. Natl. Acad. Sci. U.S.A.* **81**:4193–4197.

Lamba, D. L., Singha, P., and Chandra, S., 1974, A study of diabetes in relation to blood groups and cholesterol levels. *Humangenetik* **23**:51–58.

Langman, M. J. S., Leuthold, E., Robson, E. B., Harris, J., Luffman, J. E., and Harris, H., 1966, Influence of diet on the "intestinal" component of serum alkaline phosphatase in people of different ABO blood groups and secretor status. *Nature* **212**:41–43.

Laperche, Y., Bulle, F., Aissani, T., Chobert, M.-N., Aggerbeck, M., Hanoune, J., and Guellaen, G., 1986, Molecular cloning and nucleotide sequence of rat kidney γ-glutamyl transpeptidase cDNA. *Proc. Natl. Acad. Sci. U.S.A.* **83**:937–941.

Lemieux, R. U., Baker, D. A., Weinstein, W. M., and Switzer, C. M., 1981, Artificial antigens, antibody preparations for the localization of Lewis determinants in tissues. *Biochemistry* **20**:199–205.

Le Pendu, J., 1983, α-2-L-Fucosyltransferase activities of human serum; properties of two distinct enzymes and their relationships with the genetics of H tissue antigens. In *Red Cell Membrane Glycoconjugates and Related Genetic Markers*, pp. 183–191, Librairie Arnette, Paris.

Le Pendu, J., Cartron, J. P., Lemieux, R. U., and Oriol, R., 1985a, The presence of at least two different H-blood-group-related β-D-Gal α-2-L-fucosyltransferases in human serum and the genetics of blood group H substances. *Am. J. Hum. Genet.* **37**:749–760.

Le Pendu, J. Fredman, P. Richter, N. D., Magnani, J. L., Willingham, M. C. Pastan, I., Oriol, R., and Ginsburg, V., 1985b, Monoclonal antibody 101 that precipitates the glycoprotein receptor for epidermal growth factor is directed against the Y haptan, not H type 1 antigen. *Carbohydr. Res.* **141**:347–349.

Li, S.-C., Nakamura, T., Ogamo, A., and Li, Y.-T., 1979, Evidence for the presence of two separate protein activators for the enzymic hydrolysis of G_{M1} and G_{M2} gangliosides. *J. Biol. Chem.* **254**:10592–10595.

Marcus, D. M., and Cass, L. E., 1969, Glycosphingolipids with Lewis blood group activity: Uptake by human erythrocytes. *Science* **164**:553–555.

Massey, D., and Maroux, S., 1985, The carbohydrate moiety of aminopeptidase N of rabbit intestinal brush-border membrane. *FEBS Lett.* **181**:207–210.

Mollicone, R., Bara, J., Le Pendu, J., and Oriol, R., 1985, Immunohistologic pattern of type 1 (Le^a, Le^b) and type 2 (X,Y,H) blood group-related antigens in the human pyloric and duodenal mucosae. *Lab. Invest.* **53**:219–227.

Mollicone, R., Le Pendu, J., Bara, J., and Oriol, R., 1986, Heterogeneity of the ABH antigenic determinants expressed in human pyloric and duodenal mucosae. *Glycoconjugate J.* **3**:187–202.

Oliver, M. F., Geizerova, H., Cumming, R. A., and Heady, J. A., 1969, Serum cholesterol and ABO and rhesus blood groups. *Lancet* **ii**:605–606.

Oriol, R., Danilovs, J., and Hawkins, B. R., 1981, A new genetic model proposing that the *Se* gene is a structural gene closely linked to the *H* gene. *Am. J. Hum. Genet.* **33**:421–431.

Oriol, R., Le Pendu, J., and Mollicone, R., 1986, Genetics of ABO, H, Lewis, X and related antigens. *Vox Sang.* **51**:161–171.

Parker, P. J., Young, S., Gullick, W. J., Mayes, E. L. V., Bennett, P., and Waterfield, M. D., 1984, Monoclonal antibodies against the human epidermal growth factor receptor from A431 cells. *J. Biol. Chem.* **259**:9906–9912.

Picard, J., Edward, D. W., and Feizi, T., 1978, Changes in the expression of the blood group A,B,H,Le^a, and Le^b antigens and the blood group precursor associated I(Ma) antigen in glycoprotein-rich extracts of gastric carcinomas. *J. Clin. Lab. Immunol.* **1**:119–128.

Potter, J., Ho, M.-W., Bolton, H., Furth, A. J., Swallow, D. M., and Griffiths, B., 1985, Human lactase and the molecular basis of lactase persistence. *Biochem. Genet.* **23**:423–439.

Roth, J., Lentze, M. J., and Berger, E. G., 1985, Immunocytochemical demonstration of ectogalactosyltransferase in absorptive intestinal cells. *J. Cell Biol.* **100:**118–125.

Roth, J., Taatjes, D. J., Weinstein, J., Paulson, J. C., Greenwell, P., and Watkins, W. M., 1986, Differential subcompartmentation of terminal glycosylation in the Golgi apparatus of intestinal absorptive and goblet cells. *J. Biol. Chem.* **261:**14307–14312.

Roth, J., Greenwell, P., and Watkins, W. M., 1987, Subcellular distribution of the products of the blood group A gene. *Biochem. Soc. Trans.* **15:**599–601.

Semenza, G., 1986, Anchoring and biosynthesis of stalked brush border membrane proteins: Glycosidases and peptidases of enterocytes and of renal tubuli. *Ann. Rev. Cell Biol.* **2:**255–313.

Sneath, J. S., and Sneath, P. H. A., 1955, Transformation of the Lewis groups of human red cells. *Nature* **176:**172.

Stevens, R. L., Fluharty, A. L., Kihara, H., Kaback, M. M., Shapiro, L. J., Marsh, B., Sandhoff, K., and Fischer, G., 1981, Cerebroside sulfatase activator deficiency induced metachromatic leukodystrophy. *Am. J. Hum. Genet.* **33:**900–906.

Strecker, G., Trentesaux-Chauvet, C., Poitau, A., and Montreuil, J., 1976, Etudes des oligosaccharides riches en fucose presents dans les urines de sujets sains et melituriques de groupes sanguins A, B et O. *Biochimie* **58:**805–814.

Swallow, D. M., Potter, J., Green, F., and Ho, M. W., 1985, Preparation of a monoclonal antibody against human lactase. *J. Immunol. Methods* **77:**139–145.

Swallow, D. M., Griffiths, B., Bramwell, M., Wiseman, G., and Burchell, J., 1986, Detection of the urinary "PUM" polymorphism by the tumour-binding monoclonal antibodies Ca1, Ca2, Ca3, HMFG1, and HMFG2. *Disease Markers* **4:**247–254.

Szulman, A. E., 1960, The histological distribution of blood group substances A and B in man. *J. Exp. Med.* **111:**785–799.

Szulman, A. E., 1962, The histological distribution of the blood group substances in man as disclosed by immunofluorescence. *J. Exp. Med.* **115:**977–995.

Takasaki, S., Yamashita, K., and Kobata, A., 1978, The sugar chain structures of ABO blood group active glycoproteins obtained from human erythrocyte membrane. *J. Biol. Chem.* **253:**6086–6089.

Triadou, N., Audran, E., Rousset, M., Zweibaum, A., and Oriol, R., 1983, Relationship between the secretor status and the expression of ABH blood group antigenic determinants in human intestinal brush-border membrane hydrolases. *Biochim. Biophys. Acta* **761:**231–236.

Vauhkonen, M., Viitala, J., Parkkinen, J., and Rauvala, H., 1985, High-mannose structure of apolipoprotein-B from low-density lipoproteins of human plasma. *Eur. J. Biochem.* **152:**43–50.

Vauhkonen, M., 1986, Complex carbohydrates of apolipoprotein-B of human plasma low density lipoprotein. *Glycoconjugate J.* **3:**35–43.

Watkins, W. M., 1959, Some genetical aspects of human blood group substances. In *Ciba Foundation Symposium on Biochemistry of Human Genetics* (G. E. W. Wolstenholme and C. M. O'Connor, eds.), pp. 217–238, Churchill, London.

Watkins, W. M., 1972, Blood-group specific substances. In *Glycoproteins: Their Composition, Structure and Function* (A. Gottschalk. ed.), pp. 830–891, Elsevier, Amsterdam.

Watkins, W. M., 1980, Biochemistry and genetics of the ABO, Lewis, and P blood group systems. In *Advances in Human Genetics* (H. Harris and K. Hirschhorn, eds.), pp. 1–136, Plenum Press, New York.

Watkins, W. M., and Morgan, W. T. J., 1959, Possible genetical pathways for the biosynthesis of blood group mucopolysaccharides. *Vox Sang.* **4:**97–119.

Watkins, W. M., Yates, A. D., Greenwell, P., Bird, G. W. G., Gibson, M., Roy, T. C. F., Wingham, J., and Loeb, W., 1981, A human dispermic chimaera first suspected from analyses of the blood group gene specified transferases. *J. Immunogenet.* **8:**113–128.

Wilczynska, Z., Miller-Podraza, H., and Koscielak, J., 1980, The contribution of different glycoconjugates to the total ABH blood group activity of human erythrocytes. *FEBS Lett.* **112**:277–279.

Wiley, E. L., Murphy, P., Mendelsohn, G., and Eggleston, J. C., 1981, Distribution of blood group substances in normal human colon. Use of the unlabeled antibody (PAP) immunoperoxidase technique to identify A and B blood group substances. *Am. J. Clin. Pathol.* **76:** 806–809.

Zweibaum, A., and Bouhou, E., 1973, Studies on digestive groups. I. The A allo-antigen–allo-antibody system in rabbits. *Transplantation* **15**:291–293.

Chapter 5

Biogenesis and Intracellular Transport of Intestinal Brush Border Membrane Hydrolases

Use of Antibody Probes and Tissue Culture

Hans-Peter Hauri

1. INTRODUCTION

The brush border membrane (also designated microvillar, luminal, or apical membrane) is a highly ordered surface domain of transporting epithelial cells like small-intestinal enterocytes, large-intestinal colonocytes, or kidney proximal tubular cells. In the small intestine, the brush border membrane faces the intestinal lumen. Its microvilli provide a surface enlargement of approximately 20-fold as compared to a flat luminal cell surface. This membrane amplification serves the purpose of increasing the capacity of membrane-mediated digestion and carrier-mediated uptake.

In the last 20 years, a number of brush border membrane components have been identified and characterized that are involved in these processes, that is, disaccharidases and peptidases, as well as carriers for sugars, peptides, amino acids, bile, and ions. Among these components the disaccharidases and peptidases have been characterized most completely (see Gray, 1981; Kenny and

Hans-Peter Hauri Department of Pharmacology, Biocenter of the University of Basel, CH-4056 Basel, Switzerland.

Maroux, 1982; Norén *et al.* 1986b; Semenza, 1981a, 1986, for reviews). Three characteristic features render these hydrolases particularly amenable for experimentation: (1) they are major proteins of the luminal membrane; (2) they can easily be solubilized because of their mode of anchoring; and (3) they can readily be followed by means of their enzyme activities during purification. Studies on the biogenesis and the intracellular transport of individual brush border enzymes started in the late 1970s and were made possible by the development of improved subcellular fractionation techniques for isolated enterocytes (Colas and Maroux, 1980; Freedman *et al.*, 1977; Hopfer *et al.*, 1973; Murer *et al.*, 1974, 1976; Schmitz *et al.*, 1973; Weiser *et al.* 1978; see Weiser *et al.*, 1986 for review), by the production of specific antibodies and by the establishment of intestinal organ culture (Browning and Trier, 1969; see Trier, 1980 for review) and most recently cell culture techniques (Pinto *et al.*, 1982, 1983; see Zweibaum *et al.*, 1986 for review). Previous excellent reviews including aspects of biogenesis have focused on the topology (Kenny and Maroux, 1982) and post-translational modifications of brush border enzymes (Danielsen *et al.*, 1984a) and on membrane anchoring and early biosynthetic events (Semenza, 1986).

The aim of the present chapter is to review recent data on the biogenesis and the intracellular transport of the most extensively characterized brush border hydrolases with particular reference to the use of antibody probes and tissue culture. The review does not discuss the appearance of brush border hydrolases during ontogeny and their regulation by substrates and hormones (see Henning, 1985, for review) as most of these studies deal with enzyme activities rather than with the direct measurement of biosynthetic events.

Some of the hydrolases, like a number of different peptidases, are also found in kidney, placenta, and hepatocyte (bile canalicular) brush border membranes, and therefore knowledge on the biogenesis of the microvillar enzymes in these organs is also included when available. The introduction of the monoclonal antibody technique by Köhler and Milstein (1975) has opened new perspectives in intestinal cell biology relating to the identification and characterization of individual microvillar proteins and to studies of their maturation. On the other hand, the establishment of novel intestinal cell and tissue culture systems have significantly contributed to the understanding of biogenetic processes in epithelial cells. Particularly noteworthy is the fact that it is now possible to manipulate these events easily in differentiated adenocarcinoma cell lines (Zweibaum *et al.*, 1986). The relevance of information obtained by the use of cell culture can subsequently be tested for at a more complex level, that is, in organ culture, a system that is in many, but of course not all, respects close to the in vivo situation. Furthermore, a comparison of brush border enzymes synthesized in tissue culture with those synthesized in vivo provides insights into the role of pancreatic proteases in the post-transitional cleavage of some brush border enzymes.

2. PROTEIN COMPOSITION OF THE SMALL INTESTINAL BRUSH BORDER MEMBRANE

A highly purified brush border membrane fraction analyzed by SDS–PAGE exhibits a minimum of 40–50 individual protein bands. However, some of these proteins are not integral components of the microvillar membrane but derive from the cytoskeleton of the microvillus core, the most prominent of which is actin (see Bretscher, 1983, and Mooseker, 1985, for recent reviews on cytoskeletal proteins of brush borders). Cytoskeletal elements can most completely, but not entirely, be removed by treatment with 1 M Tris at high pH (Eichholz and Crane, 1965) or by chaotropic agents (Hopfer *et al.*, 1983; Riendeau *et al.*, 1986). Another possible source of cross-contamination is the basolateral membrane. The purest microvillar membrane preparations known are obtained from rat tissue. They are based on procedures originally designed by Crane's group for the guinea pig intestine (Miller and Crane, 1961) and further developed by Forstner *et al.* (1968) and Hopfer *et al.* (1973). In these procedures, the apical portion of the enterocytes is sheared off in a blender and purified as virtually intact brush border caps. The caps are then disrupted and the membranes are recovered by density gradient centrifugation in vesicular form. Since the shear plane through the cells is below the tight junction, the brush border caps comprise tags of lateral membranes, which upon further subfractionation can only be removed with great difficulties. Because of these technical problems, it is presently not possible to give an accurate estimate for the number of different proteins in the microvillar membrane. However, it appears that the major components have been identified as brush border enzymes. These are listed in Table I together with their respective precursor forms. The M_r values for the individual mature enzymes are remarkably similar in the different mammalian species with the exception of maltase–glucoamylase, which is considerably larger in the human than in the pig.

Minor components of the brush border membrane include carriers for glucose (Semenza and Corcelli, 1986), amino acids and dipeptides (Adibi and Kim, 1981; Semenza and Corcelli, 1986), and, in some restricted intestinal segments, receptors for iron, bile acids, and the intrinsic factor–vitamin B_{12} complex (Seetharam *et al.*, 1981). Knowledge on a molecular level of these physiologically most important systems is scant and therefore these proteins are not dealt with in this chapter. Recent immunohistochemical studies suggest that the human intestinal brush border membrane may also contain glycosyltransferases, in particular galactosyltransferase (Pestalozzi *et al.*, 1982; Roth et al., 1985a), that is, enzymes that are normally associated with the Golgi apparatus, before being released from the cells. However, this provocative finding is controversial. Boyle *et al.* (1986) were unable to measure galactosyltransferase activity in purified brush border membranes, which may suggest that the immu-

Table I
Molecular Forms of Microvillar Hydrolases During Biosynthesis[a]

Enzyme	Species	Tissue	Primary translation product, nonglycosylated form	High-mannose, endo-H-sensitive transient form	Complex-glycosylated, endo-H-resistant mature form[b]	References
Aminopeptidase N	Pig	Intestine	115	140	166 (123 + 62)	Danielsen (1982); Danielsen *et al.* (1982b, 1983c, 1984a); Sjöström *et al.* (1978); Sjöström and Norén (1982); Benajiba and Maroux (1981)
	Rat	Intestine	—	<130	130–140	Ahnen *et al.* (1983); Quaroni (1985a)
	Rabbit	Intestine	—	110	115–128	Feracci *et al.* (1985)
			99	106	125	Massey and Maroux (1985)
	Human	Caco-2	—	139	158	Hauri *et al.* (1985b)
	Human	Intestine	—	130	160–162	Naim *et al.* (1986); Hauri *et al.* (1985b)
	Pig	Kidney	115	140	160	Stewart and Kenny (1984a)
Aminopeptidase A	Pig	Intestine	—	140	170	Danielsen *et al.* (1983a)
	Pig	Kidney	—	145	170	Stewart and Kenny (1984a)
Dipeptidylpeptidase IV	Human	Intestine	—	100	128	Naim *et al.* (1986)
	Human	Caco-2	—	110/114	124	Hauri *et al.* (1985b)
	Pig	Intestine	—	115	137	Danielsen *et al.* (1983a)
	Pig	Kidney	—	115	130	Stewart and Kenny (1984a)
Endopeptidase 24.11	Pig	Kidney	—	88	93	Stewart and Kenny (1984a)
Angiotensin I-converting enzyme	Human	Intestine	—	174	185	Naim *et al.* (1986); Hauri *et al.* (1985b)
γ-Glutamyltransferase	Rat	Kidney	63	78	(51 + 22)	Nash and Tate (1982, 1984)
	Rat	Kidney	63	78	(50 + 23)	Kuno *et al.* (1983); Matsuda *et al.* (1983)

	Rat	Kidney	—	75	(49.5 + 29)	Capraro and Hughey (1983)
PABA–peptide hydrolase	Human	Intestine	—	90	100	Sterchi *et al.* (1986); Hauri *et al.* (1985a)
Sucrase–isomaltase	Rabbit	Intestine	200	230	275 (120 + 140)	Brunner *et al.* (1979); Wacker *et al.* (1981); Ghersa *et al.* (1986)
	Rat	Intestine	190	220	265 (130 + 150)	Hauri *et al.* (1979); Montgomery *et al.* (1981); Alpers *et al.* (1986); Quaroni and Isselbacher (1985)
	Pig	Intestine	225	240	265 (140 + 150)	Danielsen (1982); Danielsen and Cowell (1984)
	Human	Intestine	—	210	231–260 (145 + 151)	Naim *et al.* (1986); Hauri *et al.* (1985b); Skovbjerg (1982); Hauri *et al.* (1985a)
	Human	Caco-2	—	210	217	Hauri *et al.* (1985b)
Maltase–glucoamylase	Pig	Intestine	200	225	245 (125 + 135)	Danielsen *et al.* (1983a); Danielsen and Cowell (1984)
	Human	Intestine	—	290	350	H. Naim (personal communication)
Lactase–phlorizin hydrolase	Pig	Intestine	210	225	245 (160)	Danielsen *et al.* (1984b); Sjöström *et al.* (1983)
	Human	Intestine	200	215	— (160)	Naim *et al.* (1986); Naim *et al.* (1987); Skovbjerg *et al.* (1984); Skovbjerg *et al.* (1981)
	Human	Caco-2	—	200	214 (143)	Hauri *et al.* (1985b)

[a]Apparent values of M_r were estimated by SDS–PAGE under reducing conditions and are given in kilodaltons.
[b]Values in parenthesis indicate the M_r of proteolytically cleaved mature forms.
Source: Adapted from Danielsen *et al.* (1984a).

nohistochemically detected galactosyltransferase, if indeed present, may be enzymatically inactive. On the other hand, it is important to note that enzymatic activity of galactosyltransferase in brush border membranes may be impeded by the presence of glycosidases and phosphatases in the preparation (Pestalozzi *et al.*, 1982).

3. TOPOLOGY AND SUBUNIT STRUCTURE OF THE MAJOR BRUSH BORDER MEMBRANE HYDROLASES

3.1. Positioning in the Brush Border Membrane

The orientation of an integral membrane protein with respect to the phospholipid bilayer is likely to offer clues as to the mechanism of its insertion into the membrane (Sabatini *et al.*, 1982). A number of biochemical studies have led to a detailed picture on how disaccharides, in particular sucrase–isomaltase, and some peptidases are positioned in the brush border lipid bilayer (Kenny and Maroux, 1982; Semenza, 1986) and a unifying basic concept has emerged for these two protein classes, which can be summarized as follows (Figure 1). Disaccharidases (with the exception of trehalase, see below) and peptidases are integral membrane glycoproteins. The principal protein mass (up to 97%), including the active site(s) and the carbohydrate side chains, protrudes to the intestinal lumen while a short stretch of hydrophobic amino acids near the N terminus (i.e., on the cytoplasmic side) anchors the proteins in the brush border membrane. In most cases, the brush border hydrolases form dimers (Cowell *et al.*, 1986; Mosimann *et al.*, 1973; Norén *et al.*, 1986b, 1986b), although dimerization may not be needed for activity (Fulcher *et al.*, 1986a). With the exception of PABA–peptide hydrolase that is dimerized by disulfide bonds (E. E. Sterchi, personal communication), the subunits are held together by noncovalent interaction. Intestinal alkaline phosphatase of some species is believed to be anchored by means of a C-terminal rather than an N-terminal sequence (Colbeau and Maroux, 1978; Kam *et al.*, 1985; Seetharam *et al.*, 1985), while this enzyme in other species and in other organs can be solubilized by phospholipase C and therefore might be attached to the lipid bilayer via phosphoinositol (Low and Zilversmit, 1980; Takesue *et al.*, 1986). A similar mode of anchoring via phospholipids has been proposed for trehalase (Takesue *et al.*, 1986).

3.2. Topology Deduced from the Primary Sequence

For a number of brush border enzymes, N-terminal amino acid sequences are available (Figure 2). More complete sequence information is emerging from

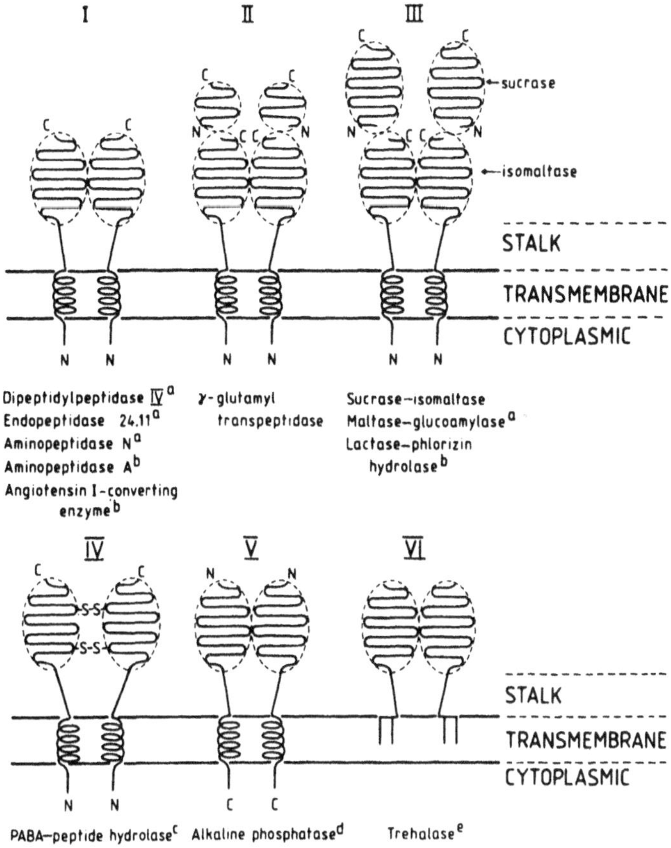

FIGURE 1. Topology and subunit composition of the microvillar hydrolases in the brush border membrane. Six different types (I–VI) are indicated. I: Dimer of a protein that is anchored N terminally and is synthesized as a one-active-site polypeptide. II: Same as I but protein is cleaved into two subunits post-translationally. III: Dimer of a protein complex that is anchored N terminally and is initially synthesized as a two-active site single polypeptide but is post-translationally cleaved intracellularly or at the cell surface to various extents into two subunits. Note that the subunit identification is known only for sucrase–isomaltase and is indicated. IV: Same as I but dimer formation involves disulfide bonds. V: Dimer of a protein that is anchored C terminally and is synthesized as a one-active-site polypeptide. VI: Dimer of a one-active-site protein that is anchored in the membrane via phospholipids rather than a hydrophobic amino acid sequence. Note that the precise topology is only known for sucrase–isomaltase and γ-glutamyltranspeptidase. [a]N-terminal anchoring is inferred from the presence of a stretch of hydrophobic residues (Figure 1). [b]N-terminal anchoring is likely on the basis of biochemical properties that are either similar to those of aminopeptidase N or sucrase–isomaltase. [c]Topology is tentative. The number of disulfide bonds is unknown. [d]At least initial C-terminal anchoring is likely on the basis of the cDNA sequences for the placental enzyme. However, this anchor may be replaced in some species by covalent binding to phosphatidylinositol as suggested by solubilization studies with phospholipase C (see text). [e]Anchoring via phosphatidylinositol is likely based on experiments with phospholipase C.

```
                   Cytoplasmic domain                          Transmembrane domain

Isomaltase  1                   5                   10                  15                  20                  25                  30
rabbit[a]   Ala Lys Arg Lys Phe Ser Gly Leu Glu Ile Thr Leu Ile Val Leu Phe Val Ile Val Phe Ile Ile Ala Ile Ala Leu Ile Ala Val Leu Ala Thr
human[b]    Ala Arg Lys Lys Phe Ser Gly Leu Glu Ile Ser Leu Ile Val Leu Phe Val Ile Val Thr Ile Ile Ala Ile Ala Leu Ile Val Val Leu Ala Thr
                                                                                                                                   ?
pig[c]      Ala Arg Lys Lys Phe Ser Gly Leu Glu Ile  X  Leu Ile Val Leu Phe Ala Ile Val Leu Ser Ile Ala Ile Ala Leu Val Val Val  X  Ala Ser
                                                                                                    ?
rat[d]      Ala Lys Lys Lys Phe Ser Ala Leu Glu Ile  X  Leu Ile Val Leu Phe Ile Ile Val Leu Ala Ile Ala Ile Ala Leu Val Leu Val

Maltase-glucoamylase
            1                   5                   10              15                  20                  25
                ?                                                       ?           ?   ?   ?   ?   ?   ?   ?
Pig[e]       X  Arg Lys Lys Leu Lys Lys Phe Thr Asp Leu Leu Met Leu Ser Val Leu Leu Leu Val Val Phe Ile Val
                                                    Glu
γ-Glutamyltransferase
                                    1                5               10                          15                  20
rat kidney [f]                      Met Lys Asn Arg Phe Leu Val Leu Gly Leu Val Ala Val Val Leu Val Phe Val Ile Ile Gly Leu Cys

Endopeptidase 24.11
pig (kidney)[g]
1         5              10              15                  20                  25
X  Pro Lys Pro Lys Lys Lys Gln Arg Trp Thr Pro Leu Glu Ile Ser Leu Glu Val Leu Val Leu Val Leu Val  X  Ile
Dipeptidylpeptidase IV                                          1               5       ?
pig (kidney)[h]                                                 Leu Gly Phe Ala Leu Ala Phe Ile
Aminopeptidase N                                1               5                   10
rabbit[i]                                       Tyr Ile Ser Lys Ala Leu Gly Ile Leu Gly Phe  X  Leu Gly
```

FIGURE 2. N-terminal amino acid sequences of some brush border hydrolases. The first residues of the transmembrane domains were aligned. Hydrophobic sequences are underlined. [a]From Frank *et al.* (1978) and Hunziker *et al.* (1986). [b]From Green *et al.* (1987). [c]From Sjöström *et al.* (1982). [d]From Hauri *et al.* (1982). [e]From Norén *et al.* (1986). [f]From Matsuda *et al.* (1983) and Laperche *et al.* (1986). [g]From Fulcher *et al.* (1986b). [h]From MacNair and Kenny (1979). [i]From Feracci *et al.* (1982a). Question marks: uncertain amino acids.

cDNA cloning. To date, this approach has been successful for rabbit sucrase–isomaltase (Hunziker *et al.*, 1986), for human sucrase–isomaltase (Green *et al.*, 1987), and for γ-glutamyltranspeptidase (Laperche *et al.*, 1986). Common to most hydrolases analyzed to date is a short hydrophilic amino acid sequence near the N terminus. For sucrase–isomaltase, this sequence is conserved; 8 of the 11 amino acid positions are identical in four species, and each of the substitutions in the remaining positions is conservative with respect to charge. The N-terminal domain is believed, but not in all cases formally proved, to be located on the cytoplasmic side. A previously proposed extracytoplasmic location of the N terminus of sucrase–isomaltase (Bürgi *et al.*, 1983) has now been ruled out on the basis of the cDNA sequence (Hunziker *et al.*, 1986). The cytoplasmic segment contains several positively charged amino acids, a frequent feature of plasma membrane proteins (Sabatini *et al.*, 1982). Dipeptidylpeptidase IV of pig kidney was reported to have no cytoplasmic extension (MacNair and Kenny, 1979). However, it cannot be excluded that this part was lost during purification owing to an unrecognized proteolytic event.

The cytoplasmic tail is followed by the membrane-spanning region. The amino acid sequence of this domain is known for sucrase–isomaltase and γ-glutamyltranspeptidase and it comprises 20 and 18 predominantly hydrophobic amino acids, respectively. These (and probably most) hydrolases span the lipid bilayer once as deduced from their cDNA (Hunziker *et al.*, 1986; Laperche *et al.*, 1986).

Immediately distal to the membrane-spanning domain of rabbit (Hunziker *et al.*, 1986) and human (Green *et al.*, 1987) sucrase–isomaltase is a sequence, the so-called stalk, that contains clusters of serine and threonine residues, some of which may carry O-linked oligosaccharide side chains. Similar threonine- and serine-rich domains are found in glycophorin (Marchesi *et al.*, 1976), the low-density lipoprotein receptor (Goldstein *et al.*, 1985), and in the receptor for interleukin-2 on T lymphocytes (Leonard *et al.*, 1984; Nikaido *et al.*, 1984). The function of the stalk in brush border hydrolases is unknown, but a more elongated rather than a globular molecular shape allows accommodation of a higher number of enzyme molecules in a given surface area. Moreover, the stalk may keep the bulk of the protein distant from the cell surface and thereby provide room for other types of proteins, including carriers, whose functional domains require a closer contact to the lipid bilayer. The stalks of the individual brush border enzymes vary somewhat in length from 2 to 5 nm (Brunner *et al.*, 1978; Cowell *et al.*, 1986; Hussain, 1985; Hussain *et al.*, 1981; Kenny and Fulcher, 1982; Kenny *et al.*, 1983; Norén *et al.*, 1986a; Spiess *et al.*, 1981).

In disaccharidases, the stalk is followed by two globular domains (Figure 1) that possess the hydrolytic activities. Electron microscopy of microvillar enzymes after their reconstitution into liposomes have indeed revealed lollipop like structures (Norén *et al.*, 1986b) resembling the knobs that have previously been observed at the luminal surface of intact brush border membrane vesicles (Johnson, 1967; Takesue and Nishi, 1978). Nishi and Takesue (1978a,b) have demonstrated that the sucrase–isomaltase complex is associated at its long axis with the membrane and that it protrudes 150 Å from the microvillar surface. In sucrase–isomaltase, the first globular domain represents the isomaltase subunit and the second one, that is more distant to the bilayer, the sucrase subunit. Assignment of enzyme activities to these two domains was possible owing to the fact that in vivo the sucrase–isomaltase precursor protein is proteolytically cleaved into an isomaltase and a sucrase subunit by pancreatic proteases (Figures 1 and 4), which remain associated by noncovalent interaction (Semenza, 1985). The subunits can be separated by citraconylation (Braun *et al.*, 1975) and recovered in an enzymatically active form. Using this procedure, scientists found that the isolated isomaltase and not the sucrase subunit carries the hydrophobic anchor (Brunner *et al.*, 1979; Frank *et al.*, 1978). Although among disaccharidases detailed structural information is only available for sucrase–isomaltase, it appears that this enzyme, in many respects, can serve as a model for the structural features of the other two major disaccharidases, lactase–phlorizin hydrolase and maltase–glucoamylase. This notion is in line with results obtained from biosynthetic studies that show that these three enzymes are synthesized as single-chain two-active site polypeptides each (see Section 6.1).

The overall topology of γ-glutamyltranspeptidase is in many respects comparable to that of sucrase–isomaltase (Figure 1). The mature enzyme is anchored to the membrane near its N terminus and has two subunits that are encoded by a single gene (Laperche *et al.*, 1986). On the basis of cDNA cloning and sequencing, the N terminus was found to reside on the cytoplasmic side (Figure 2). A short stretch of four hydrophilic amino acids constitutes the cytoplasmic tail that is followed by a sequence of 18 hydrophobic residues constituting the sole membrane-spanning domain. There are some notable differences to sucrase–isomaltase. (1) The subunits of γ-glutamyltranspeptidase are much smaller (Table I); (2) there is only one catalytic site in γ-glutamyltranspeptidase that probably encompasses domains of both subunits (Gardell and Tate, 1981, 1982); and (3) cleavage into subunits probably occurs intracellularly (see Section 6).

In summary, collective but still indirect evidence suggests that most brush border enzymes exhibit a transbilayer orientation similar to sucrase–isomaltase.

4. MONOCLONAL ANTIBODIES AS TOOLS FOR THE IDENTIFICATION OF BRUSH BORDER COMPONENTS AND FOR STUDYING THEIR BIOGENESIS

A number of brush border enzymes have been purified from various mammalian species and polyclonal antibodies produced against these proteins have proved most useful to study their biosynthesis (Danielsen *et al.*, 1984a; Kenny and Maroux, 1982; Skovbjerg *et al.*, 1978). The recent introduction of the monoclonal antibody technique has extended the experimental possibilities. The general advantages and limitations of monoclonal antibodies have been discussed in many reviews. In this chapter, I would like to focus on specific applications of monoclonal antibodies relating to intestinal brush border enzymes. Monoclonal antibodies are particularly valuable (1) as highly specific probes, (2) to identify novel brush border enzymes, (3) as subunit-specific probes and for epitope mapping, (4) to assess precursor–product relations, and (5) to unravel conformational changes during biogenesis.

A list of all the available monoclonal antibodies against brush border hydrolases is given in Table II.

4.1. Strategy for the Production of Monoclonal Antibodies

It is well known that a multitude of factors influence the immune response and the subsequent successful establishment of the desired hybridoma cell line. These factors, although determined empirically, can to a certain extent be controlled in order to increase the chances for antibodies having the desired properties. Since a single monoclonal antibody can never display all the features of a polyclonal antibody, its exact application should be known prior to the preparation of the antigen for the first immunization. The planned application will largely influence the strategy of antibody production. For instance, the use of native brush border membranes as an antigen almost exclusively lead to monoclonal antibodies against noncontinuous epitopes, and hence exclusively against the native enzymes, after 6–8 weeks immunization (Hauri *et al.*, 1980, 1985b), while the injection into Balb/c mice of SDS-denatured human sucrase–isomaltase cut out from SDS gels resulted in hybridoma lines secreting antibodies against the denatured enzyme. However, none of these antibodies was able to recognize the native enzyme (H.-P. Hauri, unpublished observation). Longer immunization schedules (up to 1 year) with the native antigen increase the chances for antibodies recognizing both the native and the denatured antigen (Hauri *et al.*. 1985b). Such antibodies are useful for many applications, in particular immunoelectron microscopy (Fransen *et al.*, 1985; Hauri *et al.*, 1985a; Sips *et al.*, 1985). Another important factor is the amount of antigen used for immunization. Usually, 200 μg of purified brush border membrane protein per injection is sufficient to obtain monoclonal antibodies against different major microvillar enzymes. With this amount of protein, the most abundant antigen in the mixture, sucrase–isomaltase, produced the greatest number of positive clones, and the number of hybridoma positive for other brush border enzymes was roughly proportional to their relative amount in the antigen mixture. Since brush border enzymes are in general equally potent antigens when isolated and

Table II
Characteristics of Monoclonal Antibodies against Microvillar Hydrolases

Enzyme	Antibody designation	Immunoglobulin class	Origin of antigen	Special properties of antibody	References
Aminopeptidase N	BB 4/33	IgG1	Rat (intestine)	—	Quaroni and Isselbacher (1985)
	YBB 2/54/3	IgG1	Rat (intestine)	—	Quaroni and Isselbacher (1985)
	HBB 3/153/63	IgG1	Human (intestine)	Higher affinity for high-mannose than for complex form; n	Hauri *et al.* (1985a)
	HA 201	—	Rat (liver)	Cross-reacts with intestinal enzyme	Bartles *et al.* (1985)
Dipeptidylpeptidase IV	HBB 3/775/42	IgG1	Human (intestine)	n	Hauri *et al.*(1985a)
	HA 301	—	Rat (liver)	—	Bartles *et al.* (1985)
Endopeptidase 24.11	GK7C2	IgG1	Pig (kidney)	Useful for affinity chromatography	Gee *et al.* (1983)
Aminopeptidase W	GK5C1	IgG1	Pig (kidney)	Aminopeptidase W was detected by means of this antibody	Gee and Kenny (1985)
γ-Glutamyltranspeptidase	—	IgG	Rat (kidney)	n	Hughey *et al.* (1986)
Angiotensin I-converting enzyme	HBB 3/264/18	IgG1	Human (intestine)	n	Hauri *et al.* (1985b)
PABA–peptide hydrolase	HBB 3/716/36	IgG2a	Human (intestine)	n	Hauri *et al.* (1985b)
Alkaline phosphatase	ALP_{P1}	—	Human (placenta)	Useful for detection of enzyme polymorphism	Slaughter *et al.* (1981)
	$ALP_{FI}a$	—	Human (intestine)		Vockley and Harris (1984)
	ALP-D98	—	D98/AH-2 cells		Wray and Harris (1983) Vockley and Harris (1984)

Sucrase–isomaltase	BB3/34/12	IgG1	Rat (intestine)	Recognize additional unknown protein of 108 kD	Hauri *et al*. (1980); Quaroni (1985a)
	BB5/8/90	IgG1			
	CC 4/11	IgG1			Hauri *et al*. (1980); Quaroni (1985a)
	CC 4/91	IgG1			
	CC 4/39/1	IgG1		Recognize two additional 110 and 108 kD proteins	Quaroni (1985a)
	BBC 1/35	IgG1			
	HBB2/614/88	IgG1	Human (intestine)	Sucrase specific; d + n	Hauri *et al*. (1985b)
	HBB2/219/20	IgG2b		n	
	HBB1/451/82	IgG1		n	
	HBB1/691/79	IgG1		n	
	HBB3/705/60	IgG1		Isomaltase specific, conformation specific	
	3/73	IgG1	Human (Caco-2)	—	Quaroni (1986)
Maltase–glucoamylase	BB 8/2	IgG1	Rat (intestine)	—	Quaroni (1985a)
	BBC 3/88	IgG1		—	
	BBC 3/90	IgG1		—	
	BBC3/91/1	IgG2b		—	
	HBB 2/143/17	IgG1	Human (intestine)	n	Hauri *et al*. (1985b)
	HBB 2/962/94	IgG1		n	
	1F12E1	IgG1	Rat (kidney)	Bind to active enzyme of young animals	Reiss and Sacktor (1983)
	8B1G6	IgG1			
	7G10H3	IgG1	(kidney)	Bind to inactive enzyme of old animals	Reiss and Sacktor (1983)
	2E1C10	IgG1			
Lactase–phlorizin hydrolase	YBB1/57	IgG2a	Rat (intestine)	—	Quaroni (1985a)
	YBB2/61	IgG1		—	
	4E1	IgG1	Human (intestine)	n	Swallow *et al*. (1985)
	HBB1/909/34/74	IgG2a		n	Hauri *et al*. (1985b)

n = antibody recognizes native enzyme only.
d + n = antibody recognizes both native and denatured enzyme.

used as antigens for the production of polyclonal antibodies in rabbits, the results with the monoclonal antibodies indicate that even for the major hydrolases the amount of antigen may be limiting in the antibody response when using 200 μg antigen protein. Therefore, I believe that the antigen amount and the conformation of the protein (native versus denatured) are the most critical factors for the successful production of monoclonal antibodies against brush border enzymes. Moreover, our experience and that of Gee *et al.* (1983) suggested that presenting the hydrolases in membrane-bound rather than in purified form may increase the immune response.

4.2. Identification and Isolation of Brush Border Enzymes

Various aspects relating to the identification of brush border components by means of monoclonal antibodies have already been discussed in previous reviews (Harris, 1983; Quaroni, 1983, 1984). Since antibodies against microvillar hydrolases are in general noninhibitory or only moderately (Feracci *et al.*, 1981; Takesue *et al.*, 1977) inhibitory, the specificity of a monoclonal antibody binding to isolated brush border membranes can easily be established by measuring enzyme activities in immunoprecipitates using common substrates for disaccharidases, peptidases, or alkaline phosphatase. This approach in conjunction with new potential substrates has also led to the identification of novel enzymes like PABA–peptide hydrolase (Hauri *et al.*, 1985b), intestinal angiotensin I-converting enzyme (Hauri *et al.*, 1985b), and kidney aminopeptidase W (Gee and Kenny, 1985). It is clear that the enzymatic identification requires antibodies recognizing the native enzyme. Analytical immunopurification is most readily achieved by immunoprecipitation of radiolabeled brush border membranes followed by SDS–PAGE (Dottavio-Martin and Ravel, 1978; Hauri *et al.*, 1985b; Quaroni and Isselbacher, 1985). Quantitative immunopurification of active enzymes by means of polyclonal antibodies is not trivial owing to the denaturing conditions required for the elution of the ligand from the affinity matrix. Although a hypotonic elution method was successfully applied to release active brush border enzymes from polyclonal antibody affinity columns (Danielsen *et al.*, 1982a), the procedure often suffers from substantial loss of protein. Monoclonal antibody affinity chromatography represents an alternative approach to solving the elution problem. Since a monoclonal antibody is specific for only a single antigenic determinant, binding of an enzyme to such a column should in principle be weaker than one made of a polyclonal antiserum that recognizes a large number of different determinants. This weaker binding makes subsequent elution of the enzyme in an active state easier as exemplified in a recent study on the immunopurification of intestinal alkaline phosphatase (Vockley and Harris, 1984). This study has shown that the relative strength of binding of an antibody to its antigen is the most important factor to consider when constructing such columns. Antibodies with intermediate avidity are preferable to those with high or low avidity.

For some applications, it is sufficient or even advantageous to isolate enzyme–immunoglobulin complexes rather than the pure enzyme. This can readily be achieved by affinity chromatography using protein A–Sepharose. If the monoclonal antibody belongs to the IgG1 class, as most murine monoclonal antibodies do, the immunocomplexes can be eluted in a native form at pH 6.0 (Ey *et al.*, 1978). In the case of PABA–peptide hyrolase, this was the only successful method for purifying the native enzyme (E. E. Sterchi, personal communication). Combined with preparative SDS–PAGE, this approach is also very useful for the analysis of the N-terminal amino acid sequences (Hauri *et al.*, 1982).

4.3. Specificity Problems

The most serious specificity problems for polyclonal antibodies against brush border hydrolases relate to anticarbohydrate specificities and the presence of contaminating proteins in the purified enzyme preparation. It is well known that intestinal hydrolases of the human (Kelly and Alpers 1973a,b; Kurisu *et al.*, 1977; Triadou, 1984; Tiradou *et al.*, 1983) and other mammals (Bernadac *et al.*, 1984; Massey and Maroux, 1985) carry highly antigenic blood group and other carbohydrate determinants that will give rise to unwanted antibodies. (See Chapter 4.) In many instances, it is extremely difficult to exclude unequivocally the contribution of such antibodies, and only recently a method has become available for the determination of anticarbohydrate specificities in polyclonal and monoclonal antibodies (Tang *et al.*, 1985). Since monoclonal antibodies are directed against a single antigenic determinant, it is much easier to establish the nature of this epitope and hence antibody specificity. It is worth noting that some carbohydrate epitopes may be present on a limited amount of brush border proteins or may even be unique to individual glycoproteins (Fredman *et al.*, 1983). Therefore, a specific immunoprecipitation of a single protein as assessed by SDS–PAGE does not necessarily exclude a carbohydrate specificity.

Another example for insufficient specificity is polyclonal antisera to sucrase–isomaltase. Such antisera, although monospecific by a number of criteria, have consistently been found to react also with the luminal membrane of intestinal crypt cells (Hauri *et al.*, 1980). However, these cells are devoid of the corresponding enzyme activities. On the basis of such an immunoreactivity and the subsequent isolation of the cross-reacting protein, Dubs *et al.* (1975) have postulated that rabbit sucrase–isomaltase is synthesized in the crypt cells as an inactive precursor that becomes activated when the crypt cells undergo differentiation at the crypt–villus junction. This concept is not compatible with monoclonal antibody data. A number of monoclonal antibodies against different epitopes of sucrase–isomaltase failed to reveal an antigenetically related precursor in crypt cells (Hauri *et al.*, 1980; Quaroni and Isselbacher, 1985). A recent study by Gorvel *et al.* (1986) on a novel 140 kD brush border protein of crypt origin may shed light on the nature of the crypt cell antigen recognized by anti-sucrase–isomaltase sera. These authors found a tight association between sucrase–isomaltase and an unrelated 140 kD protein and they speculate that conventionally purified sucrase–isomaltase may be contaminated by the latter protein, which gives rise to antibodies having both villus and crypt cell specificities.

4.4. Subunit-Specific Antibodies, Precursor–Product Relation, and Epitope Mapping

Using polyclonal antibodies, Hauri *et al.* (1979) found a high molecular weight putative precursor of rat sucrase–isomaltase in Golgi-derived membranes that was cleaved into the two subunits in the brush border. Subsequently, it was possible to immunopurify the uncleaved precursor from fetal intestinal transplants (Leapman *et al.*, 1974) by means of a monoclonal antibody specific for the sucrase subunit (Hauri *et al.*, 1980). The purified precursor displayed both sucrase and isomaltase activities, which unequivocally proved the concept of a single precursor comprising both subunits (Semenza, 1978a). Subunit-specific monoclonal antibodies are also available for the human sucrase–isomaltase (Table II; Hauri *et al.*, 1985b). Despite a substantial homology of the sucrase and the isomaltase subunit (Hunziker *et al.*, 1986), most monoclonal antibodies against human sucrase–isomaltase appear to be subunit specific or have at least a high preference for one subunit only (H.-P. Hauri, unpublished observations). Since brush border enzymes are large molecules, it is not difficult to produce monoclonal antibodies against different epitopes (Hauri *et al.*, 1985b; Quaroni and Isselbacher, 1985) that can be used for epitope mapping. For instance, epitope mapping is a powerful approach to differentiate between very similar congenital sucrase–isomaltase defects (Naim *et al.*, in press), to assess homologies among alkaline phosphatases of different tissues (Slaughter *et al.*, 1981), or to differentiate between active and inactive kidney maltase in young and old animals (Reiss and Sacktor, 1983).

4.5. Conformation-Specific Antibodies

Using highly purified native brush border membrane vesicles as an antigen, Hauri *et al.* (1985b) obtained a monoclonal antibody that was specific for an early, endoglycosidase H-sensitive, biosynthetic form of sucrase–isomaltase when the immunoprecipitation was carried out under nondenaturing conditions. The results suggested that the antigenic site for this antibody became masked during maturation by either carbohydrate chain elongation or protein folding.

Such antibodies might be extremely useful to study early biosynthetic events related to intracellular transport (Burke *et al.*, 1983; Oude Elferink *et al.*, 1984; Yewdell *et al.*, 1983).

4.6. Immunohistochemistry

Although polyclonal antibodies are in general preferable to monoclonal antibodies in histochemical studies because they give stronger signals, the above discussed specificity problems may seriously limit the application of anti-brush-border enzyme antisera. Another potential problem that has not received full attention, in my opinion, is the fact that antibody specificity is usually established by immunoblotting, that is, conditions that are drastically different from the labeling conditions used in immunohistochemistry. Thus, it appears preferable to establish antibody specificity by immunoprecipitation in the presence of nonionic detergents. These conditions more closely resemble the labeling conditions used in immunofluoresence or in the protein A–gold labeling procedure of ultrathin cryosections.

Most monoclonal antibodies that are able to immunoprecipitate their antigen under nondenaturing conditions give a positive signal with formaldehyde-fixed intestinal cryosections or intestinal cell monolayers (Hauri *et al.*, 1985b), as well as with ultrathin cryosections (Fransen *et al.*, 1985). However, these antibodies usually fail to recognize their antigen on sections of Lowicryl K4M-embedded tissue unless they bind the denatured antigen in addition to the native form. Likewise, antibodies produced against denatured sucrase–isomaltase that were unable to recognize the native enzyme were equally unreactive with Lowicryl K4M-embedded intestine (H.-P. Hauri and J. Roth, unpublished observation). It can be concluded therefore that the most valuable monoclonal probes for immunohistochemistry of brush border (and probably other) proteins are those that recognize both the native and the denatured enzyme with high avidity. At this point it may be worth noting that monoclonal antibodies may require individual blotting conditions for optimal reactions on Western blots (Hauri and Bucher, 1986). For immunohistochemical experiments, the signal can be increased by using a mixture of monoclonal antibodies against different epitopes or by linker antibodies (Fransen *et al.*, 1985).

4.7. Screening of cDNA Expression Libraries

cDNA clones encoding human sucrase–isomaltase were successfully isolated by means of a monoclonal antibody (Green *et al.*, 1987). Monoclonal antibodies may not be the first choice for directly screening a cDNA expression library. However, monoclonal antibodies are most useful to immunoisolate the corresponding antigen that, in turn, is used to produce highly specific polyclonal antibodies, preferably against the denatured protein (Timmins *et al.*, 1985). cDNAs for rabbit sucrase–isomaltase (Hunziker *et al.*, 1986), γ-glutamyltranspeptidase (Laperche *et al.*, 1986), and alkaline phosphatase (Henthorn *et al.*, 1986; Kam *et al.*, 1985; Millan, 1986; Weiss *et al.*, 1986) have been isolated by means of polyclonal antibodies. The most useful monoclonal antibodies for directly screening cDNA expression libraries are those with high affinity for the denatured antigen.

5. EXPERIMENTAL SYSTEM

The general intracellular pathway taken by newly synthesized brush border glycoproteins has initially been established by autoradiography at the ultrastructural level in pioneering studies by Neutra and Leblond (1966) and Bennett *et al.* (1974). However, only with the advent of antibodies and powerful subcellular fractionation techniques (Weiser *et al.*, 1978) was it possible to follow individual proteins through different subcellular compartments (Hauri *et al.*, 1979). Organ culture of intestinal mucosa introduced by Browning and Trier (1969) offered the possibility of manipulating the biosynthesis in vitro (Danielsen *et al.*, 1984a). Only recently, the long-lasting dream of a cell culture model became true, by the observation of Zweibaum's group that two human colon adenocarcinoma cell lines, Caco-2 and HT-29, exhibit a number of differentiation characteristics of the normal small intestine including the expression of some brush border hydrolases (Rousset, 1986; Zweibaum *et al.*, 1986).

5.1. Intestinal Tissue

The use of intact gut mucosa for biosynthesis studies has the advantage that the epithelial organization is preserved and that the cell material is not limiting for subcellular fractionation experiments. Initial procedures using scraped mucosa as a starting material, which of course comprises a number of different cell types, were improved by Weiser (1973) who introduced an enterocyte isolation procedure for rats that is based on the selective release of crypt and villus cells by EDTA. Such cell preparations are useful for the isolation of brush border, basolateral, or Golgi membranes (Weiser *et al.*, 1978, 1986). A potential redistribution of apical and basolateral membrane proteins taking place at 37°C (Ziomek *et al.*, 1980) can be reduced by a modified isolation procedure in which the EDTA incubations are carried out on ice (Bonkovsky *et al.*, 1985; Stieger *et al.*, 1986). Alternatively, intact epithelial sheets can be isolated within a few minutes by perfusing animals with an EDTA buffer through the heart (Bjerknes and Cheng, 1981). A major disadvantage of animal studies is the difficulty to manipulate experimentally the biosynthesis and intracellular transport of proteins.

5.2. Organ Culture

Mucosal explants from human (Falchuk *et al.*, 1974; Hauri *et al.*, 1975, 1977; Trier, 1980) or animal small intestine (Berteloot *et al.*, 1979; Danielsen *et al.*, 1982c; Ferland and Hugon, 1979; Howdle, 1983; Kedinger *et al.*, 1980; Malo *et al.*, 1983) do not survive much longer than 2 days in culture. However, this time lapse is sufficient for most biosynthesis experiments. Organ culture of pig intestinal mucosa has extensively been used by Danielsen and co-workers in studies on the post-translational processing of a number of brush border enzymes (Danielsen *et al.*, 1984a), and by Naim *et al.* (1986, 1987) to investigate precursor forms of human brush border enzymes in normal mucosa and in mucosa of patients with sucrase–isomaltase deficiency (Lloyd and Olsen, 1987; Naim *et al.*, in press). While this system is very useful to identify different biosynthetic precursor forms in the homogenate and the molecular forms of their mature products in the absence of pancreatic proteases, it suffers from some drawbacks. (1) Organ-cultured mucosa is a heterogeneous system containing different cell types and enterocytes at various stages of cell differentiation. (2) Because of the crypt–villus architecture and the lack of vascularization, incomplete actions of inhibitors or modulators are difficult to interpret. Is their action incomplete in all the cells or complete in only a fraction of them? This may be especially critical for substances that enter the cells through the basolateral membrane only. (3) Subcellular fractionation techniques are hampered by the low amount of tissue that can usually be placed in organ culture. A crude brush border membrane fraction can be obtained from cultured or noncultured biopsy specimens by the Ca- or Mg-precipitation technique (Danielsen *et al.*, 1981a,b; Freiburghaus *et al.*, 1978). Such preparations, particularly when performed with small amounts of tissue, are substantially contaminated by basolateral and intracellular membranes. It is therefore difficult to draw firm conclusions on intestinal membrane traffic on the basis of such an approach.

5.3. Established Epithelial Cell Lines

A number of permanent intestinal epithelial cell lines of normal intestine, that is, the IEC-6 (Quaroni and May, 1980; Quaroni *et al.*, 1979c), the IEC-16 and IEC-17 (Quaroni and Isselbacher, 1981), the IRD-98 (Negrel *et al.*, 1983), and the RIE-1 (Blay and Brown, 1984, 1985) cell lines have been established recently. Unfortunately, these cells, either rapidly loose their differentiation characteristics in culture or express only moderate amounts of brush border enzymes (Quaroni, 1985b). To date, attempts to induce differentiation in these cells in cultures has met with limiting success. This may be due to the absence of mesenchymal cells that have been shown to promote, in conjunction with glucocorticoids, the expression of brush border hydrolases of fetal endodermal cells in primary culture (Haffen *et al.*, 1986; Kedinger *et al.*, 1986a, 1987). In fact, if IEC-6 cells of rat origin are associated with mouse mesenchymal cells and grafted under the kidney capsule of nude mice, these recombinants will also give rise to intestinal structures and to the synthesis of rat brush border enzymes as assessed by rat-specific monoclonal antibodies (Kedinger *et al.*, 1986b). This shows that at least some IEC-6 cells have retained their competence for differentiation. Thus, established crypt cell lines may be valuable to study early expression of brush border enzymes in culture once the inducing mesenchymal factors have been identified. Progress in understanding the biogenesis and the intracellular transport of brush border membrane enzymes can be expected from studies using transformed intestinal cell lines. Zweibaum and collaborators described human intestinal colon adenocarcinoma cell lines that either spontaneously differentiated into highly polarized epithelial cells, like Caco-2 (Pinto *et al.*, 1983), or whose differentiation was inducible by galactose, low glucose, sodium butyrate, or inosine, like HT-29 (Augeron and Laborisse, 1984; Fantini *et al.*, 1986; Pinto *et al.*, 1982; Wice *et al.*, 1985; Zweibaum *et al.*, 1985). In particular, Caco-2 cells (Figure 3) have been found to express a number of small-intestinal brush border enzymes at high levels, like sucrase–isomaltase, dipeptidylpeptidase IV, or alkaline phosphatase, while other enzymes are moderately expressed (i.e., lactase–phlorizin hydrolase, aminopeptidase N, γ-glutamyltranspeptidase, or trehalase) and some typical brush border enzymes are absent

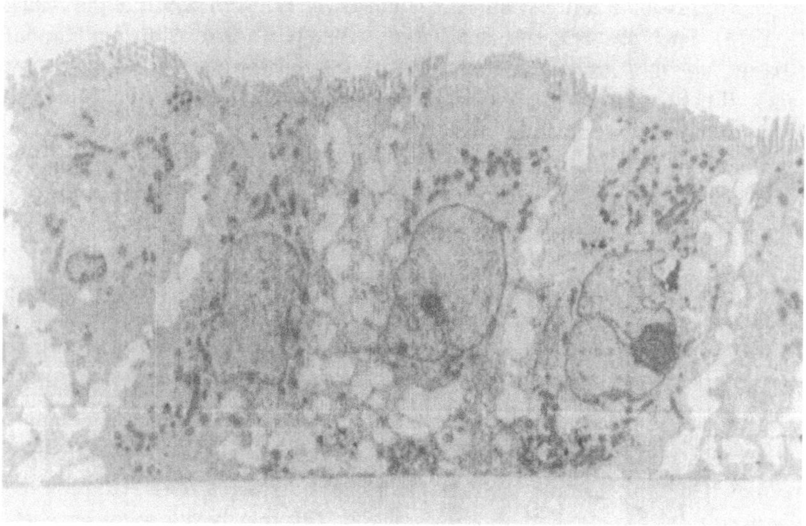

FIGURE 3. Transmission electron micrograph of a confluent Caco-2 cell monolayer cultured on a plastic dish (courtesy of Dr. A. Zweibaum).

(e.g., angiotensin I-converting enzyme and maltase–glucoamylase) (Hauri *et al.*, 1985a). Caco-2 cells grow on plastic dishes as well as on Millipore (Hauri, 1986) or Nucleopore filters (Ramond and Martinot-Peignoux, 1985). Filter-grown cells can more efficiently be labeled with ^{35}S-methionine from the basolateral surface through the filter than from the apical side since in Caco-2 as in MDCK cells (Bonsdorf *et al.*, 1985) the methionine transport system resides in the basolateral membrane (Stieger *et al.*, in press). Moreover, this culture method allows selective experimental manipulation of protein biosynthesis and transport from either the apical or the basolateral surface domain. For instance, it was found that in Caco-2 cells tunicamycin effectively inhibited N-linked glycosylation of sucrase–isomaltase only when applied from the basolateral side (H.-P. Hauri, unpublished observation). Although more difficult to achieve with cultured cells than with intact tissue, subcellular fractionation techniques can be applied to Caco-2 cells. Such an approach has, for instance, provided novel information on the kinetics of transport of newly synthesized brush border enzymes (Hauri, 1986). It is now possible to isolate, with acceptable purity, brush border membranes and Golgi membranes (Stieger *et al.*, in press), as well as lysosomes (K. Matter and H.-P. Hauri, unpublished observation) from Caco-2 cells that are minimally contaminated by membranes of other organelles.

6. BIOSYNTHESIS AND MATURATION

The synthesis and maturation of microvillar hydrolases can be accommodated within the general framework that has emerged in the past few years for a number of membrane glycoproteins (Farquhar, 1985; Kornfeld and Kornfeld, 1985; Palade, 1982; Sabatini *et al.*, 1982; Wickner and Lodish, 1985). Brush border enzymes are synthesized and core-glycosylated in the rough endoplasmic reticulum, undergo complex glycosylation in the Golgi apparatus, and are subsequently transported to the brush border membrane. An interesting aspect of the intestinal epithelial cell relates to the possibility that a number of different major glycoproteins can be studied and compared within the same cell. Biosynthetic and post-translational events are now discussed for the individual enzymes with the exception of trehalase (Galand, 1984; Nakano, 1982; Nakano and Sacktor, 1984, 1985; Nakano *et al.*, 1977; Yokota *et al.*, 1986), whose biosynthesis is unknown.

6.1. Disaccharidases

6.1.1. Sucrase–Isomaltase

This protein represents the most thoroughly studied brush border enzyme (Brunner *et al.*, 1983; Conklin *et al.*, 1975; Gray *et al.*, 1979; Hauri, 1983;

Hauser and Semenza, 1983; Ribi and Kretchmer, 1984; Rodriguez *et al.*, 1984; Semenza, 1978b, 1981a, 1985, 1986; Semenza *et al.*, 1983). Its mode of biosynthesis may be largely representative for that of the other major disaccharidases, that is, lactase–phlorizin hydrolase and maltase–glucoamylase. The observation that in the adult intestine sucrase–isomaltase is anchored to the membrane bilayer by only one of its subunits (Brunner *et al.*, 1979; Frank *et al.*, 1978; Spiess *et al.*, 1982) in conjunction with the observations that sucrase and isomaltase activities are virtually always concomitantly expressed during gut development or highly reduced in hereditary sucrase–isomaltase deficiency has led to the hypothesis that the two subunits may be proteolytic products of a single-chain precursor molecule (Semenza, 1978a). First experimental evidence for this notion has come from labeling studies in rats. It was found that short pulses with ^{3}H-fucose labeled a single high molecular weight precursor protein that was immunoprecipitable with polyclonal antibodies against sucrase–isomaltase from isolated Golgi membranes (Hauri *et al.*, 1979). In the brush border membrane this precursor protein was also found at early time points but it rapidly disappeared with the concomitant appearance of radiolabeled enzyme subunits. Since the precursor in isolated Golgi membranes was cleavable into subunitlike proteins by elastase or trypsin, it was concluded that sucrase–isomaltase was synthesized as a high molecular weight precursor protein (also termed pro-sucrase–isomaltase), comprising both the isomaltase and the sucrase subunits, which is transferred to the brush border membrane where it is immediately cleaved into its subunits by pancreatic proteases (Hauri *et al.*, 1980). This precursor–product relation was confirmed by the structural analysis of sucrase–isomaltase synthesized in the absence of pancreatic proteases, that is, in fetal intestinal transplants of rats (Hauri *et al.*, 1980, 1982; Montgomery *et al.*, 1981), in pancreatic duct-ligated pigs (Sjöström *et al.*, 1980, 1982), in pig (Danielsen, 1982; Danielsen *et al.*, 1981a,b) and human (Naim *et al.*, 1986, 1987) intestinal organ culture, in human colon cancers (Zweibaum *et al.*, 1983, 1984), in the human adenocarcinoma cell lines HT-29 (Zweibaum *et al.*, 1983, 1986) and Caco-2 (Hauri *et al.*, 1985a), and in cell-free translation (Alpers *et al.*, 1986; Wacker *et al.*, 1981). In all these cases, sucrase–isomaltase was exclusively found in the precursor form. Furthermore, in cystic fibrosis, an autosomal recessive disease, affecting the release of pancreatic zymogens, sucrase–isomaltase is less efficiently cleaved when compared to normal subjects (Sips *et al.*, 1985). All these observations do not confirm an early suggestion of Cézard *et al.* (1979) that free isomaltase and sucrase subunits may serve as intracellular precursors for the brush border sucrase–isomaltase complex.

The sequence of biogenetic events is as follows (Figure 4): Sucrase–isomaltase is synthesized as a M_r 200,000 protein with a noncleavable signal sequence near the N terminus of the isomaltase subunit that anchors the precursor protein to the lipid bilayer of the endoplasmic reticulum. This was convincingly demonstrated by in vitro translation experiments (Ghersa *et al.*, 1986; Wacker *et al.*, 1981) and by cDNA cloning (Hunziker *et al.*, 1986). The enzyme is (probably cotranslationally) core-glycosylated (Danielsen, 1982; Hauri *et al.*, 1985b). The exact number of asparagine-linked oligosaccharide side chains is unknown but a total of 19 putative N-glycosylation sites have been identified by cDNA cloning (Hunziker *et al.*, 1986). The high-mannose sucrase–isomaltase does not seem to be completely enzymatically active (Sjöström *et al.*, 1985). The trimming of glucose (and probably some mannose) residues is a rapid event since nontrimmed sucrase–isomaltase cannot be detected by SDS–PAGE even after short pulses unless glucose trimming is inhibited by castanospermine (Danielsen and Cowell, 1986) or methyl-1-deoxynojirimycin (K. Matter and H.-P. Hauri, unpublished observations) in intestinal organ or cell culture. Sucrase–isomaltase acquires its complex oligosaccharides in the Golgi apparatus as shown by subcellular fractionation (Hauri *et al.*, 1979). Besides N-linked oligosaccharides of the complex type, sucrase–isomaltase also carries O-linked sugars as shown by carbohydrate analysis (Herskovics *et al.*, 1981) or sensitivity to *O*-glycanase (K.

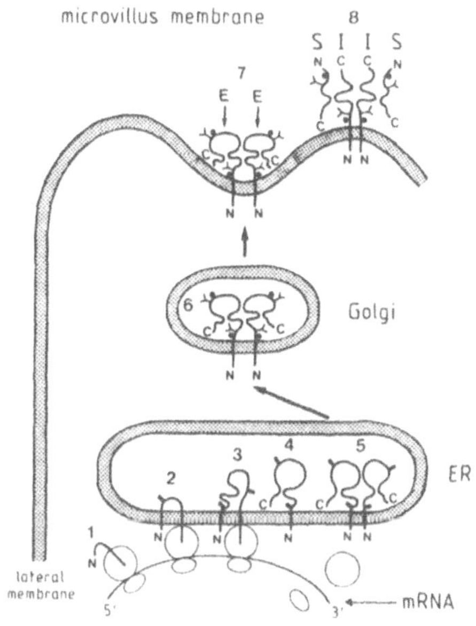

FIGURE 4. Current diagram of sequential events involved in the biogenesis of sucrase–isomaltase in vivo. Sucrase–isomaltase is synthesized starting with the N terminus of the isomaltase domain (1). Translocation through the endoplasmic reticulum membrane is mediated by a short hydrophobic sequence near the N terminus that remains uncleaved and serves as a membrane anchor (2). The rest of the isomaltase domain protrudes through the membrane (3), followed by the sucrase domain, which carries no membrane anchor (4). Concomitant with or immediately after its synthesis, the enzyme acquires N linked high-mannose oligosaccharides (❙) that are trimmed by glucosidases and mannosidases (not indicated). The enzyme dimerizes possibly in the endoplasmic reticulum (5). On its transport through the Golgi apparatus (6), further trimming of the N-linked side chains takes place, so that terminal sugars can be added (Y). Furthermore, O-linked carbohydrates are added to threonine and serine (●). Upon arrival at the brush border membrane (7), the hydrolase is cleaved into its subunits isomaltase and sucrase (8) by pancreatic enzymes (E). Sucrase remains attached to isomaltase by noncovalent interaction. Note that in the absence of pancreatic proteases, sucrase–isomaltase remains uncleaved (see text).

Matter and H.-P. Hauri, unpublished observations). A number of studies have demonstrated that human and some animal sucrase–isomaltase carries blood group antigens (Bernadac *et al.*, 1984; Kelly and Alpers, 1973b; Triadou *et al.*, 1983).

Prior to complex glycosylation, sucrase–isomaltase undergoes a conformation change as revealed by monoclonal antibody HBB 3/705/60 (Hauri *et al.*, 1985b). This antibody recognizes an epitope in the isomaltase domain of the high-mannose form immediately after its synthesis but can no longer react with either the high-mannose or complex form at later times of chase unless the protein is denatured prior to immunoprecipitation. It is possible that this antibody defines an epitope that is changed during folding or dimerization of the protein. The epitope appears to be strongly antigenic since another three independently produced monoclonal antibodies were obtained that competed with the binding of HBB3/705/60 (H.-P. Hauri, unpublished observations). No interchain disulfide bonds or fatty acid acylation were found in sucrase–isomaltase.

As discussed above, sucrase–isomaltase in vivo is cleaved into its subunits sucrase and isomaltase probably in the brush border. Although there is no doubt that sucrase–isomaltase remains uncleaved in the absence of pancreatic proteases, the precise site at which subunit cleavage occurs has not directly been visualized. Assuming that the endocytotic and the exocytotic pathways cross each other intracellularly, an issue that is unsettled at present (Farquhar, 1985; Griffiths and Simons, 1986; Marsh *et al.*, 1986; Snider and Rogers, 1986; Woods *et al.*, 1986), it would also be conceivable, although less likely, that sucrase–isomaltase is cleaved prior to its insertion into the brush border membrane by endocytosed pancreatic proteases. It is certain, however, that sucrase–isomaltase is cleaved at a later time-point than lactase. For the latter enzyme cleavage occurs immediately after its complex glycosylation at an intracellular site (see Section 6.1.3). It is not known if the splitting of the sucrase–isomaltase precursor is accompanied by the removal of a short peptide sequence at the cleavage site. Several studies have shown that the cleavage event does not modify enzyme activities (Montgomery *et al.*, 1981; Sjöström *et al.*, 1980).

6.1.2. Maltase–Glucoamylase

This enzyme complex has been purified from human (Kelly and Alpers, 1973a), pig (Sorensen *et al.*, 1982; Taravel *et al.*, 1983), rat (Flanagan and Forstner, 1978; Lee and Forstner, 1984; Lee *et al.*, 1980; Reiss and Sacktor, 1981; Schlegel-Haueter *et al.*, 1972), and rabbit (Sankaran *et al.*, 1983). Analogous to sucrase–isomaltase, the maltase–glucoamylase complex has two active sites that can be differentiated in heat-inactivation studies (Norén *et al.*, 1986a; Sorensen *et al.*, 1982). Organ culture studies using the pulse-chase technique have established that pig maltase–glucoamylase is synthesized as a single large precursor protein (Danielsen *et al.*, 1981b; Sjöström *et al.*, 1983) that in vivo is cleaved into two subunits in a way analogous to sucrase–isomaltase. The M_r 225,000 high-mannose form is slowly converted to the M_r 245,000 complex form (Danielsen *et al.*, 1983a) that persists in the intestine of pancreatic duct-ligated animals (Sorensen *et al.*, 1982). In normal pigs the complex glycosylated form is cleaved, at least in part, into two subunits (M_r 135,000 and M_r 125,000). It is not known if each subunit carries one of the two active sites. In the human intestinal mucosa, immunoprecipitated mature maltase–glucoamylase is uncleaved and has a M_r of 355,000 (Hauri *et al.*, 1985b), thus reflecting the higher resistance to proteolytic attack than the pig enzyme. Not compatible with the model of a single-chain maltase–glucoamylase with one anchor segment cleaved post-translationally is a recent report by Alpers *et al.* (1986). On the basis of in vitro translation experiments, these authors postulate that rat maltase–glucoamylase is synthesized as a M_r 145,000 subunit protein and that the other subunit (M_r 130,000) is generated by slight proteolytic modification. An alternative view is expressed by Lee and Forster (1985). Their cyanogen bromide cleavage results suggest that the two central segments of the 145,000 subunit and the 130,000 subunit of rat maltase–glucoamylase are homologous but not identical and therefore that one subunit may not be a post-translational derivative of the other. Further experiments are required to solve this controversial issue. It appears rather unlikely that maltase–glucoamylase is synthesized differently in the rat than in the pig. Moreover, maltase of the rat kidney, which is immunologically identical to the intestinal enzyme, has a large size (M_r 335,000), consistent with the lack of luminal proteases in that tissue (Reiss and Sacktor, 1981).

6.1.3. Lactase–Phlorizin Hydrolase

This third major disaccharidase has obtained considerable attention owing to the worldwide problem of lactase intolerance (Freiburghaus *et al.*, 1976; Garfield, 1980; Kretchmer, 1972; Skovbjerg *et al.*, 1980). Lactase–phlorizin hydrolase is found in the small intestine of almost all mammals during the perinatal stage of development. On weaning, the enzyme activity decreases to low levels in all adult mammals except certain human populations, for example, Caucasians and some African tribes in which lactase levels remain high (Simoons, 1978). The molecular basis of the lactase–polymorphism is unknown but some first steps have been made to understand the biogenesis of this enzyme complex in humans and in the pig. Lactase–phlorizin hydrolase has been purified from rat and human (Alpers, 1969; Birkenmeier and Alpers, 1974; Colombo *et al.*, 1987; Cousineau and Green, 1980; Schlegel-Haueter *et al.*, 1972; Skovbjerg *et al.*, 1981, 1982; see also Chapter 4 of this volume). The human enzyme has an M_r of 160,000 on SDS–polyacrylamide gels and there is evidence for two different active sites, one for lactose hydrolysis that is inhibitable by phlorizin and one for phlorizin hydrolysis. It is assumed, but not formally proved, that the two enzymatic sites reside in two different subunits that cannot be separated by SDS–PAGE. By gel filtration, a M_r of 320,000 was determined, suggesting that the native enzyme is composed of two very similar subunits. It was initially proposed that the subunits of lactase–phlorizin hydrolase may be synthesized individually (Sjöström *et al.*, 1983), but this view was soon corrected by subsequent studies (Skovbjerg *et al.*, 1984). Organ culture experiments with pig intestinal mucosa (Danielsen *et al.*, 1984b), as well as cell culture studies with Caco-2 cells (Hauri *et al.*, 1985b), have now clearly demonstrated that this enzyme complex is synthesized as a large M_r 210,000–225,000 high-mannose precursor protein that is gradually converted to the M_r 245,000 complex form and subsequent cleaved into the M_r 160,000 subunits. Interestingly, the proteolytic modification of the precursor occurs even in the absence of pancreatic proteases and is partially inhibited by leupeptin (Danielsen *et al.*, 1984b; Naim *et al.*, 1987). In organ-cultured human tissue, cleavage of lactase–phlorizin hydrolase may immediately precede complex glycosylation since neither the complex-glycosylated precursor form nor any immature (i.e., endoglycosidase-H-sensitive) subunits could be visualized (Naim *et al.*, 1987). Overall, the results suggest that the cleavage of lactase–phlorizin hydrolase is an intracellular event that may take place in the Golgi apparatus or at a late biosynthetic stage as in the case of the conversion of proinsulin to insulin (Orci *et al.*, 1984a,b, 1985) and other prohormones. Although carrying complex N-linked and probably also O-linked oligosaccharides (Naim *et al.*, 1987, and unpublished observations), mature lactase–phlorizin hydrolase still binds Con A (Potter *et al.*, 1985), suggesting the presence of oligomannose-type N-glycans. It is presently unknown if only one subunit of lactase–phlorizin hydrolase is anchored in the brush border as with sucrase–isomaltase or if both subunits have an anchoring segment as suggested by Potter *et al.* (1985).

6.2. Peptidases

As discussed in the previous section, these hydrolases, most of which are common to the kidney and the intestine (Booth *et al.*, 1979; Bruneval *et al.*, 1986; Galand, 1984; Gee *et al.*, 1983; Kenny and Maroux, 1982; Lojda and Gossrau, 1980; Marathe *et al.*, 1979; Norén *et al.*, 1979; Sacktor, 1977; Vannier *et al.*, 1976) form homodimers in the brush border membrane; that is, they are composed of identical subunits (Figure 1). The subunits, with some exceptions, are not proteolytically modified once they arrive at the brush border membrane. The biogenesis of a typical brush border peptidase can be described as follows. It is synthesized as a transmembrane protein with a noncleavable signal sequence near the N terminus, which in the mature enzyme serves as a membrane anchor. The endoglycosidase H-sensitive high-mannose form of the hydrolase is already fully enzymatically active (Sjöström *et al.*, 1985). Following its synthesis, it is rapidly transported to the Golgi apparatus (with a half-time of 15–30 min) where complex-type oligosaccharides (N- and O-linked) are attached immediately. No other post-translational modifications like fatty acid acylation or phosphorylation take place and the enzyme is transferred from the Golgi apparatus to the cell surface with a half-time of about 30–60 min. This general scheme of biosynthesis has been deduced from a number of the different peptidases of intestine and kidney. However, it is important to note that there are some variations to the common scheme and that knowledge on the different steps of biosynthesis is incomplete for a number of peptidases. Since these variations may serve as landmarks for the intracellular pathway or may put in proper perspective the presumed necessity of a post-translational modification for the intracellular transport, the enzymes are now discussed individually.

6.2.1. Dipeptidylpeptidase IV

In pig intestinal or kidney organ culture, dipeptidylpeptidase IV (Bella *et al.*, 1982; Caporale *et al.*, 1985; Erickson *et al.*, 1983; Gonschor and Scheafer, 1985; Kenny *et al.*, 1976; MacNair and Kenny, 1979; Sahara *et al.*, 1983; Svensson *et al.*, 1978) is synthesized as a single-chain high-mannose precursor form (Danielsen *et al.*, 1983d; Stewart and Kenny, 1984a), while in Caco-2 cultures two high-mannose forms can be resolved after a short pulse with ^{35}S-methionine (Hauri *et al.*, 1985b, see Table I). At present it is not certain if this precursor doublet represents two different isozymes or a proteolytic event that cannot be inhibited by commonly used protease inhibitors. It appears unlikely that the doublet reflects different stages of trimming of the N-linked oligosaccharides since no precursor–product relation was found. Fully glycosylated dipeptidylpeptidase IV appears as a single broad band on SDS–PAGE (Hauri *et al.*, 1985b; Svensson *et al.*, 1978) and the purified kidney enzyme displays five isoforms on isoelectric focusing gels (Gonschor and Schaefer, 1985) that may be due to sugar microheterogeneity. In vivo, the glycosylation of dipeptidylpeptidase IV from proximal or distal intestine is qualitatively different (Morita *et al.*, 1986).

6.2.2. Aminopeptidase N

This is the most thoroughly studied intestinal peptidase (Benajiba and Maroux, 1981; Desnuelle, 1979; Feracci and Maroux, 1980; Feracci *et al.*, 1981; Kim and Bropyh, 1976; Louvard *et al.*, 1976; Maroux and Feracci, 1983; Maroux *et al.*, 1973; Sjöström *et al.*, 1978; Svensson *et al.*, 1982). Cell-free synthesis of aminopeptidase N in the absence of pancreatic microsomes resulted in a primary translation product of M_r 115,000 that comigrated on gels with the (presumably) nonglycosylated metabolically labeled aminopeptidase N in the presence of tunicamycin (Danielsen *et al.*, 1982a, 1983c). The absence of a detectable M_r difference between the two forms provides evidence for a non-cleavable signal peptide in this enzyme. Furthermore, synchronized translation experiments with timed addition of pancreatic microsomes suggested that the signal in aminopeptidase N is located in the N-terminal part of the polypeptide. Likewise, a stretch of hydrophobic amino acids starting with position 5 from the N terminus was found that presumably is part of the membrane anchor (Feracci *et al.*, 1982b; Maroux *et al.*, 1983). Evidence for a transmembrane disposition of aminopeptidase N was first provided by macromolecular photolabeling (Louvard *et al.*, 1976). Post-translational modifications of aminopeptidase N include complex glycosylation of the N- and O-types but no acylation (Danielsen *et al.*, 1983a). Massey and Maroux (1985) have provided evidence for eight N-linked complex glycans and two O-linked side chains in rabbit aminopeptidase N. There is preliminary evidence that a proteolytic event may cleave a cytoplasmic sequence of 3 kD during maturation (Feracci *et al.*, 1985). Furthermore, Ahnen *et al.* (1982, 1983) have proposed that rat aminopeptidase N may undergo a previously unrecognized maturation step in the endoplasmic reticulum that leads to a larger, more mature species prior to transport to the Golgi apparatus. These latter two findings may be important in regard to understanding the intracellular transport mechanism. Pig, but not human, rabbit, or rat intestinal aminopeptidase N is proteolytically modified in the brush border membrane by trypsin (Banajiba and Maroux, 1981; Norén and Sjöström, 1980; Sjöström and Norén, 1982; Svensson *et al.*, 1982).

6.2.3. Aminopeptidase A, Endopeptidase-24.11, and Angiotensin I-Converting Enzyme (Peptidyl Dipeptidase A)

Information on the biosynthesis of aminopeptidase A (Benajiba and Maroux, 1980; Danielsen *et al.*, 1980a; Feracci *et al.*, 1981) is scarce. However, the available data suggest that its biosynthesis and maturation closely resemble that of aminopeptidase N (Danielsen *et al.*, 1983d; Stewart and Kenny, 1984a). Likewise, the biogenesis and maturation of endopeptidase 24.11 (Stewart and Kenny, 1984a), a short-stalked hydrolase (Danielsen *et al.*, 1980b; Fulcher and Kenny, 1983; Kenny, 1986; Kenny and Fulcher, 1983; Kenny *et al.*, 1985), and of angiotensin I-converting enzyme (Naim *et al.*, 1986), the largest peptidase of the microvillar membrane (Aoyagi *et al.*, 1985, Harris and Wilson, 1985; Hauri *et al.*, 1985b; Weare *et al.*, 1982) are similar to aminopeptidase N. For endopeptidase 24.11, molecular weight differences between the kidney and the intestinal enzyme have conclusively been attributed to differences in glycosylation (Fulcher *et al.*, 1983; Stewart *et al.*, 1984).

6.2.4. PABA-Peptide Hydrolase

The biogenesis of this recently detected intestinal peptidase, whose enzymatic activity is measured with *N*-benzoyl-L-tyrosyl-*p*-aminobenzoic acid as a substrate (PABA; Sterchi *et al.*, 1982), is somewhat exceptional since its dimerization involves the formation of disulfide bonds (E. E. Sterchi, personal communication). Dimerization precedes complex glycosylation and may therefore be an endoplasmic reticulum or *cis*-Golgi event. Meprin, another intestinal brush border peptidase (Bond and Beynon, 1986) is not related to PABA–peptide hydrolase.

6.2.5. γ-Glutamyltranspeptidase

Most information on γ-glutamyltranspeptidase originates from the kidney enzyme (Cone *et al.*, 1986; Gardell and Tate, 1981; Hughey and Curthoys, 1976; Hughey *et al.*, 1986; Matsuda *et al.*, 1983; Tate and Meister, 1976, 1977; Tate and Ross, 1977), but it is not unreasonable to assume that the intestinal enzyme is synthesized and processed in a similar manner (Norén *et al.*, 1979). γ-Glutamyltranspeptidase is the only brush border peptidase that undergoes a proteolytic cleavage even in the absence of pancreatic proteases (e.g., in the kidney), leading to a larger membrane-bound subunit and a smaller catalytic subunit. The smaller subunit does not interact directly with the lipid bilayer via noncovalent interactions (Hughey and Curthoys, 1976). Labeling experiments with ^{35}S-methionine in organ culture have provided the first evidence that γ-glutamyltranspeptidase is synthesized as a single precursor protein (Nash and Tate, 1982). Subsequently, this notion was confirmed by in vitro translation studies (Barouki *et al.*, 1984; Finidori *et al.*, 1984; Nash and Tate, 1984) and by the analysis of the amino acid sequence deduced from cDNA (Laperche *et al.*, 1986). These latter studies have also identified a single 18 residue-long hydrophobic sequence near the N terminus that is likely to serve both as an insertion signal and a membrane anchor (see Figure 2). A total of five putative N-glysolylation sites have been found. N-glycosidic side chains are different in the two subunits and carbohydrate heterogeneity is considered as the basis for the different molecular forms of the enzyme (Cone *et al.*, 1986; Matsuda *et al.*, 1980, 1983; Yamashita *et al.*, 1983a,b). Contrary to sucrase–isomaltase, potential O-glycosylation sites are not concentrated in the stalk of the heavy subunit but in the light subunit (Laperche *et al.*, 1986). However, Matsuda *et al.* (1983) have not found any galactosamine and concluded therefore that rat kidney γ-glutamyltranspeptidase is not O-glycosylated.

Controversy exists as to the subcellular site of subunit cleavage (Barouki *et al.*, 1984; Caprano and Hughey, 1983; Kuno *et al.*, 1983; Nash and Tate, 1982,

1984). The study of Barouki *et al.* (1984) on the biogenesis of γ-glutamyltranspeptidase in the rat hepatoma tissue culture cells suggests that part of the enzyme precursor is already cleaved prior to its complex glycosylation, while no uncleaved precursor was found at the cell surface. In contrast, subcellular fractionation data with rat kidney suggest predominant cleavage of kidney γ-glutamyltranspeptidase in the brush border (Kuno *et al.*, 1983). It is difficult to compare these two studies because of the differences in techniques and tissue. However, it should be noted that in the study of Barouki *et al.* (1984) the cells were solubilized at 37°C. Since the specificity of the cleaving protease is not known, the possibility must be considered that solubilization at elevated temperatures might lead to proteolytic artifacts even in the presence of protease inhibitors. A solution to the controversy might come from a recent study by Tate (1986). This author showed that a small fraction of the propeptidase escapes cleavage and persists in the kidney but has low enzyme activity. It would appear therefore that cleavage into subunits is accompanied by enzymatic activation. Furthermore, the persistence of the propeptidase in the brush border membrane indicates that cleavage is not essential for the normal intracellular transport mechanism of the enzyme.

6.3. Alkaline Phosphatase

Alkaline phosphatases exist as membrane-bound glycoproteins in eukaryotes. At least three gene loci (intestinal, placental, and liver/bone/kidney) are distinguished in humans by biochemical, immunological, and structural methods (Weiss *et al.*, 1986). Alkaline phosphatase from rat has been shown to exist in a membrane and a soluble form, the latter also being secreted into the blood stream (Yedlin *et al.*, 1981; Young *et al.*, 1981). The origin of the secreted form is not clear but its expression is believed to be independently controlled as assessed by ^{3}H-leucine incorporation and dietary manipulation (Young *et al.*, 1981). Indeed, recent in vitro translation experiments suggest the presence of two mRNA species coding for a 62 and a 65 kD protein (Sussman *et al.*, 1986). The smaller protein was tentatively equated with the secreted enzyme, and the larger one with the membrane enzyme.

Contrary to all the other brush border enzymes described above, alkaline phosphatase may be anchored to the lipid bilayer by means of its C-terminal rather than by its N-terminal portion, as indicated by the identity of the N-terminal dipeptide Phe–Ile, of the detergent- and the papain-solubilized enzymes (Colbeau and Maroux, 1978). Papain treatment solubilizes the enzyme and reduces the M_r of the porcine intestinal enzyme by 3000 daltons, which has been taken as evidence for the existence of a hydrophobic region in the intact enzyme. Controversy exists as to the precise mechanism of membrane anchoring of alkaline phosphatase. cDNA cloning of several alkaline phosphatase isozymes

(Henthorn *et al.*, 1986; Kam *et al.*, 1985; Millan, 1986; Weiss *et al.*, 1986) has identified a sequence of hydrophobic residues near the C terminus that could participate in membrane anchoring. However, there is evidence to suggest that kidney and intestinal alkaline phosphatase from rabbit (Takesue *et al.*, 1986) as well as the kidney enzyme from dog (Low and Zilversmit, 1980; Low *et al.*, 1986) are attached to the brush border membrane by covalent linkage with phosphatidylinositol, while no such role for phosphatidylinositol was found for the rat intestinal alkaline phosphatase (Seetharam *et al.*, 1985). Overall, these data may be accommodated as follows. Alkaline phosphatase may initially be attached to the bilayer by means of a C-terminal hydrophobic region that in most but not all instances is cleaved from the newly synthesized protein, leaving covalent binding with phosphatidylinositol as the only protein–membrane interaction. Such proteolytic events have been described for Thy-1 (Tse *et al.*, 1985) and the trypanosome variable surface glycoprotein (Ferguson *et al.*, 1985).

Intestinal (Sussman *et al.*, 1986), placental (Kam *et al.*, 1985; Millan, 1986), and liver/bone/kidney-type alkaline phosphatase (Weiss *et al.*, 1986) are synthesized with a cleavable signal sequence (Ito and Chou, 1984). Knowledge on the glycosylation of intestinal alkaline phosphatase is scant (Mulivor *et al.*, 1978). Placental alkaline phosphatase has two N-glycosylation sites, while liver/bone/kidney-type alkaline phosphatase has five potential N-glycosylation sites.

7. INTRACELLULAR TRANSPORT AND SORTING

The highly polarized cell surface of epithelial cells cannot be maintained without transport mechanisms that guarantee a selective targeting of surface components to the individual membrane domains. Current knowledge of these processes is largely based on studies with enveloped viruses that mature in a polarized manner from either the apical (e.g., influenza virus) or the basolateral membrane (e.g., vesicular stomatitis virus or Semliki forest virus) in infected Madine–Darby canine kidney cells (Alonso and Compans, 1981; Garoff, 1985; Gething, 1985; Griffiths and Simons, 1986; Rodriguez-Boulan, 1983, 1986; Rodriguez-Boulan *et al.*, 1985; Sabatini *et al.*, 1983; Salas *et al.*, 1985, Simons and Fuller, 1985; Simons and Warren, 1984; Willingham and Pastan, 1984) and can be summarized as follows. The vectorial transport of viral glycoproteins from the endoplasmic reticulum to the cell surface is a fairly rapid event taking 30–45 min. Protein folding and oligomerization appears to be important for protein exit from the endoplasmic reticulum (Copeland *et al.*, 1986; Gething *et al.*, 1986; Kreis and Lodish, 1986). Passage through the Golgi apparatus occurs in a cis to trans direction (Bergeron *et al.*, 1982; Bergmann and Singer, 1982;

Bergmann *et al.*, 1981; Dunphy and Rothman, 1983, 1985; Green *et al.*, 1981; Saraste and Kuismanen, 1984). Proteins destined for the apical and basolateral domains traverse the same Golgi apparatus (Fuller *et al.*, 1985; Rindler *et al.*, 1984) but their sorting out occurs before they reach the cell surface (Matlin and Simons, 1984; Misek *et al.*, 1984; Pfeiffer *et al.*, 1985; Rindler *et al.*, 1985) probably at the transside of the Golgi apparatus. It is unlikely that initial sorting of these proteins occurs at the plasma membrane although corrective rerouting by transcytosis can be detected if a protein is artificially inserted from the outside into the wrong plasma membrane domain (Matlin *et al.*, 1983; Pesonen and Simons, 1983; Pesonen *et al.*, 1984a,b). The signals mediating protein transport and sorting have not yet been elucidated (Alonso and Compans, 1981; Garoff, 1985). Studies with chimeric viral glycoprotein suggested that the information for apical or basolateral targeting is contained within the extracytoplasmic domains rather than within the transmembrane and/or cytoplasmic domains, while the signal for endocytosis of vesicular stomatitis virus G protein was associated with the transmembrane and/or cytoplasmic domains (Compton *et al.*, 1986; M. G. Roth *et al.*, 1986). To what extent do these findings apply to endogenous membrane glycoproteins like the brush border enzymes?

7.1. Intracellular Pathway of Brush Border Hydrolases

At an ultrastructural level, alkaline phosphatase activity in the duodenum of the mouse was localized in the endoplasmic reticulum, in all cisternae of the Golgi apparatus, in small 500 Å vesicles and 0.2–0.5 μm multivesicular bodies (endosomes) below the terminal web, in dense bodies (lysosomes), and in the brush border membranes (Hugon and Borgers, 1966). Most notably, the basolateral membrane was devoid of labeling. Similarly, sucrase–isomaltase of the human jejunum (Fransen *et al.*, 1985) and aminopeptidase N of pig ileum (Danielsen *et al.*, 1986a) were localized by immunoelectron microscopy to the Golgi apparatus, to apical vesicles, to the brush border (Gitzelmann *et al.*, 1970), and, in the case of sucrase–isomaltase, also to lysosomes, while the basolateral membrane was negative for both enzymes. These labeling patterns are in line with a direct delivery pathway of newly synthesized hydrolases from the endoplasmic reticulum through the Golgi apparatus to the brush border. However, biochemical studies have pointed to the possibility that the basolateral membrane may be involved in the routing of newly synthesized microvillar hydrolases to the brush border, since ^{3}H-fucosylated sucrase–isomaltase was transiently associated with the basolateral membrane fraction (Hauri *et al.*, 1979). Maximal labeling of sucrase–isomaltase in the basolateral membrane fraction occurred at a later time point than that in the Golgi fraction, suggesting that cross-contamination with Golgi membranes cannot fully explain the finding.

Cross-contamination with other membranes, however, have to be considered, for example, cofractionation with apical vesicles. Indeed, this presently seems to be a likely reason for the apparent association of sucrase–isomaltase with the basolateral membrane.

In a recent study on the subcellular localization of aminopeptidase N, the amount of this enzyme in the basolateral fraction was reported to be about 4.5% of that in brush border membrane vesicles (Moktari *et al.*, 1986). However, the same authors were unable to detect immunoreactive enzyme in the basolateral membrane by immunofluorescence (Feracci *et al.*, 1982a), and therefore the association of aminopeptidase N with the basolateral membrane remains uncertain. On the other hand, Ahnen *et al.* (1982) have claimed that in their study any incorporation of aminopeptidase N into the basolateral membrane fraction could be accounted for by contamination with endoplasmic reticulum and Golgi membranes. Moreover, Danielsen and Cowell (1985a) proposed that the basolateral membrane is unlikely to be involved in the post-Golgi transport of newly synthesized aminopeptidase N, since both the mature and the high-mannose forms of the enzyme in the Mg^{2+}-precipitated fraction (containing intracellular and basolateral membranes) were largely protected from proteolytic cleavage in the absence of detergent. However, in this study neither the cross-contamination of the microvillar fraction by basolateral membranes nor the sidedness of the basolateral membrane vesicles was assessed, making their findings inconclusive. Overall, the available data suggest a direct delivery pathway of microvillar hydrolases to the brush border, bypassing the basolateral membrane. It has not yet been possible to accumulate reversibly a brush border hydrolase in the Golgi compartment experimentally, which would permit a direct visualization of its delivery to the brush border as in the case of virus glycoproteins (Griffiths *et al.*, 1985; Matlin *et al.*, 1983; Rindler *et al.*, 1985; Ridriguez-Boulan *et al.*, 1984).

An interesting finding is the association of brush border hydrolases with lysosomes (Fransen *et al.*, 1985; Hauri *et al.*, 1985a; Hugon and Borgers, 1966). The most trivial interpretation that this localization may be due to endocytosis is challenged by the finding that (in the case of sucrase–isomaltase) the multivesicular bodies (endosomes) that represent an intermediate compartment of the endocytotic pathway (Blok *et al.*, 1981b, 1982) consistently show no labeling. Furthermore, when living monolayers of differentiated HT-29 cells were labeled with monoclonal antibodies against sucrase–isomaltase, no endocytosis of the enzyme–antibody complexes could be detected by immunoelectron microscopy, while horseradish peroxidase used as a fluid phase marker was swiftly internalized (J. Fransen, personal communication). It is therefore possible that part of newly synthesized brush border hydrolases never reach the cell surface and are directly delivered from the Golgi apparatus to the lysosomes and degraded or recycled at the Golgi apparatus as suggested by Ginsel and colleagues (Blok *et al.*, 1984). These authors speculated that this so-called crinophagic pathway may play a regulatory role for brush border enzyme expression at a post-translational level. Studies on the turnover of sucrase–isomaltase in rats indicate that the principal means of regulating the enzyme may be through changes in degradation (Kaufman *et al.*, 1980). A flow diagram depicting the routes for intracellular traffic of brush border hydrolases is given in Figure 5.

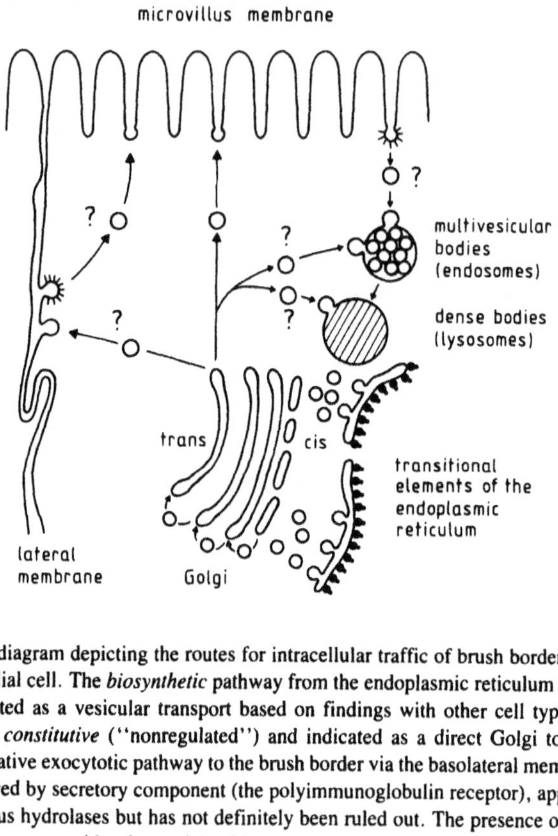

FIGURE 5. Flow diagram depicting the routes for intracellular traffic of brush border hydrolases in an intestinal epithelial cell. The *biosynthetic* pathway from the endoplasmic reticulum to and through the Golgi is indicated as a vesicular transport based on findings with other cell types. The major *exocytotic* route is *constitutive* ("nonregulated") and indicated as a direct Golgi to brush border pathway. An alternative exocytotic pathway to the brush border via the basolateral membrane, that is, for instance, followed by secretory component (the polyimmunoglobulin receptor), appears now less likely for microvillus hydrolases but has not definitely been ruled out. The presence of brush border enzymes in lysosomes may either be explained by a *crinophagic* (degradative or regulatory?) pathway in which a small fraction of newly synthesized enzyme is directly delivered to lysosomes from the Golgi apparatus or via multivesicular bodies. Alternatively, lysosomal localization of brush border hydrolases may be due to *endocytosis* from the brush border membrane. Whether all the brush border hydrolases are transported via the same vesicles is not yet known. Note that the relative distances between the Golgi apparatus and the two plasma membrane domains are not drawn to scale. In enterocytes in vivo, the Golgi apparatus is positioned in a supranuclear region close to the lateral membrane and considerably remote from the microvillar cell surface.

7.2. Kinetics of Intracellular Transport

Recent studies using intestinal organ and cell cultures suggest that transport to the cell surface is not synchronous for all brush border hydrolases. Transport of peptidases including dipeptidylpeptidase IV, aminopeptidase N, angiotensin I-converting enzyme (Danielsen and Cowell, 1985a,b; Naim *et al.*, 1986, and unpublished observation), and PABA–peptide hydrolase (E. E. Sterchi *et al.*, unpublished observation) is rapid, while that of disaccharidases including sucrase–isomaltase, lactase–phlorizin hydrolase, and maltase–glucoamylase is considerably slower. A report by Ahnen *et al.* (1983) on the biogenesis of rat aminopeptidase N is not in line with the results obtained for this enzyme in other species. These authors found a very slow transport of 180 min. The reason for this slow transport is unknown. An important factor to be considered with animal experimentation is body temperature, since intracellular protein transport slows down when the temperature falls below 37°C. On the other hand, the transport times established for the appearance of microvillar hydrolases at the cell surface by means of the organ culture methodology must be interpreted with caution. No data are available as to what extent the brush border fraction in such experiments is cross-contaminated with Golgi membranes. Such a contamination, if substantial, may lead to an overestimation of the efficiency of transport, since microvillar hydrolases in the Golgi apparatus (Pavelka and Ellinger, 1986; Roth *et al.*, 1986) can be expected to be electrophoretically indistinguishable from their counterparts in the brush border membrane.

For disaccharidases, maturation of the high-mannose to the complex-glycosylated form in Caco-2 cells has a half-time of at least 2 h as compared to approximately 20–25 min for the peptidases (Table III). Recent subcellular fractionation studies in our laboratory suggest that the asynchronous intracellular transport is due to at least two rate-limiting steps: exit from the endoplasmic reticulum and transit through the Golgi apparatus (Stieger *et al.*, in press). The molecular basis for the asynchronous transport is unknown, but it is not merely a function of molecular size, since angiotensin I-converting enzyme, which is only minimally smaller than sucrase–isomaltase is rapidly transported. Furthermore, the asynchronism is not due to differential trimming of N-linked carbohydrates since it is not affected by *N*-methyl-1-deoxynojirimycin, an inhibitor of the first trimming reaction (K. Matter and H.-P. Hauri, unpublished observation). In a recent study, Puddington *et al.* (1986) have used oligonucleotide-directed mutagenesis to study the role of the cytoplasmic domain of vesicular stomatitis G protein and cellular transmembrane proteins for efficient transport to the cell surface. On the basis of their results, the authors proposed that the cytoplasmic domain is important in signaling a normal rate of intracellular transport and that basic amino acids in this part of the molecules may slow down the process.

Table III
Apparent Intracellular Transport Rates (min) of Two Microvillar Hydrolases in Caco 2 Cells as Deduced from Pulse-Chase Experiments and Subcellular Fractionation[a]

Enzyme	ER to brush border[b]	ER to Golgi[c]	ER to site of complex glycosylation[d]	Site of complex glycosylation to brush border[e]
Sucrase–isomaltase	180–190	45	130–140	40–60
Dipeptidylpeptidase IV	80–90	<15	15–20	60–75

[a]From Stieger *et al.* (in press).
[b]Time for half-maximal appearance of mature enzyme in the brush border membrane fraction.
[c]Time for half-maximal appearance of the high-mannose form in the Golgi fraction.
[d]Time for half-maximal appearance of the complex-glycosylated form in the homogenate.
[e]Transport rate from ER to brush border minus transport rate from ER to site of complex glycosylation.

However, with brush border hydrolases there does not seem to be a correlation between basic amino acids in the cytoplasmic domains and transport kinetics. The slowly transported disaccharidases sucrase–isomaltase and maltase–glucoamylase have three and five basic amino acids, respectively, while the rapid transported peptidases γ-glutamyltranspeptidase and endopeptidase 24.11 have two and five basic amino acids, respectively (see Figure 2). Feracci *et al.* (1982a) have reported that the cytoplasmic domain of rabbit aminopeptidase N only possesses a single basic residue (see Figure 1). However, sequencing data for human aminopeptidase N indicate that the cytoplasmic portion of this enzyme may be longer and may comprise two basic amino acids (E. E. Sterchi *et al.*, unpublished observations). Overall, this sequence information clearly shows no correlation between basic residues on the cytoplasmic side and the efficiency of intracellular transport. There is even a striking homology of the N-terminal amino acid sequence of pig kidney endopeptidase 24.11 with that of isomaltase (Fulcher *et al.*, 1986b), strongly suggesting that the cytoplasmic domain is not responsible for the asynchronous transport (Figure 6). It is possible that gene duplication during evolution (Hunziker *et al.*, 1986) has lowered the efficiency of folding and hence maturation of the disaccharidases.

7.3. Experimental Manipulation of Exocytosis and Possible Signals for Transport

The intracellular transport of surface membrane proteins may be mediated by signals in the protein itself (e.g., topogenic sequences in the protein backbone

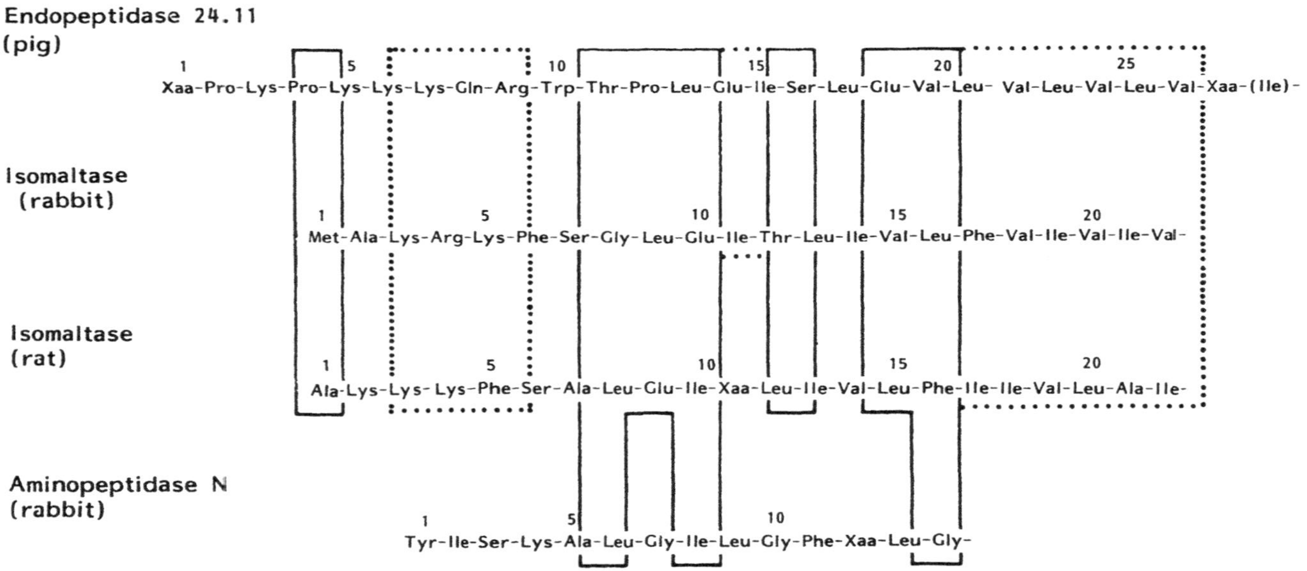

FIGURE 6. Homology of the N-terminal amino acid sequence of endopeptidase 24.11 with that of isomaltase (according to Fulcher *et al.*, 1986). Identical residues are boxed in with continuous lines. Constitutive substitutions are boxed in with dotted lines. Sequence of rabbit aminopeptidase is also given.

comparable to signal sequences for translocation; Blobel, 1980) or by covalent modifications (e.g., glycosylation, phosphorylation or sulfation, fatty acid acylation, proteolytic cleavage). Alternatively, this constitutive pathway may be nonselective and may therefore not require specific signals unless a protein has to be specifically retained in an intracellular organelle (Poruchynsky *et al.*, 1985).

In search of transport signals for brush border hydrolases, we have recently studied the role of N-linked oligosaccharides in Caco-2 cells. When glycosylation is completely prevented by tunicamycin, the transport of sucrase–isomaltase but not of dipeptidylpeptidase IV is blocked. There is twofold evidence that nonglycosylated sucrase–isomaltase is arrested in the endoplasmic reticulum. (1) The enzyme does not undergo O-glycosylation (which would not be inhibited by tunicamycin provided the protein reaches the Golgi apparatus) and (2) the protein does not undergo the early conformational change detected by monoclonal antibody HBB 3/705/60 under normal conditions (Hauri *et al.*, 1985b). This observation is in line with the results of Danielsen and Cowell (1984), who noted that (presumably) nonglycosylated aminopeptidase N, sucrase–isomaltase, and maltase–glucoamylase failed to reach the brush border membrane in tunicamycin-treated pig intestinal organ cultures. In Caco-2 cells, exit of sucrase–isomaltase from the endoplasmic reticulum was not blocked when the trimming of the N-glycosidic side chains was inhibited by *N*-methyl-1-deoxynojirimycin, an inhibitor of glucosidase I (Romero *et al.*, 1983; Schwarz and Datema, 1984). Under these conditions the enzyme was normally transported to the Golgi apparatus where it underwent O-glycosylation (K. Matter *et al.*, unpublished observation). However, this inhibitor reduced transport efficiency of both sucrase–isomaltase and dipeptidylpeptidase IV to the brush border by 60–70%. Swainsonine, an inhibitor of Golgi mannosidase II, leading to hybrid oligosaccharide side chains (Tulsiani and Touster 1983a,b), only moderately inhibited transport of aminopeptidase N to the brush border (Danielsen *et al.*, 1983a). These observations suggest that oligosaccharides may not serve as general transport signals for the endoplasmic reticulum to Golgi pathway. It appears more likely that the individual brush border hydrolases have different requirements for N-glycosylation in the folding process, leading to a conformation that permits transport. At the level of the Golgi apparatus, at least some trimming appears necessary for efficient delivery to the brush border. Much less is known on the importance of O-glycosylation for transport since no specific inhibitors exist. It has recently been suggested that pig maltase–glucoamylase may not receive O-linked carbohydrates during its transport through the Golgi apparatus and hence that O-linked glycosylation may not be of general importance as a sorting signal (Danielsen and Cowell, 1986).

Overall, the studies with brush border enzymes are in line with those of other glycoproteins, which have indicated that glycosylation most probably is of no importance as a general sorting signal for the apical pathway (Burke *et al.*,

1984). Other common post-translational protein modifications are equally unlikely to serve as transport signals since they do not occur in all the brush border enzymes (e.g., intracellular proteolytic cleavage as in lactase–phlorizinhydrolase and γ-glutamyltranspeptidase) or do not occur at all (e.g., phosphorylation).

A number of studies have now shown that the Golgi apparatus is a mildly acidic compartment owing to a proton pump (see Mellman *et al.*, 1986, for a review). There is a pH gradient in a cis to trans direction with the lowest pH on the transside and in departing vesicles destined for the cell surface (Anderson and Pathak, 1985; Orci *et al.*, 1986). It would therefore not be unreasonable to speculate that the Golgi-to-cell surface transport may be mediated by receptors that can bind to newly synthesized proteins in a pH-dependent manner (e.g., tight binding with low pH and dissociation with neutral pH).

However, surface expression of both peptidases and disaccharidases in organ culture is not impaired by chloroquine, a weak base known to raise the pH of intracellular organelles (Danielsen and Cowell, 1985b), which may indicate that exocytosis of microvillar hydrolases is less affected by changes in pH than that of other membrane proteins (Matlin, 1986; Strous *et al.*, 1985). The constitutive pathway is in general less sensitive to lysosomotropic agents than the regulated one (Kelly, 1985).

In conclusion, the concept is emerging that common post-translational modifications do not serve as obligatory transport signals for all brush border hydrolases. On the other hand, no positive signals have been uncovered to date that can mediate this transport in intestinal epithelial cells.

7.4. Sorting of Brush Border Hydrolases from Basolateral Proteins

A recent study on the surface expression of small-intestinal (Na^+ + K^+)-ATPase using subunit-specific monoclonal antibodies in conjunction with immunoelectron microscopy has demonstrated that both subunits are exclusively localized in the basolateral membrane (Marxer *et al.*, submitted). No immunoreaction was observed in the brush border membrane, while intracellular labeling was restricted to the Golgi apparatus. This labeling pattern supports the notion that intestinal (Na^+ + K^+)-ATPase is directly delivered from the Golgi to the basolateral membrane. Caplan *et al.* (1986) have recently provided evidence for such a direct biosynthetic pathway of (Na^+ + K^+)-ATPase in MDCK cells. Therefore, either apical or basolateral membrane proteins must carry sorting signals in order to guarantee polarized surface expression. While these signals have not yet been determined, a number of studies have suggested a decisive role of microtubules for the efficient sorting of brush border and basolateral membrane proteins. In the normal intestinal enterocyte, microtubules parallel the long axis of the cell and do not focus on a microtubule-organizing center (Gorbsky and Borisy, 1985; Reaven and Reaven, 1977), while the Golgi apparatus has a characteristic supranuclear structure–its stacks being connected by the transmost (and possibly additional) cisternae (Roth *et al.*, 1985b). Upon disassembly of the microtubules by colchicine, the stacks of the Golgi apparatus become dispersed throughout the entire apical, lateral, and basal cytoplasmas (Pavelka and Ellinger, 1981; Reaven and Reaven, 1977), concomitant with a loss of polarized delivery of surface–membrane glycoproteins assessed by electron microscopy (Ellinger and Pavelka, 1984; Pavelka and Gangl, 1983; Pavelka *et al.*, 1983), autoradiography (Bennett *et al.*, 1984; Blok *et al.*, 1981b; Ellinger *et al.*, 1983), and subcellular fractionation (Danielsen *et al.*, 1983d; Quaroni *et al.*, 1979a,b). Interestingly, colchicine primarily affects protein delivery to the apical cell pole. When colchicine was intravenously injected into rabbits 10 min prior to labeling with ^{3}H-fucose and the duodenal enterocytes were analyzed by autoradiography 90 min later, ^{3}H-fucose labeling of the brush border membrane was reduced by 51% with little change in labeling of the lateral plasma membrane, while labeling of the Golgi apparatus was increased (Bennett *et al.*, 1984). This suggests that labeled glycoproteins destined for the apical plasma membrane were inhibited from leaving the Golgi region, while migration to the lateral plasma membrane was not impaired. After longer exposures to colchicine in vivo (Ellinger *et al.*, 1983) or in organ culture (Blok *et al.*, 1981a), relative labeling of the brush border is further decreased but labeling of the lateral plasma membrane increases up to twofold (Blok *et al.*, 1981a). At 2.5 h after colchicine treatment, the apical microvilli were found to be reduced in length (Buschmann, 1983). At 3 h after colchicine exposure, brush border areas start to appear along the lateral and basal cell surfaces, and at 6 h, 3.5% of the basolateral cell surface exhibited microvilli (Pavelka *et al.*, 1983), suggesting a misrouting of apical membrane proteins to the basolateral membrane. Indeed, Quaroni *et al.* (1979b) showed by subcellular fractionation that the incorporation of all ^{3}H-fucosylated glycoproteins (most of which correspond to known brush border enzymes) into the microvillus membrane was greatly inhibited by colchicine given 10 min or 2 h before labeling (in vivo). In parallel, the administration of colchicine considerably increased the incorporation of radioactivity into the basolateral plasma membrane and led to an accumulation of apical glycoproteins in this membrane. Furthermore, organ culture studies by Danielsen *et al.* (1983b) demonstrated that colchicine caused a significant inhibition of the appearance of newly synthesized enzymes in the brush border fraction. Lactase appears to be particularly sensitive to colchicine, a finding that is unexplained (Herbst *et al.*, 1970; Hudson and Smith, 1986).

The most likely interpretation of all these results is as follows. Microtubules may provide "tracks" for the vectorial transport of brush border hydrolases as in other highly polarized cells like nerve cells (Allen *et al.*, 1985; McNiven and Porter, 1986; Schnapp *et al.*, 1985; Schroer and Kelly, 1985; Vale *et al.*, 1985). Only recently has it been possible to demonstrate that microtubules are indeed present in the terminal web and run into close vicinity of the microvillar membrane (Sandoz *et al.*, 1985). When the cytoskeleton is disrupted by microtubular agents, the Golgi apparatus is disturbed and the efficiency of protein exit from this organelle slows down but is not completely blocked. In the absence of assembled microtubules, departing vesicles destined for the apical pole now randomly fuse with the next target membrane, which is the basolateral membrane. Admittedly, the results with colchicine could also be accommodated within the hypothesis of Quaroni (1984), who proposes that in the normal situation newly synthesized brush border hydrolases are first inserted into the basolateral membrane and subsequently routed to the apical cell pole by transcytosis (a pathway that has been established for the polyimmunoglobulin receptor; Ahnen *et al.*, 1985; Mostov and Deitcher, 1986; Mostov and Simister, 1985). In that case, brush border enzymes would be retarded not only in the Golgi apparatus but also in the basolateral membrane, since endocytosis in the latter membrane would also be impaired by colchicine.

Although a number of studies favor a role of microtubules for directed organelle transport also in other cells (Rogalski *et al.*, 1984), evidence has recently been reported that microtubules may not be involved in the polarized maturation of enveloped viruses in MDCK cells (Salas *et al.*, 1986). Likewise, neither colchicine nor vinblastine inhibited the appearance of peptidases in pig kidney organ culture (Stewart and Kenny, 1984b). Intracellular transport in kidney epithelial cells may be less dependent on microtubules, although this discrepancy could also be attributed to differences in the experimental systems, that is, in vivo versus organ culture, slice culture, or cell monolayers.

7.5. Disruption of Transport of Sucrase–Isomaltase in Hereditary Sucrase–Isomaltase Deficiency

The analysis of naturally occurring mutations is a valuable approach to study mechanisms of intracellular protein targeting (Davis *et al.*, 1986; Tartakoff, 1983; Tolleshaug *et al.*, 1982, 1983; Yamamoto *et al.*, 1986). Hereditary sucrase–isomaltase deficiency is an autosomal recessive disease in humans leading to diarrhea upon ingestion of disaccharides owing to the absence of a functional sucrase–isomaltase complex (Kerry and Townley, 1965). On the basis of immunofluorescence with antibodies against sucrase–isomaltase, Dubs *et al.* (1973) have proposed that in some but not all patients catalytically inactive enzyme was expressed, suggesting multiple mutations. This finding was at variance with a study by Gray *et al.* (1976), who found complete absence of enzyme protein by radioimmunoassay. Biochemical studies revealed the absence of mature sucrase–isomaltase from the brush border fraction prepared from mucosal specimens of deficiency cases (Preiser *et al.*, 1974; Schmitz *et al.*, 1974). Subsequent studies have confirmed Dubs' finding that sucrase–isomaltase was not completely lacking in all the patients (Cooper *et al.*, 1979; Freiburghaus *et al.*, 1977; Hadorn *et al.*, 1981; Schmitz *et al.*, 1980; Skovbjerg and Krasilnikoff, 1980). Only recently it was found that in hereditary sucrase–isomaltase deficiency multiple mutations disrupt transport and processing of this hydrolase. In two such patients the high-mannose form of sucrase–isomaltase was overexpressed probably in the endoplasmic reticulum since no complex-glycosylated form was detectable (Naim *et al.*, in press). In two other patients, electrophoretically unchanged high-mannose form was synthesized at normal levels but the enzyme was arrested and probably degraded in the Golgi apparatus and failed to reach the brush border membrane (Hauri *et al.*, 1985a, 1986). Yet another case displayed structurally and enzymatically altered sucrase–isomaltase at the microvillar surface (Freiburghaus *et al.*, 1977). Thus, multiple genetic defects of sucrase–isomaltase can lead to a transport block at specific sites of the biosynthetic pathway. Organ culture is being used to study the biosynthesis of the mutant enzymes, and epitope mapping by means of monoclonal antibodies have revealed that the two mutations leading to a block in Golgi apparatus must be different (Hauri *et al.*, 1986). Common to all the recently analyzed cases with sucrase–isomaltase deficiency is the expression of an electrophoretically normal high-mannose precursor, suggesting that only minor alterations of the primary sequence, probably point mutations, can have a profound effect on the migration of this hydrolase.

Further characterization of mutant phenotypes at the molecular level might therefore lead to the identification of protein domains essential for the intracellular targeting of sucrase–isomaltase.

8. SUMMARY AND PERSPECTIVES

The enterocyte, the principal epithelial cell of the small-intestinal mucosa, is an attractive model for studying the biogenesis and polarized surface expression of endogenous surface-membrane proteins. Most major integral proteins of its apical brush border membrane have been identified as hydrolases, either disaccharidases or peptidases, that are involved in the terminal digestion of dietary sugars and proteins. Biochemical and more recently molecular biological studies suggest a common mechanism for initial biosynthesis of both disaccharidases and peptidases. In this process, a noncleavable hydrophobic sequence near the N terminus has a dual role as a signal for translocation and as a membrane anchor. Knowledge on the maturation and intracellular transport of brush border enzyme is increasing by studies using organ-cultured mucosal explants and the differentiated colon adenocarcinoma cell line Caco-2, which displays many structural and functional features that are remarkably similar to those of normal small-intestinal epithelial cells. In conjunction with the use of specific antibody probes, these studies have revealed interesting aspects of maturation.

The intracellular transport is considerably slower for disaccharidases than for peptidases. The application of monoclonal antibodies has been rewarding in respect to the identification of novel brush border enzymes, the detection of early conformational changes in newly synthesized enzymes, and to epitope mapping in hereditary sucrase–isomaltase deficiency, in which multiple mutations disrupt the intracellular transport of sucrase–isomaltase. Through use of differentiated adenocarcinoma lines as cell culture models, a number of questions concerning the polarized cell-surface expression of brush border hydrolases can now be approached directly, such as: Is the basolateral membrane involved in routing these enzymes to the brush border? Are all the microvillar hydrolases transported along the same pathway from the endoplasmic reticulum to the cell surface or do slowly and rapidly transported enzymes follow different constitutive routes? Does the microtubular system provide tracks for the efficient delivery of membrane proteins to the apical cell pole? Is there a crinophagic mechanism by which the surface expression of these proteins can be regulated at the level of the Golgi apparatus? What are the sorting signals for basolateral and microvillar targeting? Answers to these challenging questions that are of general interest in cell biology will undoubtedly require the development of new techniques including the in vitro reconstitution of membrane flow between the different membrane compartments.

Notes added in proof. The complete amino acid sequence of human intestinal alkaline phosphatase (Berger *et al.*, 1987, *Proc. Natl. Acad. Sci. U.S.A.* **84:**695–698) and of kidney endopeptidase 24.11 (Devault *et al.*, 1987, *EMBO J.* **6:**1317–1322) have been deduced recently from cDNA. There is evidence in support of a C terminal rather than an N terminal anchoring of pig kidney angiotensin I-converting enzyme (Hooper *et al.*, 1987, *Biochem. J.* **247:**85–93). The major brush border hydrolases have been shown to carry sulfate groups (Danielson, 1987, *EMBO J.* **6:**2891–2896).

ACKNOWLEDGMENTS. I am grateful to Dr. U. A. Meyer for his continuous support and interest, to Dr. E. G. Berger, Dr. E. E. Sterchi, Ursula Eilers, Karl Matter, and Thomas Jascur for critically reading the manuscript, to Käthy Bucher for excellent technical assistance, and to M. Liechi and J. Wittker for secretarial help. The research in the author's laboratory was supported by the Swiss National Science Foundation (grant 3.652-0.84).

9. REFERENCES

Adibi, S. A., and Kim, Y. S., 1981, Peptide absorption and hydrolysis. In *Physiology of the Gastrointestinal Tract* (L. R. Johnson, ed.), pp. 1072–1095, Raven Press, New York.

Ahnen, D. J., Santiago, N. A., Cézard, J.-P., and Gray, G. M., 1982, Intestinal aminooligopeptidase. In vivo synthesis on intracellular membranes of rat jejunum. *J. Biol. Chem.* **257:**12129–12135.

Ahnen, D. J., Mircheff, A. K., Santiago, N. A., Yoshioka, C., and Gray, G. M., 1983, Intestinal surface aminooligopeptidase. Distinct molecular forms during assembly on intracellular membranes in vivo. *J. Biol. Chem.* **258:**5960–5966.

Ahnen, D. J., Brown, W. R., and Kloppel, T. M., 1985, Secretory component: The polymeric immunoglobulin receptor. *Gastroenterology* **89:**667–682.

Allen, R. D., Weiss, D. G., Hayden, J. H., Brown, D. T., Fujiwake, H., and Simpson, M., 1985, Gliding movement of and bidirectional transport along single native microtubules in cytoplasmic transport. *J. Cell Biol.* **100:**1736–1752.

Alonso, F. V., and Compans, R. W., 1981. Differential effect of monensin on enveloped viruses that form at distinct plasma membrane domains. *J. Cell Biol.* **89:**700–705.

Alpers, D. H., 1969, Separation and isolation of rat and human intestinal beta-galactosidase. *J. Biol. Chem.* **244:**1238–1246.

Alpers, D. H., Helms, D. Seetharam, S., May V. L., and Strauss A. W., 1986, In vitro translation of intestinal sucrase–isomaltase and glucoamylase. *Biochem. Biophys. Res. Commun.* **134:**37–43.

Anderson, R. G. W., and Pathak, R. K., 1985, Vesicles and cisternae in the trans Golgi apparatus of human fibroblasts are acidic compartments. *Cell* **40:**635–643.

Aoyagi, K., Katayama Narita, M., and Tojo, S., 1985, Purification and characterization of human kidney neutral endopeptidase that hydrolyzes succinyl-4-nitroanilide and biologically active peptides. *Biochim. Biophys. Acta* **840:**211–218.

Augeron, C., and Laboisse, C. L., 1984, Emergence of permanently differentiated cell clones in a human colonic cancer cell line in culture after treatment with sodium butyrate. *Cancer Res.* **44:**3961–3969.

Barouki, R., Finidori, J., Chobert, M.-N., Aggerbeck, M., Laperche, Y., and Hanoune, J., 1984, Biosynthesis and processing of gamma-glutamyl transpeptidase in hepatoma tissue culture cells. *J. Biol. Chem.* **259:**7970–7974.

Bartles, J. R., Braiterman, L. T., and Hubbard, A. L., 1985, Biochemical characterization of domain-specific glycoproteins of the rat hepatocyte membrane. *J. Biol. Chem.* **260:**12792–12802.

Bella, A. M., Erickson, R. H., and Kim, Y. S., 1982, Rat intestinal brush border membrane dipeptidyl-aminopeptidase IV: Kinetic properties and substrate specificities of the purified enzyme. *Arch. Biochem. Biophys.* **219:**156–162.

Benajiba, A., and Maroux, S., 1980, Purification and characterization of an aminopeptidase A from hog intestinal brush border membrane. *Eur. J. Biochem.* **107:**381–388.

Benajiba, A., and Maroux, S., 1981, Subunit structure of pig small-intestinal brush border aminopeptidase N. *Biochem. J.* **197:**573–580.

Bennett, G., Leblond, C. P., and Haddad, A., 1974. Migration of glycoprotein from the Golgi apparatus to the surface of various cell types as shown by radioautography after labeled fucose injection into rats. *J. Cell Biol.* **60:**258–284.

Bennett, G., Carlet E., Wild, G., and Parsons, S., 1984, Influence of colchicine and vinblastine on the intracellular migration of secretory and membrane glycoproteins: III. Inhibition of intracellular migration of membrane glycoproteins in rat intestinal columnar cells and hepatocytes as visualized by light- and electron-microscope radioautography after ^{3}H-fucose injection. *Am. J. Anat.* **170:**545–566.

Bergeron, J. J. M., Kotwal, G. J., Levine, G., Bilan, P., Rachubinski, R., Hamilton, M. Shore, G. C., and Ghosh, H. P., 1982, Intracellular transport of the transmembrane glycoprotein G of vesicular stomatitis virus through the Golgi apparatus as visualized by electron microscope radioautography. *J. Cell Biol.* **94:**36–41.

Bergmann, J. E., and Singer, S. J., 1983, Immunoelectron microscopic studies of the intracellular transport of the membrane glycoprotein G of vesicular stomatitis virus in infected Chinese hamster ovary cells. *J. Cell Biol.* **97:**1777–1787.

Bergmann, J. E., Tokuyasu, K. T., and Singer, S. J., 1981, Passage of an integral membrane protein, the vesicular stomatitis virus glycoprotein, through the Golgi apparatus en route to the plasma membrane. *Proc. Natl. Acad. Sci. U.S.A.* **78:**1746–1750.

Bernadac, A., Gorvel, J.-P., Feracci, H., and Maroux, S., 1984, Human blood group A-like determinants as marker of the intracellular pools of glycoproteins in secretory and absorbing cells of A+ rabbit jejunum. *Biol. Cell* **50:**31–36.

Berteloot, A., Chabot, J.-G., Menard, D., and Hugon, J. S., 1979, Organ culture of adult mouse intestine. III. Behavior of the proteins, DNA content and brush border membrane enzymatic activities. *In Vitro* **15:**294–299.

Birkenmeier, E., and Alpers, D. H., 1974, Enzymatic properties of rat lactase–phlorizin hydrolase. *Biochim. Biophys. Acta* **350:**100–112.

Bjerknes, M., and Cheng, H., 1981, Methods for the isolation of intact epithelium from the mouse intestine. *Anat. Rec.* **199:**565–574.

Blay, J., and Brown, K. D., 1984, Characterization of an epitheloid cell line derived from rat small intestine: Demonstration of cytokeratin filaments. *Cell Biol. Int. Rep.* **8:**551–559.

Blay, J., and Brown, K. D., 1985, Epidermal growth factor promotes the chemotactic migration of cultured rat intestinal epithelial cells. *J. Cell Physiol.* **124:**107–112.

Blobel, G., 1980, Intracellular protein topogenesis. *Proc. Natl. Acad. Sci. U.S.A.* **77:**1496–1500.

Blok, J., Ginsel, L. A., Mulder-Stapel, A. A., Onderwater, J. J. M., and Daems, W. Th. 1981a, The effect of colchicine on the intracellular transport of ^{3}H-fucose-labelled glycoproteins in the absorptive cells of cultured human small-intestinal tissue. *Cell Tissue Res.* **215:**1–12.

Blok, J., Mulder-Stapel, A. A., Daems, W. Th., and Ginsel, L. A., 1981b, The effect of chloroquine on the intralysosomal degradation of cell-coat glycoproteins in the absorptive cells of cultured human small-intestinal tissue as shown by silver proteinate staining. *Histochemistry* **73:**429–438.

Blok, J., Scheven, B. A. A., Mulder-Stapel, A. A., Ginsel, L. A., and Daems, W. Th., 1982, Endocytosis in absorptive cells of cultured human small-intestinal tissue: Effects of cytochalasin B and D. *Cell Tissue Res.* **222:**113–126.

Blok, J., Fransen, J. A. M., and Ginsel, L. A., 1984, Turnover of brush border glycoproteins in human intestinal absorptive cells: Do lysosomes have a regulatory function? *Cell Biol. Int. Rep.* **8:**993–1014.

Bond, J. S., and Beynon, R. J., 1986, Meprin: A membrane-bound metallo-endopeptidase. *Curr. Top. Cell. Regul.* **28:**263–290.

Bonkovsky, H. L., Hauri, H. P., Marti, U., Gasser, R., and Meyer, U. A., 1985, Cytochrome P_{450} of small intestinal epithelial cells: Immunochemical characterization of the increase in cytochrome P_{450} caused by phenobarbital. *Gastroenterology* **88:**458–467.

Bonsdorff, C.-H., Fuller, S., and Simons, K., 1985, Apical and basolateral endocytosis in Madin–Darby canine kidney (MDCK) cells grown on nitrocellulose filters. *EMBO J.* **11:**2781–2792.

Booth, A. G., Hubbard, L. M. L., and Kenny, A. J. 1979, Proteins of the kidney microvillar membrane. Immunoelectrophoretic analysis of the membrane hydrolases: Identification and resolution of the detergent- and proteinase-solubilized forms. *Biochem. J.* **179:**397–405.

Boyle, F., Snape, S., Duane, P., Cook, N., and Peters, T., 1986, Galactosyltransferase activity is not localized to the brush border membrane of human small intestine. *Biosci. Rep.* **6:**171–176.

Braun, H., Cogoli, A., and Semenza, G., 1975, Dissociation of small-intestinal sucrase–isomaltase complex into enzymatically active subunits. *Eur. J. Biochem.* **52:**475–480.

Bretscher, A., 1983, Molecular architecture of the microvillus cytoskeleton. In *Brush Border Membranes* (R. Porter and G. Collins, eds.), Ciba Foundation Symposium 95, pp. 164–175, Pitman, London.

Browning, T. H., and Trier, J. S., 1969, Organ culture of mucosal biopsies of human small intestine. *J. Clin. Invest.* **48:**1423–1432.

Bruneval, P., Hinglais, N. Alhenc-Gelas, F., Tricottet, V., Corvol, P., Menard, J., Camilleri, J. P., and Bariety, J., 1986, Angiotensin I converting enzyme in human intestine and kidney. *Histochemistry* **85:**73–80.

Brunner, J., Hauser, H., and Semenza, G., 1978, Single bilayer lipid–protein vesicles formed from phosphatidylcholine and small intestinal sucrase–isomaltase. *J. Biol. Chem.* **253:**7538–7546.

Brunner, J., Hauser, H., Braun, H., Wilson, K., Wacker, H., O'Neill, B., and Semenza, G., 1979, The mode of association of the enzyme complex sucrase–isomaltase with the intestinal brush border membrane. *J. Biol. Chem.* **254:**1821–1828.

Brunner, J., Wacker, H., and Semenza, G., 1983, Sucrase–isomaltase of the small-intestinal brush border membrane: Assembly and biosynthesis. *Methods Enzymol.* **96:**368–406.

Bürgi, R., Brunner, J., and Semenza, G., 1983, A chemical procedure for determining the sidedness of the NH_2-terminus in a membrane protein. The small intestinal sucrase–isomaltase. *J. Biol. Chem.* **258:**15114–15119.

Burke, B., Walter, C., Griffiths, G., and Warren, G., 1983, Viral glycoproteins at different stages of intracellular transport can be distinguished using monoclonal antibodies. *Eur. J. Cell Biol.* **31:**315–324.

Burke, B., Matlin, K., Bause, E., Legler, G., Peyrieras, N., and Ploegh, H., 1984, Inhibition of N-linked oligosaccharide trimming does not interfere with surface expression of certain integral membrane proteins. *EMBO J.* **3:**551–556.

Buschmann, R. J., 1983, Morphometry of small intestinal enterocytes of the fasted rat and the effects of colchicine. *Cell Tissue Res.* **231:**289–299.

Caplan, M., Anderson, H. C., Palade, G. E., and Jamieson, J. D., 1986, Intracellular sorting and polarized cell surface delivery of Na-K-ATPase, an endogenous component of MDCK cell basolateral plasma membrane. *Cell* **46:**623–631.

Caporale, C., Fontanella, A., Petrilli, P., Pucci, P., Molinaro, M. F., Picone, D., and Auricchio, S., 1985, Isolation and characterization of dipeptidylpeptidase IV from human meconium. Functional role of beta-casomorphins. *FEBS Lett.* **184:**273–277.

Capraro, M. A., and Hughey, R. P., 1983, Processing of the propeptide form of rat renal gamma-glutamyl transpeptidase. *FEBS Lett.* **157:**139–143.

Cézard, J.-P., Conklin, K. A., Das, B. C., and Gray, G. M., 1979, Incomplete intracellular forms of intestinal surface membrane sucrase–isomaltase. *J. Biol. Chem.* **254:**8969–8975.

Colas, B., and Maroux, S., 1980, Simultaneous isolation of brush border and basolateral membrane from rabbit enterocytes. Presence of brush border hydrolases in the basolateral membrane of rabbit enterocytes. *Biochim. Biophys. Acta* **600:**406–420.

Colbeau, A., and Maroux, S., 1978. Integration of alkaline phosphatase in the intestinal brush border membrane. *Biochim. Biophys. Acta* **511:**39–51.

Colombo, V., Lorenz-Meyer, H., and Semenza, G., 1973, Small intestinal phlorizin hydrolase: The beta-glycosidase complex. *Biochim. Biophys. Acta* **327:**412–424.

Compton, T., Gottlieb, T., Rindler, M., Adesnik, M., and Sabatini, D. D., 1986, Polarized delivery and recycling in MDCK cells of a chimeric protein containing the luminal portion of VSV G and the cytoplasmic and transmembrane domain of influenza HA. *J. Cell Biol.* **103:**7a (abstract).

Cone, J. L., Glowinski, I. B., Wirth, P. J., Grantham, P. H., and Roller, P. P., 1986, Structural studies and two-dimensional gel electrophoresis of gamma-glutamyl transpeptidase. *Arch. Biochem. Biophys.* **247:**165–170.

Conklin, K. A., Yamashiro, K. M., and Gray, G. M., 1975, Human intestinal sucrase isomaltase. Identification of free sucrase and isomaltase and cleavage of the hybrid into active distinct subunits. *J. Biol. Chem.* **250:**5735–5741.

Cooper, B. T., Candy, D. C. A., Harries, J. T., and Peters, T. J., 1979, Subcellular fractionation studies of the intestinal mucosa in congenital sucrase–isomaltase deficiency. *Clin. Sci.* **57:**181–185.

Copeland, C. S., Doms, R. W., Bolzau, E. M., Webster, R. G., and Helenius, A., 1986, Assembly of influenza hemagglutination trimers and its role in intracellular transport. *J. Cell. Biol.* **103:**1179–1191.

Cousineau, J., and Green, J. R., 1980, Isolation and characterization of the proximal and distal forms of lactase–phlorizin hydrolase from the small intestine of the suckling rat. *Biochim. Biophys. Acta* **615:**147–157.

Cowell, G. M., Tranum-Jensen, J., Sjöström, H., and Norén, O., 1986, Topology and quaternary structure of pro-sucrase–isomaltase and final-form sucrase–isomaltase. *Biochem. J.* **137:**455–461.

Danielsen, E. M., 1982, Biosynthesis of intestinal microvillar proteins. Pulse-chase labelling studies on aminopeptidase N and sucrase–isomaltase. *Biochem. J.* **204:**639–645.

Danielsen, E. M., and Cowell, G. M., 1984, Biosynthesis of intestinal microvillar proteins. Further characterization of the intracellular processing and transport. *FEBS Lett.* **166:**28–32.

Danielsen, E. M., and Cowell, G. M., 1985a, Biosynthesis of intestinal microvillar proteins. Evidence for an intracellular sorting taking place in, or shortly after, exit from the Golgi complex. *Eur. J. Biochem.* **152:**493–499.

Danielsen, E. M., and Cowell, G. M., 1985, Biosynthesis of intestinal microvillar proteins. The intracellular transport of aminopeptidase N and sucrase–isomaltase occurs at different rates pre-Golgi but at the same rate post-Golgi. *FEBS Lett.* **190:**69–72.

Danielsen, E. M., and Cowell, G. M., 1986, Biosynthesis of intestinal microvillar proteins. Processing of N-linked carbohydrate is not required for surface expression. *Biochem. J.* **240:**777–782.

Danielsen, E. M., Norén, O., Sjöström, H., Ingram, J., and Kenny, A. J., 1980a, Proteins of the kidney microvillar membrane. Aspartate aminotransferase: Purification by immunoadsorbent chromatography and properties of the detergent- and proteinase-solubilized forms. *Biochem. J.* **189:**591–603.

Danielsen, E. M., Vyas, J. P., and Kenny, A. J., 1980b, A neutral endopeptidase in the microvillar membrane of pig intestine. Partial purification and properties. *Biochem. J.* **191:**645–648.

Danielsen, E. M., Sjöström, H., and Norén, O., 1981a, Biosynthesis of intestinal microvillar proteins. Putative precursor forms of microvillus aminopeptidase and sucrase–isomaltase isolated from Ca-precipitated enterocyte membranes. *FEBS Lett.* **127:**129–132.

Danielsen, E. M., Skovbjerg, H., Norén, O., and Sjöström, H., 1981b, Biosynthesis of intestinal microvillar proteins. Nature of precursor forms of microvillar enzymes from Ca-precipitated enterocyte membranes. *FEBS Lett.* **132:**197–200.

Danielsen, E. M., Norén, O., and Sjöström, H., 1982a, Biosynthesis of intestinal microvillar proteins. Translational evidence in vitro that aminopeptidase N is synthesized as a M_r115,000 polypeptide. *Biochem. J.* **204:**323–327.

Danielsen, E. M., Sjöström, H., and Norén, O., 1982b, Hypotonic elution, a new desorption principle in immunoadsorbent chromatography. *J. Immunol. Methods* **52:**223–232.

Danielsen, E. M., Sjöström, H., Norén, O., Bro, B., and Dabelsteen, E., 1982c. Biogenesis of intestinal microvillar proteins. Characterization of intestinal explants in organ culture and evidence for the existence of pro-forms of the microvillar enzymes. *Biochem. J.* **202:**647–654.

Danielsen, E. M., Cowell, G. M., Norén, O., Sjöström, H., and Droling, P. R., 1983a, Biosynthesis of intestinal microvillar proteins. Effect of swainsonine on post-translational processing of aminopeptidase N. *Biochem. J.* **216:**325–331.

Danielsen, E. M., Cowell, G. M., and Poulsen, S. S., 1983b, Biosynthesis of intestinal microvillar proteins. Role of the Golgi complex and microtubules. *Biochem. J.* **216:**37–42.

Danielsen, E. M., Norén, O., and Sjöström, H., 1983c, Biosynthesis of intestinal microvillar proteins. Processing of aminopeptidase N by microsomal membranes. *Biochem. J.* **212:**161–165.

Danielsen, E. M., Sjöström, H., and Norén, O., 1983d, Biosynthesis of intestinal microvillar proteins. Pulse-chase labelling studies on maltase–glucoamylase, aminopeptidase A and dipeptidyl peptidase IV. *Biochem. J.* **210:**389–393.

Danielsen, E. M., Cowell, G. M., Norén, O., and Sjöström, H., 1984a, Biosynthesis of microvillar proteins. *Biochem. J.* **221:**1–14.

Danielsen, E. M., Skovbjerg, H., Norén, O., and Sjöström, H., 1984b, Biosynthesis of intestinal microvillar proteins. Intracellular processing of lactase–phlorizin hydrolase. *Biochem. Biophys. Res. Commun.* **122:**82–90.

Danielsen, E. M., Sjöström, and Norén, O., 1985, Biosynthesis of intestinal microvillar proteins. Surface expression of aminopeptidase N is not affected by chloroquine. *FEBS Lett.* **179:**160–164.

Danielsen, E. M., Cowell, G. M., Hansen, G. H., Gorr, S.-U., Sjostrom, H., and Noren, O., 1986a, Role of the Golgi complex and characteristics of post-Golgi transport in the biosynthesis of intestinal microvillar proteins. *Biochem. Soc. Trans.* **14:**165–170.

Danielsen, E. M., Cowell, G. M., Sjöström, H., and Norén, O., 1986b, Translational control of an intestinal microvillar enzyme. *Biochem. J.* **235:**447–451.

Davis, G. C., Lehrman, M. A., Russel, D. W., Anderson, R. G. W., Brown, M. S., and Goldstein, J. L., 1986, The J. D. mutation in familial hypercholesterolemia: Amino acid substitution in cytoplasmic domain impedes internalization of LDL receptors. *Cell* **45:**15–24.

Desnuelle, P., 1979, Intestinal and renal aminopeptidase: A model of a transmembrane protein. *Eur. J. Biochem.* **101:**1–11.

Dottavio-Martin, D., and Ravel, J. M., 1978, Radiolabeling of proteins by reductive alkylation with ^{14}C-formaldehyde and sodium cyanoborohydride. *Anal. Biochem.* **87:**562–565.

Dubs, R., Steinmann, B., and Gitzelmann, R., 1973, Demonstration of an inactive enzyme antigen in sucrase–isomaltase deficiency. *Helv. Paediat. Acta* **28:**187–198.

Dubs, R., Gitzelmann, R., Steinmann, B., and Lindenmann, J., 1975, Catalytically inactive sucrase antigen of rabbit small intestine: The enzyme precursor. *Helv. Paediat. Acta* **30:**89–102.

Dunphy, W. G., and Rothman, J. E., 1983, Compartmentation of asparagine-linked oligosaccharide processing in the Golgi apparatus. *J. Cell. Biol.* **97:**270–275.

Dunphy, W. G., and Rothman, J. E., 1985, Compartmental organization of the Golgi stack. *Cell* **42:**13–21.

Eichholz, A., and Crane, R. K., 1965, Studies on the organization of the brush border in intestinal epithelial cells. I. Tris disruption of isolated hamster brush borders and density gradient centrifugation of fractions. *J. Cell Biol.* **26:**687–691.

Ellinger, A., and Pavelka, M., 1984, Colchicine-induced tubular, vesicular and cisternal organelle aggregates in absorptive cells of the small intestine of the rat. Morphology and phosphatase cytochemistry. *Biol. Cell* **52:**43–52.

Ellinger, A., Pavelka, M., and Gangl, A., 1983, Effect of colchicine on rat small intestinal absorptive cells. II. Distribution of label after incorporation of ^{3}H-fucose into plasma membrane glycoproteins. *J. Ultrastruct. Res.* **85,** 260–271.

Erickson, R. H., Bella, A. M., Brophy, E. J., Kobata, A., and Kim, Y. S., 1983, Purification and molecular characterization of rat intestinal brush border membrane dipeptidyl aminopeptidase IV. *Biochim. Biophys. Acta* **756:**258–265.

Ey, P. L., Prowse, S.J., and Jenkin, C. R., 1978, Isolation of pure IgG1, IgG2a and IgG2b immunoglobulins from mouse serum using protein A–Sepharose. *Immunochemistry* **15:**429–436.

Falchuk, Z. M., Gebhard, R. L., Sessoms, C., and Strober, W., 1974, An in vitro model of gluten-sensitive enteropathy. Effect of gliadin on intestinal epithelial cells of patients with gluten-sensitive enteropathy in organ culture. *J. Clin. Invest.* **53:**487–500.

Fantini, J., Abadie, B., Tirard, A., Remy, L., Ripert, J. P., Elbattari, E., and Marvaldi, J., 1986,

Spontaneous and induced dome formation by two clonal cell populations derived from a human adenocarcinoma cell line. HT-29. *J. Cell Sci.* **83**:235–249.

Farquhar, M. G., 1985, Progress in unraveling pathways of Golgi traffic. *Annu. Rev. Cell Biol.* **1**:447–488.

Feracci, H., and Maroux, S., 1980, Rabbit intestinal aminopeptidase N. Purification and molecular properties. *Biochim. Biophys. Acta* **599**:448–463.

Feracci, H., Benajiba, A., Gorvel, J. P., Doumeng, Ch., and Maroux, S., 1981, Enzymatic and immunological properties of the protease forms of aminopeptidases N and A from pig and rabbit intestinal brush border. *Biochim. Biophys. Acta* **658**:148–157.

Feracci, H., Bernadac, A., Gorvel, J. P., and Maroux, S., 1982a, Localization by immunofluorescence and histochemical labeling of aminopeptidase N in relation to its biosynthesis in rabbit and pig enterocytes. *Gastroenterology* **82**:317–324.

Feracci, H., Maroux, S., Bonicel, J., and Desnuelle, P., 1982b, The amino acid sequence of the hydrophobic anchor of rabbit intestinal brush border aminopeptidase N. *Biochim. Biophys. Acta* **684**:133–136.

Feracci, H., Rigal, A., and Maroux, S., 1985, Biosynthesis and intracellular pool of aminopeptidase N in rabbit enterocytes. *J. Membr. Biol.* **83**:139–146.

Ferguson, M. A. J., Low, M. G., and Cross, G. A. M., 1985, Glycosyl-*sn*-1,2-dimyristylphosphatidylinositol is covalently linked to *Trypanosoma brucei* variant surface glycoprotein. *J. Biol. Chem.* **260**:14547–14555.

Ferland, S., and Hugon, J. S., 1979, Organ culture of adult mouse intestine. I. Morphological results after 24 and 48 hours of culture. *In Vitro* **15**:278–287.

Finidori, J., Laperche, Y., Haguenauer-Tsapis, R., Barouki, R., Guellaen, G., and Hanoune, J., 1984, In vitro biosynthesis and membrane insertion of gamma-glutamyl transpeptidase. *J. Biol. Chem.* **259**:4687–4690.

Flanagan, P., and Forstner, G. G., 1978, Purification of rat intestinal maltase–glucoamylase and its anomalous dissociation either by heat or by low pH. *Biochem. J.* **173**:553–563.

Forstner, G. G., Sabesin, S. M., and Isselbacher, K. J., 1968, Rat intestinal microvillus membranes. Purification and biochemical characterization. *Biochem. J.* **106**:381–390.

Frank, G., Brunner, J., Hauser, H., Wacker, H., Semenza, G., and Zuber, H., 1978, The hydrophobic anchor of small-intestinal sucrase–isomaltase. N-terminal sequence of the isomaltase subunit. *FEBS Lett.* **96**:183–188.

Fransen, J. A. M., Ginsel, L. A., Hauri. H. P., Sterchi, E. E., and Blok, J., 1985, Immunoelectronmicroscopical localization of a microvillus membrane disaccharidase in the human small-intestinal epithelium with monoclonal antibodies. *Eur. J. Cell Biol.* **38**:6–15.

Fredman, P., Richert, N. D., Magnani, J. L., Willingham, M. C., Pastan, I., and Ginsburg, V., 1983, A monoclonal antibody that precipitates the glycoprotein receptor for epidermal growth factor is directed against human blood group H Type 1 antigen. *J. Biol. Chem.* **258**:11206–11210.

Freedman, R. A., Weiser, M. M., and Isselbacher, K. H., 1977, Calcium translocation by Golgi and lateral-basel membrane vesicles from rat intestine: Decrease in vitamin D-deficient rats. *Proc. Natl. Acad. Sci. U.S.A.* **74**:3612–3616.

Freiburghaus, A. U., Schmitz, J., Schindler, M., Rotthauwe, H. W., Kuitunen, P., Launiala, K., and Hadorn, B., 1976, Protein patterns of brush border membrane fragments in congenital lactose malabsorption and in specific hypolactasia of the adult. *N. Engl. J. Med.* **294**:1030–1032.

Freiburghaus, A. U., Dubs, R., Hadorn, B., Gaze, H., Hauri, H. P., and Gitzelmann, R., 1977, The brush border membrane in hereditary sucrase–isomaltase deficiency: Abnormal protein pattern and presence of immunoreactive enzyme. *Eur. J. Clin. Invest.* **7**:455–459.

Freiburghaus, A. U., Hauri, H. P., Green, J. R., and Hadorn, B., 1978, A micromethod for the separation and identification of digestive enzymes in brush border membrane fragments of single human intestinal biopsies. *Clin. Chim. Acta* **86:**227–234.

Fulcher, I. S., and Kenny, A. J., 1983, Proteins of the kidney microvillar membrane. The amphipathic forms of endopeptidase purified from pig kidneys. *Biochem. J.* **211:**743–753.

Fulcher, I. S., Chaplin, M. F., and Kenny, A. J., 1983, Endopeptidase-24.11 purified from pig intestine is differently glycosylated from that in kidney. *Biochem. J.* **215:**317–323.

Fulcher, I. S., Ingram, J., and Kenny, A. J., 1986a, Radiation inactivation analysis of kidney microvillar peptidases. *FEBS. Lett.* **205:**323–327.

Fulcher, I. S., Pappin, D. J. C., and Kenny, A. J., 1986b, The N-terminal amino acid sequence of pig kidney endopeptidase-24.11 shows homology with pro-sucrase–isomaltase. *Biochem. J.* **240:**305–308.

Fuller, S. D., Bravo, R., and Simons, K., 1985, An enzymatic assay reveals that proteins destined for the apical or basolateral domains of an epithelial cell line share the same late Golgi compartments. *EMBO J.* **4:**297–307.

Galand, G., 1984, Purification and characterization of kidney and intestinal brush border membrane trehalase from the rabbit. *Biochim. Biophys. Acta* **789:**10–19.

Gardell, S., and Tate, S. S., 1981, Subunit interactions in gamma-glutamyl transpeptidase. Reconstitution of the active species from isolated subunits. *J. Biol. Chem.* **256:**4799–4804.

Gardell, S. J., and Tate, S. S., 1982, Modification of the acceptor binding site of gamma-glutamyl transpeptidase by the diazonium derivatives of *p*-aminohippurate and *p*-aminobenzoate. *Arch. Biochem. Biophys.* **216:**719–726.

Garfield, E., 1980, The worldwide problem of lactose intolerance. *Curr. Contents, Life Sci.* **49:**5–8.

Garoff, H., 1985. Using recombinant DNA techniques to study protein targeting in the eukaryotic cell. *Annu. Rev. Cell Biol.* **1:**403–445.

Gee, N. S., and Kenny, A. J., 1985, Proteins of the kidney microvillar membrane. The 130 kDa protein in pig kidney, recognized by monoclonal antibody GK5C1, is an ectoenzyme with aminopeptidase activity. *Biochem. J.* **230:**753–764.

Gee, N. S., Matsas, R., and Kenny, A. J., 1983, A monoclonal antibody to kidney endopeptidase-24.11. Its application in immunoadsorbent purification and immunofluorescent microscopy of kidney and intestine. *Biochem. J.* **214:**377–386.

Gething, M.-J., 1985, Protein transport and secretion. In *Current Communications in Molecular Biology*, Cold Spring Harbor Laboratory, Cold Spring Harbor, NY.

Gething, M.-J., McCammon, K., and Sambrook, J., 1986, Expression of wild-type and mutant forms of influenza hemagglutinin: The role of folding in intracellular transport. *Cell* **46:**939–950.

Ghersa, P., Huber, P., Semenza, G., and Wacker, H., 1986, Cell-free synthesis, membrane integration, and glycosylation of pro-sucrase–isomaltase. *J. Biol. Chem.* **261:**7969–7974.

Gitzelmann, R., Bächi, Th., Binz, H., Lindenmann, J., and Semenza, G., 1970, Localization of rabbit intestinal sucrase with ferritin–antibody conjugates. *Biochim. Biophys. Acta* **196:**20–28.

Goldstein, J., Brown, M. S., Anderson, R. G. W., Russell, D. W., and Schneider, W. J., 1985, Receptor-mediated endocytosis: Concepts emerging from the LDL receptor system. *Annu. Rev. Cell Biol.* **1:**1–39.

Gonschor, H., and Schaefer, W., 1985, Dipeptidyl-peptidase IV. Reinigung für die Verwendung zur Peptidsequenzierung. *Biol. Chem. Hoppe Seyler* **366:**157–165.

Gorbsky, G., and Borisy, G. G., 1985, Micortubule distribution in cultured cells and intact tissues: Improved immunolabeling resolution through the use of reversible embedment cytochemistry. *Proc. Natl. Acad. Sci. U.S.A.* **82:**6889–6893.

Gorvel, J. P., Rigal, A., Bernadac, A., and Maroux, S., 1986, Further characterization of an early expressed glycoprotein of the rabbit small intestinal brush border: Its interaction with some hydrolases. *Eur. J. Biochem.* **159:**499–505.

Gray, G. M., 1981, Carbohydrate absorption and malabsorption. In *Physiology of the Gastrointestinal Tract* (L. R. Johnson, ed.), pp. 1063–1072, Raven Press, New York.

Gray, G. M., Conklin, K. A., and Townley, R. R. W., 1976, Sucrase–isomaltase deficiency: Absence of an inactive enzyme variant. *N. Engl. J. Med.* **294:**750–753.

Gray, G. M., Lally, B. C., and Conklin, K. A., 1979, Action of intestinal sucrase–isomaltase and its free monomers on an alpha-limit dextrin. *J. Biol. Chem.* **254:**6038–6043.

Green, F., Edwards, Y., Hauri, H. P., Povey, S., Ho, M. W., Pinto, M., and Swallow, D., 1987, Isolation of a cDNA probe for a human jejunal brush border hydrolase, sucrase–isomaltase, and assignment of the gene locus to chromosome 3. *Gene* **57:**101–110.

Green, J., Griffiths, G., Louvard, D., Quinn, P., and Warren, G., 1981, Passage of viral membrane proteins through the Golgi complex. *J. Mol. Biol.* **152:**663–698.

Griffiths, G., and Simons, K., 1986, The trans Golgi network: Sorting at the exit site of the Golgi complex. *Science* **234:** 438–443.

Griffiths, G., Pfeiffer, S., Simons, K., and Matlin, K., 1985, Exit of newly synthesized membrane proteins from the trans cisterna of the Golgi complex to the plasma membrane. *J. Cell. Biol.* **101:**949–964.

Hadorn, B., Green, J. R., Sterchi, E. E., and Hauri, H. P., 1981. Biochemical mechanisms in congenital enzyme deficiencies of the small intestine. *Clin. Gastroenterol.* **10:**671–690.

Haffen, K., Kedinger, M., and Lacroix, B., 1986, Cytodifferentiation of the intestinal villus epithelium. In *Molecular and Cellular Basis of Digestion* (P. Desnuelle, H. Sjöström, and O. Norén, eds.), pp. 311–322, Elsevier, Amsterdam.

Harris, H., 1983, Applications of monoclonal antibodies in enzyme genetics. *Annu. Rev. Genet.* **17:**279–314.

Harris, R. B., and Wilson, I. B., 1985, Sequencing of an active-site peptide of angiotensin I-converting enzyme containing an essential glutaminc residue. *J. Biol. Chem.* **260:**2208–2211.

Hauri, H. P., 1983, Biosynthesis and transport of plasma membrane glycoproteins in the rat intestinal epithelial cell: Studies with sucrase–isomaltase. In *Brush Border Membranes* (R. Porter and G. Collins, eds.), Ciba Foundation Symposium 95, pp. 132–147, Pitman, London.

Hauri, H. P., 1986, Use of monoclonal antibodies to investigate the intracellular transport and biogenesis of intestinal brush border proteins. *Biochem. Soc. Trans.* **14:**161–163.

Hauri, H. P., and Bucher, K., 1986, Immunoblotting with monoclonal antibodies: Importance of the blocking solution. *Anal. Biochem.* **159:**386–389.

Hauri, H. P., Kedinger, M., Haffen, K., Grenier, J. F., and Hadorn, B., 1975, Organ culture of human duodenum and jejunum. *Biol. Gastroenterol.* **8:**307–319.

Hauri, H. P., Kedinger, M., Haffen, K., Freiburghaus, A., Grenier, J. F., and Hadorn, B., 1977, Biosynthesis of brush border glycoproteins by human small intestinal mucosa in organ culture. *Biochim. Biophys. Acta* **467:**327–339.

Hauri, H. P., Quaroni, A., and Isselbacher, K. J., 1979, Biogenesis of intestinal plasma membrane: Posttranslational route and cleavage of sucrase–isomaltase. *Proc. Natl. Acad. Sci. U.S.A.* **76:**5183–5186.

Hauri, H. P., Quaroni, A., and Isselbacher, K. J., 1980, Monoclonal antibodies to sucrase–isomaltase: Probes for the study of postnatal development and biogenesis of the intestinal microvillus membrane. *Proc. Natl. Acad. Sci. U.S.A.* **77:**6629–6633.

Hauri, H. P., Wacker, H., Rickli, E. E., Bigler-Meier, B., Quaroni, A., and Semenza, G., 1982, Biosyntehsis of sucrase–isomaltase. Purification and NH_2-terminal amino acid sequence of the rat sucrase–isomaltase precursor (pro-sucrase–isomaltase) from fetal intestinal transplants. *J. Biol. Chem.* **257:**4522–4528.

Hauri, H. P., Roth, J., Sterchi, E., and Lentze, M., 1985a, Transport to cell surface of intestinal sucrase–isomaltase is blocked in the Golgi apparatus in a patient with congenital sucrase–isomaltase deficiency. *Proc. Natl. Acad. Sci. U.S.A.* **82:**4423–4427.

Hauri, H. P., Sterchi, E. E., Bienz, D., Fransen, J. A. M., and Marxer, A., 1985b, Expression and intracellular transport of microvillus membrane hydrolases in human intestinal epithelial cells. *J. Cell Biol.* **101**:838–851.

Hauri, H. P., Roth, J., and Kedinger, M., 1986, Congenital sucrase–isomaltase deficiency due to hold-up at the Golgi apparatus. *Experientia* **42**:634 (abstract).

Hauser, H., and Semenza, G., 1983, Sucrase–isomaltase: A stalked intrinsinc protein of the brush border membrane. *CRC Crit. Rev. Biochem.* **14**:319–345.

Henning, S. J., 1985, Ontogeny of enzymes in the small intestine. *Annu. Rev. Physiol.* **47**:231–245.

Henthorn, B. S., Knoll, B. J., Raducha, M., Rothblum, K. N., Slaughter, C., Weiss, M., Lafferty, M. A., Fischer, T., and Harris, H., 1986, Products of two common alleles at the locus for human placental alkaline phosphatase differ by seven amino acids. *Proc. Natl. Acad. Sci. U.S.A.* **83**:5597–5601.

Herbst, J. J., Hurwitz, R., Sunshine, P., and Kretchmer, N., 1970, Effect of colchicine on intestinal disaccharidases: Correlation with biochemical aspects of cellular renewal. *J. Clin. Invest.* **49**:530–536.

Herskovics, A., Quaroni, A., Bugge, B., and Kirsch, K., 1981, Partial characterization of the carbohydrate units of rat intestinal sucrase–isomaltase. *Biochem. J.* **197**:511–514.

Hopfer, Y., Nelson, K., Perotto, J., and Isselbacher, K. J., 1973, Glucose transport in isolated brush border membrane from rat small intestine. *J. Biol. Chem.* **248**:25–32.

Hopfer, U., Crowe, T. D., and Tandler, B., 1983, Purification of brush border membrane by thiocyanate treatment. *Anal. Biochem.* **131**:447–452.

Howdle, P. D., 1983, Organ culture in the study of the gastrointestinal tract in health and disease. *Clin. Sci.* **65**:105–110.

Hudson, D. A., and Smith, M. W., 1986, Colchicine selectively inhibits lactase expression by rat enterocytes. *Biochem. Pharmacol.* **35**:3929–3934.

Hughey, R. P., and Curthoys, N. P., 1976, Comparison of the size and physical properties of gamma-glutamyltranspeptidase purified from rat kidney following solubilization with papain or with Triton X-100. *J. Biol. Chem.* **251**:7863–7880.

Hughey, R. P., Altman, R. A., Wells, W. J., and Curto, K. A., 1986, Evidence for stable homodimers and heterodimers of gamma-glutamyl transpeptidase subunits under protein-denaturing conditions. *Biochim. Biophys. Acta* **874**:150–159.

Hugon, J., and Borgers, M., 1966, Ultrastructural localization of alkaline phosphatase activity in the absorbing cells of the duodenum of mouse. *J. Histochem. Cytochem.* **14**:629–640.

Hunziker, W., Spiess, M., Semenza, G., and Lodish, H. F., 1986, The sucrase–isomaltase complex: Primary structure, membrane orientation, and evolution of a stalked, intrinsic brush border protein. *Cell* **46**:227–234.

Hussain, M. M., 1985, Reconstitution of purified dipeptidyl peptidase IV. A comparison with aminopeptidase N with respect to morphology and influence of anchoring peptide on function. *Biochim. Biophys. Acta* **815**:306–312.

Hussain, M. M., Tranum-Jensen, J., Norén, O., Sjostrom, H., and Christiansen, K., 1981, Reconstitution of purified amphiphilic pig intestinal microvillus aminopeptidase. *Biochem. J.* **199**:179–186.

Ito, F., and Chou, J. Y., 1983, Biosynthesis and processing of placental alkaline phosphatase. *Biochem. Biophys. Res. Commun.* **111**:611–618.

Ito, F., and Chou, J. Y., 1984, Suppression of placental alkaline phosphatase biosynthesis by tunicamycin. *J. Biol. Chem.* **259**: 14997–14999.

Johnson, C. F., 1967, Disaccharidase: Localization in hamster intestine brush borders. *Science* **155**:1470–1472.

Kam, W., Clauser, E., Kim, Y. S., Kan, Y. W., and Rutter, W. J., 1985, Cloning, sequencing, and chromosomal localization of human term placental alkaline phosphatase cDNA. *Proc. Natl. Acad. Sci. U.S.A.* **82**:8715–8719.

Kaufman, M. A., Korsmo, H. A., and Olsen, W. A., 1980, Circadian rhythm of intestinal sucrase activity in rats. Mechanism of enzyme change. *J. Clin. Invest.* **65:**1174–1181.

Kedinger, M., Simon, P. M., Raul, F., Grenier, J. F., and Haffen, K., 1980, The effect of dexamethasone on the development of rat intestinal brush border enzymes in organ culture. *Dev. Biol.* **74:**9–21.

Kedinger, M., Haffen, K., and Simon-Assmann, P., 1986a, Control mechanisms in the ontogenesis of villus cells. In *Molecular and Cellular Basis of Digestion* (P. Desnuelle, H. Sjöström, and O. Norén, eds.), pp. 323–334, Elsevier, Amsterdam.

Kedinger, M., Simon-Assmann, P. M., Lacroix, B., Marxer, A., Hauri, H. P., and Haffen, K., 1986b. Fetal gut mesenchyme induces differentiation of cultured intestinal endodermal and crypt cells. *Dev. Biol.* **113:**474–483.

Kedinger, M., Simon-Assmann, P. Alexandre, E., and Haffen, K., 1987, Importance of a fibroblastic support for in vitro differentiation of intestinal endodermal cells and for their response to glucocorticoids. *Cell Differ.* **20:**171–182.

Kelly, J. J., and Alpers, D. H., 1973a, Blood group antigenicity of purified human intestinal disaccharidases. *J. Biol. Chem.* **248:**8216–8221.

Kelly, J. J., and Alpers, D. H., 1973b, Properties of human intestinal glucoamylase. *Biochim. Biophys. Acta* **315:**113–120.

Kelly, R. B., 1985, Pathways of protein secretion in eukaryotes. *Science* **230:**25–33.

Kenny, A. J., and Fulcher, I. S., 1983, Microvillar endopeptidase, an enzyme with special topological features and wide distribution. In *Brush Border Membranes* (R. Porter and G. Collins, eds.), pp. 12–25, Ciba Foundation Symposium 95, Pitman, London.

Kenny, A. J., Booth, A. G., George, S. G., Ingram, J., Kershaw, D., Wood, E. J., and Young, A. R., 1976, Dipeptidyl peptidase IV, a kidney brush border serine peptidase. *Biochem. J.* **155:**169–192.

Kenny, A. J., Bowes, M. A., Gee, N. S.,and Matsas, R., 1985, Endopeptidase-24.11: A cell-surface enzyme for metabolizing regulatory peptides. *Biochem. Soc. Trans.* **13:**293–295.

Kenny, A. J., Fulcher, I. S., McGill, K. A., and Kershaw, D., 1983, Proteins of the kidney microvillar membrane. Reconstitution of endopeptidase in liposomes shows that it is a short-stalked protein. *Biochem. J.* **211:**755–762.

Kenny, J., 1986, Cell surface peptidases are neither peptide- nor organ-specific. *Trends Pharmacol. Sci.* **11:**40–42.

Kenny, J., and Maroux, S., 1982, Topology of microvillar membrane hydrolases of kidney and intestine. *Physiol. Rev.* **62:**91–128.

Kerry, K. R., and Townley, R. R. W., 1965, Genetic aspects of intestinal sucrase–isomaltase deficiency. *Aust. Paediatr. J.* **1:**223–235.

Kim, Y. S., and Brophy, E. J., 1976, Rat intestinal brush border membrane peptidases. I. Solubilization, purification and physicochemical properties of two different forms of the enzyme. *J. Biol. Chem.* **251:**3199–3205.

Köhler, G., and Milstein, C., 1975, Continuous cultures of fused cells secreting antibody of predefined specificity. *Nature* **256:**495–497.

Kornfeld, R., and Kornfeld, S., 1985, Assembly of asparagine-linked oligosaccharides. *Annu. Rev. Biochem.* **54:**631–664.

Kreis, T., and Lodish, H., 1986, Oligomerization is essential for transport of vesicular stomatitis viral glycoprotein to the cell surface. *Cell* **46:**929–937.

Kretchmer, N., 1972, Lactose and lactase. *Sci. Am.*, October, 71–78.

Kuno, T., Matsuda, Y., and Katunuma, N., 1983, The conversion of the precursor form of gamma-glutamyltranspeptidase to its subunit form takes place in brush border membranes. *Biochem. Biophys. Res. Commun.* **114:**889–895.

Kurisu, M., Numanyu, N., Kawachi, T., and Sugimura, T., 1977, Blood group activity of human sucrase from intestinal metaplasia. *J. Biol. Chem.* **252**:3277–3280.

Laperche, Y., Bulle, F., Aissani, T., Chobert, M.-N., Aggerbeck, M., Hanoune, J., and Guellaen, G., 1986, Molecular cloning and nucleotide sequence of rat kidney gamma-glutamyl transpeptidase cDNA. *Proc. Natl. Acad. Sci. U.S.A.* **83**:937–941.

Leapman, S. B., and Deutsch, A. A., Grand, R. J., and Folkman, J., 1974, Transplantation of fetal intestine: Survival and function in a subcutaneous location in adult animals. *Ann. Surg.* **179**:109–114.

Lee, L., and Forstner, G., 1984, Rat intestinal maltase–glucoamylase. Purification of the detergent-solubilized enzyme in the presence of protease inhibitors: Properties and identification of a protease-sensitive subunit. *Can. J. Biochem. Cell Biol.* **62**:36–43.

Lee, L., and Forstner, G., 1985, Mapping of proteolytic and cyanogen bromide peptides from subunits of intestinal maltase–glucoamylase: Evidence for significant homology. *Can. J. Biochem. Cell Biol.* **63**:257–262.

Lee, L. M. Y., Salvatore, A. K., Flanagan, P. R., and Forstner, G. G., 1980, Isolation of a detergent-solubilized maltase–glucoamylase from rat intestine and its comparison with a maltase–glucoamylase solubilized by papain. *Biochem. J.* **187**:437–446.

Leonard, W. J., Depper, J. M., Crabtree, G. R., Rudikoff, S., Pumphrey, J., Robb, R. J., Kronkë, M., Svetlik, P. B., Pfeffer, N. J., Waldmann, T. A., and Greene, W. C., 1984, Molecular cloning and expression of cDNAs for the human interleukin-2 receptor. *Nature* **311**:626–631.

Lloyd, M. L., and Olsen, W. A., 1987, A study of the molecular pathology of sucrase–isomaltase deficiency. A defect in the intracellular processing of the enzyme. *New Engl. J. Med.* **316**:438–442.

Lodish, H. F., Braell, W. A., Schwartz, A. L., Strous, G. J. A. M., and Zilberstein, A., 1981, Synthesis and assembly of membrane and organelle proteins. *Int. Rev. Cytol. Suppl.* **12**:247–307.

Lojda, Z., and Gossrau, R., 1980, Study on aminopeptidase A. *Histochemistry* **67**:267–290.

Louvard, D., Semeriva, M., and Maroux, S., 1976, The brush border intestinal aminopeptidase, a transmembrane protein as probed by macromolecular photolabelling. *J. Mol. Biol.* **106**:1023–1035.

Low, M. G., and Zilversmit, D. B., 1980, Role of phosphatidylinositol in attachment of alkaline phosphatase to membranes. *Biochemistry* **19**:3913–3918.

Low, M. G., Ferguson, M. A. J., Futerman, A. H., and Silman, I., 1986, Covalently attached phosphatidylinositol as a hydrophobic anchor for membrane proteins. *Trends Biochem. Sci.* **11**:212–215.

MacNair, R. D. C., and Kenny, A. J., 1979, Proteins of the kidney microvillar membrane. The amphiphilic form of dipeptidyl peptidase IV. *Biochem. J.* **179**:379–395.

Malo, C., Arsenault, P. and Menard, D., 1983, Organ culture of the small intestine of the suckling mouse in a serum-free medium. *Cell Tissue Res.* **228**:75–84.

Marathe, G. V., Nash, B., Haschemeyer, R. H., and Tate, S. S., 1979, Ultrastructural localization of gamma-glutamyl transpeptidase in kidney and jejunum. *FEBS Lett.* **107**:436–440.

Marchesi, V., Furthmayr, H., and Tomita, M., 1976, The red cell membrane. *Annu. Rev. Biochem.* **45**:667–698.

Maroux, S., and Feracci, H., 1983, Structure and biosynthesis of aminopeptidases. *Methods Enzymol.* **96**:406–423.

Maroux, S., Louvard, D., and Baratii, J., 1973, The aminopeptidase from hog intestinal brush border. *Biochim. Biophys. Acta* **321**:282–295.

Maroux, S., Feracci, H., Gorvel, J. P., and Benajiba, A., 1983, Aminopeptidase and proteolipids of intestinal brush border. In *Brush Border Membranes* (R. Porter and G. Collins, eds.), Ciba Foundation Symposium 95, pp. 34–44, Pitman, London.

Marsh, M., Griffiths, G., Dean, G. E., Mellman, I., and Helenius, A., 1986, Three-dimensional structure of endosomes in BHK-21 cells. *Proc. Natl. Acad. Sci. U.S.A.* **83**:2899–2903.

Massey, D., and Maroux, S., 1985, The carbohydrate moiety of aminopeptidase N of rabbit intestinal brush-border membrane. *FEBS Lett.* **181**:207–210.

Matlin, K. S., 1986, Ammonium chloride slows transport of the influenza virus hemagglutinin but does not cause mis-sorting in a polarized epithelial cell line. *J. Biol. Chem.* **261**:15172–15178.

Matlin, K. S., and Simons, K., 1983, Reduced temperature prevents transfer of a membrane glycoprotein to the cell surface but does not prevent terminal glycosylation. *Cell* **34**: 233–243.

Matlin, K. S., and Simons, K., 1984, Sorting of an apical plasma membrane glycoprotein occurs before it reaches the cell surface in cultured epithelial cells. *J. Cell Biol.* **99**:2131–2139.

Matlin, K., Bainton, D. F., Pesonen, M., Louvard, D., Genty, N., and Simons, K., 1983, Transepithelial transport of a viral membrane glycoprotein implanted into the apical plasma membrane of Madin–Darby canine kidney cells. I. Morphological evidence. *J. Cell Biol.* **97**:627–637.

Matsuda, Y., Tsuji, A., and Katunuma, N., 1980, Studies on the structure of gamma-glutamyl transpeptidase. I. Correlation between sialylation and isozymic forms. *J. Biochem.* **87**:1243–1248.

Matsuda, Y., Tsuji, A., and Katunuma, N., 1983, Membrane bound gamma-glutamyl transpeptidase: Its structure, biosynthesis and degradation. In *Advances of Enzyme Regulation* (G. Weber, ed.), Vol. 21, pp. 103–119, Pergamon Press, New York.

McNiven, M. A., and Porter, K. R., 1986, Microtubule polarity confers direction to pigment transport in chromatophores. *J. Cell. Biol.* **103**:1547–1555.

Mellman, I., Fuchs, R., and Helenius, A., 1986, Acidification of the endocytotic and exocytotic pathways. *Annu. Rev. Biochem.* **55**:663–700.

Millan, J. L., 1986. Molecular cloning and sequence analysis of human placental alkaline phosphatase. *J. Biol. Chem.* **261**:3112–3115.

Miller, D., and Crane, R. K., 1961. The digestive function of the epithelium of the small intestine. II. Localization of disaccharide hydrolysis in the isolated brush border portion of intestinal epithelial cells. *Biochim. Biophys. Acta* **52**:293–298.

Misek, D. E., Bard, E., and Rodriguez-Boulan, E., 1984, Biogenesis of epithelial cell polarity: Intracellular sorting and vectorial exocytosis of an apical plasma membrane glycoprotein. *Cell* **39**:537–546.

Moktari, S., Feracci, H,. Gorvel, J. P., Mishal, Z., Rigal, A., and Maroux, S., 1986, Subcellular fractionation and subcellular localization of aminopeptidase N in the rabbit enterocytes. *J. Membr. Biol.* **89**:53–63.

Montgomery, R. K., Sybicki, M. A., Forcier, A. G., and Grand, R. J., 1981, Rat intestinal microvillus membrane sucrase–isomaltase is a single high molecular weight protein and fully active enzyme in the absence of luminal factors. *Biochim. Biophys. Acta* **661**:346–349.

Mooseker, M. S., 1985, Organization, chemistry, and assembly of the cytoskeletal apparatus of the intestinal brush border. *Annu. Rev. Cell Biol.* **1**:209–241.

Morita, A., Miura, S., Erickson, R., Sleisenger, M. H., and Kim, Y. S., 1986, Comparison of brush border membrane glycoproteins and glycoenzymes in the proximal and distal rat small intestine. *Biochim. Biophys. Acta* **883**:506–516.

Mosimann, H., Semenza, G., and Sund, H., 1973, Hydrodynamic properties of the sucrase–isomaltase complex from rabbit small intestine. *Eur. J. Biochem.* **36**:489–494.

Mostov, K., and Deitcher, D. L., 1986, Polymeric immunoglobulin receptor expressed in MDCK cells transcytoses IgA. *Cell* **46**:613–621.

Mostov, K. E., and Simister, N. E., 1985, Transcytosis. *Cell* **43**:389–390.

Mulivor, R. A., Hanning, V., and Harris, H., 1978. Developmental changes in human intestinal alkaline phosphatase. *Proc. Natl. Acad. Sci. U.S.A.* **78**:3909–3912.

Murer, H., Hopfer, U., Kinne-Saffran, E., and Kinne, R., 1974, Glucose transport in isolated brush border and basal-lateral membrane vesicles from intestinal epithelial cells. *Biochim. Biophys. Acta* **345**:170–179.

Murer, H., Hopfer, Y., and Kinne, R., 1976, Sodium/proton antiport in brush border membrane vesicles isolated from rat small intestine and kidney. *Biochem. J.* **154**:597–604.

Naim, H. Y., Sterchi, E. E., Hauri, H. P., and Lentze, M. J., 1986, Purification and biosynthesis of the brush border hydrolases of the human small intestine. In *Ion Gradient-Coupled Transport* (F. Alvarado and C. H. van Oss, eds.), pp. 71–74, Elsevier, Amsterdam.

Naim, H. Y., Sterchi, E. E., and Lentze, M. J., 1987, Biosynthesis and maturation of lactase–phlorizin hydrolase of the human small intestinal epithelial cells. *Biochem. J.* **241**:427–434.

Naim, H. Y., Roth, J., Sterchi, E., Lentze, M., Milla, P., Schmitz, J., and Hauri, H.-P., 1988, Sucrase–isomaltase deficiency in man. Different mutations disrupt intracellular transport, processing and function of an intestinal brush border enzyme. *J. Clin. Invest.* (in press).

Nakano, M., 1982, Localization of renal and intestinal trehalase with immunofluorescence- and enzyme-labeled antibody techniques. *J. Histochem. Cytochem.* **30**:1243–1248.

Nakano, M., and Sacktor, B., 1984, Renal trehalase: Two subsites at the substrate-binding site. *Biochim. Biophys. Acta* **791**:45–49.

Nakano, M., and Sacktor, B., 1985, Isolation and characterization of four forms of trehalase from rabbit kidney cortex. *J. Biochem.* **97**:1329–1335.

Nakano, M., Sumi, Y., and Mikakawa, M., 1977, Purification and properties of trehalase from rat intestinal mucosal cells. *J. Biochem.* **81**:1041–1049.

Nash, B., and Tate, S. S., 1982, Biosynthesis of rat renal gamma-glutamyl transpeptidase. *J. Biol. Chem.* **257**:585–588.

Nash, B., and Tate, S. S., 1984, In vitro translation and processing of rat kidney gamma-glutamyl transpeptidase. *J. Biol. Chem.* **259**:678–685.

Negrel, R., Rampal, P., Nano, J.-L., Cavenel, C., and Ailhaud, G., 1983, Establishment and characterization of an epithelial intestinal cell line from rat fetus. *Exp. Cell Res.* **143**:427–437.

Neutra, M., and Leblond, C. P., 1966, Radioautographic comparison of the uptake of galactose-H_3 and glucose-H_3 in the Golgi region of various cells secreting glycoproteins or mucopolysaccharides. *J. Cell Biol.* **30**:137–140.

Nikaido, T., Shimizu, A., Ishida, N., Sabe, H., and Teshigawara, K., 1984, Molecular cloning of cDNA encoding human interleukin-2 receptors. *Nature* **311**:631–635.

Nishi, Y., and Takesue, Y., 1978a, Localization of intestinal sucrase–isomaltase complex on the microvillus membrane by electron microscopy using nonlabeled antibodies. *J. Cell Biol.* **79**:516–525.

Nishi, Y., and Takesue, Y., 1978b, Electron microscope studies on Triton-solubilized sucrase from rabbit small intestine. *J. Ultrastruct. Res.* **62**:1–12.

Norén, O., and Sjöström, H., 1980, The insertion of pig microvillus aminopeptidase into the membrane as probed by (^{125} I) iodonaphthylazide. *Eur. J. Biochem.* **104**:25–31.

Norén, O., Sjöström, H., Danielsen, M. E., Staun, M., Jeppesen, L., and Svensson, B., 1979, Comparison of two pig intestinal brush border peptidases with the corresponding renal enzymes. *Hoppe Seylers Z. Physiol. Chem.* **360**:151–157.

Norén, O., Sjöström, H., Cowell, G. M., Tranum-Jensen, J., Hansen, O. C., and Welinder, K. G., 1986a, Pig intestinal microvillar maltase–glucoamylase. Structure and membrane insertion. *J. Biol. Chem.* **261**:12306–12309.

Norén, O., Sjöström, H., Danielsen, E. M., Cowell, G. M., and Skovbjerg, H., 1986b, The enzymes of the enterocyte plasma membrane. In *Molecular and Cellular Basis of Digestion* (P. Desnuelle, H. Sjöström, and O. Norén, eds.), pp. 335–365, Elesevier, Amsterdam.

Orci, L., Halban, P., Amherdt, M., Ravazzola, M., Vasalli, J.-D., and Perrelet, A., 1984a, A clathrin-coated, Golgi-related compartment of the insulin secreting cell accumulates proinsulin in the presence of monensin. *Cell* **39**:39–47.

Orci, L., Halban, P., Amherdt, M., Ravazzola, M., Vassalli, D., and Perrelet, A., 1984b, Nonconverted, amino acid analog-modified proinsulin stays in a Golgi-derived clathrin-coated membrane compartment. *J. Cell Biol.* **99:**2187–2192.

Orci, L., Ravazzola, M., Amherdt, M., Madsen, O., Vassalli, J. D., and Perrelet, A., 1985, Direct identification of prohormone conversion site in insulin-secreting cells. *Cell* **42:**671–681.

Orci, L., Ravazzola, M., Amherdt, M., Madsen, O., Vassali, J. D., and Anderson, R. G. W., 1986, Conversion of proinsulin occurs coordinately with acidification of maturing secretory vesicles. *J. Cell Biol.* **103:**2273–2281.

Oude Elferink, R. P. J., Strijland, A., Surya, I., Brouwer-Kelder, E. M., Kroos, M., Hilkens, J., Hilgers, J., Reuser, A. J. J., and Tager. J. M., 1984, Use of a monoclonal antibody to distinguish between precursor and mature forms of human lysosomal alpha-glucosidase. *Eur. J. Biochem.* **139:**497–502.

Palade, G. E., 1982, Problems in intracellular membrane traffic. In *Membrane Recycling.* (D. Evered and G. M. Collins, eds.), Ciba Foundation Symposium 92, pp. 1–28, Pitman, London.

Pan, Y. T., Hori, H., Saul, R., Sanford, B. A., Molyneux, R. J., and Elbein, A. D., 1983, Castanospermine inhibits the processing of the oligosaccharide portion of the influenza viral hemagglutinin. *Biochemistry* **22:**3975–3984.

Pavelka, M., and Ellinger, A., 1981, Effects of colchicine on the Golgi apparatus and on GERL of rat jejunal absorptive cells. Ultrastructural localization of thiamine pyrophosphatase and acid phosphatase activity. *Eur. J. Cell Biol.* **24:**53–61.

Pavelka, M., and Ellinger, A., 1986, RCA I-binding patterns of the Golgi apparatus. *Eur. J. Cell Biol.* **41:**270–278.

Pavelka, M., and Gangl, A., 1983, Effects of colchicine on the intestinal transport of endogenous lipid. Ultrastructural, biochemical and radiochemical studies in fasting rats. *Gastroenterology* **84:**544–555.

Pavelka, M., Ellinger, A., and Gangl, A., 1983, Effects of colchicine on rat small intestinal absorptive cells. I. Formation of basolateral microvillus borders. *J. Ultrastruct. Res.* **85:**249–259.

Pesonen, M., and Simons, K., 1983, Transepithelial transport of a viral membrane glycoprotein implanted into the apical plasma membrane of Madin–Darby canine kidney cells. II. Immunological quantitation. *J. Cell Biol.* **97:**638–643.

Pesonen, M., Ansorge, W., and Simons, K., 1984a, Transcytosis of the G protein of vesicular stomatitis virus after implantation into the apical plasma membrane of Madin–Darby canine kidney cells. I: Involvement of endosomes and lysosomes. *J. Cell Biol.* **99:**796–802.

Pesonen, M., Bravo, R., and Simons, K., 1984b, Transcytosis of the G protein of vesicular stomatitis virus after implantation into the apical membrane of Madin–Darby canine kidney cells. II. Involvement of the Golgi complex. *J. Cell Biol.* **99:**803–809.

Pestalozzi, D. M., Hess, M., and Berger, E. G., 1982, Immunohistochemical evidence for cell surface and Golgi localization of galactosyltransferase in human stomach, jejunum, liver and pancreas. *J. Histochem. Cytochem.* **30:**1146–1152.

Pfeiffer, S., Fuller, S. D., and Simons, K., 1985, Intracellular sorting and basolateral appearance of the G protein of vesicular stomatitis virus in Madin–Darby canine kidney cells. *J. Cell Biol.* **101:**470–476.

Pinto, C., Robine-Leon, S., Appay M.-D., Kedinger, M., Triadou, N., Dassaulx, E., Lacroix, B., Simon-Assmann, P., Haffen, K., Fogh, J., and Zweibaum, A., 1983, Enterocyte-like differentiation and polarization of the human colon carcinoma cell line Caco-2 in culture. *Biol. Cell.* **47:**323–330.

Pinto, M., Appay, M.-D., Simon-Assmann, P., Chevalier, G., Dracopoli, N., Fogh, J. and Zweibaum, A., 1982, Enterocyte differentiation of cultured human colon cancer cells by replacement of glucose by galactose in the medium. *Biol. Cell* **44:**193–196.

Poruchynsky, M. S., Tyndall, C., Both, G. W., Sato, F., Bellamy, A. R., Atkinson, P. H., 1985,

Deletions into an NH_2-terminal hydrophobic domain result in secretion of rotavirus VP7, a resident endoplasmic reticulum membrane glycoprotein. *J. Cell Biol.* **101**:2199–2209.

Potter, J., Ho, M.-W., Bolton, H., Furth, A. J., Swallow, D. M., and Griffiths, B., 1985, Human lactase and molecular basis of lactase persistence. *Biochem. Genet.* **23**:423–439.

Preiser, H., Menard, D., Crane, R. K., and Cerda, J. J., 1974, Deletion of enzyme protein from the brush border membrane in sucrase–isomaltase deficiency. *Biochim. Biophys. Acta* **363**:279–282.

Puddington, L., Machamer, C. E., and Rose, J. K., 1986, Cytoplasmic domains of cellular and viral integral membrane proteins substitute for the cytoplasmic domain of the vesicular stomatitis virus glycoprotein in transport to the plasma membrane. *J. Cell Biol.* **102**:2147–2157.

Quaroni, A., 1983, Use of monoclonal antibodies in the study of intestinal structure and function. In *Brush Border Membranes* (R. Porter and G. Collins, eds.), Ciba Foundation Symposium 95, pp. 113–127, Pitman, London.

Quaroni, A., 1984. Identification and biogenesis of intestinal cell surface components. *Prog. Cancer Res. Therapy* **29**:267–293.

Quaroni, A., 1985a, Crypt cell development in newborn rat small intestine. *J. Cell Biol.* **100**:1601–1610.

Quaroni, A., 1985b, Development of fetal rat intestine in organ and monolayer culture. *J. Cell Biol.* **100**:1611–1622.

Quaroni, A., 1986. Crypt cell antigen expression in human colon tumor cell lines: Analysis with a panel of monoclonal antibodies to Caco-2 luminal membrane components. *J. Natl. Cancer Inst.* **76**:571–585.

Quaroni, A., and Isselbacher, K. J., 1981, Cytotoxic effects and metabolism of benzo[*a*]pyrene and 7,12-dimethylbenz[*a*]anthracene in duodenal and ileal epithelial cell cultures. *J. Natl. Cancer Inst.* **67**:1353–1362.

Quaroni, A., and Isselbacher, K. J., 1985, Study of intestinal cell differentiation with monoclonal antibodies to intestinal cell surface components. *Dev. Biol.* **111**:267–279.

Quaroni, A., and May, R., 1980, Establishment and characterization of intestinal epithelial cell cultures. *Methods Cell Biol.* **21b**:403–427.

Quaroni, A., Kirsch, K., and Weiser, M. M., 1979a, Synthesis of membrane glycoproteins in rat small-intestinal villus cells. Redistribution of L-(1,5,6-^{3}H)fucose-labeled membrane glycoproteins among Golgi, lateral basal and microvillus membranes in vivo. *Biochem. J.* **182**:203–212.

Quaroni, A., Kirsch, K., and Weiser, M. M., 1979b, Synthesis of membrane glycoproteins in rat small-intestinal villus cells. Effect of colchicine on the redistribution of L-(1,5,6-^{3}H)fucose-labeled membrane glycoproteins among Golgi, lateral basal and microvillus membranes. *Biochem. J.* **182**:213–221.

Quaroni, A., Wands, J., Trelstad, R. L., and Isselbacher, K. J., 1979c, Epitheloid cell cultures from rat small intestine. Characterization by morphologic and immunologic criteria. *J. Cell Biol.* **80**:248–265.

Ramond, M.-J., and Martinot-Peignoux, M., 1985, Dome formation in the human colon carcinoma cell line Caco-2 in culture. Influence of ouabain and permeable supports. *Biol. Cell* **54**:89–92.

Reaven, E. P., and Reaven, G. M., 1977, Distribution and content of microtubules in relation to the transport of lipid. *J. Cell Biol.* **75**:559–572.

Reiss, U., and Sacktor, B., 1981, Kidney brush border membrane maltase: Purification and properties. *Arch. Biochem. Biophys.* **209**:342–348.

Reiss, U., and Sacktor, B., 1983, Monoclonal antibodies to renal brush border mambrane maltase: Age-associated antigenic alterations. *Proc. Natl. Acad. Sci. U.S.A.* **80**:3255–3259.

Ribi, J. E., and Kretchmer, N., 1984, Effect of dietary sucrose on synthesis and degradation of intestinal sucrase. *Am. J. Physiol.* **246**:757–763.

Riendeau, D., Lemaire, J., Maestracci, D., and Lessard, L., 1986, Selective release of inner core proteins from intestinal microvillus membrane by lithium diiodosalicylate. *Mol. Cell. Biochem.* **71**:45–52.

Rindler, M. J., Ivanov, I. E., Plesken, H., Rodriguez-Boulan, E., and Sabatini, D. D., 1984, Viral glycoproteins destined for apical or basolateral plasma membrane domains traverse the same Golgi apparatus during their intracellular transport in doubly infected Madin–Darby canine kidney cells. *J. Cell Biol.* **98**:1304–1319.

Rindler, M. J., Ivanov, I. E., Plesken, H., and Sabatini, D. D., 1985, Polarized delivery of viral glycoproteins to the apical and basolateral plasma membranes of Madin–Darby canine kidney cells infected with temperature-sensitive viruses. *J. Cell Biol.* **100**:136–151.

Rodriguez, I. R., Taravel, F. T., and Whelan, W. J., 1984, Characterization and function of pig intestinal sucrase–isomaltase and its separate subunits. *Eur. J. Biochem.* **143**:575–582.

Rodriguez-Boulan, E., 1983, Polarized assembly of enveloped viruses from cultured epithelial cells. *Methods Enzymol.* **98**: 486–501.

Rodriguez-Boulan, E., 1986, Genesis of polarity in renal tubular cells. *Mineral Electrolyte Metab.* **12**:20–24.

Rodriguez-Boulan, E., Paskiet, K. T., Salas, P. J. I., and Bard, E., 1984, Intracellular transport of influenza virus hemaglutinin to the apical surface of Madin–Darby canine kidney cells. *J. Cell Biol.* **98**:308–319.

Rodriguez-Boulan, E., Misek, D., DeSalas, D. V., Salas, J. I., and Bard, E., 1985, Protein sorting in the secretory pathway. *Curr. Top. Membr. Transport* **24**:251–294.

Rogalski, A. A., Bergmann, J. E., and Singer, S. J., 1984, Effect of microtubule assembly status on the intracellular processing and surface expression of an integral protein of the plasma membrane. *J. Cell Biol.* **99**:1101–1109.

Romero, P. A., Datema, R., and Schwarz, R. T., 1983, *N*-methyl-1-deoxynojirimycin, a novel inhibitor of glycoprotein processing and its effect on fowl plague virus maturation. *Virology* **130**:238–242.

Roth, J., Lentze, M. J., and Berger, E. G., 1985a, Immunocytochemical demonstration of ecto-galactosyltransferase in absorptive intestinal cells. *J. Cell Biol.* **100**:118–125.

Roth, J., Taatjes, D. J., Lococq, J. M., Weinstein, J., and Paulson, J. C., 1985b, Deomonstration of an extensive trans-tubular network continuous with the Golgi apparatus stack that may function in glycosylation. *Cell* **43**:287–295.

Roth, J., Taatjes, D. J., Weinstein, J., Paulson, J. C., Greenwell, P., and Watkins, W. M., 1986, Differential subcompartmentation of terminal glycosylation in the Golgi apparatus of intestinal absorptive and goblet cells. *J. Biol. Chem.* **261**:14307–14312.

Roth, M. G., Doyle, C., Sambrook, J., and Gething, M. J., 1986, Heterologous transmembrane and cytoplasmic domains direct functional chimeric influenza virus hemagglutinins into the endocytic pathway. *J. Cell Biol.* **102**:1271–1283.

Rousset, M., 1986, The human colon carcinoma cell lines HT-29 and Caco-2: Two in vitro models for the study of intestinal differentiation. *Biochimie* **68**:1035–1040.

Rousset, M., Trugnan, G., Brun, J.-L., and Zweibaum, A., 1986, Inhibition of post-translational processing of microvillar hydrolases is associated with a specific decreased expression of sucrase–isomaltase and an increased turnover of glucose in Caco-2 cells. *FEBS Lett.* **208**:34–38.

Sabatini, D. D., Kreibich, G., Morimoto, T., and Adesnik, M., 1982, Mechanisms for the incorporation of proteins in membranes and organelles. *J. Cell Biol.* **92**:1–22.

Sabatini, D. D., Griepp, E. B., Rodriguez-Boulan, E. J., Dolan, W. J., Robbins, E. S., Papadopoulos, S., Ivanov, I. E., and Rindler, M. J., 1983. Biogenesis of epithelial cell polarity. *Mod. Cell Biol.* **2**:419–450.

Sacktor, B., 1977, The brush border of the renal proximal tubule and the intestinal mucosa. In *Mammalian Cell Membranes* (G. A. Jamieson and D. M. Robinson, eds.), Vol. 4, pp. 221–254, Butterworth, London.

Sahara, N., Fukasawa, K., Harada, M., and Suzuki, K., 1983, Immunohistochemical localization of dipeptidylpeptidase IV in rat digestive organs. *Acta Histochem. Cytochem.* **16**:494–501.

Salas, P. J. I., Vega-Salas, D., Misek, D., and Rodriguez-Boulan, E., 1985, Intracellular routes of apical and basolateral plasma membrane proteins to the cell surface in epithelial cells. *Pflugers Arch.* **405**:S152–S157.

Salas, P. J. I., Misek, D. E., Vega-Salas, D. E., Gundersen, D., Cereijido, M., and Rodriguez-Boulan, E., 1986, Microtubules and actin filaments are not critically involved in the biogenesis of epithelial cell surface polarity. *J. Cell Biol.* **102**:1853–1867.

Sandoz, D., Laine, M.-C., and Nicolas, G., 1985, Distribution of microtubules within the intestinal terminal web as revealed by quick-freezing and cryosubstitution. *Eur. J. Cell Biol.* **39**:481–484.

Sankaran, K., Sivakami, S., and Radhakrishnan, A. N., 1983, Studies on the size and shape of rabbit intestinal glucoamylase–maltase complex. *Biochem. J.* **213**:719–725.

Saraste, J., and Kuismanen, E., 1984, Pre- and post-Golgi vacuoles operate in the transport of Semliki Forest virus membrane glycoproteins to the cell surface. *Cell* **38**:535–549.

Schlegel-Haueter, S., Kerry, K. R., and Semenza, G., 1972, The preparation of lactase and glucoamylase of rat small intestine. *Biochim. Biophys. Acta* **259**:506–519.

Schmitz, J., Preiser, H., Maestracci, D., Ghosh, B. K., Cerda, J. J., and Crane, R. K., 1973, Purification of the human intestinal brush border membrane. *Biochim. Biophys. Acta* **323**:98–112.

Schmitz, J., Commegrain, D., Maestracci, D., and Rey, J., 1974, Absence of brush border sucrase–isomaltase complex in congenital sucrose intolerance. *Biomedicine* **21**:440–443.

Schmitz, J., Bresson, J. L., Triadou, N., Bataille, J., and Rey, J., 1980, Analyse electrophorese sur gel de polyacrylamide des proteines de la membrane microvillositaire et d'une fraction cytoplasmique das 8 cas d'intolerance congenitale au saccharose. *Gastroenterol. Clin. Biol.* **4**:251–256.

Schnapp, B. J., Vale, R. D., Sheetz, M. P., and Reese, T. S., 1985, Single microtubules from squid axoplasm support bidirectional movement of organelles. *Cell* **40**:455–462.

Schroer, T. A., and Kelly, R. B., 1985, In vitro translocation of organelles along microtubules. *Cell* **40**:729–730.

Schwarz, R. T., and Datema, R., 1984, Inhibitors of trimming: New tools in glycoprotein research. *Trends Biochem. Sci.*, **January**, 32–34.

Seetharam, B., Alpers, D. H., and Allen, R. H., 1981, Isolation and characterization of the ileal receptor for intrinsic factor-cobalamin. *J. Biol. Chem.* **256**:3785–3790.

Seetharam, B., Tiruppathi, C., and Alpers, D. H., 1985, Membrane interactions of rat intestinal alkaline phosphatase: Role of polar head groups. *Biochemistry* **24**:6603–6608.

Semenza, G., 1978a, The mode of anchoring of sucrase–isomaltase to the small intestinal brush border membrane and its biosynthetic implications. In *Proceedings of the 12th FEBS Meeting* (S. Rapoport and T. Schewe, eds.), Pergamon, Oxford.

Semenza, G., 1978b, Intestinal oligosaccharidases and disaccharidases. In *Handbook of Physiology* (C. F. Code, ed.), Sec. 6, Vol. V, pp. 2543–2566, American Physiology Society, Washington, DC.

Semenza, G., 1981a, Intestinal oligo- and disaccharidases. In *Carbohydrate Metabolism and Its Disorders* (P. J. Randle, D. F. Steiner, and W. J. Whelan, eds.), Vol. 3, pp. 425–479, Academic Press, London.

Semenza, G., 1981b, Molecular pathophysiology of small-intestinal sucrase–isomaltase. In *Clinics in Gastroenterology*, Vol. 10, pp. 691–706, W. B. Saunders, Philadelphia.

Semenza, G., 1985, Anchoring and biosynthesis of a major intrinsic plasma membrane protein: The sucrase–isomaltase complex of the small-intestinal brush border. In *The Enzymes of Biological Membranes* (A. Martonosi, ed.), Vol. 2, pp. 331–348, Plenum Press, New York.

Semenza, G., 1986, Anchoring and biosynthesis of stalked brush border membrane proteins: Glycosidase and peptidase of the enterocytes and renal tubuli. *Ann. Rev. Cell Biol.* **2**:255–313.

Semenza, G., and Corcelli, A., 1986, The absorption of sugars and amino acids across the small intestine. In *Molecular and Cellular Basis of Digestion* (P. Desnuelle, H. Sjöström, and O. Norén, eds.), pp. 381–412, Elsevier, Amsterdam.

Semenza, G., Brunner, J., and Wacker, H., 1983, Biosynthesis and assembly of the largest and major intrinsic polypeptide of the small intestinal brush borders. In *Brush Border Membranes* (R. Porter and G. Collins, eds.), Ciba Foundation Symposium 95, pp. 92–107, Pitman, London.

Simons, K., and Fuller, S. D., 1985, Cell surface polarity in epithelia. *Annu. Rev. Cell Biol.* **1**:243–288.

Simons, K., and Warren, G., 1984, Semliki Forest Virus: A probe for membrane traffic in the animal cell. *Adv. Protein Chem.* **36**:79–132.

Simoons, F. J., 1978, The geographic hypothesis and lactose malabsorption. A weighing of the evidence. *Digest. Dis.* **23**:963–980.

Sips, H. J., Claass, A. H. W., van Dongen, J. M., Willemsen, R., Hoogeveen, A. T., Galjaard, H., Siinaasappel, M., Hauri, H. P., and Sterchi, E. E., 1985, Sucrase–isomaltase and cystic fibrosis. *J. Inherited Metab. Dis.* **8**:163–168.

Sjöström, H., and Norén, O., 1982. Changes in the quarternary structure of microvillus aminopeptidase in the membrane. *Eur. J. Biochem.* **122**:245–250.

Sjöström, H., Norén, O., Jeppesen, L., Staun, M., Svensson, B., and Christiansen, L., 1978, Purification of different amphiphilic forms of a microvillus aminopeptidase from pig small intestine using immunoadsorbent chromatography. *Eur. J. Biochem.* **88**:503–511.

Sjöström, H., Norén, O., Christiansen, L., Wacker, H., and Semenza, G., 1980, A fully active, two-active-site, single-chain sucrase–isomaltase from pig small intestine. Implications for the biosynthesis of a mammalian integral stalked membrane protein. *J. Biol. Chem.* **255**:11332–11338.

Sjöström, H., Norén, O., Christiansen, L., Wacker, H., and Semenza, G., 1982, Membrane anchoring and partial structure of small-intestinal pro-sucrase–isomaltase. A possible biosynthetic mechanism. *FEBS Lett.* **148**:321–325.

Sjöström, H., Norén, O., Danielsen, E. M., and Skovbjerg, H., 1983, Structure of microvillar enzymes in different phases of their life cycles. In *Brush Border Membranes* (R. Porter and G. Collins, eds.), Ciba Foundation Symposium 95, pp. 50–69, Pitman, London.

Sjöström, H., Norén, O., and Danielsen, E. M., 1985, Enzymatic activity of "high mannose" glycosylated forms of intestinal microvillar hydrolases. *J. Pediatr.* **4**:980–983.

Skovbjerg, H., 1982, High molecular weight pro-sucrase–isomaltase in human fetal intestine. *Pediatr. Res.* **16**:948–949.

Skovbjerg, H., and Krasilnikoff, P. A., 1980, Immunoelectrophoretic studies on human small intestinal brush border proteins. Residual isomaltase in sucrose intolerant patients. *Pediatr. Res.* **15**:214–218.

Skovbjerg, H., Norén, O., and Sjöström, H., 1978, Immunoelectrophoretic studies on human small intestinal brush border proteins. *Scand. J. Clin. Lab. Invest.* **38**:723–729.

Skovbjerg, H., Gudmand-Hoyer, E., and Fenger, H. J., 1980. Immunoelectrophoretic studies on human small intestinal brush border proteins—amount of lactase protein in adult-type hypolactasia. *Gut* **21**:360–364.

Skovbjerg, H., Sjöström, H., and Norén, O., 1981, Purification and characterization of amphiphilic lactase–phlorizin hydrolase from human small intestine. *Eur. J. Biochem.* **114**:653–661.

Skovbjerg, H., Norén, O., Sjöström, H., Danielsen, E. M., and Enevoldsen, B. S., 1982, Further characterization of intestinal lactase–phlorizin hydrolase. *Biochim. Biophys. Acta* **707**:89–97.

Skovbjerg, H., Danielsen, E. M., Norén, O., and Sjöström., H., 1984, Evidence for biosynthesis of lactase–phlorizin hydrolase as a single-chain high-molecular weight precursor. *Biochim. Biophys. Acta* **798:**247–251.

Slaughter, C. A., Coseo, M. C., Cancro, M. P., and Harris, H., 1981, Detection of enzyme polymorphism by using monoclonal antibodies. *Proc. Natl. Acad. Sci. U.S.A.* **78:**1124–1128.

Snider, M. D., and Rogers, O. C., 1986, Membrane traffic in animal cells: Cellular glycoproteins return to the site of Golgi mannosidase I. *J. Cell Biol.* **103:**265–275.

Solari, R., and Kraehenbuhl, J.-P., 1984, Biosynthesis of the IgA antibody receptor: A model for the transepithelial sorting of a membrane glycoprotein. *Cell* **36:**61–71.

Sorensen, S. H., Norén, O., Sjöström, H., and Danielsen, M., 1982, Amphiphilic pig intestinal microvillus maltase–glucoamylase. Structure and specificity. *Eur. J. Biochem.* **126:**559–568.

Spiess, M., Hauser, H., Rosenbusch, J. P., and Semenza, G., 1981, Hydrodynamic properties of phospholipid vesicles and of sucrase–isomaltase–phospholipid vesicles. *J. Biol. Chem.* **256:**8977–8982.

Spiess, M., Brunner, J., and Semenza, G., 1982, Hydrophobic labeling, isolation and partial characterization of the NH_2-terminal membranous segment of sucrase–isomaltrase complex. *J. Biol. Chem.* **257:**2370–2377.

Sterchi, E. E., Green, J. R., and Lentze, M. J., 1982, Non-pancreatic hydrolysis of *N*-benzoyl-L-tyrosyl-*p*-aminobenzoic acid (PABA–peptide) in the human small intestine. *Clin Sci.* **62:**557–560.

Sterchi, E. E., Naim, H., Hauri, H. P., and Lentze, M., 1986, PABA–peptide hydrolase is a human brush border membrane-bound endopeptidase distinct from enkephalinase. *Eur. J. Pediatr.* **144:**319 (abstract).

Stewart, J. R., and Kenny, A. J., 1984a, Proteins of the kidney microvillar membrane. Biosynthesis of endopeptidase-24.11, dipeptidylpeptidase IV and aminopeptidase N and A in pig kidney slices. *Biochem. J.* **224:**549–558.

Stewart, J. R., and Kenny, A. J., 1984b, Proteins of the kidney microvillar membrane. Effects of monensin, swainsonine and glucosamine on the processing and assembly of endopeptidase-24.11 and dipeptidyl peptidase IV in pig kidney slices. *Biochem. J.* **224:**559–568.

Stewart, J. R., Chaplin, M. F., and Kenny, A.J., 1984, Deglycosylation by trifluoromethanesulphonic acid of endopeptidase-24.11 purified from pig kidney and intestine. *Biochem. J.* **221:**919–922.

Stieger, B., and Murer, H., 1983, Heterogeneity of brush border membrane vesicles from rat small intestine prepared by a precipitation method using Mg/EGTA. *Eur. J. Biochem.* **135:**95–101.

Stieger, B., Marxer, A.. and Hauri, H. P., 1986, Isolation of brush border membranes from rat and rabbit colonocytes: Is alkaline phosphatase a marker enzyme? *J. Membr. Biol.* **91:**19–31.

Stieger, B., Matter, K., Baur, B., Bucher, K., Höchli, M., and Hauri, H.-P., 1988, Dissection of the asynchronous transport of intestinal microvillar hydrolases to the cell surface. *J. Cell Biol.* (in press).

Strous, G. J., DuMaine, A., Zijderhand-Bleekemolen, J. E., Slot, J. W., and Schwartz, A. L., 1985, Effect of lysosomotropic amines on the secretory pathway and on the recycling of the asialoglycoprotein receptor in human hepatoma cells. *J. Cell Biol.* **101:**531–539.

Sussman, N. L., Seetharam, S., Blaufuss, M. C., and Alpers, D. H., 1986, Translation of rat intestinal RNA yields two alkaline phosphatases. *Biochem. J.* **234:**563–568.

Svensson, B., Danielsen, M., Staun, M., Jeppesen, L., Norén, O., and Sjöström, H., 1978, An amphiphilic form of dipeptidylpeptidase IV from pig small intestinal brush border membrane. Purification by immunoadsorbent chromatography and some properties. *Eur. J. Biochem.* **90:**489–498.

Svensson, B., Sjöström, H., and Norén, O., 1982, Structural studies on microvillus aminopeptidase from pig small intestine. *Eur. J. Biochem.* **126:**481–488.

Swallow, D. M., Potter, J., Green, F., and Ho, M. W., 1985, Preparation of a monoclonal antibody against human lactase. *J. Immunol. Methods* **77:**139–145.

Takesue, Y., and Nishi, Y., 1978, Topographical studies on intestinal microvillus leucine beta-naphthylamidase on the outer membrane surface. *J. Membr. Biol.* **39**:285–296.

Takesue, Y., Tamura, R., and Nishi, Y., 1977, Immunohistochemical studies on the subunits of rabbit-intestinal sucrase–isomaltase complex. *Biochim. Biophys. Acta* **483**:375–385.

Takesue, T., Yokota, K., Nishi, Y., Taguchi, T., and Ikezawa, H., 1986, Solubilization of trehalase from rabbit renal and intestinal brush-border membranes by a phosphatidylinositol-specific phospholipase C. *FEBS Lett.* **201**:5–8.

Tang, P. W., Gool, G. C., Hardy, M., Lee, Y. C., and Feizi, T., 1985, Novel approach to the study of the antigenicities and receptor functions of carbohydrate chains of glycoproteins. *Biochem. Biophys. Res. Comm.* **132**:474–480.

Taravel, F. R., Datema, R., Woloszczuk, W., Marshall, J. J., and Whelan, W. J., 1983, Purification and characterization of a pig intestinal alpha-limit dextrinase. *Eur. J. Biochem.* **130**:147–153.

Tartakoff, A. M., 1983, Mutations that influence the secretory path in animal cells. *Biochem. J.* **216**:1–9.

Tate, S. S., 1986, Single-chain precursor of renal gamma-glutamyl transpeptidase. *FEBS Lett.* **194**:33–38.

Tate, S. S., and Meister, A., 1976, Subunit structure and isozymic forms of gamma-glutamyl transpeptidase. *Proc. Natl. Acad. Sci. U.S.A.* **73**: 2599–2603.

Tate, S. S., and Meister, A., 1977, Affinity labeling of gamma-glutamyl transpeptidase and location of the gamma-glutamyl binding site on the light subunit. *Proc. Natl. Acad. Sci. U.S.A.* **74**:931–935.

Tate, S. S., and Ross, M. E., 1977, Human kidney gamma-glutamyl transpeptidase. Catalytic properties, subunit structure, and localization of the gamma-glutamyl binding site on the light subunit. *J. Biol. Chem.* **252**:6042–6045.

Thorens, B., and Vassalli, P., 1986, Chloroquine and ammonium chloride prevent terminal glycosylation of immunoglobulins in plasma cells without affecting secretion. *Nature* **321**:618–620.

Timmins, J. G., Petrovskis, E. A., Marchioli, C. C., and Post, L. E., 1985, A method for efficient gene isolation from phage lambda-gt11 libraries: Use of antisera to denatured, acetone-precipitated proteins. *Gene* **39**:89–93.

Tobey, N., Heizer, W., Yeh, R., Huang, T.-I., and Hofner, C., 1985, Human intestinal brush border peptidases. *Gastroenterology* **88**:913–926.

Tolleshaug, H., Goldstein, J. L., Schneider, W. J., and Brown, M. S., 1982, Posttranslational processing of the LDL receptor and its genetic disruption in familial hypercholesterolemia. *Cell* **30**:715–724.

Tolleshaug, H., Hobgood, K. K., Brown, M. S., and Goldstein, J. L., 1983, The LDL receptor locus in familial hypercholesterolemia: Multiple mutations disrupt transport and processing of a membrane receptor. *Cell* **32**:941–951.

Triadou, N., 1984, Antigenic cross-reactions among human intestinal brush border enzymes revealed by the immunoblotting method and rabbit anti-enzyme sera. *J. Immunol. Methods* **73**:283–291.

Triadou, N., and Zweibaum, A., 1985, Maturation of sucrase–isomaltase complex in human fetal small and large intestine during the gestation. *Pediatr. Res.* **19**:136–138.

Triadou, N., Audran, E., Rousset, M., Zweibaum, A., and Oriol, R., 1983, Relationship between the secretor status and the expression of ABH blood group antigenic determinants in human intestinal brush border membrane hydrolases. *Biochim. Biophys. Acta* **761**:231–236.

Trier, J. S., 1980, Organ culture of the mucosa of human small intestine. *Methods Cell Biol.* **21B**:365–384.

Trugnan, G., Rousset, M., Chantret, I., Barbat, A., and Zweibaum, A., 1987, The posttranslational processing of sucrase–isomaltase in HT-29 cells is a function of their state of enterocytic differentiation. *J. Cell Biol.* **104**:1199–1205.

Tse, A. G. D., Barclay, A. N., Watts, A., and Williams, A. F., 1985, A glycosphingolipid tail at the carboxyl terminus of the Thy-1 glycoprotein of neurons and thymocytes. *Science* **230:**1003–1008.

Tsuji, A., Matsuda, Y., and Katanuma, N., 1980, Studies on the structure of gamma-glutamyl transpeptidase. II. Location of the segment anchoring gamma-glutamyl transpeptidase to the membrane. *J. Biochem.* **87:**1567–1571.

Tulsiani, D. R. P., and Touster, O., 1983a. Swainsonine, a potent mannosidase inhibitor, elevates rat liver and brain lysosomal alpha-D-mannosidase, decreases Golgi alpha-D-mannosidase II, and increases the plasma levels of several acid hydrolases. *Arch. Biochem. Biophys.* **224:**594–600.

Tulsiani, D. R. P., and Touster, O., 1983b, Swainsonine causes the production of hybrid glycoproteins by human skin fibroblasts and rat liver Golgi preparations. *J. Biol. Chem.* **258:**7578–7585.

Vale, R. D., Schnapp, B. J., Mitchinson, T., Steuer, E., Reese, T. S., and Sheetz, P., 1985, Different axoplasmic proteins generate movement in opposite directions along microtubules in vitro. *Cell* **43:**623–632.

Vannier, Ch., Louvard, D., Maroux, S., and Desnuelle, P., 1976, Structural and topological homology between porcine intestinal and renal brush border aminopeptidase. *Biochim. Biophys. Acta* **445:**185–199.

Vockley, J., and Harris, H., 1984, Purification of human adult and foetal intestinal alkaline phosphatases by monoclonal antibody immunoaffinity chromatography. *Biochem. J.* **217:**535–541.

Vora, S., 1985, Monoclonal antibodies in enzyme research: Present and potential applications. *Anal. Biochem.* **144:**307–318.

Wacker, H., Jaussi, R., Sonderegger, P., Dokow, M., Gersa, P., Hauri, H. P., Christen, P., and Semenza, G., 1981, Cell-free synthesis of the one-chain precursor of a major intrinsic protein complex of the small-intestinal brush border membrane (pro-sucrase–isomaltase). *FEBS Lett.* **136:**329–332.

Wacker, H., Aggeler, R., Kretchmer, N., O'Neill, B., Takesue, Y., and Semenza, G., 1984, A two-active site one-polypeptide enzyme: The isomaltase from sea lion small intestinal brush border membrane. Its possible phylogenetic relationship with sucrase–isomaltase. *J. Biol. Chem.* **259:**4878–4884.

Weare, J. A., Gafford, J. T., Lu, H. S., and Erdös, E. G., 1982, Purification of human kidney angiotensin I converting enzyme using reverse-immunoadsorption chromatography. *Anal. Biochem.* **123:**310–319.

Weiser, M. M., 1973, Intestinal epithelial cell surface membrane glycoprotein synthesis. An indicator of cellular differentiation. *J. Biol. Chem.* **248:**2536–2541.

Weiser, M. M., Neumeier, M. M., Quaroni, A., and Kirsch, K., 1978, Synthesis of plasmalemmal glycoproteins in intestinal epithelial cells. Separation of Golgi membranes from villus and crypt cell surface membranes: Glycosyltransferase activity of surface membrane. *J. Cell. Biol.* **77:**722–734.

Weiser, M. M., Walters, J. R. F., and Wilson, J. R., 1986, Intestinal cell membranes. *Int. Rev. Cytol.* **101:**1–57.

Weiss, M. J., Henthorn, P. S., Lafferty, M. A., Slaughter, C., Raducha, M., and Harris, H., 1986, Isolation and characterization of a cDNA encoding a human liver/bone/kidney-type alkaline phosphatase. *Proc. Natl. Acad. Sci. U.S.A.* **83:**7182–7186.

Wice, B. M., Trugnan, G., Pinto, M., Rousset, M., Chevalier, G., Dussaulx, E., Lacroix, B., and Zweibaum, A., 1985, The intracellular accumulation of UDP-*N*-acetylhexosamines is concomitant with the inability of human colon cancer cells to differentiate. *J. Biol. Chem.* **260:**139–146.

Wickner, W., and Lodish, H. F., 1985, Multiple mechanisms of protein insertion into and across membranes. *Science* **230:**400–407.

Willingham, M., and Pastan, I., 1984, Endocytosis and exocytosis: Current concepts of vesicle traffic in animal cells. *Int. Rev. Cytol.* **92:**51–92.

Woods, J. W., Doriaux, M., and Farquhar, M. G., 1986, Transferrin receptors recycle to cis and middle as well as trans Golgi cisternae in Ig-secreting myeloma cells. *J. Cell Biol.* **103:**277–286.

Wray, L. K., and Harris, H., 1983, Monoclonal antibodies to an ectopically expressed alkaline phosphatase in a human malignant cell line. *Cancer Res.* **43:**758–762.

Yamamoto, T., Bishop, R. W., Brown, M. S., Goldstein, J. L., and Russel, D. W., 1986, Deletion in cysteine-rich region of LDL receptor impedes transport to cell surface in WHHL rabbit. *Science* **232:**1230–1237.

Yamashita, K., Hitoi, A., Matsuda, Y. X., Tsuji, A., Katunuma, N., and Kobata, A., 1983a, Structural studies of the carbohydrate moieties of rat kidney gamma-glutamyltranspeptidase. An extremely heterogeneous pattern with nonreducing terminal *N*-acetylglucosamine residues. *J. Biol. Chem.* **258:**1098–1107.

Yamashita, K., Tachibana, Y., Hitoi, A., Matsuda, Y., Tsuji, A., Katunuma, N., and Kobata, A., 1983b, Difference in the sugar chains of two subunits and of isozymic forms of rat kidney gamma-glutamyl transpeptidase. *Arch. Biochem. Biophys.* **227:**225–232.

Yedlin, S. T., Young, G. P., Seetharam, B., Seetharam, S., and Alpers, D. H., 1981, Characterization and comparison of soluble and membranous forms of intestinal alkaline phosphatase from suckling rat. *J. Biol. Chem.* **256:**5620–5626.

Yeh, K.-Y., and Moog, F., 1984, Biosynthesis and transport of glycoproteins in the small intestinal epithelium of rats. I. Developmental change and effect of hypophysectomy. *Dev. Biol.* **101:**446–462.

Yewdell, J. W., Gerhard, W., and Bachi, T., 1983, Monoclonal anti-hemagglutinin antibodies detect irreversible antigenic alterations that coincide with the acid activation of influenza virus A/PR/834-mediated hemolysis. *J. Virol.* **48:**239–248.

Yokota, K., Nishi, Y., and Takesue, Y., 1986, Purification and characterization of amphiphilic trehalase from rabbit small intestine. *Biochim. Biophys. Acta* **881:**405–414.

Young, G. P., Yedlin, S. T., and Alpers, D. H., 1981, Independent biosynthesis of soluble and membrane-bound alkaline phosphatases in the suckling rat ileum. *Biochem. J.* **200:**645–654.

Yusufi, A. N. K., Low, M. G., Turner, S. T., and Dousa, T. P., 1983, Selective removal of alkaline phosphatase from renal brush border membrane and sodium-dependent brush border membrane transport. *J. Biol. Chem.* **258:**5695–5701.

Ziomek, C. A., Schulman, S., and Edidin, M., 1980, Redistribution of membrane proteins in isolated rat mouse intestinal epithelial cells. *J. Cell Biol.* **86:**849–857.

Zweibaum, A., Triadou, N., Kedinger, M., Augeron, C., Robine-Leon, S., Pinto, M., Rousset, M., and Haffen, K., 1983, Sucrase–isomaltase: A marker of foetal and malignant epithelial cells of the human colon. *Int. J. Cancer* **32:** 407–412.

Zweibaum, A., Hauri, H. P., Sterchi, E., Chantret, I., Haffen, K., Bamat, J., and Sordat, B., 1984, Immunohistological evidence, obtained with monoclonal antibodies, of small-intestinal brush border hydrolases in human colon cancers and fetal colons. *Int. J. Cancer* **34:**591–598.

Zweibaum, A., Pinto, M., Chevalier, G., Dussaulx, E., Triadou, N., Lacroix, B., Haffen, K., Brun, J.-L., and Rousset, M., 1985, Enterocytic differentiation of a subpopulation of the human colon tumor cell line HT-29 selected for growth in sugar-free medium and its inhibition by glucose. *J. Cell. Physiol.* **122:**21–29.

Zweibaum, A., Laburthe, M., Grasset, E., and Louvard, D., 1987, The use of cultured lines in studies of intestinal cell differentiation and function. In *Handbook of Physiology* (in press).

Chapter 6

Immunoaffinity Purification of Membrane Fractions from Mammalian Cells

Peter J. Richardson and J. Paul Luzio

1. INTRODUCTION

A wide variety of techniques have been developed for the subfractionation of mammalian cell membranes and organelles. The most widely used are density gradient centrifugation, free-flow electrophoresis, polymer phase partition, fluorescent activated "cell-sorting," and immunoaffinity purification. While the immunoaffinity approach relies totally on antibody–antigen reactions, the specificity of the other systems, which depend largely on physical differences between particles, can be increased by the selective use of affinity procedures. This chapter, after discussing the relative merits of other membrane separation techniques, is concerned primarily with the design, application, and advantages of immunoaffinity purification using a solid matrix.

The correct identification of fractions is essential to the success of all subcellular fractionation techniques and relies on the use of marker enzymes and chemicals. Two postulates, biochemical homogeneity and single location (DeDuve, 1975), have widely been applied to data interpretation using such markers. The postulates state that members of a given subcellular particle population

Peter J. Richardson and J. Paul Luzio Department of Clinical Biochemistry, University of Cambridge, Cambridge CB2 2QR, United Kingdom.

have the same biochemical composition and that each biochemical marker is restricted to a single subcellular site. On the basis of these postulates, the presence of specific enzymes such as cytochrome oxidase is used to define mitochondria and acid phosphatase to define lysosomes. The purification of any given particle is determined by following the increase in specific activity (usually with reference to protein) of a given marker enzyme. The realization that subcellular structures might also have specific antigenic determinants led to Kamat and Wallach (1965) using antibodies to monitor the preparation of plasma membranes from Ehrlich ascites carcinoma cells. The cells were disrupted and the membranes of the microsomal fraction separated by means of Ficoll density gradients. The distribution of cell surface antigens and thus of plasma membrane on the gradients was determined by the ability of a gradient fraction to absorb antibody capable of agglutinating intact tumor cells. A simplification of this approach is to bind radioactively labeled antibodies to the surface of intact cells and follow the isolation of plasma membranes by measuring the radioactivity of the various fractions (Boone *et al.*, 1969; Gahmberg and Simons, 1970; Luzio *et al.*, 1976). It should be noted, however, that some subcellular particles thought to be homogeneous may in fact differ in their biochemical composition, for example, mitochondria (Reijnierse *et al.*, 1975). Similarly, plasma membrane markers, for example, 5′-nucleotidase, are frequently found in more than one membrane compartment (Stanley *et al.*, 1982). It is therefore prudent to measure more than one "specific marker" when attempting to purify subcellular fractions.

The most commonly used method for separating cell membranes is centrifugation, in which the membranes and particles are separated on the basis of differences in their size, shape, and buoyant density. Two basic procedures are used—differential centrifugation and density gradient centrifugation. The former involves centrifugation at different speeds such that large dense particles sediment at slow speeds, while the smaller and less dense membranes require higher speeds before they sediment. Density gradient centrifugation can be used in two ways—rate separations that are dependent on the size and shape of the particles and equilibrium (or isopycnic) separations that are dependent on the buoyant density of the particles. A number of different media have been used to generate the density gradients, the most common being sucrose. The use of density gradients is limited by the inherent differences in density of the membranes being separated; that is, it is useful for the separation of dense membranes such as intestinal brush borders (George and Kenny, 1973) but less useful for the separation of less dense membranes such as endosomes (Mullock *et al.*, 1987). One procedure that increases the utility of density gradients is the use of hyperosmotic media, for example, sucrose (Whittaker, 1984). These result in the dehydration and consequent increase in density of osmotically sensitive particles—thus removing them from other membranous particles. (*Note:* This can also be a drawback since the membranes are then exposed to unphysiological media.) To avoid the osmotic problems presented by low molecular weight solutes in density gradients, a number of high molecular weight solutes have been used, for example, Ficoll, Percoll, and metrizamide. Variations in the solvent composition (e.g., the use of D_2O, which selectively increases the density of osmotically sensitive particles) have also been useful in organelle purification such as coated vesicles (Pearse, 1980) and chromaffin granules (Morris and Schovanka, 1977). Other procedures that have been used to increase the applications of density gradient separation techniques include density shifts induced by the use of cytochemical reactions (Davis and Bloom, 1973; Leskes *et al.*, 1981; Porter-Jordan *et al.*, 1986; Ryan and Smith 1971). A similar approach was used by Courtoy *et al.* (1984), who isolated endosome membranes after endocytosis of peroxidase-labeled ligand into the liver endocytic compartment. They utilized the insoluble reaction product of diaminobenzidine to increase the density of the compartments containing the labeled ligand. Despite these modifications, density gradient centrifugation is seldom capable of producing pure membrane fractions. However, it is still the method of choice for initial separations and analysis (Evans, 1987).

Free-flow electrophoresis relies on membranes exhibiting quite distinct differences in charge (Hannig and Heidrich, 1977). In this technique, the membranes to be separated are injected as a fine stream into a separation buffer moving through an electric field. Membranes bearing different electrical charges will consequently migrate different distances across the separation chamber, for example, the cis and trans faces of the Golgi apparatus (Morré *et al.*, 1983). This approach, while separating membranes on a different basis from that of centrifugation, also relies primarily on physical differences between membranes.

Polymer two-phase separations are based on differences in hydrophobicity between different populations of membranes (Albertsson, 1986). In this procedure, the membranes to be separated are distributed between the two phases produced upon mixing solutions of water-soluble polymers—usually polyethylene glycol and dextran. To maximize the separation, automated countercurrent distribution devices are often used. Because the extraction times can be very long with aqueous polymers, thin layers are also used. Alternatively, one phase may be kept stationary in a centrifugal field, while the other phase (bearing the membrane components) is pumped through it. Any given membrane will then pass through the stationary phase in a time inversely proportional to the membrane's affinity for this phase (Ito and Bowman, 1978). Potential increases in the applicability of two-phase separations have been demonstrated by covalent labeling of one polymer with a specific ligand, for example, hexamethonium for the purification of nicotinic acetylcholine receptor bearing membranes (Flanagan *et al.*, 1984). More recently, it has been shown that specific antibodies can be used in these systems for the immunoaffinity purification of cells (Sharp *et al.*, 1986).

The fluorescence activated cell sorting (FACS) approach is apparently well suited for the separation of organelles and membranes since unphysiological media (e.g., sucrose and dextran) are not used, and the specificity can be determined by the antibody–antigen reaction. The principle of this procedure is that the membranes to be separated are introduced as a fine stream into a laser beam. Those membranes that are fluorescently labeled are separated from unlabeled ones by charging the membrane-containing droplets and electrically deflecting them into separate collection chambers. For example, Gorvel and Maroux (1987) separated plasma membrane vesicles from epithelial cells into two populations—basolateral and apical—using fluorescent-labeled antibodies specific to the apical surface. The required specificity can also be obtained by the use of fluorescent-labeled ligands. Thus, Murphy (1985) demonstrated the presence of three distinct endocytic compartments using FITC-labeled dextran as a marker. The FACS procedure is not useful for preparative subcellular fractionation, since long separation times are required to obtain enough material even for SDS–polyacrylamide gel analysis. The principal advantages of the free-flow electrophoresis, polymer two-phase, and FACS systems are that the membranes are not subjected to hyperosmotic media, nor are they irretrievably bound to a solid matrix. Immunoaffinity techniques have the advantages that they are based on the selectivity of the antibody–antigen reaction and can be used for both analytical and preparative procedures. They can be carried out in physiological salt solutions, do not require unusual or hyperosmotic media, are applicable to small amounts of tissue (e.g., tissue culture cells), and are rapid (purification being within 1 h of homogenization). The possibilities of applying immunoaffinity procedures to the separation of subcellular particles were first discussed by Hales (1972). Some successful applications were soon reported (Kawajiri *et al.*, 1977; Luzio *et al.*, 1974, 1976), but it is only in the past decade that the technique has gained wider popularity. This was in part due to the difficulty in eluting membrane fractions from solid supports without structural damage. However, since most membrane properties can be studied while still attached to the solid matrix (see Section 5), this is not a great drawback. The other major hindrance to the exploitation of this approach has been the absence of suitable membrane-specific antibodies. Since the advent of monoclonal antibodies (Kohler and Milstein, 1975), this hindrance has also been removed (see Section 2). The consequent increase in popularity of this approach to subcellular fractionation is illustrated in Table I.

Immunoaffinity techniques for the separation of organelles and membranes can broadly be defined into two methods—the use of direct or indirect immunoadsorbents. The principles behind these methods are shown in Figures 1 and 2 with respect to the affinity purification of fat cell plasma membranes (Luzio *et al.*, 1976). In this study, polyclonal antibodies were bound to intact cells prior to

Table I
Subcellular Particles Isolated Using Immunoaffinity Techniques

Year	Cell type	Particle or membrane	Reference
1974	Adipocyte	Plasma membrane	Luzio *et al.*
1976	Adipocyte	Plasma membrane	Luzio *et al.*
	Lymphocyte	Plasma membrane	Walsh *et al.*
1977	Hepatocyte	Microsomes	Kawajiri *et al.*
	Pancreatic islets	Plasma membrane	Luzio
1978	Hepatocyte	Golgi membrane	Ito and Palade
1979	Hepatocyte	Plasma membrane	Westwood *et al.*
1981	Neurone	Synaptic vesicles	Matthew *et al.*
	PC12 culture cells	Adrenergic vesicles	Matthew *et al.*
1982	Neural cells	Clathrin-coated vesicles	Merisko *et al.*
	Neurone	Presynaptic plasma membrane	Miljanich *et al.*
1983	Hepatocyte	Endosome-derived vesicles	Luzio and Stanley
1984	Hepatocyte	Plasma membrane domains	Roman and Hubbard
	Neurone	Cholinergic nerve terminals	Richardson *et al.*
	Hepatocyte	Prelysosomes	Debanne *et al.*
	Hepatocyte	Lysosomes	Pontremoli *et al.*
1985	Erythrocyte	Plasma membrane vesicles	Gruenberg and Howell
	MDCK culture cells	Apical plasma membrane	Gruenberg and Howell
	Fao culture cells	Plasma membrane	Devaney and Howell
	Neural cells	Clathrin-coated vesicles	Pfeffer and Kelly
1986	Hepatocyte	Endosomes (asialoglycoprotein receptor-positive/negative)	Mueller and Hubbard
	BHK culture cells	Endosomes	Gruenberg and Howell
	Neurones	Cholinergic and GABAergic nerve terminals	Docherty *et al.*

homogenization, thus avoiding the complications caused by the presence of certain antigens being present on many different types of membrane. The direct method (Figure 1) involves the covalent coupling of the anti-membrane-component antibody directly to the solid matrix, followed by the incubation of the heterogeneous membrane preparation with the matrix. Most workers do not use this approach, although two successful applications have been reported (Kawajiri *et al.*, 1977; Pontremoli *et al.*, 1984). The indirect method (Figure 2), in which an antibody-binding reagent is coupled to the matrix and mixed with the antibody-coated membrane preparation, is more commonly used. This approach may have been more successful because it provides a longer and more flexible spacer arm between the membrane particle and the matrix.

There are two major requirements in the design of a successful immu-

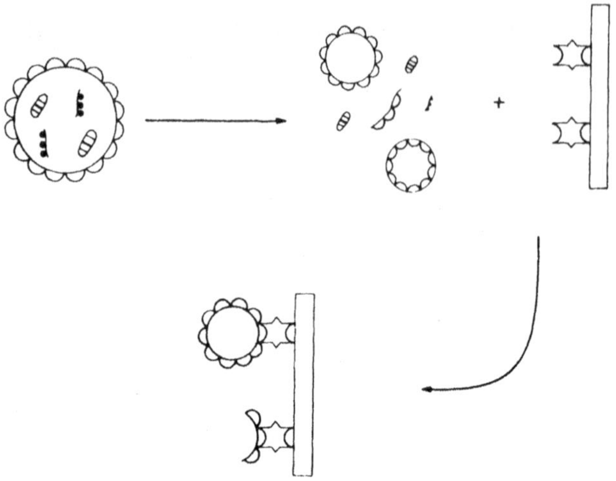

FIGURE 1. Schematic representation of direct immunoaffinity techniques for the preparation of plasma membrane. Cells are homogenized and the homogenate incubated with an antiplasma membrane immunoadsorbent.

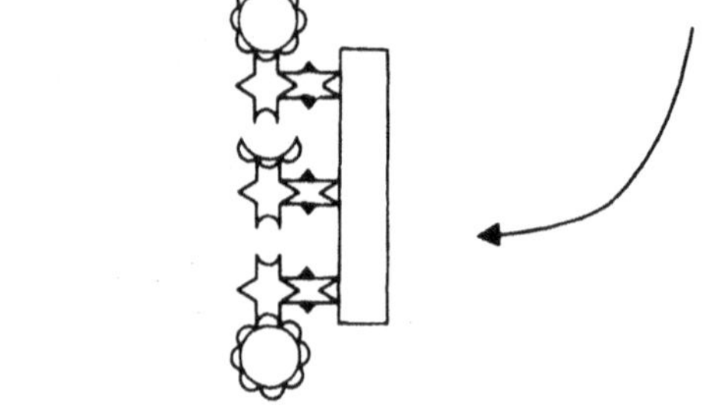

FIGURE 2. Schematic representation of indirect immunoaffinity techniques for the preparation of plasma membrane. Cells are coated with antibodies and then homogenized and the homogenate incubated with an anti-immunoglobulin immunoadsorbent.

noaffinity separation procedure—a specific primary antibody that recognizes the membrane of interest and a high-capacity solid phase. The desired nature and properties of these two components are discussed in the next two sections.

2. ANTIBODY REQUIREMENTS

2.1. Primary Antibody

Most of the properties required of a primary antibody have been determined, although the choice is often dictated by availability rather than design. In practice, polyclonal antibodies are seldom ideal since they are rarely specific to one membrane since most organelle preparations used for immunization are impure, and, as mentioned in Section 1, many antigens are present on more than one membrane. A number of different strategies have been developed to overcome these drawbacks. For example, polyclonal antisera can be used in the isolation of plasma membrane, if the antibodies are added to intact cells (Luzio *et al.*, 1976). After washing excess antibody away, the cells are then homogenized. If the operations are performed at low temperatures to prevent endocytosis (Luzio and Stanley, 1983), the only membrane labeled by the antibody will be the plasma membrane.

Polyclonal antisera have also been used when specific alterations in the antigenic structure of membranes have been undertaken. There have been two such procedures used: haptenization of cell membranes and incorporation of viral coat proteins. One example of the former was trinitrophenylation of membrane proteins (Kaplan *et al.*, 1979) and subsequent affinity purification (using anti-dinitrophenol antibodies) of individual plasma membrane components. An example of the latter was the incorporation of the surface G-protein of vesicular stomatis virus into the plasma membrane of red blood cells (Gruenberg and Howell, 1985) and subsequent isolation of the plasma membrane using polyclonal anti-G-protein antibodies. These authors have since modified the procedure by permitting endocytosis of the incorporated G-protein into BHK cells and then isolating endosomes using antibodies to the cytoplasmic tail of the G-protein (Gruenberg and Howell, 1986). By using this approach of incorporating G-protein into cell membranes, it has been estimated that for indirect immunoaffinity purification of membranes an antigen (i.e., G-protein) with a density of more than 50 molecules/μm^2 is necessary. This suggests that even when two different membranes bear the same antigen it will be possible to design affinity purification procedures that purify only the membrane fraction particularly enriched in the antigen.

One obvious approach to improving the utility of polyclonal antibodies is to affinity purify them on a specific antigen prior to use. Alternatively, they may be prepared with highly purified membrane components. The former may be especially useful since the affinity purification procedure removes most of the unwanted immunoglobulin, thus preventing it from competing with membrane-bound immunoglobulin for sites on the immunoadsorbent. The use of very small amounts of specific immunoglobulin may also preclude the necessity of washing the labeled membranes to remove unwanted antibody. This is particularly appropriate when affinity purifying low-density membranes that cannot be centrifuged (and so washed) rapidly. It may also allow single-step affinity purifications since the primary antibody can be prebound to the anti-immunoglobulin immunoadsorvent (Ito and Palade, 1978).

Monoclonal antibodies, by their very nature, also confer many of the advantages of affinity-purified polyclonal antibodies. Monoclonal antibodies are pure reagents, so avoiding the problems posed by the presence of the excess non-specific immunoglobulin present in polyclonal sera. One drawback with monoclonal antibodies is that they are often of relatively low affinity ($K_a < 0.5$ nM^{-1}), since in any set of cell fusions the range of antibodies produced reflects the range present in a polyclonal antiserum where it is unusual to have more than 25% that have $K_a > 10$ nM^{-1} (Siddle and Soos, 1981). It is these high-affinity antibodies that are likely to be of most use in subcellular fractionation, unless it is essential to elute the purified membranes from the solid matrix, in which case low-affinity antibodies may be more appropriate. It is not known whether the fact that monoclonal antibodies each react with only one antigenic site affects their use for immunoaffinity purification. If it does, it may be that in some cases tailored mixtures of monoclonal antibodies should be used. The other main drawback encountered with monoclonal antibodies is devising a suitable screening procedure for selecting hydridomas. Proof that an antibody is specific for a given organelle requires immunoelectron microscopy, although for screening purposes binding assays to crude membrane preparations must suffice. In conclusion, the primary antibody of choice should be either an affinity-purified polyclonal antibody or a high-affinity monoclonal antibody that recognizes a suitably specific membrane antigen.

Potential Sources of Specific Antigens

It can be seen from the above discussion that the correct choice of antigen and primary antibody is helpful to the success of immunoaffinity purification techniques. A number of strategies have been used, other than those involving changes in the antigenic structure of specific membranes, which are discussed above. For example, polyclonal antibodies raised against rat erythrocyte membranes were used to affinity purify rat fat-cell plasma membranes (Luzio *et al.*, 1976). Similarly, antibodies were raised to a lymphocyte plasma membrane fraction that did not bind to a concanavilin-A column (i.e., inside-out vesicles) to purify inside-out vesicles from a crude plasma membrane preparation (Walsh *et al.*, 1976). A rather different strategy was used in the search for antisera specific for cholinergic nerve terminals in mammalian brain. It was thought likely that only cholinergic-specific antigens would be conserved between such disparate organs as the *Torpedo* electric organ and the mammalian brain. Therefore, there would be few antibodies in an anti-*Torpedo*-synaptic-membrane that recognized other (unwanted) subcellular fractions in a mammalian brain. These antisera have been used successfully in the affinity purification of mammalian cholinergic nerve terminals (Richardson *et al.*, 1984). It may therefore be that, at least for highly conserved antigens, immunization with membrane preparations of widely divergent tissues will produce antisera and monoclonal antibodies (Phillips and Richardson, 1986) of the required specificity.

It is probable that in the future, further development in the generation of organelle and membrane-specific antibodies will occur. One very promising approach is the use of recombinant DNA methods, in which antibodies can be raised against antigens generated as fusion proteins in expression vectors (Stanley and Luzio, 1984; Young and Davis, 1983). The impurities present in such an immunogen are derived from *E. coli* proteins and consequently do not usually present a problem. Fusion proteins are also an ideal source of material for the affinity purification of polyclonal antibodies (Tooze and Stanley, 1986). Another potentially powerful approach is the raising of antibodies to synthetic peptides (Atassi, 1984; Lerner, 1982), which are by design directed at specific epitopes.

2.2. Second Antibody

The main requirement of the solid-phase immunoadsorbent is that it should have a high binding capacity. This depends both on the chemical nature of the support (see Section 3) and the nature of the antibody-binding reagent used. The latter may be either an anti-immunoglobulin second antibody, or Protein A from *Staphylococcus aureus*, which binds to the Fc portion of many immunoglobulin molecules (Kronvall *et al.*, 1970). When using an anti-immunoglobulin second antibody, this should be of high affinity, react with many immunoglobulin subclasses, and be available in large amounts. Appropriately selected monoclonal antibodies are useful since essentially unlimited quantities of antibody can be produced from a single hybridoma cell line. For some species of polyclonal first antibody (e.g., sheep and rabbit), suitable mouse monoclonal second antibodies are available, and if mouse monoclonal first antibodies are used rat antimouse Ig monoclonal antibodies may be coupled to the solid support. A variation of this is to make a potentially "universal immunoadsorbent" bearing avidin and using a panel of biotinylated second antibodies. The high affinity of biotin for avidin (K_a

= $10^{15}M^{-1}$) permits very tight binding, while the ability of avidin to bind four biotin molecules potentially increases the capacity of the immunoadsorbent. The use of staphylococcal Protein A on the solid support is limited by the fact that it binds to some species of immunoglobulin better than others and is especially poor at binding sheep and mouse IgG, though binding of the latter can be improved by using protein alkaline pH (Ey *et al.*, 1978). Since Protein A binds to the Fc portion of immunoglobulins, directional single-step immunoadsorbents may be prepared by reacting with affinity-purified polyclonal antibodies (Schneider *et al.*, 1982).

3. NATURE OF THE SOLID SUPPORT

The ideal solid support should have a high capacity for immunoglobulin while exhibiting a very low nonspecific binding of subcellular particles. In addition, it should be nonporous, have flexible spacer arms, and be easily separated from unbound particles.

The high binding capacity is necessary for two main reasons. First, after sensitizing the membrane to be purified with the primary antibody, some free immunoglobulin will still be present despite extensive washing. This free immunoglobulin will then compete effectively with membrane-bound antibody for binding sites on the solid phase. In addition, the binding of the sensitized membrane to the solid support is greatly facilitated by the presence of potential binding sites over the entire surfaces of both. The binding capacities of immunoadsorbents are determined primarily by the number of attachment sites for the second antibody, or equivalent immunoglobulin binding molecule, and the degree of inactivation incurred during the covalent bonding to the solid support. One of the most common mechanisms of second antibody inactivation is a consequence of multisite attachment of the antibody molecules to the support. To avoid this problem and to ensure the optimal orientation of the second antibody, Schneider *et al.* (1982) covalently bound Protein A to a Sepharose solid support. This was followed by reaction with the second antibody, and the Protein A–antibody complex covalently coupled with a bifunctional reagent. The capacity of this immunoadsorbent was therefore limited primarily by the number of reactive groups on the solid support. This number varies widely between different types of matrix. For example, diazotized cellulose (Hales and Woodhead, 1980) has one reactive group every 41 glucose residues, which amounts to 150 μequiv/g (Addison, 1971). This can be compared with approximately 15 μequiv/g of cyanogen-bromide-activated Sepharose. In contrast, if it is essential for the affinity-purified membrane to be eluted from the solid support, the low-capacity immunoadsorbents may be more advantageous since extreme conditions, for example, of pH or salt concentration, may not be required (Ghetie *et al.*, 1978).

Table II illustrates the binding capacities of a number of different immunoadsorbents using a variety of solid phases. The highest binding capacity published was that obtained using diazotized cellulose with a monoclonal second antibody (Richardson *et al.*, 1984). One other aspect of the solid support that may influence its efficacy is its flexibility; that is, the more flexible the immunoadsorbent, the greater the chance of it binding a large subcellular particle. In this context, it is worth noting that cellulose has considerable flexibility (Krassig, 1985) when compared to Sepharose or polyacrylamide.

Nonspecific binding of membranes to the solid phase is a major limiting factor in affinity purification procedures. The amount of nonspecific binding is affected by both the quantity of solid phase and the amount of second antibody bound to the immunoadsorbent. In general, the degree of nonspecific binding falls with decreases in the amount of solid phase used (Miljanich *et al.*, 1982); therefore, high-capacity immunoadsorbents are capable of reducing the contamination of affinity-purified membranes. Probably, the most informative way of determining the degree of contamination is to estimate the ratio of specific to nonspecific binding. Some examples of this are shown in Table III.

The most commonly used supports for immunoaffinity purification are nonporous so as to maximize the surface binding capacity (e.g., Sepharose 6MB, polyacrylamide, and cellulose). Porous supports give greater total binding capacities and are useful for the purification of molecules, but most of the binding sites

Table II
Immunoadsorbent Binding Capacities

Anti-immunoglobulin binding reagent	Matrix	Immunoglobulin binding capacity[a] (μg/mg)	Reference
Protein A	Sepharose 6MB	14	Ghetie *et al.* (1978)
	Polyacrylamide (magnetic)	27.2	Docherty *et al.* (1987)
	Polyacrylamide	75	Matthew *et al.* (1981)
Polyclonal	Cellulose	1	Luzio *et al.* (1976)
Affinity-purified polyclonal	Polyacrylamide	4–6	Au and Varon (1979); Ito and Palade (1978); Gruenberg and Howell (1985)
	Polyacrolein (magnetic)	80	Howell *et al.* (1985)
Monoclonal	Sepharose 4B	6.6	de Kretser *et al.* (1980)
	Cellulose	250	Richardson *et al.* (1984)

[a]When the immunoglobulin binding capacity was not quoted, affinity purified polyclonal antibodies and monoclonal antibodies were assumed to bind 2 μg IgG/μg anti-IgG, polyclonal antibodies were assumed to bind 0.02 μg IgG/μg anti-IgG, and Protein A 5 μg/μg Protein A. For reference 2 it was assumed that 1 mL of Magnogel contained 250 μg beads.

Table III
Immunoadsorbent Purification Efficiencies

Matrix	Percent binding: Specific	Percent binding: Nonspecific	Binding ratio[a]	Reference
Polyacrylamide	22	3.2	6.9	Gruenberg and Howell (1986)
	64	7	9.1	Ito and Palade (1978)
	70	4.1	17.0	Gruenberg and Howell (1986)
	76	2.0	38.0	Debanne *et al.* (1984)
	7	0.1	70.0	Gruenberg and Howell (1985)
Sepharose	21	0.7	30.0	Pontremoli *et al.* (1984)
S. aureus cells	73	5.0	14.6	Roman and Hubbard (1984)
	98	5.0	19.6	Mueller and Hubbard (1986)
	68	2.2	31.0	Gruenberg and Howell (1985)
Cellulose	35	4.0	8.8	Luzio *et al.* (1976)
	41	4.0	10.2	Westwood *et al.* (1979)
	48	4.0	12.0	Luzio (1977)
	17	0.2	85.0	Richardson *et al.* (1984)
	70	0.6	116.7	Devaney and Howell (1985)

[a]Binding ratio = % yield of specific component/% yield of nonspecific component. In cases where more than one marker was measured, the major one of interest is quoted. It should be noted that in some of the above-mentioned publications alternative criteria of purification have been used by the original authors.

are localized inside the beads and so are unavailable for the binding of the relatively large membranes and organelles. However, two porous supports have been used, one based on Sepharose 4B (de Kretser *et al.*, 1980) and one on AcA 44 (Docherty *et al.*, 1987). Both these supports had high binding capacities—the first bearing a monoclonal antibody and the second Protein A, so a sufficiently large number of binding sites were available on the surface of the beads to bind the particles being separated.

The method used to separate the immunoadsorbent from unbound membranes also influences the ratio of specific to nonspecific binding, mostly by affecting the latter. Three separation methods have been used—column, centrifugation, and magnetism. Separation of bound from free particles on columns (Au and Varon, 1979) is severely restricted because many particles become nonspecifically trapped on the column matrix. This drawback has been overcome to some extent by washing the immunoadsorbent using both column and batch methods (Antoine *et al.*, 1978). Centrifugation methods utilize the fact that most of the solid supports commonly used are denser than free subcellular organelles. Rapid separations using low-speed centrifugation have therefore been widely used. Centrifugation protocols, however, do have one disadvantage. Thus, when optimizing an affinity purification procedure it is tempting to aim for very-high-capacity immunoadsorbents and low solid support/organelle ratios. This can lead to very high binding of organelle to solid support, thus considerably reducing the support's density and destroying the separation protocol (P. J. Richardson and R. E. H. Phillips, unpublished observations). Such problems can be avoided by either increasing the solid support/organelle ratio (which would increase the nonspecific binding) or using an alternative separation procedure (e.g., magnetism).

Magnetic immunoaffinity purification of subcellular membranes may be achieved either by using solid supports containing magnetic iron or by ferritin labeling of the membranes, a technique successfully used for cell separation (Odette *et al.*, 1984). An additional advantage of magnetic-based separation is that it can be performed in a continuous flow system in which the magnetic particles can be kept in suspension by a magnetic field (Howell *et al.*, 1985). Such systems should then give a further reduction in nonspecific binding since the solid support and the membranes are not forced together as they are during column and centrifugation procedures. A variety of magnetic solid phases have been developed including polyacrolein (Howell *et al.*, 1985; Ugelstadt *et al.*, 1983), dextran (Molday and Molday, 1984), polyacrylamide–agarose (Antoine *et al.*, 1978), and cellulose (Jackson *et al.*, 1984). However, magnetic separation in free flow may require strong magnetic fields (1 tesla; Molday and Molday, 1984) or specifically designed magnetic "bottles" (Howell *et al.*, 1985). Even the use of weak magnets in a batch process can still be advantageous since density differences between particle and matrix are unnecessary.

4. ASSESSMENT OF PURIFICATION

As mentioned in Section 3, it is imperative that the degree of contamination (and therefore of purification) is determined when using immunoaffinity techniques. Table III illustrates one method of assessing the purification achieved, that is, the percent yield of specific component per percent yield of nonspecific component. There are, however, two ways of assessing the nonspecific component: one is the nonspecific binding of organelles to the solid support, either in the absence of the primary antibody or with a nonspecific immunoadsorbent, while the second is the amount of a given contaminant that copurifies with the desired membrane. Under some circumstances, the latter can also give an indication of the specificity of the primary antibody.

An alternative method of assessing the purification process is to use the efficiency (Howell *et al.*, 1985), which is defined as ratio of specific to nonspecific efficiency indices where the efficiency index (e.g., EI specific) represents the yield of specific component at saturation per amount of specific component required for half saturation. The efficiency index for nonspecific adsorption is also calculated using, for instance, the immunoadsorbent without primary antibody. The ratio of the two indices is then taken as the efficiency.

A third way of evaluating the purification is the enrichment of the desired

Table IV
Immunoaffinity Purification of Rat Fat Cell Plasma Membrane

Plasma membrane markers	Percent homogenate activity recovered		Binding ratio[b]
	Anti-IgG ImAd	NI ImAd[a]	
5′-Nucleotidase	35.3	4.2	8.4
Adenylate cyclase (NaF stimulated)	36.4	1.6	22.8
Nonspecific markers			
Cytochrome oxidase	5.8		6.2
NADH-cytochrome c reductase	7.0		5.1
Lactate dehydrogenase	4.2		8.5

[a]NI ImAd, nonimmune Ig coupled to cellulose (immunoadsorbent). Data taken from Luzio *et al.* (1976).
[b]The binding ratios presented are either the recovery of plasma membrane marker on anti-IgG ImAd/NI ImAd or plasma membrane marker/contaminant marker on anti-IgG ImAd.

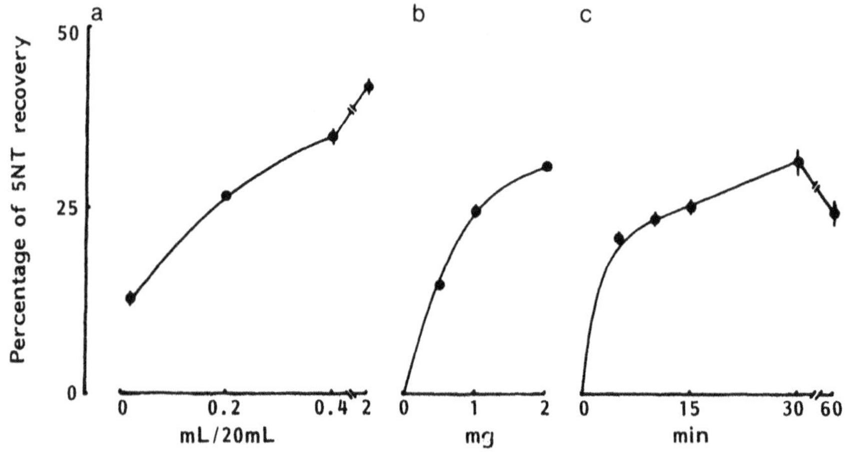

FIGURE 3. Factors affecting the yield of plasma membrane immunoaffinity purified on cellulose immunoadsorbent from collagenase-isolated rat fat cells. The fat cells from eight epididymal fat pads were incubated with (a) different amounts of anticell antiserum and then the homogenate was incubated with 2 mg of anti-Ig immunoadsorbent for 15 min at 4°C; (b) 1:100 anticell antiserum and then the homogenate was incubated with different amounts of anti-Ig immunoadsorbent for 15 min at 4°C; (c) 1:100 anticell antiserum and then the homogenate was incubated with 2 mg of anti-Ig immunoadsorbent for different times at 4°C. Plasma membrane yield was assessed by measuring 5′-nucleotidase (5NT) recovery. Data redrawn from Luzio *et al.* (1976).

component, that is, the specific activity of purified component per specific activity of component in the original fraction. One problem with this indicator is that when using antibody-coated immunoadsorbents it is difficult to obtain accurate values for the protein content of the purified fraction. Of the three indicators mentioned, the first (i.e., percent yield of specific component per percent yield of nonspecific component) is the easiest to perform and most informative. Ideally, the nonspecific component should be assessed in both the ways mentioned. Table IV illustrates the results obtained in the first successful immunoaffinity purification of subcellular membranes (Luzio *et al.*, 1974, 1976). It is apparent that a reliance on one method of assessing nonspecific binding could be misleading.

The most important factors affecting the yield are the amount of primary antibody, the amount of immunoadsorbent, and the time allowed for binding to the immunoadsorbent. Consequently, in any specific immunoaffinity protocol for isolating an organelle or membrane fragment, the time course and dose–response curve for binding of first antibody and immunoadsorbent must be determined experimentally to find the most appropriate conditions for specific isolation and high yield in the shortest possible time (Figure 3) (Luzio *et al.*, 1976; Richardson *et al.*, 1984).

5. FUNCTIONAL INTEGRITY OF PURIFIED MEMBRANES

As mentioned earlier, most immunoaffinity-purified subcellular fractions maintain their predicted properties while attached to the solid phase. For example, the affinity-purified fat cell plasma membrane contains adenylate cyclase, which is hormonally sensitive (Figure 4). Similarly, the affinity-purified cholinergic synaptosome preparation is osmotically sensitive and metabolically active while still attached to the solid support (Richardson, 1986; Richardson *et al.*, 1984). It has recently been shown that the release of acetylcholine from these terminals is regulated via an adenosine-mediated feedback mechanism involving presynaptically released ATP and ectophosphohydrolases present at the synaptic cleft (P. J. Richardson, unpublished observations). One further example of the functional integrity of affinity-purified membranes has been demonstrated by Gruenberg and Howell (1986). These authors isolated endosomes on an immunoadsorbent and elegantly demonstrated the reconstitution of endocytic vesicle fusions using these endosomes. It is therefore concluded that elution of membranes is not necessary for the study of their function. It may, however, be necessary in order to study their protein and lipid constituents. Elution protocols to recover protein antigens from immunoadsorbent have been reviewed by Bailyes *et al.* (1987).

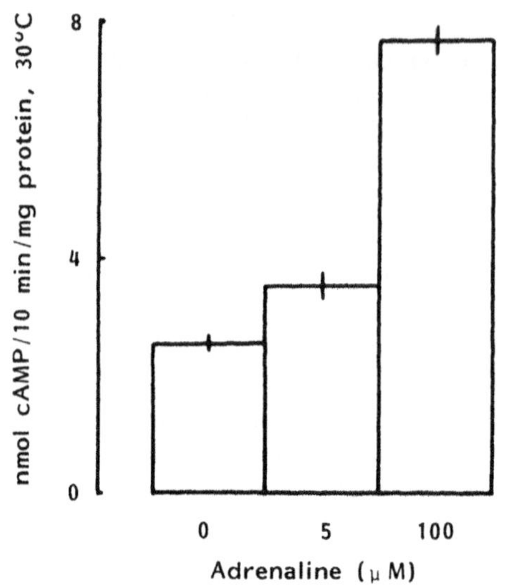

FIGURE 4. Hormonal sensitivity of affinity purified rat fat cell plasma membrane adenylate cyclase. Data from Luzio *et al.* (1976).

6. CONCLUSIONS

Immunoaffinity purification techniques have been applied to the subcellular fractionation of a wide variety of systems (Table I). Their increasing popularity is a consequence of their inherent advantages, that is, speed, applicability to small amounts of tissue, and the ability to prepare fractions that cannot be separated by conventional techniques. Some striking examples of the success of these techniques include the selection of right-side-out versus inside-out vesicles (Gruenberg and Howell, 1985; Walsh *et al.*, 1976), asialoglycoprotein receptor-positive versus receptor-negative endosomes (Mueller and Hubbard, 1986), the subfractionation of clathrin-coated vesicles (Pfeffer and Kelly, 1985), and the purification of cholinergic and other nerve terminals present in the complex homogenates of mammalian brain (Docherty *et al.*, 1987; Richardson *et al.*, 1984). The versatility of immunoadsorption methods has been extended to include the removal of contaminants from fractions prepared by classical subcellular fractionation techniques (Luzio and Stanley, 1983).

A sufficiently large number of affinity purification procedures have been published to permit a number of general recommendations to be made. A specific, high-avidity, antiorganelle antibody (either monoclonal or affinity-purified polyclonal) directed to an antigen present at high density on the membrane of interest should be used. The immunoadsorbent should be of high capacity with a solid support capable of covalently binding large quantities of protein and, if a second antibody is used, this should also be of high avidity and available in large amounts. The solid support should demonstrate low levels of nonspecific binding. Incubation and washing conditions are of importance and should be gentle to prevent disruption of organelles and vesiculation of membranes (Howell *et al.*, 1985).

Table V shows the experimental protocol used in the affinity purification of cholinergic nerve terminals from rat cerebral cortex. In this example, the above recommendations are satisfied; that is, the primary polyclonal antibody is effectively affinity purified on the surface of the nerve terminal, the immunoadsorbent has a binding capacity of approximately 250 μg sheep IgG/mg cellulose, and the avidity of the second antibody (monoclonal) is high (K_a is approximately 10^{11} M^{-1}). It should also be noted that the incubation of the sensitized terminals with the immunoadsorbent is performed in a small volume.

It is probable that as more organelle and vesicle-specific antibodies become available, immunoaffinity purification will become the method of choice for separating membranes of similar physical properties but different biological

Table V
Immunoaffinity Purification of Cholinergic Nerve Terminals

Approximately 1 g scraped rat cerebral cortex

1. Make a 10% homogenate in sucrose solution—12 up-and-down strokes in rotating (640 rpm) teflon/glass (loose-fitting) homogenizer. Keep at 4°C.
2. Centrifuge at 1000*g* for 10 min; take supernatant and centrifuge at 18,000*g* for 20 min. Make pellet to 10 mL with sucrose solution.
3. Split into two 5-mL portions. Incubate one with sheep anti-Chol-1 serum (1:25 to 1:200 dilution), the other with a similar dilution of nonimmune sheep serum, for 40 min at 4°C with periodic shaking. Dilute to 30 mL with sucrose.
4. Centrifuge at 18,000*g* for 15 min (Sorvall SS-34, 14,000 rpm).
5. Wash pellets three times with 30 mL of sucrose, centrifuging between washes.
6. Resuspend pellets in 3 mL of Percoll, centrifuge at 9000*g* for 2 min. Take top layer of synaptosomes, dilute with KRH + EDTA solution, and centrifuge again.
7. Incubate 200-μL aliquots with 1–2 mg of immunoadsorbent in a total volume of 500 μL. Ensure good mixing. Incubate 40 min at 4°C with occasional swirling.
8. Separate bound and free by addition of 2.5 mL of Percoll, mix, and centrifuge at 9000*g* for 2 min.
9. Remove supernatant. Wash immunoadsorbent three times with 2.5 mL of Percoll and once with KRH solution.

Solutions

Sucrose	0.32 M in 1 mM EDTA, 5 mM Hepes pH 7.4
KRH	140 mM NaCl, 5 mM KCl, 2.6 mM CaCl, 2 mM $MgCl_2$, 5 mM glucose, 10 mM Hepes pH 7.4
KRH + EDTA	As KRH + 1 mM EDTA pH 7.4, without Ca^{2+} and Mg^{2+}
Percoll	45% in KRH + EDTA solution (v/v)
Immunoadsorbent	Cellulose bearing monoclonal mouse antisheep-IgG antibody (approximately 0.3 mg mouse IgG/mg cellulose)

functions. Such a development will be facilitated by further improvements in the design of the solid phase, that is, monoclonal antibodies to a wide range of immunoglobulins (especially mouse IgG) and high-capacity supports for magnetic separation.

ACKNOWLEDGMENTS. We thank Professor C. N. Hales, Dr. K. Siddle, Dr. K. Howell, and Dr. J. Gruenberg for many useful and stimulating discussions. Experimental work by the authors reported in this chapter was supported by grants from the Medical Research Council, the British Diabetic Association, the Beit Foundation, St. John's College Cambridge, and Muscular Dystrophy Group of Great Britain and Northern Ireland.

7. REFERENCES

Addison, G. M., 1971, Preparation and properties of labelled antibodies. Ph.D. thesis, University of Cambridge, U.K.

Albertson, P.-A., 1986, *Partition of Cell Particles and Macromolecules*, Wiley, New York.

Antoine, J.-C., Ternynck, T., Rodrigot, M., and Avrameas, A., 1978, Lymphoid cell fractionation on magnetic polyacrylamide–agarose beads. *Immunochemistry* **15**:443–452.

Atassi, M. Z., 1984, Antigenic structures of proteins. *Eur. J. Biochem.* **145**:1–20.

Au, M. M.-J., and Varon, S., 1979, Neural cell sequestration on immunoaffinity columns. *Exp. Cell Res.* **120**:269–276.

Bailyes, E. M., Richardson, P. J., and Luzio, J. P., 1987, Immunological methods applicable to membranes. In *Biological Membranes: A Practical Approach* (J. Findlay and W. H. Evans, eds.), IRL Press, Oxford, U.K.

Boone, C. W., Ford, L. E., Bond, H. E., Stuart, D. C., and Lorenz, D., 1969, Isolation of plasma membrane fragments from Hela cells. *J. Cell Biol.* **41**:378–392.

Courtoy, P. J., Quintart, J., and Baudhuin, P., 1984, Shift of equilibrium density induced by 3,3′-diamino-benzidine cytochemistry: A new procedure for the analysis and purification of peroxidase containing organelles. *J. Cell Biol.* **98**:870–876.

Davis, G. A., and Bloom, F. E., 1973, Subcellular particles separated through a histochemical reaction. *Anal. Biochem.* **51**:429–435.

Debanne, M. T., Bolyos, M., Gauldie, J., and Regoeczi, E., 1984, Two populations of prelysosomal structures transporting asialoglycoproteins in rat liver. *Proc. Natl. Acad. Sci. U.S.A.* **81**:2995–2999.

DeDuve, C., 1975, Exploring cells with a centrifuge. *Science* **189**:186–194.

de Kretser, T. A., Bodmer, J. G., and Bodmer, W. F., 1980, The separation of cell populations using monoclonal antibodies attached to sepharose. *Tissue Antigens* **16**:317–325.

Docherty, M., Bradford, H. F., and Wu, J.-Y., 1987, Purification and characterization of mammalian cholinergic and GABAergic synaptosomes. *Biochem. Soc. Trans.* **15**:638–641.

Evans, W. H., 1987, Isolation of membranes. In *Biological Membranes: A Practical Approach* (J. Findlay and W. H. Evans, eds.), IRL Press, Oxford, U.K.

Ey, P. L., Prowse, S. J., and Jenkin. C. R., 1978, Isolation of pure IgG_1, IgG_{2a} and IgG_{2b} immunoglobulins from mouse serum using protein A–Sepharose. *Immunochemistry* **15**:429–436.

Flanagan, S. D., Johansson, G., Yost, B., Ito, Y., Sutherland, I. A., 1984, Toroidal coil countercurrent chromatography in the affinity partitioning of nicotinic cholinergic receptor enriched membranes. *J. Liquid Chromatogr.* **7**:385–402.

Gahmberg, C. G., and Simons, K., 1970, Isolation of plasma membrane fragments from BHK 21 cells. *Acta Pathol. Microbiol. Scand. B* **78**:176–182.

George, S. G., and Kenny, A. J., 1973, Studies on the enzymology of purified preparations of brush border from rabbit kidney. *Biochem. J.* **134**:43–57.

Ghetie, V., Mota, A., and Sjoquist, J., 1978, Separation of cells by affinity chromatography on SpA–sepharose 6MB. *J. Immunol. Methods* **21**:133–141.

Gorvel, J.-P., and Maroux, S., 1987, Characterization of intestinal membrane vesicles with FACS analysis. In *Cell Free Analysis of Membrane Traffic* (J. Moore, ed.), Alan R. Liss, New York.

Gruenberg, J., and Howell, K. E., 1985, Immunoisolation of vesicles using antigenic sites either located on the cytoplasmic or exoplasmic domain of an implanted viral protein. A quantitative analysis. *Eur. J. Cell Biol.* **38**:312–321.

Gruenberg, J., and Howell, K. E., 1986, Reconstitution of vesicle fusions occurring in endocytosis with a cell-free system. *EMBO J.* **5**:3091–3101.

Hales, C. N., 1972, Immunological techniques in diabetes research. *Diabetologia* **8**:229–235.

Hales, C. N., and Woodhead, J. S., 1980, Labelled antibodies and their use in the immunoradiometric assay. *Methods Enzymol.* **70**:334–355.

Hannig, K., and Heidrich, H.-G., 1977, Continuous freeflow electrophoresis and its application in biology. In *Cell Separation Methods Part IV Electrophoretic Methods* (H. Bloomendal, ed.), pp. 93–116, Elsevier, Amsterdam.

Howell, K. E., Ansorge, W., and Gruenberg, J., 1985, Immunoisolation system using beads maintained in free flow within a magnetic field. In *Microspheres: Medical and Biological Applications* (A. Rembaum and Z. Tokes, eds.), CRC Press, Boca Raton, FL.

Ito, Y., and Bowman, R. L., 1978, Countercurrent chromatography with the flow-through centrifuge without rotating seals. *Anal. Biochem.* **85**:614–617.

Ito, A., and Palade, G. E., 1978, Presence of NADPH-cytochrome P-450 reductase in rat liver golgi membranes. Evidence obtained by immunoadsorption method. *J. Cell Biol.* **79**:590–597.

Jackson, A. P., Siddle, K., and Thompson, R. J., 1984, Two site monoclonal antibody assays for human heart- and brain-type creatine kinase. *Clin. Chem.* **30**:1157–1162.

Jones, R. T., Walker, J. H., Richardson, P. F., Fox, G. W., and Whittaker, V. P., 1981, Immunohistochemical localization of cholinergic nerve terminals. *Cell Tissue Res.* **218**: 355–373.

Kamat, V. B., and Wallach, D. F. H., 1965, Separation and partial purification of plasma membrane fragments from Ehrlich ascites carcinoma microsomes. *Science* **148**:1343–1345.

Kaplan, G., Unkeless, J. C., and Cohn, Z. A., 1979, Insertion and turnover of macrophage plasma membrane proteins. *Proc. Natl. Acad. Sci. U.S.A.* **76**:2824–2828.

Kawajiri, K., Ito, A., and Omura, T., 1977, Subfractionation of rat liver microsomes by immunoprecipitation and immunoadsorbtion methods. *J. Biochem.* **81**:779–789.

Kohler, G., and Milstein, C., 1975, Continuous cultures of fused cells secreting antibody of predefined specificity. *Nature* **256**:495–497.

Krassig, H., 1985, Structure of cellulose and its relation to properties of cellulose fibres. In *Cellulose and Its Derivatives: Chemistry, Biochemistry and Applications* (J. F. Kennedy, G. O. Phillips, D. J. Wedlock, and P. A. Williams, eds.), pp. 3–25, Ellis Horwood, Chichester, U.K.

Kronvall, G., Seal, V. S., Finstad, J., and Williams, R. C., 1970, Phylogenetic insight into evolution of mammalian Fc fragment of γG globulin using staphylococcal protein A. *J. Immunol.* **104**:140–147.

Lerner, R. A., 1982, Tapping the immunological repertoire to produce antibodies of predetermined specificity. *Nature* **299**:592–596.

Leskes, A., Siekewitz, P., and Palade, G. E., 1971, Differentiation of endoplasmic reticulum in hepatocytes II. Glucose-6-phosphatase in rough microsomes. *J. Cell Biol.* **49**:288–302.

Luzio, J. P., 1977, Immunological approaches to the study of membrane features in adipocytes. In *Methodological Surveys in Biochemistry* (E. Reid, ed.), Vol 6, pp. 131–142, Ellis Horwood, Chichester, U.K.

Luzio, J. P., and Stanley, K. K., 1983, The isolation of endosome derived vesicles from rat hepatocytes. *Biochem. J.* **216**:27–36.

Luzio, J. P., Newby, A. C., and Hales, C. N., 1974, Immunological isolation of rat fat cell plasma membranes. *Biochem. Soc. Trans.* **2**:1385–1386.

Luzio, J. P., Newby, A. C., and Hales, C. N., 1976, A rapid immunological procedure for the isolation of hormonally sensitive rat fat cell plasma membrane. *Biochem. J.* **154**:11–21.

Matthew, W. D., Tsavaler, L., and Reichardt, L. F., 1981, Identification of a synaptic vesicle-specific membrane protein with a wide distribution in neuronal and neurosecretory tissue. *J. Cell Biol.* **91**:257–269.

Merisko, E. M., Farquhar, M. G., and Palade, G. E., 1982, Coated vesicle isolation by immunoadsorption on staphylococcal aureus cells. *J. Cell Biol.* **92**:846–857.

Miljanich, G. P., Brasier, A. R., and Kelly, R. B., 1982, Partial purification of presynaptic plasma membrane by immunoadsorption. *J. Cell Biol.* **84**:88–96.

Molday, R. A., and Molday, L. L., 1984, Separation of cells labelled with immunospecific iron dextran microspheres using high gradient magnetic chromatography. *FEBS Lett.* **170**:232–238.

Mooré, D. J., Moore, D. M., and Heidrich, H.-G., 1983, Subfractionation of rat liver Golgi apparatus by free-flow electrophoresis. *Eur. J. Cell Biol.* **31**:263–274.

Morris, S. J., and Schovanka, I., 1977, Some physical properties of adrenal medulla chromaffin granules isolated by a new continuous iso-osmotic density gradient method. *Biochim. Biophys. Acta* **464**: 53–64.

Mueller, S. C., and Hubbard, A. L., 1986, Receptor-mediated endocytosis of asialoglycoproteins by rat hepatocytes: Receptor-positive and receptor-negative endosomes. *J. Cell Biol.* **102**:932–942.

Mullock, B. M., Hinton, R. H., Peppard, J. V., Slot, J. W., and Luzio, J. P., 1987, The preparative isolation of endosome fractions. *Cell Biochem. Function* **5**:235–243.

Murphy, R. F., 1985, Analysis and isolation of endocytic vesicles by flow cytometry and sorting: Demonstration of three kinetically distinct compartments involved in fluid phase endocytosis. *Proc. Natl. Acad. Sci. U.S.A.* **82**:8523–8526.

Odette, L. L., McCloskey, M. A., and Young, S. H., 1984, Ferritin conjugates as specific magnetic labels—implications for cell separation. *Biophys. J.* **45**:1219–1222.

Pearse, B., 1980, Coated vesicles. *Trends Biochem. Sci.* **5**:131–134.

Pfeffer, S. R., and Kelly, R. B., 1985, The subpopulation of brain coated vesicles that carries synaptic vesicle proteins contains two unique polypeptides. *Cell* **40**:949–957.

Phillips, R. E. H., and Richardson, P. J., 1986, The detection of the cholinergic ganglioside Chol-1 in rat brain using monoclonal antibodies. *Biochem. Soc. Trans.* **14**:765.

Pontremoli, S., Diamine, G., Michetti, M., Salamino, F., Sparatore, B., and Honecker, B. L., 1984, Binding of monoclonal antibody to Cathepsin M located on the external surface of rabbit lysosomes. *Arch. Biochem. Biophys.* **233**:267–271.

Porter-Jordan, K., Benson, R. J. J., Buoniconti, P., and Fine, R. E., 1986, An acetylcholinesterase mediated density shift technique demonstrates that coated vesicles from chick myotubes may contain both newly synthesised acetylcholinesterase and acetylcholine receptors. *J. Neurosci.* **6**:3112–3119.

Reijnierse, L. A., Velstra, H., and Van den Berg, C. J., 1975, Subcellular localization of γ-aminobutyrate transaminase and glutamate dehydrogenase in adult rat brain. *Biochem. J.* **152**:469–475.

Richardson, P. J., 1986, Choline uptake and metabolism in affinity purified cholinergic nerve terminals from rat brain. *J. Neurochem.* **46:**1251–1255.

Richardson, P. J., Siddle, K., and Luzio, J. P., 1984, Immunoaffinity purification of intact, metabolically active, cholinergic nerve terminals from mammalian brain. *Biochem. J.* **219:**647–654.

Roman, L. R., and Hubbard, A. L., 1984, A domain specific marker for the hepatocyte plasma membrane. III. Isolation of bile canalicular membrane by immunoadsorption. *J. Cell Biol.* **98:**1497–1504.

Ryan, J. W., and Smith, U., 1971, A rapid simple method for isolating pinocytotic vesicles and plasma membrane of lung. *Biochim. Biophys. Acta* **249:**177–180.

Schneider, C., Newman, R. A., Sutherland, D. R., Asser, V., and Greaves, M. F., 1982, A one step purification of membrane proteins using a high efficiency immunomatrix. *J. Biol. Chem.* **257:**10766–10769.

Sharp, K. A., Yalpani, M., Howard, S. J., and Brooks, D. E., 1986, Synthesis and application of a polyethylene-glycol-antibody affinity ligand for cell separations in aqueous two phase systems. *Anal. Biochem.* **154:**110–117.

Siddle, K., and Soos, M., 1981, Monoclonal antibodies for human pituitary. In *Monoclonal Antibodies and Developments in Immunoassay* (A. Albertini and R. Ekins, eds.), pp. 53–66, Elsevier, Amsterdam.

Stanley, K. K., and Luzio, J. P., 1984, *EMBO J.* **3:**1429–1434.

Stanley, K. K., Newby, A. C., and Luzio, J. P., 1982, What do ectoenzymes do? *Trends Biochem. Sci.* **7:**145–147.

Tooze, S. A., and Stanley, K. K., 1986, The identification of two epitopes in the carboxyterminal 15 amino acids of the E1 glycoprotein of MHV-A59 using hybrid proteins. *J. Virol.* **60:**928–934.

Ugelstadt, J., Soderberg, L., Berge, A., and Bergstrom, J., 1983, Monodisperse polymer particles—a step forward for chromatography. *Nature* **303:**95–96.

Walsh, F. S., Barber, B. H., and Crumpton, M. J., 1976, Preparation of inside-out vesicles of pig lymphocyte plasma membrane. *Biochemistry* **15:**3557–3563.

Westwood, S. A., Luzio, J. P., Flockhart, D. A., and Siddle, K., 1979, Investigation of the subcellular distribution of cyclic AMP phosphodiesterase in rat hepatocytes, using a rapid immunological procedure for the isolation of plasma membrane. *Biochim. Biophys. Acta* **583:**454–466.

Whittaker, V. P., 1984, The synaptosome. In *Handbook of Neurochemistry* (A. Lajtha, ed.), Vol. 7, pp. 1–30, Plenum Press, New York.

Young, R. A., and Davis, R. Z., 1983, Efficient isolation of genes by using antibody probes. *Proc. Natl. Acad. Sci. U.S.A.* **80:**1194–1198.

Chapter 7

The Use of Antibodies for Analysis of the Secretory and Endocytic Paths of Eukaryotic Cells

Yves Chicheportiche and Alan Michael Tartakoff

1. INTRODUCTION

The complex machinery of the eukaryotic cell cytoplasm coordinates multiple simultaneous events of vesicular transport. The best characterized are those pertaining to transport along the secretory and endocytic paths.

This massive membrane flux involves organelles of complex composition and organization; in particular, the endoplasmic reticulum, the Golgi complex, secretion vesicles or granules, the plasma membrane, endocytic vesicles, endosomes, and lysosomes. Several of these are compositionally and functionally heterogeneous. For example, the Golgi complex is composed of a stack of distinct cisternae (proximal, medial, distal), the plasma membrane is composed of distinct though anatomically continuous domains, and endosomes bear diverticulae that appear to be intimately related to their ability to sequester ligands from receptors.

All these compartments are involved in repeated rapid events of membrane–

Yves Chicheportiche Department of Pathology, University Medical Center, 1211 Geneva 4, Switzerland. **Alan Michael Tartakoff** Institute of Pathology, Case Western Reserve University, Cleveland, Ohio 44106.

membrane recognition, fusion, and fission. Nevertheless, they each appear to maintain a distinct identity.

To unravel the events that lead to macromolecular secretion or internalization, it is important (1) to isolate and characterize all the vesicular carriers involved, (2) to assign functions and enzyme activities to each of them, and (3) ultimately to reconstitute in vitro these transport paths.

In this chapter, we concentrate on the smooth membrane-bounded compartments that participate in intracellular transport: the Golgi complex, lysosomes, and all the smooth vesicles derived from or destined to fuse with the plasma membrane. We especially emphasize the importance and utility of polyclonal and monoclonal antibodies for such analysis.

2. THE SECRETORY AND ENDOCYTIC PATHWAYS: AN OVERVIEW

The well-accepted secretory pathway conducts secretory, lysosomal, and cell surface proteins as follows: after synthesis and vectorial release to the cisternal space of the rough endoplasmic reticulum, these proteins encounter transitional elements, from which vesicles bud that enclose and convey quanta of content and membrane to the proximal Golgi cisternae. These newly arrived proteins then pass successively from one to the next cisterna, often while undergoing post-translational modifications: trimming, maturation and addition of sugar moieties, acylation, sulfation, and acquisition of a mannose-6-phosphate recognition signal, in the case of lysosomal enzymes. After having reached the distal cisternae, two pathways are possible.

In *regulated* secretory cells, proteins destined for export are packaged into immature granules (condensing vacuoles) that subsequently transform into mature secretion granules whose content is exocytosed upon stimulation. Examples are endocrine and exocrine cells of the pancreas and pituitary cells. In *constitutive* secretory cells, secretory proteins are continuously discharged without being stored in granules. The vesicular carriers that mediate their delivery to the plasma membrane are essentially uncharacterized. Such behavior is illustrated by hepatocytes, fibroblasts, and plasma cells.

In both regulated and constitutive secretory cells, membrane proteins may use a constitutive pathway(s) to reach the plasma membrane. Lysosomal enzymes are deviated from the secretory path upon exit from the Golgi complex.

The uptake of extracellular solutes also takes two forms: (1) constitutive fluid-phase pinocytosis, in which vesicles invaginate from the plasma membrane, and (2) receptor-mediated endocytosis, which apparently may be either constitutive or controlled by the presence of ligand. In many cases, the ligands are soluble, for example, low-density lipoprotein (LDL), transferrin, asialoglycoproteins, insulin, epidermal growth factor, and immunoglobulin A (IgA). These ligands are internalized via coated pits, which give rise to coated vesicles that lose their clathrin coat within less than 1 min. The newly formed endocytic vesicles or endosomes contain an electrogenic proton pump that acidifies their content to pH 5–6.

After arrival in endosomes, any of several pathways can be taken: (1) the ligand may dissociate from its receptor, the receptor recycle back to the plasma membrane, and the ligand pass on to the lysosome (LDL, asialoglycoproteins, acid hydrolases bearing the mannose-6-phosphate signal); (2) the ligand and the receptor may both be conveyed to lysosomes (EGF, insulin); (3) the ligand and the receptor may return to the plasma membrane. (In the case of transferrin, iron dissociates from the complex in endosomes. In transcytosis of IgA, the ligand–receptor complex is released at a different domain of the plasma membrane from the domain at which IgA was internalized.)

In addition, a little understood path of transport from the plasma membrane to the Golgi has been documented. This route may include the recapture of membrane delivered by exocytosis to the cell surface and the path followed by receptors that mediate receptor-mediated endocytosis (Farquhar, 1985; Snider and Rogers, 1985, 1986; Woods *et al.*, 1986).

The paths outlined above are well documented by numerous criteria. Nevertheless, and this is the central concern of this chapter, many of the smooth-membrane bounded compartments are anatomically similar to each other, coexist within the same region of the cell cytoplasm, and cannot readily be distinguished from each other.

For details and references on the overall performance of the secretory and endocytic paths, the reader is referred to the recent publications on endocytosis (Pastan and Willingham, 1985; Wileman *et al.*, 1985) and on protein secretion (Dunphy and Rothman, 1985; Farquhar, 1985; Kelly, 1985; Tartakoff, 1980, 1983a).

We use the following terminology: the *endodomain* of a membrane is the surface in contact with the cell sap or ground substance of the cytoplasm. The *ectodomain* faces a fluid volume that is topologically equivalent to the extracellular space.

3. AVAILABLE METHODS FOR DEFINING COMPARTMENTS ALONG THE SECRETORY AND ENDOCYTIC PATHWAYS

3.1. Subcellular Fractionation

Differential and/or isopycnic sedimentation and free-flow electrophoresis have often been used to recover subcellular fractions highly enriched in components of the secretory and endocytic paths. Thus, rough microsomes (Kreibich *et al.*, 1973), lysosomes (Harms *et al.*, 1980), zymogen granules (Arvan *et al.*, 1983; Meldolesi, 1983; Paquet *et al.*, 1982) coated vesicles (Pearse, 1983; Simion *et al.*, 1983), and multivesicular bodies (Hornick *et al.*, 1985) have been obtained at near homogeneity.

With regard to the membranes usually recovered in a total smooth microsomal fraction, however, the situation is more difficult. Nevertheless, the resolution of sucrose or Percoll gradients has been sufficient to document, for example, subcompartmentalization of enzyme activities of the Golgi complex and to resolve endosomes from lysosomes (Evans and Flint, 1985; Storrie *et al.*, 1984; Wall and Hubbard, 1985).

With regard to Golgi-associated activities, N- and O-glycan processing enzymes, including the enzymes responsible for the addition of the mannose-6-phosphate signal of lysosomal enzymes, have partially been resolved from each other and from rough microsomes. Indeed, three subcompartments of overlapping density have been identified. These are thought to correspond to proximal, medial, and distal elements of Golgi (Deutscher *et al.*, 1983; Dunphy and Rothman, 1983; Dunphy *et al.*, 1981; Elhammer and Kornfeld, 1984; Goldberg and Kornfeld, 1983). There is no reason to believe, however, that such methods come close to yielding homogeneous fractions. As an alternate approach to the study of the Golgi complex, "stacked" Golgi fractions almost devoid of enzyme activities known to be associated with other organelles have been prepared from liver and CHO cells (Balch *et al.*, 1984a; Bergeron *et al.*, 1982; Fleischer, 1983; Morré, 1971).

Differential perturbation of organelle density may heighten isopycnic resolution. For example, digitonin may be used to increase the density of the cholesterol-rich membranes (Amar-Costesec *et al.*, 1974; Borgese and Meldolesi, 1980). Moreover, density-shift experiments employing endocytosed peroxidase, colloidal gold, and endogenous or exogenous cholinesterase have been used to purify endosomes (Courtoy *et al.*, 1984), to discriminate between endocytic and exocytic coated vesicles (Helmy *et al.*, 1986), and to clean Golgi fractions from plasma membrane contaminants (D. K. Gupta and A. M. Tartakoff, unpublished observations).

3.2. Electron Microscopic Cytochemistry

To relate isolated cell fractions to their in situ cytologic counterpart(s), one must use electron microscopy to evaluate the distribution of anatomic, cytochemical, or immunocytochemical markers. Unfortunately, only a few cytochemical markers are known. Nevertheless, these markers have historically been of critical importance. Thus, acid phosphatase has been used to identify lysosomes and GERL, a compartment at the distal side of the Golgi stack

(Novikoff *et al.*, 1971); nucleoside diphosphatase (Goldfisher *et al.*, 1971) and NADPase (Smith, 1980; Smith *et al.*, 1986) are markers for distal and medial to distal Golgi cisternae, respectively; and overimpregnation of cells with osmium tetroxide leads to massive blackening of the cisternal space of proximal Golgi cisternae (Friend and Murray, 1965).

A modest augmentation of this limited repertoire of cytochemical markers has been made possible by the use of lectins. These data reinforce the reasonable postulate of the colinearity of the secretory path with the sequence of steps of glycoprotein oligosaccharide maturation. The lectins may be adsorbed to colloidal gold or covalently coupled to peroxidase (whose activity survives aldehyde fixation). Thus, concanavilin A (specific for oligosaccharides bearing terminal nonreducing glucosyl, mannosyl, and *N*-acetylglucosamine residues) labels the cisternal space of the RER and proximal (and, in certain cases, medial and distal) Golgi cisternae (Griffiths *et al.*, 1982; Tartakoff and Vassalli, 1983). In contrast, *Ricinus communis* lectin (specific for terminal galactose and *N*-acetylgalactosamine) and wheat germ agglutinin (specific for clustered sialic acid or *N*-acetylglucosamine) can be used to label the cisternal space of more distal Golgi cisternae (Griffiths *et al.*, 1982; Tartakoff and Vassalli, 1983). O-glycosylation has been demonstrated to initiate in proximal Golgi cisternae of goblet cells by use of *Helix pomatia* lectin (specific for *N*-acetylgalactosamine) (Roth, 1984).

To trace fluid-phase pinocytosis and membrane recapture following exocytosis, soluble or particulate tracers (e.g., peroxidase, colloidal gold, and ferritin) have often been used (Farquhar, 1978; Herzog and Farquhar, 1985). For tracing receptor-mediated endocytosis, ligands coupled to comparable tracers may be employed (Dunn *et al.*, 1986; Hopkins, 1985; Miller *et al.*, 1986; Willingham and Pastan, 1982). The use of such tracers provides the opportunity to establish a satisfactory correlation between anatomically defined compartments and isolated subcellular fractions.

3.3. Perturbation by Drugs

Several types of drugs have been used to perturb vesicular traffic and therefore discriminate among segments of either the secretory or endocytic paths: (1) acidotropic weak bases (e.g., ammonium chloride, chloroquine, primaquine, and methylamine), (2) carboxylic ionophores (e.g., nigericin and monensin A23187), (3) drugs that interact directly with enzymes responsible for oligosaccharide addition or maturation, and (4) drugs that interact with the cytoskeleton (e.g., colchicine and cytochalasins).

1. The weak bases (reviewed by Dean *et al.*, 1984; deDuve *et al.*, 1974; Maxfield, 1985) accumulate in and therefore increase the pH of acidic compartments. They provoke reversible massive dilation of lysosomes. Treatment of cells with these agents leads to a reduced rate of pinocytosis and to trapping of

receptors and ligands within endosomes (Harford *et al.*, 1983a; Wileman *et al.*, 1985). Hence, receptor-mediated endocytosis is inhibited. Moreover, the alkalinization of endosomes and lysosomes results in a reduction in toxicity of diptheria toxin and modeccin and reduction of infectivity of those enveloped viruses whose surface glycoproteins become fusogenic at acid pH (e.g., VSV and Sindbis). These agents also perturb oligosaccharide maturation and secretion of newly synthesized proteins in a number of cells (Oda *et al.*, 1983; Thorens and Vassalli, 1986). This observation may be related to the suggestion that distal compartments of the Golgi are slightly acidic (Anderson and Pathak, 1985; Schwartz *et al.*, 1985).

2. Carboxylic ionophores (reviewed by Tartakoff, 1983b). The monovalent ionophores (e.g., monensin) lead to a partial Na^+/K^+ equilibration and also, at high doses, reversibly raise the pH of acidic compartments. Their main site of action on the secretory path is the Golgi complex where they provoke a rapid dilation of cisternae, slow secretion, and cause overaccumulation of secretory proteins. Furthermore, in a number of monensin-treated cells, (a) newly synthesized glycoproteins do not acquire their complex type asparagine-linked oligosaccharides, (b) newly synthesized lysosomal enzymes acquire the mannose-6-phosphate recognition marker, and (c) viral spike glycoproteins become acylated. These observations are consistent with other data suggesting that (a), but not (b) or (c), occurs in a distal Golgi subcompartment. The impact of monensin on the endocytic path is much like that of the acidotropic weak bases (Dean *et al.*, 1984; Harford *et al.*, 1983b; Wileman *et al.*, 1985).

3. A number of enzyme inhibitors interrupt the normal progress of oligosaccharide addition and maturation (Datema *et al.*, 1982). The transport of selected glycoproteins may be blocked (Elbein, 1981; Lodish and Kong, 1984); however, these data cannot at present be generalized.

4. The literature pertaining to drugs that interact with identified components of the cytoskeleton is by no means clear. The interested reader should consult the recent text of Schliwa (1986) and Allan and Kreis (1986) for summaries of certain dramatic effects that have been observed when cells are treated with microtubule-active drugs. The review of Mollenhauer and Morre (1978) also provides valuable electron microscopic documents.

3.4. Limitations and New Prospects

Although unprecedented progress has been achieved by these several approaches, the isolation of the compartments of the secretory and endocytic paths is not complete. The area of greatest obscurity includes Golgi subcompartments, endosomes, and related structures. Moreover, almost nothing is known of the mechanisms of vesicle–vesicle recognition that lead to membrane fusion and, ultimately, to a protein being sorted to a specific destination. The critical membrane determinants involved must lie on the endodomain of this set of membranes, that is, in contact with the cell sap.

In vitro reconstitution of vesicular transport is potentially of great value for analyzing the underlying mechanisms. The progress that has resulted from studies of in vitro protein synthesis and vectorial transport has motivated these studies.

As a first step toward establishing an in vitro system that accomplishes intracellular transport, Rothman and co-workers (1984a,b) fused VSV-infected mutant CHO cells (that lack characteristic sugar transferase activities) with wild-type CHO cells. After fusion, although the two Golgi complexes appear to remain distinct, the G protein originating in the mutant Golgi acquires terminal sugars characteristic of the wild type, thus demonstrating complementation between the two cell types and showing that transport between two stacks of Golgi cisternae is both possible and efficient. This suggested that intercompartmental transport (proximal-to-medial or medial-to-distal cisternae) is a vesicular "dissociative" process.

These same authors extended these results to establish a cell-free system (Balch *et al.*, 1984a,b; Braell *et al.*, 1984). By mixing, in the presence of ATP and cytosol, stacked Golgi preparations of both wild-type cells and VSV-infected CHO 15B mutant cells (which lack GlcNAc transferase I activity), they were able to follow the incorporation of labeled *N*-acetylglucosamine into the G protein. Further analysis of this process by Balch *et al.* (1984b) gave evidence for the existence of three sequential intermediates in the transfer from one cisterna to the next.

Haselbeck and Schekman (1986) have used yeast to demonstrate in vitro complementation between a mutant deficient in 1–3 mannosyl transferase and a mutant deficient in invertase synthesis. In this case, endoplasmic reticulum-to-Golgi transfer appears to account for the observed invertase glycosylation. ATP, but not supernatant proteins, are essential.

Davey *et al.* (1985) have reconstituted an endocytic event in a cell-free system. The protocol makes use of two postnuclear supernatants of BHK21 cells—either from (a) cells in an early stage (15 min at 37°C) of infection with fowl plague virus or (b) cells infected 2 h at 20°C with ^{3}H-sialic acid labeled Semliki forest virus (SFV). Under these circumstances, SFV accumulates in "late" endosomes. In the presence of ATP, fusion between the two populations of endocytic vesicles is reflected by the release of [^{3}H]-sialic acid from SFV since one of the envelope glycoproteins of fowl plague virus is a neuraminidase.

These experiments have shown that only early endocytic vesicles can fuse with the late endocytic vesicles: neither the plasma membrane nor lysosomes participate in this process. A related cell-free investigation (Gruenberg and Howell, 1986) of endosome fusion, making use of immuno-isolated endosomes, has also called attention to the relation between the age of endosomes and their ability to fuse. In an independent line of investigation, Altsteil and Branton (1983) have demonstrated in vitro that coated vesicles stripped of their clathrin coat were able to fuse with lysosomes.

Most recently, the "cell-free" approach for study of the secretory path has been transposed to a compromise strategy in which opened, but still partly intact adherent cells have been shown to be able to accomplish transport (Beckers *et al.*, 1987; Simons and Virta, 1987). In principle, these several cell-free systems are ideally suited for deeper analysis of membrane recognition and fusion events. For example, one would like to know the makeup of the endodomain of the membranes that interact. A major step in this direction could be made by generating endodomain-specific antibodies that inhibit the in vitro events or block transport after microinjection into living cells.

4. IMPORTANCE OF POLYCLONAL AND MONOCLONAL ANTIBODIES

4.1. Defining the Paths of Intracellular Transport with Antibodies

The basic information describing the paths of vesicular transport has relied primarily on subcellular fractionation, cytochemistry, and autoradiography. Immunoelectron microscopy has confirmed these data. Moreover, recently, the immunocytochemical approach has made major new contributions, for example, (1) in showing that multiple exocrine pancreatic secretory proteins use the same pathway (Bendayan *et al.*, 1980; Geuze *et al.*, 1979; Kraehenbuhl *et al.*, 1977; Posthuma *et al.*, 1984), (2) in showing that the site of conversion of proinsulin to insulin is in condensing vacuoles (Orci *et al.*, 1985), (3) in documenting the proximal-to-distal passage of viral membrane glycoproteins across the Golgi stack (Bergmann and Singer, 1983; Bergmann *et al.*, 1981; Green *et al.*, 1981; Saraste and Hedman, 1983), and (4) in detecting antigenic (conformational?) determinants of proteins undergoing transport, which are manifest only within selected compartments (Burke *et al.*, 1983; Schmaljohn *et al.*, 1983).

Equally significant advances have been made with reference to the endocytic paths. For example, Geuze *et al.* (1983a) followed the fate of asialoglycoproteins and the asialoglycoprotein receptor by double-label immunoelectron microscopy. They observed that the receptor and the ligand are associated in the same coated and peripheral vesicles until they reach a "compartment of uncoupling of receptor and ligand" (CURL). At this point, the receptors are segregated in tubular extensions of CURL and the ligands (now soluble) are retained in the body of these vacuoles. These and related observations have led to the suggestion that endosomes (CURL) play an essential role in delivering content to the lysosomes while the tubular extensions recycle receptors back to the plasma membrane. The same group (Geuze *et al.*, 1983b) showed in immunocytochemical studies that CURL was also the site of segregation of the asialoglycoprotein receptor from the polymeric-IgA-receptor.

Elegant electron microscopic studies have also delineated the path of diacytosis of transferrin and its receptor (Hopkins, 1983; Hopkins and Trowbridge, 1983) and the path of endocytosis of several other receptors.

Immunocytochemical studies also have helped define the events of sorting of newly synthesized acid hydrolases. These studies make use of several polyclonal antibodies specific for the 215,000-dalton mannose-6-phosphate receptor.

For the moment, different laboratories are not in agreement. Willingham *et al.* (1983) found by immunoferritin electron microscopy of fixed and saponin-permeabilized CHO cells that this receptor is concentrated in the rough endoplasmic reticulum and a distal Golgi compartment tentatively identified as GERL. Brown and Farquhar (1984), using aldehyde-fixed saponin-permeabilized hepatocytes, exocrine pancreatic cells, and cultured hepatoma cells have found the receptor concentrated in proximal Golgi cisternae, associated vesicles, endosomes, and at the cell surface. Geuze *et al.* (1984) found that in hepatocytes the receptor can be detected throughout the Golgi stack and possibly in CURL. These researchers used an immunogold technique applied to ultrathin frozen sections. These discrepancies may be due to the different fixatives and staining methods, the different cell types studied, or the different polyclonal antisera used.

4.2. Defining Organelles and Subcompartments with Antibodies

At least three classes of resident proteins may exist in a given membrane-bounded compartment:

1. Enzymes responsible for post-translational modification of passenger macromolecules (e.g., for glycosylation of secretory proteins).
2. Proteins responsible for the recognition, fusion, and targeting of vesicles (e.g., to the apical versus basolateral cell surface).
3. Proteins with other metabolic responsibilities (e.g., ion ports).

For class 1, the active site is expected to be on the ectodomain of the membrane. For class 2, the active site should be on the endodomain. Such proteins need not be integral membrane proteins.

In recent years, polyclonal and monoclonal antibodies have been raised to known and unknown proteins of diverse membranes. We will now describe in detail these new markers, giving particular emphasis to antigens that may belong to class 2. (An inclusive list of the available antibodies is given in Table II.)

4.2.1. Endoplasmic Reticulum

A number of polyclonal and monoclonal antibodies have been produced against enzymes and proteins of the rough endoplasmic reticulum including HMGCoA reductase (Clark *et al.*, 1982), NADH-cytochrome b_5 oxidoreductase (Borgese and Gaetani, 1983; Okada *et al.*, 1982), cytochrome b_5, cytochrome P-450 reductase (Okada *et al.*, 1982), docking protein or SRP receptor (Hortsch *et al.*, 1985a, b), ribophorins (Hortsch *et al.*, 1986; Kreibich *et al.*, 1983), protein disulfide isomerase (Dawson *et al.*, 1984; Kaetzel *et al.*, 1986; Morin *et al.*, 1983; Ohba *et al.*, 1981; Roth and Mesirow, 1984; Varandani *et al.*, 1978), ERp60, ERp72, and ERp99 (Lewis *et al.*, 1985), and four proteins of 29,000, 58,000, 66,000, and 91,000 (Louvard *et al.*, 1982). Two enzymes, epoxide hydrolase (Galteau *et al.*, 1985; Oesch and Bentley, 1976) and glucosidase II (Lucocq *et al.*, 1986), have been demonstrated by electron microscopy to reside in the smooth endoplasmic reticulum but only epoxide hydrolase is missing from rough endoplasmic reticulum.

No antibodies have been observed to react with transitional elements of the endoplasmic reticulum.

4.2.2. Golgi Complex

4.2.2.1. Enzymes Involved in N-Glycosylation. Subcellular fractionation has demonstrated that several of the enzymes responsible for trimming and elongation of N-glycans, O-glycans, and proteoglycans are recovered primarily in Golgi-enriched fractions. With regard to N-glycan biosynthesis, isopycnic centrifugation allows partial resolution of Golgi subcompartments differentially enriched in mannosidases, *N*-acetylglucosamine transferase I, or galactosyl and sialyl transferase activities (Dunphy and Rothman, 1983). A corresponding cytological demonstration of this heterogeneity of the Golgi comes from the use of monoclonal and polyclonal antibodies for electron microscopic immunocytochemistry.

4.2.2.2. Mannosidase II. α-D-mannosidase II was purified from rat liver by Novikoff *et al.* (1983). A rabbit antiserum was prepared and tested for lack of cross reactivity for other mannosidases (Golgi mannosidases IA and IB, lysosomal mannosidase, cytosolic mannosidase). Immunoperoxidase electron microscopy in the presence of saponin showed, surprisingly, that all Golgi cisternae are stained in hepatocytes. It is conceivable that only the enzyme in the medial cisternae is active and that the antigen detected elsewhere is destined for return to medial cisternae.

4.2.2.3. N-Acetylglucosamine Transferase I. Dunphy *et al.* (1985) purified *N*-acetylglucosamine transferase I from rabbit liver. After immunization of mice, monoclonal antibodies were produced that immunoprecipitated the 55,000-dalton enzyme. Immunoperoxidase studies of saponin-permeabilized rabbit cells (LLC-RK1 kidney and hepatocytes) showed that staining is restricted to medial Golgi cisternae, thus confirming the subcellular fractionation experiments and indicating that, along with NADPase (Smith, 1980), this enzyme may be considered a marker of medial cisternae in selected cells.

4.2.2.4. Galactosyl Transferase. β(1–4) Galactosyl transferase (GT) localization was studied by Berger *et al.* (1981) and Roth and Berger (1982) using an affinity-purified polyclonal rabbit antiserum that recognizes human milk GT. At the ultrastructural level, cytochemical staining for nucleoside diphosphatase (TPPase) (a classical cytochemical marker of distal cisternae) followed by postembedding immunogold staining of thin sections revealed that GT colocalizes with TPPase activity in the two most distal Golgi cisternae of HeLa cells. In selected tissues, cell surface (Roth *et al.*, 1985) or secretory (Strous and Berger, 1982) galactosyltransferase has also been documented. Recently, monoclonal antibodies to different domains of the bovine enzyme have been produced (Ulrich *et al.*, 1986).

4.2.2.5. Sialyl Transferase. Roth *et al.* (1985) have immunized rabbits with purified (α2,6) sialyl transferase (ST). The affinity-purified polyclonal antibody recognized by immunoblotting a protein of 47,000 dalton from a rat liver Golgi preparation as well as purified sialyltransferase. ST was localized in rat hepatocytes by immunogold postembedding staining. The distal cisternae of the Golgi complex and a network of tubules in continuity with the distal Golgi cisternae were positive. The same elements were reactive for TPPase but a more restricted reaction was seen for acid phosphatase (CMPase).

4.2.2.6. Golgi Proteins of Unknown Function. Louvard *et al.* (1982) have produced a rabbit antiserum against rat liver Golgi membranes freed of their content. The serum was sequentially absorbed on fixed rat plasma proteins, rough and smooth microsomal fractions, purified rat liver plasma membranes, and intestinal microvillar vesicles. The resulting polyclonal antiserum immunoprecipitated a protein of 135,000 dalton from a rat liver Golgi preparation and NRK cells. At the ultrastructural level, by using immunoperoxidase staining of saponin-permeabilized NRK cells, the reaction product was restricted to the cisternae on one side of the Golgi, to nearby smooth vesicles, and some coated vesicles. In prolactin-secreting cells, the staining was observed in medial cisternae and with lesser intensity in distal cisternae (Tougard *et al.*, 1983). It also stained a number of small vesicles and lysosomelike structures.

A monoclonal antibody has been obtained by Burke *et al.* (1982) after immunization of mice with rat liver Golgi membranes. This antibody immunoprecipitated the same 135,000-dalton protein as the polyclonal mentioned in the preceding paragraph. In NRK cells, the monoclonal antibody stained only one or two cisternae on one side of the Golgi stack. This 135,000-dalton protein is a detergent-binding membrane protein. The epitope recognized by the monoclonal antibody is on the ectodomain of Golgi cisternae. The polyclonal but not the monoclonal antibody is suitable for immunoblotting.

Smith *et al.* (1984) raised monoclonal antibodies by immunization of mice with plasmalemma and microsomal membranes of pancreatic acinar cells. In immunoblotting, one monoclonal stained 103,000–108,000-dalton proteins and another one stained an additional 180,000-dalton membrane protein. Gold immunostaining of sections of preembedded rat pancreas showed staining by both antibodies of the ectodomain of the trans-most cisternae of the Golgi. These monoclonal antibodies have a restricted tissue specificity.

Chicheportiche *et al.* (1984) and Chicheportiche and Tartakoff (1987) immunized mice with galactosyltransferase-rich microsomal fractions of rat myeloma cells and produced monoclonal antibodies against proteins of 54,000 and 86,000 daltons (GCI), 44,000 daltons (GCII), 130,000 daltons (GCIII), and 140,000 daltons (SMII). The former two react with antigens that are exposed on the endodomain of Golgi membranes and are eluted with modest concentrations of salt. The latter two are detergent-binding glycoproteins. Immunoperoxidase electron microscopy of saponin-permeabilized myeloma cells demonstrated that the first three antibodies react with medial Golgi cisternae. Anti-GCI stains three cisternae (Figure 1), whereas GCII and GCIII appear to have a more restricted specificity. GCI is present in a broad variety of cell types and species (Table I). SMII has the striking property of reacting with multivesicular bodies in addition to Golgi cisternae.

Allen and Kreis (1986) have raised a monoclonal antibody to a Golgi-associated microtubule-binding protein (MAP). Both this protein and the superficial Golgi proteins identified by Chicheportiche *et al.* (1984) and Chicheportiche and Tartakoff (1987) may constitute part of the "zone of exclusion" and "intercisternal elements" that are often characteristic of the Golgi complex (Mollenhauer and Morre, 1978). The finding of a MAP may be related to the long-recognized sensitivity of the structure of the Golgi complex to microtubule-active drugs.

Yuan *et al.* (1987) have described a pair of monoclonal antibodies which detect two integral membrane glycoproteins of the Golgi complex. The 130,000-dalton protein is phosphorylated and is found in proximal and medial cisternae. The 100,000-dalton protein is concentrated in distal cisternae and more distal structures.

Smith *et al.* (1984) have reported the presence of antibodies that recognize Golgi membrane proteins in sera of nonimmunized mice. Other studies (Fritzler *et al.*, 1984; Rodriguez *et al.*, 1982) have provided immunofluorescent documentation of the existence of human autoimmune anti-Golgi antibodies.

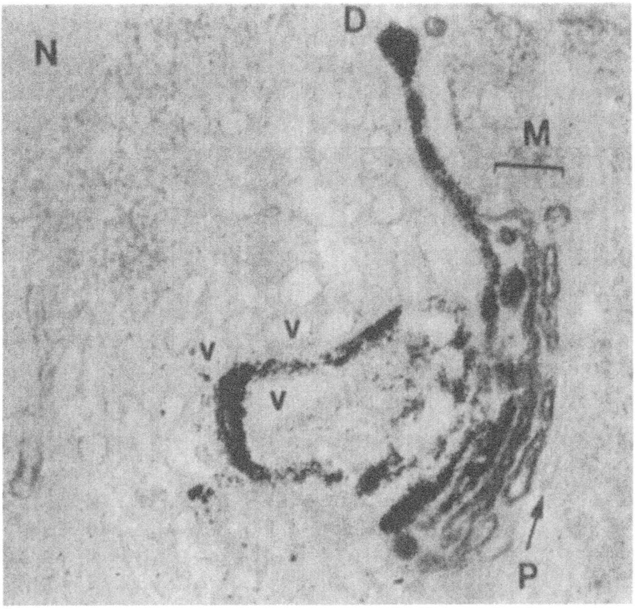

FIGURE 1. Double-staining electron microscopy of IR202 rat myeloma cells and localization of GCI to medial cisternae of the Golgi stack. Cells in suspension were fixed with 4% paraformaldehyde in PBS for 1 h at room temperature, permeabilized with 0.05% saponin, and incubated with biotinylated $6F_4C_5$ Mab (GCI) followed by avidin-peroxidase (Chicheportiche *et al.*, 1984). The cells were then washed, refixed with 0.2% glutaraldehyde/2% paraformaldehyde, and stained for thiamine pyrophosphatase activity 40 min at 37°C. After washing, the cells were then incubated with peroxidase substrates, postfixed with 1% osmium tetroxide, dehydrated, and embedded in Epon. Unstained thin sections were examined in a Phillips 300 electron microscope. (N) Nucleus, (V) Ig transport vesicles; arrow indicates an unstained proximal cisterna (P). Note that the three medial cisternae (M) are peroxidase stained and that TPPase activity is abundant in distal cisternae (D). ×67,000.

4.2.3. Secretion Granules and Vesicles; Coated Vesicles

4.2.3.1. Secretion Granules and Vesicles. Synapsin I (a and b) are proteins of 86,000 and 80,000 daltons that are major endogenous substrates for cAMP-dependent and Ca^{2+}- and calmodulin-dependent protein kinases in neurons of the central and peripheral nervous systems.

De Camilli *et al.* (1983) used purified synapsin I to immunize rabbits. The resulting antiserum reacted with only synapsin I in rat and bovine brain homogenates after radioimmunostaining of a SDS–polyacrylamide gel. Immunoferritin

Table I
Presence of GCI in Different Cell Types and Species Determined by Immunoblot Analysis[a] and/or Immunofluorescence[b]

Species	Cell type	
Mouse	Sp2/0 MOPC-21	Myeloma
	6F4C5	Hybridoma
	L cells 3T3	Fibroblast
	SIA/B (Thy-1$^-$ mutant)	Lymphoma
Rat	IR983F IR202 Y3	Myeloma
	IEC-18	Epithelial ileum
	NRK-52E	Epithelial kidney
	FR3T3	Fibroblast
		Liver
		Neuron
Hamster	BHK21	Fibroblast kidney
	CHO-K1	Epithelial ovary
Monkey	CV1	Fibroblast kidney
Human	CHANG	Liver
	HEP-2	Hepatoma
	HT-29	Adenocarcinoma
	HL-60	Promyelocytic leukemia
	HeLa	Epitheloid carcinoma

[a] 2×10^6 cells were lysed in 1% NP40/PBS. After a low speed centrifugation, the clarified lysates were mixed with an equal volume of SDS–PAGE sample buffer concentrated two times, containing β-mercaptoethanol and boiled. The samples were run on 7.5–20% SDS–polyacrylamide gels, and the resolved proteins transferred to nitrocellulose 3 h at 200 mM in the presence of 0.01% SDS in the transfer buffer containing 10% methanol. The nitrocellulose was incubated for 30 min in 10% N rabbit serum in Tris buffered saline (NRS/TBS). It was then incubated overnight at 4°C with $6F_4C_5$ Mab (anti-GCI). After several washes, the replica was incubated for 1 h at room temperature with an HRP-conjugated rabbit IgG antimouse Ig at 1 : 500 dilution in NRS/TBS. The staining was done by incubation for a few minutes in 0.5 mg/mL chloronaphthol in TBS in the presence of 0.05% H_2O_2 (see Chicheportiche *et al.*, 1984).

[b] Cells grown on coverslips were fixed with 3% paraformaldehyde in PBS, permeabilized with 0.2% Triton X-100 in PBS for 4 min, and then washed with 0.2% gelatin/PBS. The coverslips were turned upside down on a drop (50 L deposited on Parafilm) of $6F_4C_5$ (50 μg/mL) for 1 h in a humid chamber. The coverslips were returned to Petri dishes and washed. The incubation with the rabbit antimouse second antibody conjugated to rhodamine was done as for the first incubation. After washes, the coverslips were mounted on slides with glycerol and examined in an epifluorescence microscope.

and immunoperoxidase electron microscopy on fixed and agarose-embedded synaptosomes showed that synapsin I is associated with the cytoplasmic surface of small synaptic vesicles (Navone *et al.*, 1984). Huttner *et al.* (1983) showed that synapsin I is a superficial protein that represents 6% of the protein of highly purified synaptic vesicles. This preparation contains a Ca^{2+}- and calmodulin-dependent protein kinase that phosphorylates and triggers the release of synapsin I from synaptic vesicles. Recent experiments suggest that synapsin I and its kinase are also associated with the cytoskeleton and that synapsin I may play a role as a link between cytoskeleton and synaptic vesicles.

Matthew *et al.* (1981) raised monoclonal antibodies after immunization of mice with rat brain synaptic junctional complexes. Two of these antibodies immunoprecipitate a protein of 65,000 daltons in synaptic membranes and react with antigenic sites on the endodomain of synaptic vesicles. Immunoperoxidase electron microscopy on sections of rat tissues revealed the presence of this protein on synaptic vesicles of a variety of neurons. This protein and the epitope in question are widely conserved through evolution.

Jahn *et al.* (1985) prepared a rabbit antiserum to a major protein of 38,000 daltons of synaptic vesicles purified after SDS–gel electrophoresis. They also raised monoclonal antibodies after immunization of mice with synaptic vesicles of rat brain. Four of the antibodies reacted with the 38,000-dalton protein on immunoblots. This protein is a membrane protein whose epitope is on the endodomain of synaptic vesicles. The 38,000-dalton protein was also characterized by Wiedenmann and Franke (1985) with monoclonal antibodies raised after immunization of mice with bovine brain crude coated vesicle fractions. These authors showed that this protein is glycosylated. Immunogold electron microscopy documented the presence of the 38,000-dalton protein on synaptic vesicles of mouse and rat neuronal tissues and synaptosomes.

Buckley and Kelly (1985) produced monoclonal antibodies by immunization of mice with highly purified cholingergic synaptic vesicles from the electric organ of elasmobranchs. Immunoblotting experiments showed that some of the monoclonal antibodies react with a protein (SV2) of 100,000 daltons. SV2 is a detergent-binding protein and is synthesized as a nonglycosylated precursor of 62,000 daltons. Immunoperoxidase and immunogold electron microscopy showed the specificity of the antibody for the endodomain of synaptic vesicles. SV2 is present on secretion granules in a variety of species from fish to mammals, in both neural and endocrine cells.

4.2.3.2. Coated Vesicles. The clathrin light (LC) and heavy chains (HC) of coated vesicles have been studied extensively. A number of polyclonal antibodies have been raised against clathrin (Bloom *et al.*, 1980; Keen *et al.*, 1981; Louvard *et al.*, 1983). Moreover, a number of monoclonal antibodies are now available that recognize HC (Brodsky, 1985; Kelly *et al.*, 1983), free HC (Brodsky, 1985; Louvard *et al.*, 1983), and LC associated with HC (Brodsky, 1985). Recently, antisera were raised against the 100,000-dalton proteins present in coated vesicles (Robinson and Pearse, 1986).

No monoclonal and only a limited number of polyclonal antibodies have been raised to membrane components of secretory granules of exocrine cells

(Fisher-Colbrie *et al.*, 1984; Geuze *et al.*, 1981). Antibodies that recognize dopamine β-hydroxylase recognize the ectodomain of the membrane of chromaffin cell granules (Phillips *et al.*, 1983).

4.2.4. Lysosomes/Endocytic Vesicles

In the study of Reggio *et al.* (1984), rat liver lysosomes were isolated by differential centrifugation followed by isopycnic centrifugation and free-flow electrophoresis. The hydrolase content was eliminated by repeated freezing and thawing followed by sedimentation of a purified lysosomal membrane fraction that was used to immunize rabbits. The resulting sera were sequentially absorbed on aldehyde-fixed mitochondrial and microvillar fractions to eliminate unwanted specificities. The purified antiserum defined by immunoblotting a 100,000-dalton membrane protein and cross-reacted with several purified ATPases. Judging from immunoperoxidase staining of saponin-permeabilized NRK cells, the authors concluded that the antigen is present in the membrane of lysosomes, endosomes, coated vesicles, proximal Golgi cisternae, and nearby vesicles and vacuoles. In prolactin-secreting cells, endocytic vesicles, lysosomes, and cisternae on one side of the Golgi stack were stained (Tougard *et al.*, 1985). These observations suggest that the antigen might be responsible for the acidification of prelysosomal and lysosomal compartments.

Lewis *et al.* (1985a,b) have generated mouse monoclonal antibodies against rat liver lysosomal membranes purified on Percoll gradients and by free-flow electrophoresis. Three membrane glycoproteins of 120,000 (lgp120), 100,000 (lgp100), and 80,000 (lgp80) daltons were identified by immunoprecipitation of lysates from [^{35}S]-methionine-labeled NRK cells. Double immunofluorescence observations show that these proteins colocalize with acid phosphatase.

Rabbit polyclonal antibodies have been prepared that recognize lgp120 since the monoclonal antibodies react only with rat cells. At the ultrastructural level, through use of an immunoperoxidase technique on saponin-permeabilized NRK cells, the staining with the polyclonal was restricted to lysosomes. Digestion by endoglycosidase H and neuraminidase and the use of tunicamycin showed that these proteins have a high content in N-linked and possibly some O-linked oligosaccharides: lgp120 appears to have a peptide backbone of 42,000 daltons and ten sialated asparagine-linked oligosaccharides. Since sialyl transferase is found only in distal Golgi cisternae and neighboring tubules (Roth *et al.*, 1985), the existence of complex sugar chains on lysosomal membrane proteins argues strongly that these proteins pass through the distal portion of the Golgi stack before reaching lysosomes.

Granger *et al.* (1985) have produced rat monoclonal antibodies to mouse liver lysosomes. Two of these antibodies immunoprecipitate a membrane glycoprotein of 110,000–120,000 daltons that has a small protein core and a large number of complex type oligosaccharides. Furthermore, unlike most acid hydrolases, this glycoprotein does not bear a mannose-6-phosphate recognition signal. The absence of this signal suggests that a different targeting mechanism may govern the transport of lysosomal membrane proteins versus acid hydrolases.

Croze *et al.* (1985) used rat liver lysosomal membranes as immunogen to raise mouse monoclonal antibodies that recognize by immunoblotting glycoproteins of 150,000, 105,000 and 160,000 and 105,000 daltons. By immunofluorescence and double-label immunoelectron microscopy of ultrathin frozen sections, the epitope of the 150,000-dalton glycoprotein has been shown to be on the ectodomain of lysosomal membranes.

Monoclonal antibodies have been prepared by Chen *et al.* (1985) after immunization of rats with lentil-lectin binding proteins of a 3T3 BALB/c mouse cell lysate. Two proteins of 105,000–115,000 (LAMP-1) and 100,000–110,000 daltons (LAMP-2) were identified by immunoprecipitation of biosynthetically labeled lysates. These proteins are membrane glycoproteins with complex type oligosaccharides. Studies of the heterogeneity of LAMP-1 in two-dimensional gel electrophoresis showed that at least 16 sialic acid residues are present. On saponin-permeabilized cells, double-label immunoelectron microscopy showed that these glycoproteins are present on the ectodomain of endocytic vesicles containing endocytosed α-2 macroglobulin. These antigens showed cross-reactivity with the macrophage differentiation antigen (MAC-3).

Lippincott-Schwartz and Fambrough (1986, 1987) have immunized mice with chicken coated vesicles to produce a monoclonal antibody that immunoprecipitates a 95,000–105,000-dalton protein from biosynthetically labeled lysates of chicken fibroblasts. Metabolic labeling and glycosidase treatment reveal that this membrane protein has a large number of complex N-linked oligosaccharides and a protein core of 48,000 daltons. Immunoperoxidase electron microscopy of saponin-permeabilized chicken fibroblasts showed that the epitope is concentrated on the ectodomain of lysosomes but is also present elsewhere along the endocytic pathway and at the plasma membrane. Study of the kinetics of internalization of the monoclonal antibody by living cells demonstrated a rapid and continuous movement of the protein between the plasma membrane endosomes and lysosomes.

Chicheportiche and Tartakoff (1987) have produced monoclonal antibodies after immunization of mice with wheat germ agglutinin (WGA) binding microsomal proteins of rat myeloma cells. Immunoblotting experiments show that two of the antibodies recognize proteins of 50,000–92,000 (SM-I) and 140,000 (SM-II) daltons. SM-I and SM-II are detergent-binding glycoproteins (judging by TX-114 partitioning). Since they bind to WGA, they probably contain a number of complex type oligosaccharide chains. Immunocytochemical studies of saponin-permeabilized myeloma cells demonstrated that SM-I is concentrated in coated pits and in vesicles along the endocytic pathway. Anti-SM-I is rapidly endocytosed by living cells. Comparable immunocytochemical studies show that SM-II is concentrated in distal Golgi cisternae and multivesicular bodies. No other structures are stained.

No markers for endosomes, as opposed to lysosomes and coated pits, are known.

4.3. Antigen Characterization: Functional Implications

The antibodies we have described (Table II) reinforce the current concept of compositional differences along the paths of transport. Moreover, they may give new insight into the underlying mechanisms of intracellular transport, both because (1) they define and allow the characterization of otherwise unrecognized membrane components and (2) they can be exploited in a functional context to analyze vesicular traffic.

For example, a number of proteins of lysosomal membranes have been described. Two common features have emerged: They all have an unusually high content of complex sialic acid-containing oligosaccharides. These may be important for their protection from the action of acid hydrolases and maintenance of protein conformation at acid pH. Moreover, to the extent that they have been studied, these proteins do not possess the mannose-6-phosphate recognition marker of lysosomal hydrolases, suggesting that their mechanism of delivery to lysosomes is different from that of acid hydrolases. The fact that they acquire sialic acid demonstrates that they traverse distal Golgi cisternae and/or associated tubules known to house sialyl transferase.

For the Golgi complex, a number of enzymes responsible for the maturation of N-glycans have now been localized in situ. These data argue strongly that this sequence of enzymatically catalyzed steps is cytologically distributed. They do not all occur within a single compartment. Moreover, localization of sialyl transferase has revealed a tubular network in continuity with the Golgi stack that may be the site of convergence of the endocytic and exocytic pathways. This structure may include the critical compartment studied by Regoeczi *et al.* (1982) and by Snider and Rogers (1985) in their investigation of the resialylation of desialated transferrin and its desialated receptor upon endocytosis. It may also be the compartment where viral membrane proteins are arrested when infected cells are incubated at 20°C (Griffiths *et al.*, 1985).

The subcompartmentalization of the Golgi stack has also been demonstrated with antibodies that detect components of unknown function. One of these, known as GCI or 6F4 (Chicheportiche *et al.*, 1984), is a superficial 54,000-dalton protein of the endodomain of medial cisternae and has a further curious property. At millimolar concentration of Ca^{2+} it is modified in such a way that it is no longer detected by immunoblotting. This modification occurs only when the protein is associated to Golgi membranes (not after elution from the mem-

Table II
Markers of Cytoplasmic Membranes Identified with Antibodies

Organelle	Activity of the antigen	Molecular weight[a] (daltons)	Oligosaccharide[b] content	Topology[c]	References
Endoplasmic reticulum	HMG–CoA reductase	58,000	High mannose	Integral	Clark *et al.* (1982)
	NADH–cytb$_5$ reductase	33,000	None	Integral	Borgese and Gaetani (1983); Okada *et al.* (1982)
	Cytochrome b$_5$	18,000	None	Extrinsic, endodomain	Okada *et al.* (1982)
	Cytochrome P-450 reductase	78,000	None	Integral	Okada *et al.* (1982)
	SRP receptor/docking protein	73,000	None	Integral	Hortsch *et al.* (1985a,b)
	Ribophorins	65,000/63,000	High mannose	Integral	Kreibich *et al.* (1983); Hortsch *et al.* (1986)
	Protein disulfide isomerase	53,000	+	Extrinsic, ectodomain	Varandani *et al.* (1978); Ohba *et al.* (1981); Morin *et al.* (1983); Dawson *et al.* (1984); Roth and Mesirow (1984); Kaetzel *et al.* (1986)
	Epoxide hydrolase	50,000	None	Integral	Oesch and Bentley (1976); Galteau *et al.* (1985)
	Glucosidase II	100,000	High mannose	Integral	Lucocq *et al.* (1986)
		29,000	N.D.[d]	N.D.	Louvard *et al.* (1982)
		58,000	N.D.	N.D.	Louvard *et al.* (1982)
		66,000	N.D.	N.D.	Louvard *et al.* (1982)
		91,000	N.D.	N.D.	Louvard *et al.* (1982)
		72,000	—	Integral	Lewis *et al.* (1985a,b)
		99,000	High mannose	Integral	Lewis *et al.* (1985a,b)
		60,000	—	Integral	Lewis *et al.* (1985a,b)

(continued)

Table II (*Continued*)

Organelle	Activity of the antigen	Molecular weight[a] (daltons)	Oligosaccharide[b] content	Topology[c]	References
Golgi complex	Galactosyl transferase	48,000	Complex	Integral	Berger *et al.* (1981); Strous and Berger (1982); Ulrich *et al.* (1986)
	Mannosidase II	75,000	+	Integral	Novikoff *et al.* (1983)
	N-acetyl glucosamine transferase I	55,000	N.D.	Integral	Dunphy *et al.* (1985)
	Sialyl transferase	47,000	N.D.	Integral	Roth *et al.* (1985)
	cAMP-dependent protein kinase type II	52,000/56,000 (RII)	N.D.	N.D.	Nigg *et al.* (1985a,b); De Camilli *et al.* (1986); Kuettel *et al.* (1985)
		40,000(C)	N.D.	Associated, endodomain	
	Heat shock	100,000	N.D.	Associated endodomain	Lin and Queally (1982); Lin *et al.* (1982)
		103,000–108,000	N.D.	Integral	Smith *et al.* (1984)
		180,000/103,000–108,000	N.D.	Integral	Smith *et al.* (1984)
	src Gene product	60,000	None	Associated, endodomain	Resh and Erikson (1985)
		135,000	None	Associated endodomain	Louvard *et al.* (1982); Burke *et al.* (1982)
	Microtubule–associated protein	110,000	None	Associated	Allan and Kreis (1986)
		54,000/86,000	None	Associated, endodomain	Chicheportiche *et al.* (1984)
		45,000	None	Associated endodomain	Chicheportiche and Tartakoff (1987)
		130,000	Complex	Integral	Chicheportiche and Tartakoff (1987)
		100,000	Complex	Integral	Yuan *et al.* (1987)
		130,000	Complex	Integral	Yuan *et al.* (1987)

Granules and vesicles		86,000/80,000	None	Associated, endodomain	De Camilli *et al.* (1983); Huttner *et al.* (1983)
		65,000	N.D.	Integral, endodomain	Matthew *et al.* (1981)
		38,000	+	Integral	Jahn *et al.* (1985); Wiedenmann and Franke (1985)
		100,000	+	Integral	Buckley and Kelly (1985)
	Clathrin light chain	30,000/35,000	None	Associated endodomain	Keen *et al.* (1981); Louvard *et al.* (1983); Bloom *et al.* (1980); Kelly *et al.* (1983); Brodsky (1985)
	heavy chain	180,000	None	Associated endodomain	
		100,000	N.D.	Integral	Robinson and Pearse (1986)
Lysosome/endosome	ATPase?	100,000	N.D.	Integral	Reggio *et al.* (1984)
		80,000	Complex	Integral	Lewis *et al.* (1985)
		100,000	Complex	Integral	Lewis *et al.* (1985)
		120,000	Complex	Integral	Lewis *et al.* (1985)
		110,000–120,000	Complex	Integral	Granger *et al.* (1985)
		110,000–115,000	Complex	Integral	Chen *et al.* (1985)
		100,000–110,000	Complex	Integral	Chen *et al.* (1985)
		95,000–105,000	Complex	Integral	Lippincott-Schwartz and Fambrough (1986)
		58,000–90,000	Complex	Integral	Chicheportiche *et al.* (1984)
		140,000	Complex	Integral	Chicheportiche *et al.* (1984)
		150,000/105,000	Complex	Integral	Croze *et al.* (1985)
		160,000/105,000	Complex	Integral	Croze *et al.* (1985)

[a]Molecular weights are determined after SDS–PAGE.

[b]Carbohydrate content is defined either by incorporation of radioactive sugar in a protein, by study of biosynthetic labeling in the absence and presence of tunicamycin, by susceptibility to neuraminidase or endoglycosidase H treatment, or by adsorption to lectins. Complex type oligosaccharides are those that include sialic acid.

[c]*Integral* proteins are those that have been proved to be transmembrane proteins, proteins with detergent-binding properties, or proteins nonextractable with salt (see below). *Extrinsic* proteins are membrane-associated proteins that can be extracted with low detergent concentrations but not by salt. They may simply be anchored to the bilayer without traversing it. *Associated* proteins are extractable from membranes with salt (1 M Tris, 0.5 M NaCl, 0.5M KCl, 0.1 M Na_2CO_3, etc.). For the extrinsic and associated classes, the terms endodomains and ectodomains refer to the orientation of the epitope or active site of the protein.

[d]N.D. means not determined.

branes with salt) and is specific for the 54,000-dalton versus the 86,000-dalton form of the antigen (both bind the same Mab). This modification is inhibited by *N*-ethylmaleimide or iodoacetamide (but not leupeptin or antipain), suggesting that a free thiol (not a thiol protease) is involved. Furthermore, a protein comigrating with and possibly identical to this 54,000-dalton protein (but different from tubulin) is phosphorylated in the presence of Ca^{2+} (Y. Chicheportiche, unpublished observations). These observations may be related to the Golgi phosphoproteins and protein kinases, which are also inhibited by NEM (Capasso *et al.*, 1985). The cell-free system that reconstitutes proximal-to-medial Golgi transport is also NEM sensitive (Balch *et al.*, 1984b).

A further striking characteristic of some antibodies that react with the Golgi complex is that they also stain multivesicular bodies (Chicheportiche and Tartakoff, 1987) or "lysosomelike" structures (Tougard *et al.*, 1983). The corresponding antigens, like the 215,000 mannose-6-phosphate receptor, may transit to and from Golgi cisternae (Brown *et al.*, 1984).

Regulation of Golgi function may involve several "transiently resident" proteins. For example, in 1982 Lin and Queally described a monoclonal antibody recognizing a Golgi protein of 110,000 daltons. Later Lin *et al.* (1982) demonstrated that this protein is a heat shock protein that, remarkably, appears to be translocated to the nucleus when cells are incubated at 43°C. A second example of transient Golgi association is that of cAMP-dependent protein kinase II. Antibodies against the catalytic and regulatory subunits of this enzyme have allowed Nigg *et al.* (1985a) to localize these proteins by immunofluorescence to the Golgi region and centrosomes. De Camilli *et al.* (1986) have localized the RII subunit, by immunoperoxidase electron microscopy of rat neurons, to the distal region of the Golgi in nonstimulated cells. After adenylate cyclase activation, the catalytic subunit is rapidly and reversibly translocated to the nucleus, while the regulatory subunit stays in the Golgi (Kuettel *et al.*, 1985; Nigg *et al.*, 1985b).

The Rous sarcoma virus src gene product, $pp60^{v\text{-}src}$ (the tyrosine-specific protein kinase), and its cellular counterpart, $pp60^{c\text{-}src}$, have been demonstrated to be concentrated in the plasma membrane and the Golgi area, thus suggesting a possible role of the Golgi complex in cell transformation (Resh and Erikson, 1985). Considering that this protein is synthesized on free polysomes, any Golgi association is presumably with the endodomain of Golgi membranes.

5. APPLICATIONS AND PERSPECTIVES

5.1. Affinity Isolation of Organelles

When antibodies are available that recognize epitopes on the endodomain of organelles, immunopurification of these membranes may be possible. Thus, by using antibodies bound to *Staphylococcus aureus*, Ito and Palade (1978) have shown that Golgi subfractionation may be achieved. Similarly, Pfeffer and Kelly (1985) have used immunoprecipitation to purify two subpopulations of brain coated vesicles, and Mueller and Hubbard (1986) used antibodies to the asialoglycoprotein receptor and *Staphylococcus aureus* to adsorb receptor-bearing endosomes. Immunoadsorbates have also been used to isolate the plasma membrane or the endosomes of VSV-infected cells (Gruenberg and Howell, 1985a,b). (See also Chapter 6 in this volume.)

5.2. Functional Studies

Proteins on the endodomain of vesicles are believed to be important in the events of membrane–membrane recognition, membrane fusion, and interaction with the cytoskeleton. These possibilities may be explored by microinjection of an appropriate antibody into living cells. Thus, Wehland *et al.* (1981, 1982) have demonstrated that microinjected polyclonal anticlathrin antibodies, although they react with clathrin, do not interfere with the endocytosis of EGF or with transport of the G protein of VSV to the cell surface. This apparent nonimportance of clathrin in vesicular transport was reinforced by Payne and Schekman (1985) in their studies of yeast, which lack an intact clathrin gene.

In the case of antigens whose epitope is on the ectodomain, mRNA of the corresponding hybridoma cells should be injected if antibody is to gain access to the appropriate membrane surface. This approach has been used by Burke and Warren (1984) to interrupt intracellular transport of the G protein by VSV-infected BHK cells. These investigators have injected mRNA from hybridoma cells whose secretion product recognizes a luminal epitope of a 135,000-dalton Golgi protein. The BHK cells then synthesize this specific Ig and transport it along their own secretory path. In VSV-infected microinjected BHK cells, the G protein overaccumulates in the Golgi complex rather than reaching the plasma membrane.

In a related line of investigation (Arnheiter *et al.*, 1984; Kreis, 1986), polyclonal antibodies that recognize the cytoplasmic domain of the G protein were microinjected into VSV-infected cells. In this condition, the G protein no longer moved to the plasma membrane. Injection of the monovalent Fab fragment of these antibodies does not result in arrest.

A particularly striking use of antibodies that react with the ectodomain of synaptic vesicles and secretory granules has been to document the events of exocytosis. Such antigenic sites become exposed at the cell surface when exocytosis occurs (Lingg *et al.*, 1983; Phillips *et al.*, 1983; von Wedel *et al.*, 1981).

5.3. Perspectives

Although new markers identified by antibodies have clarified and extended our view of compartmentation of the cell cytoplasm, more such antibodies and further applications are needed to dissect the mechanisms involved in intracellular transport. Thus, with the exception of clathrin-coated vesicles, all the transport vesicles in the cell cytoplasm are almost uncharacterized. These include endocytic vesicles, endosomes, the carriers between transitional elements of the endoplasmic reticulum and proximal Golgi cisternae, those apparently responsible for transport from one cisterna of the Golgi to the next, and those transporting mature proteins from the distal Golgi to lysosomes or to the cell surface. Even for the coated vesicles, a distinction remains to be made between those used in endocytosis and those involved in protein secretion (Helmy *et al.*, 1986; Kedersha *et al.*, 1986). The Golgi complex remains equally obscure; with the exception of OsO_4 impregnation, no cytochemical markers are available for proximal cisternae and, although the medial region of the Golgi contains several cisternae, no markers of individual cisternae are known.

5.4. Note of Caution

Although antibodies are of great value in the definition and analysis of membranes, one should bear in mind that the characterization of antigens often makes use of antibodies that are poorly defined with regard to the epitopes recognized in a given protein. Thus, carbohydrate epitopes may be recognized. In this case, the use of antibodies for cytologic studies may define a step in glycoprotein maturation, but not the location of an altogether unique protein. One should also be aware of the *nonequivalence* of the methods used for characterization of an antigen. For example, blotting and immunoprecipitation are frequently used to identify the proteins recognized by a given antibody, but blotting, in the majority of cases, results in the denaturation and often the reduction of the protein. Conversely, immunoprecipitation allows the recovery of native antigens but is representative only of the labeled pool of molecules. In immunofluorescence and immunoelectron microscopy, there is even more methodological diversity; detection always follows a fixation step and in many cases the tissue is first dehydrated in organic solvents and embedded before processing. There is no reason to assume that epitopes characteristic of an immunogen remain unaltered by either the current blotting or immunocytochemical procedures.

ACKNOWLEDGMENTS. We wish to thank S. Cechner and A. Gutierrez for typing the manuscript. This work was supported by NIH grants AM27651 to Y. Chicheportiche and AI-21269 to A. M. Tartakoff.

6. REFERENCES

Allan, V., and Kreis, T., 1986, A microtubule-binding protein associated with membranes of the Golgi apparatus. *J. Cell Biol.* **103**:2229–2239.

Altstiel, L., and Branton, D., 1983, Fusion of coated vesicles with lysosomes: Measurement with a fluorescence assay. *Cell* **32**:921–929.

Amar-Costesec, A., Wibo, M., Thines-Sempoux, D., Beaufay, H., and Berthet, J., 1974, Analytical study of microsomes and isolated subcellular membranes from rat liver. IV. Biochemical, physical and morphological modifications of microsomal components induced by digitonin, EDTA and pyrophosphate. *J. Cell Biol.* **62**:717–745.

Anderson, R. G. W., and Pathak, R. K., 1985, Vesicles and cistrnae in the *Trans* Golgi apparatus of human fibroblasts are acidic compartments. *Cell* **40**: 635–643.

Arnheiter, H., Dubois-Dalcq, M., and Lazzarini, R. A., 1984, Direct visualization of protein transport and processing in the living cell by microinjection of specific antibodies. *Cell* **39**:99–109.

Arvan, P., Cameron, R. S., and Castle, J. D., 1983, Secretory membranes of rat parotid gland: Preparation and comparative characterization. *Methods Enzymol.* **98**:75–87.

Balch, W. E., Dunphy, W. G., Braell, W. A., and Rothman, J. E., 1984a, Reconstitution of the transport of protein between successive compartments of the Golgi measured by the coupled incorporation of *N*-acetylglucosamine. *Cell* **39**:405–416.

Balch, W. E., Glick, B. S., and Rothman, J. E., 1984b, Sequential intermediates in the pathway of intercompartmental transport in a cell-free system. *Cell* **39**:525–536.

Beckers, C., Keller, D., and Balch, W., 1987, Semi-intact cells permeable to macromolecules. Use in reconstitution of protein transport from the endoplasmic reticulum to the Golgi complex. *Cell* **50**:523–534.

Bendayan, M., Roth, J., Perrelet, A., and Orci, L., 1980, Quantitative immunocytochemical localization of pancreatic secretory proteins in subcellular compartments of the rat acinar cell. *J. Histochem. Cytochem.* **28**:149–160.

Berger, E. G., Mandel, T., and Schilt, U., 1981, Immunohistochemical localization of galactosyltransferase in human fibroblasts and HeLa cells. *J. Histochem. Cytochem.* **29**:364–370.

Bergeron, J. J. M., Rachubinski, R. A., Sikstrom, R. A., Posner, B. I., and Paiement, J., 1982, Galactose transfer to endogenous acceptors within Golgi fractions of rat liver. *J. Cell Biol.* **92**:139–146.

Bergmann, J. E., and Singer, S. J., 1983, Immunoelectron microscopic studies of the intracellular transport of the membrane glycoprotein (G) of vesicular stomatitis virus in infected Chinese hamster ovary cells. *J. Cell Biol.* **97**:1777–1787.

Bergmann, J. E., Tokuyasu, K. T., and Singer, S. J., 1981, Passage of an integral membrane protein, the vesicular stomatitis virus glycoprotein, through the Golgi apparatus en route to the plasma membrane. *Proc. Natl. Acad. Sci. U.S.A.* **78**:1746–1750.

Bloom, W. S., Fields, K. L., Yen, S. H., Haver, K., Schook, W., and Puszkin, S., 1980, Brain clathrin: Immunofluorescent patterns in cultured cells and tissues. *Proc. Natl. Acad. Sci. U.S.A.* **77**:5520–5524.

Borgese, N., and Gaetani, S., 1983, *In vitro* synthesis and post-translational insertion into microsomes of the integral membrane protein, NADH–cytochrome b_5 oxidoreductase. *EMBO J.* **2**:1263–1269.

Borgese, N., and Meldolesi, J., 1980, Localization and biosynthesis of NADH–cytochrome b_5 reductase, an integral membrane protein, in rat liver cells. I. Distribution of the enzyme activity in microsomes, mitochondria, and Golgi complex. *J. Cell Biol.* **85**:501–515.

Braell, W. A., Balch, W. E., Dobbertin, D. C., and Rothman, J. E., 1984, The glycoprotein that is transported between successive compartments of the Golgi in a cell-free system resides in stacks of cisternae. *Cell* **39**:511–524.

Brodsky, F. M., 1985, Clathrin structure characterized with monoclonal antibodies. I. Analysis of multiple antigenic sites. *J. Cell Biol.* **101**:2047–2054.

Brown, W. J., and Farquhar, M. G., 1984, The mannose-6-phosphate receptor for lysosomal enzymes is concentrated in *cis* Golgi cisternae. *Cell* **36**:295–307.

Brown, W. J., Constantinescu, E., and Farquhar, M. G., 1984, Redistribution of mannose-6-phosphate receptors induced by tunicamycin and chloroquine. *J. Cell Biol.* **99:**320–326.

Buckley, K., and Kelly, R. B., 1985, Identification of a transmembrane glycoprotein specific for secretory vesicles of neural and endocrine cells. *J. Cell Biol.* **100:**1284–1294.

Burke, B., and Warren, G., 1984, Microinjection of mRNA coding for an anti-Golgi antibody inhibits intracellular transport of a viral membrane protein. *Cell* **36:**847–856.

Burke, B., Griffiths, G., Reggio, H., Louvard, D., and Warren, G., 1982, A monoclonal antibody against a 135-K Golgi membrane protein. *EMBO J.* **1:**1621–1628.

Burke, B., Walter, C., Griffiths, G., and Warren, G., 1983, Viral glycoproteins at different stages of intracellular transport can be distinguished using monoclonal antibodies. *Eur. J. Cell Biol.* **31:**315–324.

Capasso, J. M., Abeijon, C., and Hirschberg, C. B., 1985, Phosphoproteins and protein kinases of the Golgi apparatus membrane. *J. Biol. Chem.* **260:**14879–14884.

Chen, J. W., Murphy, T. L., Willingham, M. C., Pastan, I., and August, J. T., 1985, Identification of two lysosomal membrane glycoproteins. *J. Cell Biol.* **101:**85–95.

Chicheportiche, Y., and Tartakoff, A., 1987, Monoclonal antibodies as markers of the endocytic and secretory pathways. *Eur. J. Cell Biol.* **44:**135–143.

Chicheportiche, Y., Vassalli, P., and Tartakoff, A. M., 1984, Characterization of cytoplasmically oriented Golgi proteins with a monoclonal antibody. *J. Cell Biol.* **99:**2200–2210.

Clark, R. E., Martin, G. G., Barton, M. C., and Shapiro, D. J., 1982, Production and characterization of monoclonal antibodies to rat liver microsomal 3-hydroxyl-3-methylglutaryl-coenzyme A reductase. *Proc. Natl. Acad. Sci. U.S.A.* **79:**3734–3738.

Courtoy, P. J., Quintart, J., and Baudhuin, P., 1984, Shift of equilibrium density induced by 3,3′-diaminobenzidine cytochemistry: A new procedure for the analysis and purification of peroxidase-containing organelles. *J. Cell Biol.* **98:**870–876.

Croze, E., Ivanov, I., Snitkin, H., Kreibich, G., Sabatini, D., and Rosenfeld, M., 1985, Identification of lysosomal membrane proteins using specific monoclonal antibodies, Abstract 198. *J. Cell Biol.* **101:**53a.

Datema, R., Romero, P. A., Legler, G., and Schwarz, R. T., 1982, Inhibition of formation of complex oligosaccharides by the glucosidase inhibitor bromoconduritol. *Proc. Natl. Acad. Sci. U.S.A.* **79:**6787–6791.

Davey, J., Hurtley, S. M., and Warren, G., 1985, Reconstitution of an endocytic fusion event in a cell-free system. *Cell* **43:**643–652.

Dawson, D. B., Smith, R. A., and Varandani, P. T., 1984, Production of a monoclonal antibody directed against rat liver glutathione–insulin transhydrogenase. *Biochem. Biophys. Res. Commun.* **125:**668–674.

Dean, R. T., Jessup, W., and Roberts, C. R., 1984, Effects of exogenous amines on mammalian cells, with particular reference to membrane flow. *Biochem. J.* **217:**27–40.

De Camilli, P., Harris, S. M., Jr., Huttner, W. B., and Greengard, P., 1983, Synapsin I (Protein I), a nerve terminal-specific phosphoprotein. II. Its specific association with synaptic vesicles demonstrated by immunocytochemistry in agarose-embedded synaptosomes. *J. Cell Biol.* **96:**1355–1373.

De Camilli, P., Moretti, M., Donini, S. D., Walter, U., and Lohmann, S. M., 1986, Heterogeneous distribution of the cAMP receptor protein RII in the nervous system: Evidence for its intracellular accumulation on microtubules, microtubule-organizing centers, and in the area of the Golgi complex. *J. Cell Biol.* **103:**189–203.

deDuve, C., De Barsy, T., Poole, B., Trouet, A., Tulkens, P., and Van Hoof, F., 1974, Lysosomotropic agents. *Biochem. Pharmacol.* **23:**2495–2531.

Deutscher, S. L., Creek, K. E., Merion, M., and Hirschberg, C. B., 1983, Subfractionation of rat liver Golgi apparatus: Separation of enzyme activities involved in the biosynthesis of the

phosphomannosyl recognition marker in lysosomal enzymes. *Proc. Natl. Acad. Sci. U.S.A.* **80**:3938–3942.
Dunn, W. A., Connolly, T. P., and Hubbard, A. L., 1986, Receptor-mediated endocytosis of epidermal growth factor by rat hepatocytes: Receptor pathway. *J. Cell Biol.* **102**:24–36.
Dunphy, W. G., and Rothman, J. E., 1983, Compartmentation of asparagine-linked oligosaccharide processing in the Golgi apparatus. *J. Cell Biol.* **97**:270–275.
Dunphy, W. G., and Rothman, J. E., 1985, Compartmental organization of the Golgi stack. *Cell* **42**:13–21.
Dunphy, W. G., Fries, E., Urbani, L. J., and Rothman, J. E., 1981, Early and late functions associated with the Golgi apparatus reside in distinct compartments. *Proc. Natl. Acad. Sci. U.S.A.* **78**:7453–7457.
Dunphy, W. G., Brands, R., and Rothman, J. E., 1985, Attachment of terminal *N*-acetylglucosamine to asparagine-linked oligosaccharides occurs in central cisternae of the Golgi stack. *Cell* **40**:463–472.
Elbein, A. D., 1981, The tunicamycins—useful tools for studies on glycoproteins, *TIBS* **6**:219–223.
Elhammer, A., and Kornfeld, S., 1984, Two enzymes involved in the synthesis of O-linked oligosaccharides are localized on membranes of different densities in mouse lymphoma BW5147 cells. *J. Cell Biol.* **98**:327–331.
Evans, W. H., and Flint, N., 1985, Subfractionation of hepatic endosomes in Nycodenz gradients and by free-flow electrophoresis. *Biochem. J.* **232**:25–32.
Farquhar, M. G., 1978, Recovery of surface membrane in anterior pituitary cells. Variations in traffic detected with anionic and cationic ferritin. *J. Cell Biol.* **77**:R35–R42.
Farquhar, M. G., 1985, Progress in unraveling pathways of Golgi traffic. *Annu. Rev. Cell Biol.* **1**:447–488.
Fischer-Colbrie, R., Zangerle, R., Frischenschlager, I., Weber, A., and Winkler, H., 1984, Isolation and characterization of a glycoprotein from adrenal chromaffin granules. *J. Neurochem.* **42**:1008–1016.
Fleischer, B., 1983, Isolation of Golgi apparatus from rat liver using D_2O-sucrose gradients. *Methods Enzymol.* **98**:60–67.
Friend, D. S., and Murray, M. J., 1965, Osmium impregnation of the Golgi apparatus. *Am. J. Anat.* **117**:135–150.
Fritzler, M., Etherington, J., Sokoluk, C., Kinsella, T., and Valencia, D., 1984, Antibodies from patients with autoimmune disease react with a cytoplasmic antigen in the Golgi apparatus. *J. Immunol.* **132**:2904–2908.
Galteau, M.-M., Antoine, B., and Reggio, H., 1985, Epoxide hydrolase is a marker for the smooth endoplasmic reticulum in rat liver. *EMBO J.* **4**:2793–2800.
Geuze, H. J., Slot, J. W., and Tokuyasu, K. T., 1979, Immunocytochemical localization of amylase and chymotrypsinogen in the exocrine pancreatic cell with special attention to the Golgi complex. *J. Cell Biol.* **82**:697–707.
Geuze, H. J., Slot, J. W., van Derley, P. A., and Scheffer, R. C. T., 1981, Use of colloidal gold particles in double-labeling immunoelectron microscopy of ultrathin frozen tissue sections. *J. Cell Biol.* **89**:653–665.
Geuze, H. J., Slot, J. W., Strous, G. J. A. M., Lodish, H. F., and Schwartz, A. L., 1983a, Intracellular site of asialoglycoprotein receptor–ligand uncoupling: Double-label immunoelectron microscopy during receptor-mediated endocytosis. *Cell* **32**:277–287.
Geuze, H. J., Slot, J. W., Strous, G. J. A. M., Peppard, J., von Figura, K., Hasilik, A., and Schwartz, A. L., 1983b, Intracellular receptor sorting during endocytosis: Comparative immunoelectron microscopy of multiple receptors in rat liver. *Cell* **37**:195–204.
Geuze, H. J., Slot, J. W., Strous, G. J. A. M., Hasilik, A., and von Figura, K., 1984, Ultrastructural localization of the mannose-6-phosphate receptor in rat liver. *J. Cell Biol.* **98**:2047–2054.

Goldberg, D. E., and Kornfeld, S., 1983, Evidence for extensive subcellular organization of asparagine-linked oligosaccharide processing and lysosomal enzyme phosphorylation. *J. Biol. Chem.* **258**:3159–3165.

Goldfischer, S., Essner, E., and Schiller, B., 1971, Nucleoside diphosphatase and thiamine pyrophosphatase activities in the endoplasmic reticulum and Golgi apparatus. *J. Histochem. Cytochem.* **19**:349–360.

Gonatas, N. K., Stieber, A., Hickey, W. F., Herbert, S. H., and Gonatas, J. O., 1984, Endosomes and Golgi vesicles in adsorptive and fluid phase endocytosis. *J. Cell Biol.* **99**:1379–1390.

Granger, B. J., Plutner, H., Green, S. A., Mellman, I., and Helenius, A., 1985, Defining the lysosome membrane: Characterization of specific glycoproteins, Abstract 199. *J. Cell Biol.* **101**:54a.

Green, J., Griffiths, G., Louvard, D., Quinn, P., and Warren, G., 1981, Passage of viral membrane proteins through the Golgi complex. *J. Mol. Biol.* **152**:663–698.

Griffiths, G., Brands, R., Burke, B., Louvard, D., and Warren, G., 1982, Viral membrane proteins acquire galactose in *trans* Golgi cisternae during intracellular transport. *J. Cell Biol.* **95**:781–792.

Griffiths, G., Pfeiffer, S., Simons, K., and Matlin, K., 1985, Exit of newly synthesized membrane proteins from the *trans* cisterna of the Golgi complex to the plasma membrane. *J. Cell Biol.* **101**:949–964.

Gruenberg, J., and Howell, K. E., 1985a, Immuno-isolation of the plasma membrane and the endosomal compartment using antigenic sites on opposite domains of a foreign (viral) protein, Abstract 1594. *J. Cell Biol.* **101**:423a.

Gruenberg, J., and Howell, K. E., 1985b, Immuno-isolation of vesicles using antigenic sites either located on the cytoplasmic or the exoplasmic domain of an implanted viral protein. A quantitative analysis. *Eur. J. Cell Biol.* **38**:312–321.

Gruenberg, J., and Howell, K., 1986, Reconstitution of vesicle fusions occurring in endocytosis with a cell-free system. *EMBO J.* **5**:3091–3101.

Harford, J., Bridges, K., Ashwell, G., and Klausner, R. D., 1983a, Intracellular dissociation of receptor bound asialoglycoproteins in cultured hepatocytes. A pH mediated non-lysosomal event. *J. Biol. Chem.* **258**:3191–3197.

Harford, J., Wolkoff, A. W., Ashwell, G., and Klausner, R. D., 1983b, Monensin inhibits intracellular dissociation of asialoglycoproteins from their receptors. *J. Cell Biol.* **96**:1824–1828.

Harms, E., Kern, H., and Schneider, J. A., 1980, Human lysosomes can be purified from diploid skin fibroblasts by free-flow electrophoresis. *Proc. Natl. Acad. Sci. U.S.A.* **77**:6139–6143.

Haselbeck, A., and Schekman, R., 1986, Interorganelle transfer and glycosylation of yeast invertase *in vitro*. *Proc. Natl. Acad. Sci. U.S.A.* **83**:2017–2021.

Helmy, S., Porter-Jordan, K., Dawidowicz, E. A., Pilch, P., Schwartz, A. L., and Fine R. E., 1986, Separation of endocytic from exocytic coated vesicles using a novel cholinesterase mediated density shift technique. *Cell* **44**:497–506.

Herzog, V., and Farquhar, M. G., 1983, Use of electron opaque tracers for studies on endocytosis and membrane recycling. *Methods Enzymol.* **98**:203–225.

Hopkins, C. R., 1983, Intracellular routing of transferrin and transferrin receptors in epidermoid carcinoma A 431 cells. *Cell* **35**:321–330.

Hopkins, C. R., 1985, The appearance and internalization of transferrin receptors at the margins of spreading human tumor cells. *Cell* **40**:199–208.

Hopkins, C. R., and Trowbridge, I. S., 1983, Internalization and processing of transferrin and the transferrin receptor in human carcinoma A 431 cells. *J. Cell Biol.* **97**:508–521.

Hornick, C. A., Hamilton, R. L., Spaziani, E., Enders, G. H., and Havel, R. J., 1985, Isolation and characterization of multivesicular bodies from rat hepatocytes: An organelle distinct from secretory vesicles of the Golgi apparatus. *J. Cell Biol.* **100**:1558–1569.

Hortsch, M., Avossa, D., and Meyer, D. I., 1985a, A structural and functional analysis of the docking protein. *J. Biol. Chem.* **260:**9137–9145.

Hortsch, M., Griffiths, G., and Meyer, D. I., 1985b, Restriction of docking protein to the rough endoplasmic reticulum: Immunocytochemical localization in rat liver. *Eur. J. Cell Biol.* **38:**271–279.

Hortsch, M., Avossa, D., and Meyer, D. I., 1986, Characterization of secretory protein translocation: Ribosome–membrane interaction in endoplasmic reticulum. *J. Cell Biol.* **103:**241–253.

Huttner, W. B., Schiebler, P., Greengard, P., and De Camilli, P., 1983, Synapsin I (Protein I), a nerve terminal-specific phosphoprotein. III. Its association with synaptic vesicles studied in a highly purified synaptic vesicle preparation. *J. Cell Biol.* **96:**1374–1388.

Ito, A., and Palade, G. E., 1978, Presence of NADPH–cytochrome P-450 reductase in rat liver Golgi membranes. Evidence obtained by immunoadsorption method. *J. Cell Biol.* **79:**590–597.

Jahn, R., Schiebler, W., Ouimet, C., and Greengard, P., 1985, A 38,000-dalton protein (p38) present in synaptic vesicles. *Proc. Natl. Acad. Sci. U.S.A.* **82:**4137–4141.

Kaetzel, C. S., Rao, C. K., and Lamm, M. E., 1987, Protein disulphide isomerase from human placenta and rat liver: Purification and immunological characterization with monoclonal antibodies. *Biochem. J.* **241:**39–47.

Kedersha, N. L., Hill, D. F., Kronquist, K. E., and Rome, L. H., 1986, Subpopulations of liver coated vesicles resolved by preparative agarose gel electrophoresis. *J. Cell Biol.* **103:**287–297.

Keen, J. H., Willingham, M. C., and Pastan, I., 1981, Clathrin and coated vesicle proteins. Immunological characterization. *J. Biol. Chem.* **256:**2538–2544.

Kelly, R. B., 1985, Pathways of protein secretion in eukaryotes. *Science* **230:**25–32.

Kelly, W. G., Passaniti, A., Woods, J. W., Daiss, J. L., and Roth, T. F., Tubulin as a molecular component of coated vesicles, 1983, *J. Cell Biol.* **97:**1191–1199.

Kirchhausen, T., Harrison, S. C., Parham, P., and Brodsky, F. M., 1983, Location and distribution of the light chains in clathrin trimers. *Proc. Natl. Acad. Sci. U.S.A.* **80:**2481–2485.

Kraehenbuhl, J. P., Racine, L., and Jamieson, J. D., 1977, Immunocytochemical localization of secretion proteins in bovine pancreatic exocrine cells. *J. Cell Biol.* **72:**406–423.

Kreibich, G., Debey, P., and Sabatini, D. D., 1973, Selective release of content from microsomal vesicles without membrane disassembly. I. Permeability changes induced by low detergent concentrations. *J. Cell Biol.* **58:**436–462.

Kreibich, G., Marcantonio, E. E., and Sabatini, D. D., 1983, Ribophorins I and II: Membrane proteins characteristic of the rough endoplasmic reticulum. *Methods Enzymol.* **96:**520–530.

Kreis, T. E., 1986, Microinjected antibodies against the cytoplasmic domain of vesicular stomatitis virus glycoprotein block its transport to the cell surface. *EMBO J.* **5:**931–941.

Kuettel, M. R., Squinto, S. P., Kwast-Welfeld, J., Schwoch, G., Schweppe, J. S., and Jungmann, R. A., 1985, Localization of nuclear subunits of cyclic AMP-dependent protein kinase by the immunocolloidal gold method. *J. Cell Biol.* **101:**965–975.

Lewis, M. J., Mazzarella, R. A., and Green, M., 1985a, Structure and assembly of the endoplasmic reticulum. The synthesis of three major endoplasmic reticulum proteins during lipopolysaccharide-induced differentiation of murine lymphocytes. *J. Biol. Chem.* **260:**3050–3057.

Lewis, M. J., Turco, S. J., and Green, M., 1985b, Structure and assembly of the endoplasmic reticulum. Biosynthetic sorting of endoplasmic reticulum proteins. *J. Biol. Chem.* **260:**6926–6931.

Lewis, V., Green, S. A., Marsh, M., Vihko, P., Helenius, A., and Mellman, I., 1985, Glycoproteins of the lysosomal membrane. *J. Cell. Biol.* **100:**1839–1847.

Lin, J. J.-C., and Queally, S. A., 1982, A monoclonal antibody that recognizes Golgi-associated protein of cultured fibroblast cells. *J. Cell Biol.* **92:**108–112.

Lin, J. J.-C., Welch, W. J., Garrels, J. I., and Feramisco, J. R., 1982, The association of the 100-kD heat-shock protein with the Golgi apparatus. In *Heat Shock from Bacteria to Man* (M. J.

Schlesinger, M. Ashburner, and A. Tissieres, eds.), pp. 267–273, Cold Spring Harbor Laboratory, Cold Spring Harbor, NY.

Lingg, G., Fisher-Colbrie, R., Schmidt, W., and Winkler, H., 1983, Exposure of an antigen of chromaffin granules on cell surface during exocytosis. *Nature* **301**:610–611.

Lippincott-Schwartz, J., and Fambrough, D. M., 1986, Lysosomal membrane dynamics: Structure and interorganellar movement of a major lysosomal membrane glycoprotein. *J. Cell Biol.* **102**:1593–1605.

Lippincott-Schwartz, J., and Fambrough, D. M., 1987, Cycling of the integral membrane glycoprotein LEP100 between plasma membrane and lysosomes: Kinetic and morphologic analysis. *Cell* **49**:669–677.

Lodish, H. F., and Kong, N., 1984, Glucose removal from N-linked oligosaccharides is required for efficient maturation of certain secretory glycoproteins from the rough endoplasmic reticulum to the Golgi complex. *J. Cell Biol.* **98**:1720–1729.

Louvard, D., Reggio, H., and Warren, G., 1982, Antibodies to the Golgi complex and the rough endoplasmic reticulum. *J. Cell Biol.* **92**:92–107.

Louvard, D., Morris, C., Warren, G., Stanley, K., Winkler, F., and Reggio, H., 1983, A monoclonal antibody to the heavy chain of clathrin. *EMBO J.* **2**:1655–1664.

Lucocq, J. M., Brada, D., and Roth, J., 1986, Immunolocalization of the oligosaccharide trimming enzyme glucosidase II. *J. Cell Biol.* **102**:2137–2146.

Matthew, W. D., Tsavaler, L., and Reichardt, L. F., 1981, Identification of a synaptic vesicle-specific membrane protein with a wide distribution in neuronal and neurosecretory tissue. *J. Cell Biol.* **91**:257–269.

Maxfield, F. R., 1985, Acidification of endocytic vesicles and lysosomes. In *Endocytosis* (I. Pastan and M. C. Willingham, eds.), pp. 235–257, Plenum Press, New York.

Meldolesi, J., 1983, Membranes of pancreatic zymogen granules. *Methods Enzymol.* **98**:67–75.

Miller, K., Beardmore, J., Kanety, H., Schlessinger, J., and Hopkins, C. R., 1986, Localization of the epidermal growth factor (EGF) receptor within the endosome of EGF-stimulated epidermoid carcinoma (A431) cells. *J. Cell Biol.* **102**:500–509.

Mollenhauer, H., and Morré, D. J., 1987, Structural compartmentation of the cytosol; zones of exclusion, zones of adhesion, cytoskeletal and intercisternal elements. *Subcell. Biochem.* **5**:327–359.

Morin, J. E., Dixon, J. E., Chang, P. P., and Moss, J., 1983, Identification of thiol:protein disulfide oxidoreductase activity in cultured human fibroblasts: Dependence of enzyme activity on growth conditions. *Biochem. Biophys. Res. Commun.* **111**:872–877.

Morré, J., 1971, Isolation of Golgi apparatus. *Methods Enzymol.* **22**:130–148.

Mueller, S. C., and Hubbard, A. L., 1986, Receptor-mediated endocytosis of asialoglycoproteins by rat hepatocytes: Receptor-positive and receptor-negative endosomes. *J. Cell Biol.* **102**:932–942.

Navone, F., Greengard, P., and De Camilli, P., 1984, Synapsin I in nerve terminals: Selective association with small synaptic vesicles. *Science* **226**:1209–1211.

Nigg, E. A., Schafer, G., Hilz, H., and Eppenberger, H. M., 1985a, Cyclic-AMP-dependent protein kinase type II is associated with the Golgi complex and with centrosomes. *Cell* **41**:1039–1051.

Nigg, E. A., Hilz, H., Eppenberger, H. M., and Dutly, F., 1985b, Rapid and reversible translocation of the catalygic subunit of cAMP-dependent protein kinase type II from the Golgi complex to the nucleus. *EMBO J.* **4**:2801–2806.

Novikoff, P. M., Novikoff, A. B., Quintana, N., and Hauw, J.-J., 1971, Golgi apparatus, GERL and lysosomes of neurons in rat dorsal root ganglia studied by thick section and thin section cytochemistry. *J. Cell Biol.* **50**:859–886.

Novikoff, P. M., Tulsiani, D. R. P., Touster, O., Yam, A., and Novikoff, A. B., 1983, Immunocytochemical localization of α-D-mannosidase II in the Golgi apparatus of rat liver, *Proc. Natl. Acad. Sci. U.S.A.* **80:**4364–4368.

Oesch, F., and Bentley, P., 1976, Antibodies against homogeneous epoxide hydratase provide evidence for a single enzyme hydrating styrene oxide and benz[*a*]pyrene 4,5-oxide. *Nature* **259:**53–55.

Ohba, H., Harano, T., and Omura, T., 1981, Intracellular and intramembranous localization of a protein disulfide isomerase in rat liver. *J. Biochem.* **89:**889–900.

Oda, K., Misumi, Y. and Ikehara, Y., 1983, Disparate effects of monensin and colchicine on intracellular processing of secretory proteins in cultured rat hepatocytes. *Eur. J. Biochem.* **135:**209–216.

Okada, P., Frey, A. B., Guenthner, T. M., Oesch, F., Sabatini, D. D., and Kreibich, G., 1982, Studies on the biosynthesis of microsomal membrane proteins: Site of synthesis and mode of insertion of cytochrome b_5, cytochrome b_5 reductase, cytochrome P-450 reductase, and epoxide hydrolase. *Eur. J. Biochem.* **122:**393–402.

Orci, L., Ravazzola, M., Amherdt, M., Madsen, O., Vassalli, J.-D., and Perrelet, A., 1985, Direct identification of prohormone conversion site in insulin-secreting cells. *Cell* **42:**671–681.

Paquet, M. R., St.-Jean, P., Roberge, M., Beaudoin, A. R., 1982, Isolation of zymogen granules from rat pancreas and characterization of their membrane proteins. *Eur. J. Cell Biol.* **28:**20–26.

Pastan, I., and Willingham, M. C., 1985, The pathway of endocytosis. In *Endocytosis* (I. Pastan and M. C. Willingham, eds.), pp. 1–44, Plenum Press, New York.

Payne, G. S., and Schekman, R., 1985, A test of clathrin function in protein secretion and cell growth. *Science* **230:**1009–1014.

Pearse, B. M. F., 1983, Isolation of coated vesicles. *Methods Enzymol.* **98:**320–326.

Pfeffer, S. R., and Kelly, R. B., 1985, The subpopulation of brain coated vesicles that carries synaptic vesicle proteins contains two unique polypeptides. *Cell* **40:**949–957.

Phillips, J. H., Burridge, K., Wilson, S. P. and Kirshner, N., 1983, Visualization of the exocytosis/endocytosis secretory cycle in cultured adrenal chromaffin cells. *J. Cell Biol.* **97:**1906–1917.

Posthuma, G., Slot, J. W., and Geuze, H. J., 1984, Immunocytochemical assays of amylase and chymotrypsinogen in rat pancreas secretory granules. Efficacy of using immunogold-labeled ultrathin cryosections to estimate relative protein concentrations. *J. Histochem. Cytochem.* **32:**1028–1034.

Reggio, H., Bainton, D., Harms, E., Coudrier, E., and Louvard, D., 1984, Antibodies against lysosomal membranes reveal a 100,000-mol-wt protein that cross-reacts with purified H^+,K^+ ATPase from gastric mucosa. *J. Cell Biol.* **99:**1511–1526.

Regoeczi, E., Chindemi, P. A., Debanne, M. T., and Charlwood, P. A., 1982, Partial resialylation of human asialotransferrin type 3 in the rat. *Proc. Natl. Acad. Sci. U.S.A.* **79:**2226–2230.

Resh, M. D., and Erikson, R. L., 1985, Highly specific antibody to rous sarcoma virus src gene product recognizes a novel population of $pp60^{v\text{-}src}$ and $pp60^{c\text{-}src}$ molecules. *J. Cell Biol.* **100:**409–417.

Robinson, M. S., and Pearse, B. M. F., 1986, Immunofluorescent localization of 100K coated vesicle proteins. *J. Cell Biol.* **102:**48–54.

Rodriguez, R., Gelpi, C., Thompson, T., Real, F., and Fernandez, J., 1982, Anti-Golgi complex antibodies in a patient with Sjogren's syndrome and lymphoma. *Clin. Exp. Immunol.* **49:**579–586.

Roth, J., 1984, Cytochemical localization of terminal *N*-acetyl-D-galactosamine residues in cellular compartments of intestinal goblet cells: Implications for the topology of O-glycosylation. *J. Cell Biol.* **98:**399–406.

Roth, J., and Berger, E. G., 1982, Immunocytochemical localization of galactosyltransferase in

HeLa cells: Codistribution with thiamine pyrophosphatase in *trans*-Golgi cisternae. *J. Cell Biol.* **92:**223–229.

Roth, J., Taatjes, D. J., Lucocq, J. M., Weinstein, J., and Paulson, J. C., 1985, Demonstration of an extensive *trans*-tubular network continuous with the Golgi apparatus stack that may function in glycosylation. *Cell* **43:**287–295.

Roth, R., and Mesirow, M. L., 1984, Production and characterization of a monoclonal antibody to rat liver thiol: Protein–disulfide oxidoreductase/glutathione–insulin transhydrogenase. *Biochim. Biophys. Acta* **788:**189–192.

Rothman, J., Urbani, L. J., and Brands, R., 1984a, Transport of protein between cytoplasmic membranes of fused cells: Correspondence to processes reconstituted in a cell-free system. *J. Cell Biol.* **99:**248–259.

Rothman, J., Miller, R. L., and Urbani, L. J., 1984b, Intercompartmental transport in the Golgi complex is a dissociative process: Facile transfer of membrane protein between two Golgi populations. *J. Cell Biol.* **99:**260–271.

Saraste, J., and Hedman, K., 1983, Intracellular vesicles involved in the transport of Semliki Forest virus membrane proteins to the cell surface. *EMBO J.* **2:**2001–2006.

Schliwa, M., 1986, *The Cytoskeleton, An Introductory Survey,* Springer-Verlag, New York.

Schmaljohn, A. L., Kokubun, K. M., and Cole, G. A., 1983, Protective monoclonal antibodies define maturational and pH dependent changes in Sindbus virus E1 glycoprotein. *Virology* **130:**144–154.

Schwartz, A. L., Strous, G. J. A. M., Slot, J. W., and Geuze, H. J., 1985, Immunoelectron microscopic localization of acidic intracellular compartments in hepatoma cells. *EMBO J.* **4:**899–904.

Simion, F. A., Winek, D., Brandan, E., Fleischer, B., and Fleischer, S., 1983, Isolation and characterization of coated vesicles from rat liver. *Methods Enzymol.* **98:**326–336.

Simons, K., and Virta, H., 1987, Perforated MDCK cells support intracellular transport. *EMBO J.* **6:**2241–2247.

Smith, C. E., 1980, Ultrastructural localization of nicotinamide adenine dinucleotide phosphatase (NADPase) activity to the intermediate saccules of the Golgi apparatus in rat incisor ameloblasts. *J. Histochem. Cytochem.* **28:**16–26.

Smith, C. E., Paiement, J., and Bergeron, J. J. M., 1986, Subcellular distribution of acid NADPase activity within the parenchymal cells of rat liver. *J. Histochem. Cytochem.* **34:**649–658.

Smith, Z. D. J., D'Eugenio-Gumkowski, F., Yanagisawa, K., and Jamieson, J. D., 1984, Endogenous and monoclonal antibodies to the rat pancreatic acinar cell Golgi complex. *J. Cell Biol.* **98:**2035–2046.

Snider, M. D., and Rogers, O. C., 1985, Intracellular movement of cell surface receptors after endocytosis: Resialylation of asialo-transferrin receptor in human erythroleukemia cells. *J. Cell Biol.* **100:**826–834.

Snider, M. D., and Rogers, O. C., 1986, Membrane traffic in animal cells: Cellular glycoproteins return to the site of Golgi mannosidase I. *J. Cell Biol.* **103:**265–275.

Storrie, B., Pool, R. R., Jr., Sachdeva, M., Maurey, K. M., and Oliver, C., 1984, Evidence for both prelysosomal and lysosomal intermediates in endocytic pathways. *J. Cell Biol.* **98:**108–115.

Strous, G. J. A. M., and Berger, E. G., 1982, Biosynthesis, intracellular transport, and release of the Golgi enzyme galactosyltransferase (lactose synthetase A protein) in HeLa cells. *J. Biol. Chem.* **257:**7623–7628.

Tartakoff, A. M., 1980, The Golgi complex: Crossroads for vesicular traffic. *Int. Rev. Exp. Pathol.* **22:**227–251.

Tartakoff, A. M., 1983a, The confined function model of the Golgi complex: Center for ordered processing of biosynthetic products of the rough endoplasmic reticulum. *Int. Rev. Cytol.* **85:**221–252.

Tartakoff, A. M., 1983b, Perturbation of the structure and function of the Golgi complex by monovalent carboxylic ionophores. *Methods Enzymol.* **98**:47–59.

Tartakoff, A. M., and Vassalli, P., 1983, Lectin-binding sites as markers of Golgi subcompartments: Proximal-to-distal maturation of oligosaccharides. *J. Cell Biol.* **97**:1243–1248.

Thorens, B., and Vassalli, P., 1986, Chloroquine and ammonium chloride prevent terminal glycosylation of immunoglobulins in plasma cells without affecting secretion. *Nature* **321**:618–620.

Tougard, C., Louvard, D., Picart, R., and Tixier-Vidal, A., 1983, The rough endoplasmic reticulum and the Golgi apparatus visualized using specific antibodies in normal and tumoral prolactin cells in culture. *J. Cell Biol.* **96**:1197–1207.

Tougard, C., Louvard, D., Picart, R., and Tixier-Vidal, A., 1985, Antibodies against a lysosomal membrane antigen recognize a prelysosomal compartment involved in the endocytic pathway in cultured prolactin cells. *J. Cell Biol.* **100**:786–793.

Ulrich, J. T., Schenck, J. R., Rittenhouse, H. G., Shaper, N. L., and Shaper, J. H., 1986, Monoclonal antibodies to bovine UDP–galactosyltransferase. Characterization, cross-reactivity, and utilization as structural probes. *J. Biol. Chem.* **261**:7975–7981.

van Deurs, B., Tønnessen, T. I., Petersen, O. W., Sandvig, K., and Olsnes, S., 1986, Routing of internalized ricin and ricin conjugates to the Golgi complex. *J. Cell Biol.* **102**:37–47.

Varandani, P. T., Raveed, D., and Nafz, M. A., 1978, Insulin degradation. XXIII. Distribution of glutathione–insulin transhydrogenase in isolated rat hepatocytes as studied by immuno-ferritin and electron microscopy. *Biochim. Biophys. Acta* **538**:343–353.

von Wedel, R. J., Carlson, S. S., and Kelly, R., 1981, Transfer of synaptic vesicle antigens to the presynaptic plasma membrane during exocytosis. *Proc. Natl. Acad. Sci. U.S.A.* **78**:1014–1018.

Wall, D. A., and Hubbard, A. L., 1985, Receptor-mediated endocytosis of asialoglycoproteins by rat liver hepatocytes: Biochemical characterization of the endosomal compartments. *J. Cell Biol.* **101**:2104–2112.

Wehland, J., Willingham, M. C., Dickson, R., and Pastan, I., 1981, Microinjection of anticlathrin antibodies into fibroblasts does not interfere with the receptor-mediated endocytosis of α_2-macroglobulin. *Cell* **25**:105–119.

Wehland, J., Willingham, M. C., Gallo, M., and Pastan, I., 1982, The morphologic pathway of exocytosis of the VSV G protein in cultured fibroblasts. *Cell* **28**:831–841.

Wiedenmann, B., and Franke, W. W., 1985, Identification and localization of synaptophysin, an integral membrane glycoprotein of M_r 38,000 characteristic of presynaptic vesicles. *Cell* **41**:1017–1028.

Wileman, T., Harding, C., and Stahl, P., 1985, Receptor-mediated endocytosis. *Biochem. J.* **232**:1–14.

Willingham, M. C., and Pastan, I., 1982, Transit of epidermal growth factor through coated pits of the Golgi system. *J. Cell Biol.* **94**:207–212.

Willingham, M. C., Pastan, I., and Sahagian, G. G., 1983, Ultrastructural immunocytochemical localization of the phosphomannosyl receptor in Chinese hamster ovary (CHO) cells. *J. Histochem. Cytochem.* **31**:1–11.

Woods, J. W., Doriaux, M. and Farquhar, M. G., 1986, Transferrin receptors recycle to cis and middle as well as trans Golgi cisternae in Ig-secreting myeloma cells- *J. Cell Biol.* **103**:277–286.

Yuan, L., Barriocanal, J., Bonifacino, J., and Sandoval, I., 1987, Two integral membrane proteins located in the cis-middle and trans-part of the Golgi system acquire sialylated N-linked carbohydrates and display different turnovers and sensitivity to cAMP-dependent phosphorylation. *J. Cell. Biol.* **105**:215–227.

Chapter 8

Monoclonal Antibodies in Investigations on Astrocytes

S. K. Malhotra

1. INTRODUCTION

The impact of a new concept or a new technological advance usually results in accelerated pace in the application of the newly acquired concept or technique that produces advances of knowledge in the field. In cell biology in general, the repercussions in cell research have been spectacular following Abbé's theory of image formation and following development of commercial transmission electron microscopes. These advances in microscopy have had parallel advances in preparative techniques suitable for examination of biological specimens by light microscopy and electron microscopy (Palade, 1967). Such advances in microscopy coupled with coincidental advances in biochemical and biophysical approaches for cell research have been instrumental in bringing our knowledge of the cell to its current phenomenal level. The functioning of cell organelles and of cells themselves, their interactions with other cells and with matrix of tissues are determined by the molecular architecture. Toward this end, hybridoma technology and consequently the ability to generate monoclonal antibodies (Mabs) (Köhler and Milstein, 1975) have provided cell biologists with means, hitherto unavailable to them, to probe the uniqueness of cells and tissues and map the distribution of their antigenic components. Monoclonal antibodies enable us to

S. K. Malhotra Department of Zoology, University of Alberta, Edmonton, Alberta, Canada T6G 2E9.

study functional homologies in cells and tissues between diverse species. They enable us to detect sparsely distributed and unknown antigens that could not be discovered by routine biochemical techniques. These antigenic molecules can then be purified and subjected to further analyses, such as amino acid sequence, isolation of mRNA, and cDNA cloning. In the case of glycoproteins and glycolipids, characterization of the carbohydrate moiety can be achieved. The variations in the spatial and temporal distributions of molecules of interest in understanding the differentiation process provide insight into growth and development. An example of this sort is the knowledge gained in the role of cell adhesion molecules (CAMs) in the development of the nervous system (Bunge and Waksman, 1985; Rutishauser, 1984, 1986). A comparative study of the gene coding for particular antigenic molecules can lead to understanding of the biological diversity and maturation process. Gene cloning has provided insight, for example, into differences between the fetal and adult acetylcholine receptor (AchR) molecules in which one of the five subunits (γ subunit) of the fetal AchR is replaced by the ϵ subunit in the adult. The channel conductance properties of the AchR differ in the fetal and adult skeletal muscle and this difference has now been attributed to the difference in the subunit structure of the receptor molecule in developing muscle. These results also explain the differences in the channel conductance properties of the nonjunctional receptors in the denervated muscle and junctional receptors at the neuromuscular junction. The nonjunctional Ach receptors are similar to the fetal Ach receptors in their channel conductance properties (reviewed by Mishina *et al.*, 1986). By cDNA cloning of the muscarinic Ach receptor, the homology between this receptor, the β-adrengeric receptor, and rhodopsin in amino acid sequence and transmembrane topography has been revealed (Kubo *et al.*, 1986). Also by cDNA cloning the similarity in the myelin basic proteins (MBP) in CNS and PNS has been revealed and these proteins are likely to be encoded by a single gene expressed in both myelin-forming cells, that is, oligodendrocytes (CNS) and Schwann cells (PNS)(Lemke, 1986; Mentaberry *et al.*, 1986). It is also of interest that a Mab (M6C5) raised against MBP cross-reacts with oligodendrocytes and Schwann cells, as well as astrocytes (Brenner *et al.*, 1986). Astrocytes are not known to produce MBP, which is a well-known component of the myelin sheath; therefore, the identity of the molecule in astrocytes cross-reacting with the Mab to MBP should be of interest when revealed. The sharing of an epitope between these three glial cell types may point to a common neuroglial progenitor cell (Brenner *et al.*, 1986). Alternatively, such a similarity may simply reside in the sharing of this particular epitope carried by a molecule in astrocytes distinct from MBP in myelin-forming cells.

Monoclonal antibodies provide an incisive means to study the anatomical distribution of antigens without the need to purify antigenic molecules (Milstein, 1981). Their application in the study of CNS in particular has already produced vast information on the spatial and temporal distributions of antigens, yielding cell subtypes and compartmentilization of antigens where none was suspected to exist. Such morphological knowledge on the distribution of antigenic molecules may seem confusing because it is difficult at present to assign functional significance to various patterns of distribution. However, this can only facilitate advances in understanding of the molecular mechanisms underlying functions that cells and tissues perform in the normal state or fail to perform in the abnormal or diseased state. By application of Mabs and other approaches now available to probe the uniqueness of the CNS, advances have been made toward understanding the molecular organization of neurons and glia and the interactions between cells and their matrices, as well as developmental changes (Bunge and Waksman, 1985).

This chapter is a brief account of some of the current work aimed at understanding the structure and functioning of astrocytes in CNS (Figure 1). The rationale for this choice has a personal bias. These cells constitute a predominant single cell type in the vertebrate CNS and are implicated in disease and injury to the CNS although the understanding of their biology is incomplete. This is by no means a comprehensive review of what is known about astrocytes, because only some of the current work of general significance in understanding the biology of this glial cell type in vertebrates is included. Thus, I have often avoided mention of particular animal or human material used for investigations. The application of Mabs has been one of the stimuli for current activity on astrocytes and therefore such work carried out by immunocytochemical techniques is emphasized.

To the great many authors whose important work is not cited in the references, my sincere apology. In order to limit references, I have cited only recent reviews and articles, often by authors who previously published on the same subject.

2. DEFINITION OF ASTROCYTES

Astrocytes have come to be operationally defined as glial fibrillary acidic protein (GFAP) positive cells (Figure 2). Such an operational definition should not obscure the vast literature, extending over nearly a century, using special staining procedures for light microscopy and electron microscopy. Such studies provided valuable knowledge on the distribution and morphological diversity of this glial cell type in the CNS (Duffy, 1983; Palay and Chan-Palay, 1974). A reliable method of identifying astrocytes became feasible when GFAP was identified as a marker protein for astrocytes (Bignami and Dahl, 1973). The use of antibodies to GFAP spread quickly, and the term GFAP positive cells in CNS is used as a synonym for astrocytes. This operational definition should cover not only commonly recognized protoplasmic and fibrillary astrocytes (Duffy, 1983)

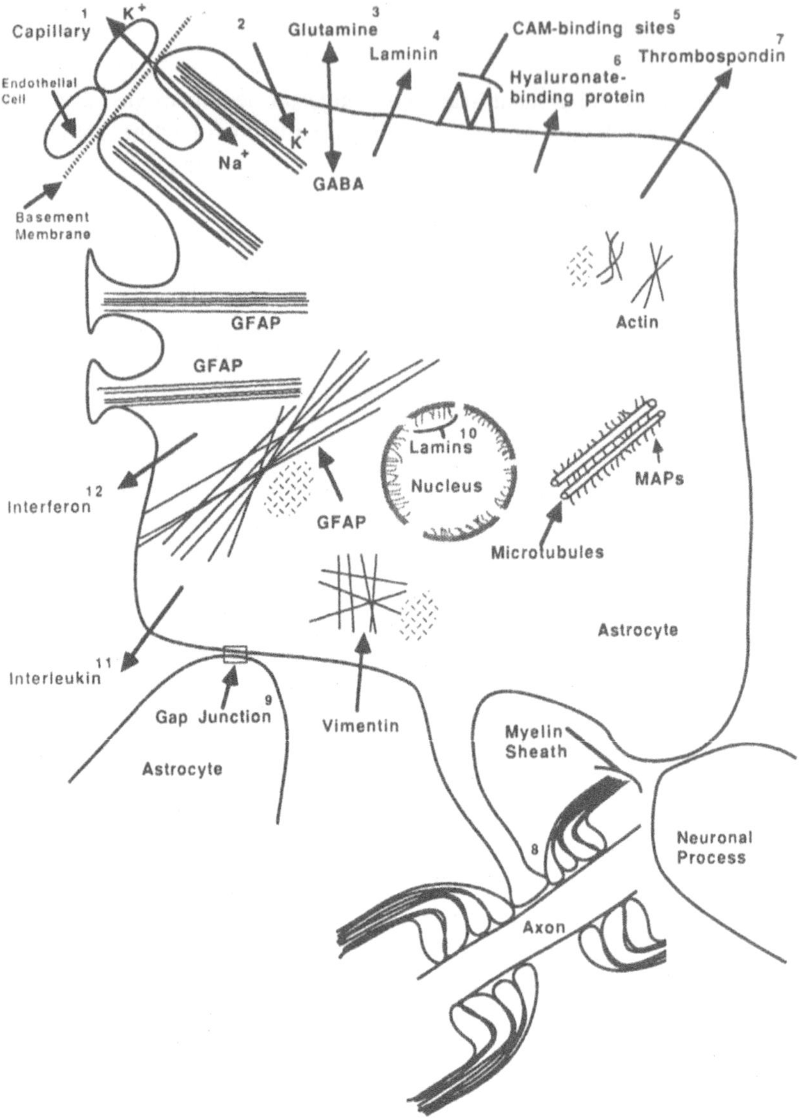

FIGURE 1. A highly diagrammatic representation of some of the functions served by astrocytes compiled from publications indicated by numerals 1–12. No attempt has been made to subclassify astrocytes on the basis of morphology, location, or functions. Both GFAP and vimentin are shown as homopolymer even though the two core IF proteins may form heteropolymers. (Not to scale.) References: 1. Goldstein and Betz (1986). 2. Newman (1986); Kuffler and Nicholls (1976). 3. Duffy (1983); Kuffler and Nicholls (1976). 4. Liesi *et al.* (1983). 5. Cunningham (1986). 6. Bignami and Dahl (1986). 7. Asch *et al.* (1986). 8. ffrench-Constant and Raff (1986). 9. Brightman and Reese (1969). 10. Harris and Marshall (1981); Aebi *et al.* (1986); Burke and Gerace (1986). 11. Frei *et al.* (1985). 12. Tedeschi *et al.* (1986).

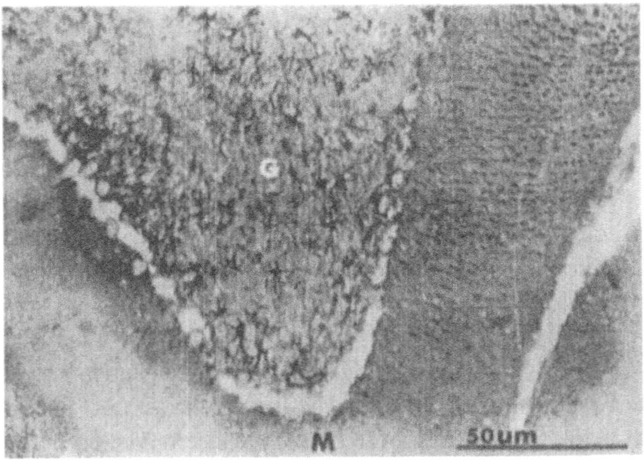

FIGURE 2. Rat cerebellum, fixed in paraformaldehyde and embedded in paraffin; section stained with antiserum to GFAP, using PAP technique of Sternberger (Malhotra, 1984b) to show astrocytes in the granular layer (G) and Bergmann fibers in the molecular layer (M) running perpendicular to the surface of the folium. (The tissue was processed by Dr. Manickvale in T. K. Shnitka's laboratory.)

and type I and type II astrocytes (Miller *et al.*, 1986; Raff *et al.*, 1983) but also the specialized form of radially oriented glia Müller cells of the retina (Björklund *et al.*, 1985), tanycytes lining the ventricles (Flament-Durand and Brion, 1985), and Bergmann fibers that terminate beneath the pia in the cerebellum (Palay and Chan-Palay, 1974). This definition also includes progenitor cells that give rise to GFAP positive astrocytes (Raff *et al.*, 1983) as well as tumors of astrocytic origin (malignant astrocytomas), some of which have been found to have sparse or no detectable GFAP (Duffy, 1983; Jacque *et al.*, 1979), as there appears to be reduction in levels of GFAP as the astrocytomas become more malignant, that is, less differentiated (Duffy, 1983). The operational definition of astrocytes may be extended to include ciliated ependymal cells; in humans they have been reported to express GFAP transiently during development (Roessmann *et al.*, 1980). However, this does seem to be an exception as the ependymal cells, other than tanycytes, are not generally known to express GFAP in the mature state (Shaw *et al.*, 1981).

The value of GFAP as a marker for astrocytes in CNS should not be detracted by recent findings that GFAP or GFAP-like immunoreactivity is present in non-CNS tissues such as Schwann cells, lens epithelium, and Kupffer cells (Yen and Fields, 1985). Whether the nonastrocytic GFAP-like protein is associated with 10-nm IFs does not seem to have been determined thus far.

3. SUBTYPES OF ASTROCYTES

There are two generally recognized subpopulations of astrocytes—protoplasmic (type 1) and fibrous (type 2) based on their morphology (Duffy, 1983) and reactivity to a Mab (Raff *et al.*, 1983). Type 1 in optic nerve, for example, is located at the periphery and forms the glial limiting membrane, whereas type 2 constitute the majority of the astrocytes in the optic nerve. Both types are GFAP positive, although the relative amounts may differ in the two types (Raff *et al.*, 1983). It is quite likely that further subpopulations of astrocytes based on functional characteristics may emerge; for example, certain astrocytes are involved in immunological response (expression of H-2 antigens in mice in response to interferon treatment; Tedeschi *et al.*, 1986). No such subpopulation study has been emphasized in this article, even though it is more than likely that various astrocytic functions (Figure 1) are served by a subpopulation only.

It is Bergmann-like fibers (protoplasmic or type-1 astrocytes) along which neurons migrate, not along the processes of star-shaped (fibrous or type-2) astrocytes in culture (Hatten and Mason, 1986). There is also some evidence that type-1 astrocytes proliferate in response to injury; type-2 astrocytes in the region of the wound eventually die off (Miller *et al.*, 1986). Also, processes of type-2 astrocytes (not type-1) contribute to the structure of nodes of Ranvier. This has been determined by immunofluorescence staining with a Mab (NSP-4, which recognizes several glycoproteins) in cultured optic nerve cells of perinatal rat, in which perinodal processes of astrocytes are NSP-4 positive (ffrench-Constant and Raff, 1986). These type-2 astrocytes are also positive for Mab A2B5 and arise from the same progenitor as oligodendrocytes. Type-1 astrocytes are A2B5 negative (see ffrench-Constant and Raff, 1986). However, type-1 astrocytes (protoplasmic) proliferate and accumulate glial filaments (IFs) in reaction to injury (Duffy, 1983).

Reactive astrocytes develop not only in the proximity of the lesion but also further away from the lesion. Although GFAP decorates all the reactive astrocytes, only those astrocytes that are at the edge of the lesion begin to show expression of vimentin (Schiffer *et al.*, 1986). Therefore, it is apparent that the development of astrocytes is dependent on outside factors. When neurons are added to a culture of pure astrocytes, they undergo division and differentiate into various shapes; in the absence of neurons, the astrocytes proliferate and have flat (protoplasmic) morphology (Hatten and Mason, 1986). The expression of vimentin by astrocytes has been regarded as a sign of immaturity, and a transition from vimentin to GFAP is thought to be related to their state of differentiation (Bignami and Dahl, 1986). Vimentin, however, is demonstrable particularly in radially oriented glia in the adult animal (Müller cells, tanycytes, and Bergmann fibers).

Astrocytes play a role in uptake of the neurotransmitter GABA, but there also appear to be astrocytes that are GABA-uptake negative; the two subtypes of astrocytes differentiate from different precursors. GABA-uptake positive cells are Mab A2B5 positive, whereas GABA-uptake negative cells are A2B5 negative (Levi *et al.*, 1986).

Some of the antigens associated with astrocytes or their subpopulations are listed in Table I.

4. ORIGIN OF ASTROCYTES

Astrocytes (and other glial cell types, including ependymal cells) differentiate from a ventricular zone of the neural tube from which neurons also arise (Duffy, 1983; Goldman *et al.*, 1986). Astrocytes, in contrast to neurons, continue to divide and proliferate even after birth; they also retain the capacity to undergo further division when stimulated, as in gliosis. In the adult, proliferation of neuroglial cells may take place throughout the parenchyma of the CNS, but there appear to be exclusive foci in the brain for the formation of glial cells. The subependymal plate is thought to be a potential source of glia for surrounding regions of the brain (Bologa, 1985).

Reyners *et al.* (1986) have attempted a characterization of yet another type of glial cell—the beta astrocyte—in adult rat cerebral cortex, using [^{3}H]thymidine incorporation. A high proportion of the beta astrocytes were observed to be labeled (92% of cells labeled). The identification of astrocytes was carried out by electron and light microscopy. Such cells represent approximately 10% of the total glia in the adult rat brain and are thought to represent a constant population of a glial progenitor cell in the grey matter of the adult rat. Some of these beta astrocytes transform into astrocytes but most of them differentiate into oligodendrocytes. In this respect, the beta astrocytes are comparable to the bipotential glial progenitor in the optic nerve of the newborn rat, where the glial precursor may develop into an oligodendrocyte or an astrocyte (see below).

At least two progenitors of astrocytes have been recognized through application of Mabs (particularly Mab A2B5) to cultured cells from rat optic nerve (Raff *et al.*, 1983) and rat cerebellum (Levi *et al.*, 1986). One type of precursor cell gives rise to astrocytes (type-1 or protoplasmic) that are Mab A2B5 negative, whereas the second precursor is bipotential; it gives rise to oligodendrocytes or type-2 (fibrous) astrocytes that are A2B5 positive (ffrench-Constant and Raff, 1986). Using another Mab to GD3 ganglioside and cultures of neonatal rat forebrain, Goldman *et al.* (1986) found that GD3 positive cells can give rise to both A2B5 negative (type-1) and A2B5 positive (type-2) astrocytes and thus cells bearing GD3 antigen represent an earlier stage in the development of astrocytes than that represented by Mab A2B5 binding. GFAP is first detected in cells that

Table I
Some of the Antigens Associated with Astrocytes

	Antibody	Molecule	Distribution	Reference
1.	Anti-GFAP	50K protein	Intermediate filaments (10-nm IFs) and soluble fraction; persists in mature astrocytes, except in some malignant tumors	Bignami *et al.* (1972); Schachner *et al.* (1977); Duffy (1983)
2.	Antivimentin	55K protein	IFs and soluble fraction; abundant in immature astrocytes; may persist in mature astrocytes particularly in radially oriented glia (Müller cells, tanycytes, and Bergmann fibers)	Bignami and Dahl (1985); Shaw and Weber (1983); Bovolenta *et al.* (1984)
3.	Mab J1-31	30K protein	Intracellular, soluble, and possibly in association with IFs; persists in mature astrocytes; first appears after birth but later than GFAP in cerebellum; also expressed in Müller glia and ependymal cells	Malhotra *et al.* (1984a); Schröder and Malhotra (1987); Singh *et al.* (1986); Predy *et al.* (1987)
4.	Mab A_2B_5	Ganglioside	Cell surface; neurones, immature astrocytes, and oligodendrocytes; in rat optic nerve it recognizes a subpopulation of astrocytes (type 2)	Eisenbarth *et al.* (1979); Schachner (1982); ffrench-Constant and Raff (1986); Kim *et al.* (1986)
5.	Mab J1-NSP_4	Glycoproteins	Cell surface; differential staining of astrocytes type 2	ffrench-Constant and Raff (1986)
6.	Mab 8C7	Glycolipids and glycoprotein	Radial glia	Misson *et al.* (1986); Edwards *et al.* (1986)
7.	Mab.M1	?	Astrocytes in White matter; Purkinje perikaryon	Lagenaur *et al.* (1980)
8.	Mab 54-3C4	?	Cell surface; subpopulation of astrocytes	Wujek and Akeson (1986)
9.	Mab M1/N1 308	?	Cell surface; subpopulation of astrocytes	Dickson *et al.* (1983)
10.	Mab 5E10	High molecular weight, resembling actin-binding protein filamin	All neuroepithelial cells in developing chick; after embryonic day 10, restricted to Müller cells in retina and radial glia in optic tectum	Lemmon (1986)
11.	Mab Ran-2	Protease sensitive	Rat cell surface, astrocytes, ependymal cells, and leptomeningeal cells	Bartlett *et al.* (1981)
12.	Mab G.3.5	Protease insensitive	Astrocytes; specificity undetermined	Malhotra *et al.* (1984b)

also express GD3 ganglioside; the latter is eventually lost so that the mature astrocytes do not express this ganglioside.GD3 positive cells also give rise to oligodendrocytes that can be recognized by their characteristic galactocerebroside expression (Goldman *et al.*, 1986).

With respect to the functioning of astrocytes, type-1 astrocytes (GFAP positive, A2B5 negative, RAN-2 positive) have been reported to produce a soluble factor(s) that stimulates the oligodendrocyte type-2 astrocyte (O-2A) progenitor cells (A2B5 positive, GFAP negative) to divide in cultures (Noble and Murray, 1984; Raff *et al.*, 1985). In the absence of purified type-1 astrocytes, the O-2A progenitor cells differentiated within 3 days without undergoing division. However, with the addition of purified type-1 astrocytes to the culture, the O-2A progenitors (A2B5 positive) manifested incorporation of [^{3}H]thymidine and their number increased four- to fivefold over a 2-day period. These observations raise the possibility that the type-1 astrocytes may play a role in the development of CNS.

5. GENERAL REMARKS ON RESULTS FROM IMMUNOCYTOCHEMICAL STUDIES

A great deal of information on the distribution of various antigens in CNS has been obtained by use of immunocytochemical techniques, and therefore some of the difficulties encountered in such an approach are now mentioned. A problem may arise in the results obtained by using polyclonal antisera in immunocytochemistry. For example, there are inconsistent reports on the presence of GFAP in the normal Müller glia of the rat retina as well as human retina. Several studies report that there is no expression of GFAP in the normal Müller glia (Bartlett *et al.*, 1981); but it becomes detectable after injury to the eye (Bignami and Dahl, 1979). However, the presence of GFAP in normal Müller glia of the rat has been noted in a recent report (Björklund *et al.*, 1985), which has been confirmed in my laboratory (Predy *et al.*, 1987). On similar lines, the distribution of three neurofilament (NF) subunits has been visualized differently in different parts of neurons in mature CNS and PNS. Using Mabs specific for each bovine NF subunit, Trojanowski *et al.* (1986) report that all three subunits are expressed by most CNS and PNS neurons including their processes. Those neurons that are devoid of immunodetectable NF antigens are generally small and have low levels of NF proteins. Such a low level of antigens could result in an apparent absence of reactivity owing to NF proteins in some neurons. Some of these differences in results may arise because of differences in fixation protocols, in antisera, and in immunodetective reagents; some others may reflect differences in the chemical nature of antigens. A protein similar to GFAP, the well-known marker for astrocytes, is now known to be present in non-CNS tissues such as Schwann cells, lens epithelium, and Kupffer cells in liver (Eng *et al.*, 1985), and fixation in formaldehyde and embedding in paraffin does not preserve GFAP-like antigen in Schwann cells, whereas similar fixation reveals substantial amount of GFAP reactivity in CNS (Yen and Fields, 1985). Another reason for inconsistency in results from different laboratories may be the differences in culture condition. The expression of GFAP in astrocytes is subject to regulation by hormones and growth factors, and it is commonly believed that rat Müller cells in cultures of retina are GFAP negative (Bartlett *et al.*, 1981). As mentioned above, recent studies have revealed immunostaining for GFAP in normal Müller glia of the rat (Björklund *et al.*, 1985; Predy *et al.*, 1987).

The value of correlative study on in vitro and in vivo systems is illustrated in the critical analysis of a Mab (7-1.1) raised against a bovine brain microtubule-associated protein (MAP1) (Asai *et al.*, 1985). Mab 7-1.1 stains reconstituted brain microtubules in vitro, and it was determined that this Mab recognizes MAP1 on the microtubules. In contrast, in cultured cells (fixed and permeabilized embryonic fibroblasts and brain presumptive astrocytes) Mab 7-1.1 did not stain microtubules (detectable by an antitubulin antibody) but stress fibers (actin). (In immunoblot experiments, Mab 7-1.1 did not react with actin.) The authors concluded that, while the MAP1 reacts with microtubules in vitro, it could represent in vivo a family of high molecular weight proteins, the precise characterization of which remains to be achieved. In support of such a conclusion, Asai *et al.* cite several instances of distinctive immunostaining by antibodies to MAP1 such as nuclei, dendrites, and uniform staining of brain tissue.

It should be mentioned that there can be problems with the localization of antigens by the PAP technique because the immunoperoxidase reaction product may diffuse over a short distance, thus giving a nonspecific localization (Landis, 1985). This is exemplified by the distribution of synapsin, a nerve terminal phosphoprotein, which by immunoferritin is found to be associated with synaptic vesicles. By immunoperoxidase, the reaction is also found on membranes other than the synaptic vesicles (de Camilli *et al.*, 1983). [Synapsin I is a substrate for cAMP-dependent and Ca/calmodulin-dependent protein kinases and is highly concentrated in a great majority of the synapses and may function in the release of the neurotransmitter since this process is regulated by Ca and by cAMP (de Camilli *et al.*, 1983).] Also, soluble antigens may precipitate onto other cellular structures during fixation and thus produce artificial labeling with immunoreagents.

Besides the above limitations on results from immunocytochemical studies, many molecules in the nervous system may exist at such low concentrations that the current immunocytochemical techniques are not sensitive enough to reveal their distributions. Such molecules are exemplified by nerve growth factor and Ca^{2+}-dependent K^+ channels (Valentino *et al.*, 1985).

6. GENERAL REMARKS ON THE CYTOSKELETON IN ASTROCYTES

Recent studies on the cytoskeleton have advanced our knowledge of the molecular constituents that make up the cytoskeletal framework in eukaryotic cells. In astrocytes, most of the advances have been associated with intermediate filaments (IFs) (see below). Very little seems to be known about 5-nm microfilaments made up of actin and actin-binding proteins (Lemmon, 1986), or about microtubules, made up of tubulin, and microtubule-associated proteins (MAPs). Microtubules are thought to be rare in mature astrocytes (Duffy, 1983). However, astrocytes in vitro from cerebellum cultures of neonatal mice contain a heterogeneous population of proteins that cross-react with antisera raised against the two brain MAPS (I and II). Also, marked quantitative changes were observed during the differentiation of astrocytes. These changes are thought to be related to a buildup of a denser microtubule network in the differentiated cells than in the dividing immature astroblasts (Couchie *et al.*, 1985). Based on existing knowledge of the differential expression of MAPs during development and distinctive distribution of different MAPs in different regions of neurons (Huber *et al.*, 1986), several types of MAPs may be found in astrocytes that manifest differential expression during maturation and distinctive distribution associated with different parts of the CNS. The evidence that one category of MAP interacts with stress fibers in cultured fibroblasts has been presented by Asai *et al.* (1985) (see p. 286). Differences in distribution of a MAP in astrocytes in different regions of the CNS have been reported (Papasozomenos and Binder, 1986). A great deal of information relevant to the cytoskeleton and its interactions with other cell components is needed to comprehend some of the functions that astrocytes perform during their life and are associated with their cytoskeletal framework. GFAP (MW 50K) is the principal constituent of the 10-nm IFs of astrocytes and one of the core IF proteins (Schachner *et al.*, 1977; Steinert and Steven, 1985). The other IF types are represented by keratins (40–70K, epithelial cells), desmin (52K, muscle cells), vimentin (55K, fibroblasts, see below), neuronal filaments (NF 68K, 150K, and 200K), and the new lamin-type forming nuclear lamina (Aebi *et al.*, 1986; see below). All these types of IFs share a similar structure represented by a well-conserved α-helical secondary structure in the central domain, which is flanked by end domains. The latter vary in size and chemical nature in different types of IFs and provide them their distinctive nature (Ip *et al.*, 1985; Steinert and Parry, 1985; Weber and Geisler, 1985). Based on in vitro assembly of vimentin, IFs are made up of 2 nm × 3 nm × 48 nm rods (protofilaments) and there are eight protofilaments in each IF (10–12-nm diameter referred to here as 10-nm IFs). The protofilaments are arranged in such a way that there is an ~22-nm axial stagger between neighboring protofilaments, which corresponds to the 20–22-nm axial repeat seen in IFs (Ip *et al.*, 1985). Despite the structural similarities between IFs (Pruss *et al.*, 1981), there is no cross-hybridization at high stringency between various IF protein-specific cDNA probes. Thus, significant divergence has occurred among members of the multigene family encoding these IF polypeptides. However, within a given type of IF, some cross-hybridization is detectable (Cowan *et al.*, 1985). In respect to GFAP, a mouse cDNA probe has been found to cross-hybridize weakly at high stringency with DNA from rat, human, and chicken, and there appears to be a single gene encoding GFAP (Cowan *et al.*, 1985).

In addition to the five well-recognized core IF proteins represented by keratins, desmin, vimentin, NF, and GFAP, a new category of IFs appears to have been established. This is represented by lamins A, B, and C (MW 60–70K), which form the filamentous meshwork (nuclear lamina) lining the surface of the inner nuclear membrane in somatic cells (Aebi *et al.*, 1986). A similarity between lamins and IF proteins has been revealed by protein sequence derived from cDNA clones of human lamins A and C, which shows a region of ~350 amino acids (McKeon *et al.*, 1986). This region resembles the central rod domain of IF proteins, which shows a coiled-coil α helical configuration (Steinert and Parry, 1985). Aebi and others have demonstrated that lamins A and C isolated from rat liver can be reconstituted into typical 10-nm IFs. Lamin B differs from the other two lamins but has properties of IF proteins. The authors speculate that the lamins are assembled into heteropolymers containing all three lamins, in making up the nuclear lamina. It should not be long before the presence of lamin-type IFs becomes widely known throughout eukaryotes, since antisera and Mabs are available for rapid screening. (A nuclear lamina interacts with both chromatin and the inner nuclear membrane.)

7. INTERMEDIATE FILAMENTS (IFs) OF ASTROCYTES

Mature astrocytes contain their characteristic 10-nm IFs, which are made up of GFAP, as their principal core protein. As mentioned elsewhere in this article, another IF core protein (vimentin) may also be present simultaneously with GFAP as in Bergmann fibers in the cerebellum (Bovolenta *et al.*, 1984), Müller cells in the retina, and tanycytes lining the ventricles of the brain (Shaw *et al.*, 1981). Vimentin is particularly predominant in immature astrocytes and a vimentin to GFAP transition may occur in glial differentiation (Bignami and Dahl, 1985). The first expression of GFAP-positive cells is usually detectable around the time of birth in rat (Raju *et al.*, 1981). When vimentin persists in adult astrocytes, this may reflect a state of immaturity of these particular cells (Schachner *et al.*, 1984). However, the presence of heteropolymeric IFs made up of GFAP and vimentin has been demonstrated in glioma cells (Wang, 1985).

Apart from the general cytoskeletal function of these IF proteins, their precise role in the biology of astrocytes is not known, and therefore the relevance of relative contributions of GFAP and vimentin to the cytoskeleton is difficult to assess. Also the presence of GFAP and vimentin may not necessarily be associated with the assembly of IFs, because there are relatively few filaments in immature astrocytes although vimentin in particular can be demonstrated by immunofluorescence; the IFs become abundant as this glia matures (Vaughn and Peters, 1967).

Both these proteins exist in soluble form as well as polymeric IFs and IFs may assemble and disassemble in the living cell (Rueger *et al.*, 1979). A distinct pool of tetramers of vimentin has been shown to exist in cultured cells which could exchange with the polymeric IF form (Soellner *et al.*, 1985). Notwithstanding the relative contributions made by GFAP and vimentin to IFs, their assembly into IFs and bundling into macromolecular assemblies in vivo presumably requires involvement of other proteins referred to as intermediate filament associated proteins (IFAPs) (Osborn, 1985; Wang, 1985). These proteins are not integrated into the IF themselves but bind to the filaments; they may function to link IFs and serve as spacers between IFs or as branching points for individual filaments (Wang, 1985). They may also serve in interactions between IFs and microtubules (Lawson, 1984), as a microtubule associated protein (MAP) does (Leterrier *et al.*, 1982). They could also serve in interactions between membranes and IFs (Georgatos and Marchesi, 1985; Wang, 1985). Just as there are several categories of MAPs, there could be several IFAPs but "the field of IF-associated proteins is as yet relatively unexplored" (Osborn, 1985). While epinemin (P50, 50K) associates specifically with vimentin type IFs (Lawson, 1984; Wang, 1985), filaggrins (30–48K), which are species specific, have been reported to interact in vitro with IFs of different types of cells (e.g., keratin, vimentin, and desmin) to form large fibers or macrofibrils (Steinert *et al.*, 1981; see Table II). There are discrete sites expressed asymmetrically on the IF, where one (occasionally two or three) antiepinemin molecule binds as observed by electron microscopy of deep-etched preparations of fibroblast cytoskeletons (Lawson, 1984).

Table II
Some of the Proteins Associated with IFs Relevant to CNS

300K	IFAP	Lieska *et al.* (1985)
66K-α-Internexin	IFAP	Pachter and Liem (1985)
70K-β-Internexin	IFAP	Napolitano *et al.* (1985)
50K-(P50) (epinemin)	IFAP	Wang (1985); Lawson (1983)
30K (J1-31)	IFAP?	Singh *et al.* (1986)
30–48K (filaggrins)	IFAP	Steinert *et al.* (1981)

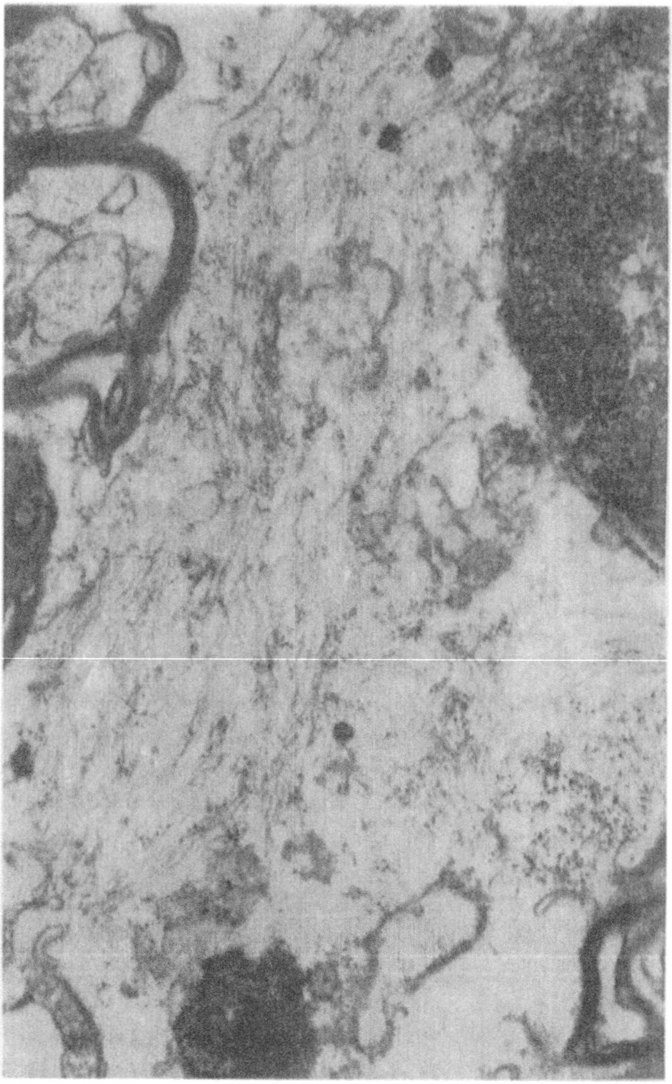

FIGURE 3. Cerebellum from adult rat, fixed in 4% paraformaldehyde, and sectioned in an Oxford vibratome. Sections 25–50 μm thick were incubated with Mab J1-31 followed by immunogold (5-nm particles) labeling. The sections were subsequently fixed in glutaraldehyde followed by Os04 and processed for electron microscopy. Gold particles are associated with IFs. Note the IFs are not uniformly labeled with gold particles.

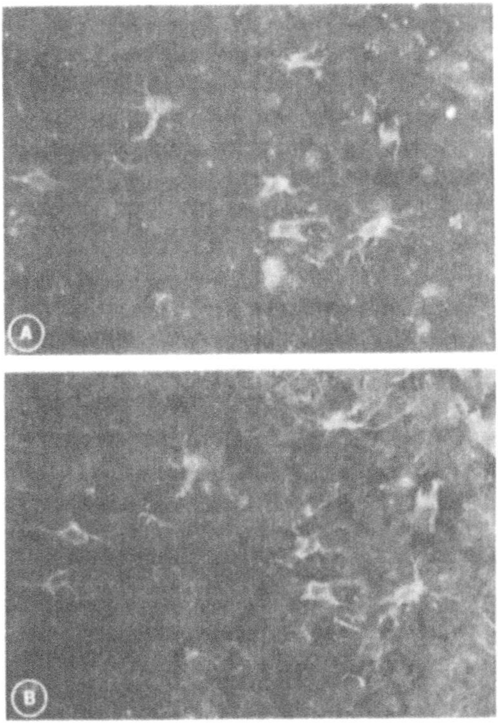

FIGURE 4. Human brain from MS patient (autopsy sample) fixed in 4% paraformaldehyde in phosphate buffer, cryoprotected in sucrose; cryostat sections were double-labeled with Mab J1-31 (A) and anti-GFAP antiserum (B) to show a general overlap in staining owing to Mab J1-31 and GFAP. Several GFAP-positive processes do not show J1-31 antigen, which could be due to low fluorescence signal. It is also conceivable that J1-31 antigen is present in only a subpopulation of astrocytic processes.

Another protein (J1-31) tentatively described as an IFAP has been discovered in my laboratory. That J1-31 antigen may be an IFAP is based on immunoelectron microscopy, using PAP and colloidal gold labeling (Figure 3), and the staining appears in association with IFs (Malhotra *et al.*, 1986; Singh *et al.*, 1986). A mouse Mab raised by using crude homogenate from autopsy samples of multiple sclerosis (MS) patients (Malhotra *et al.*, 1984b), J1-31 has been found to stain GFAP-positive cells in cerebellum (Malhotra *et al.*, 1984a; Singh *et al.*, 1986), cerebrum (Figure 4A,B), spinal cord, hypothalamus, retina, and pineal (Schröder and Malhotra, 1986).

By double-label immunofluorescence microscopy on cryostat sections, the staining resulting from J1-31 antigen mostly overlaps with that resulting from GFAP (Figure 4). Sometimes the impression is given that only a subpopulation of GFAP-positive cell processes are stained, but this has not yet been determined. It is more likely that the overlap in staining owing to J1-31 antigen and GFAP is not complete, or there is such a low concentration of J1-31 antigen, particularly in the finer processes of GFAP-positive cells, that it is not revealed by immunofluorescence microscopy. However, J1-31 antigen appears to be restricted to the CNS, as no immunostaining has been detected outside the CNS.

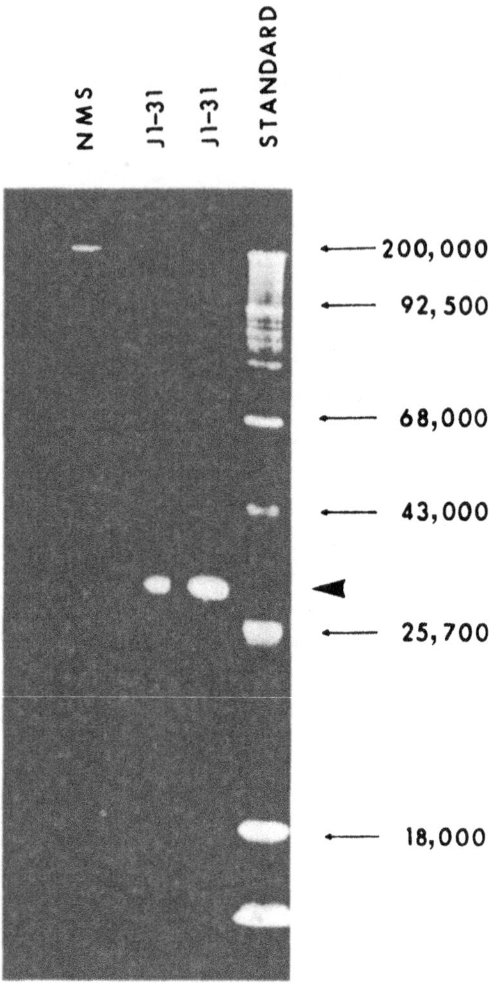

FIGURE 5. Autoradiograph of the immunoprecipitation of J1-31 antigen. ^{125}I-labeled human brain homogenate was immunoprecipitated with various antibodies. The precipitates were separated on 10% SDS–PAGE under reducing conditions and developed after 2 weeks of exposure. ^{14}C-labeled protein standards were used for molecular weight determinations (Singh *et al.*, 1986). Unlabeled arrow indicates a band at ~30K reactive with Mab J1-31. NMS, normal mouse serum.

By immunoprecipitation followed by SDS–gel electrophoresis (reducing conditions) and radioautography, Mab J1-31 reacts with a single band at approximately MW 30K and is thus distinct from GFAP (MW 50K) (Singh *et al.*, 1986; Figure 5). J1-31 antigen is readily extracted in aqueous buffers and is also present in detergent extracts of human CNS homogenate; thus, it appears that the antigen exists in a soluble form in the cytosol as well as in a form bound to the cytoskeleton and/or to a membrane. In view of the soluble nature of the antigen, it is conceivable that during fixation the antigen could precipitate onto the cytoskeleton and thus provide artificial staining. Thus, the characterization of J1-31 antigen as an IFAP remains to be established.

8. COEXPRESSION OF VIMENTIN AND GFAP IN ASTROCYTES

Vimentin, the well-known core protein of IFs first described from fibroblasts and cells of mesenchymal nature, has come to be known as the nonspecific IF protein (Bignami *et al.*, 1982), because it is also expressed in a variety of nonmesenchymal cells, such as epithelial cells (Lazarides, 1980), neurons (Drager, 1983; Lee and Page, 1984; Bignami and Dahl, 1985), and glial cells in CNS (Bignami and Dahl, 1985) and PNS (Yen and Fields, 1985). Both in the neurons and astrocytes, the expression of vimentin may be transitory (Bignami and Dahl, 1985; Lee and Page, 1984) or it may persist along with the characteristic IF, that is, NF or GFAP, in the adult, as in axonless horizontal cells in the adult mouse retina (Drager, 1983). Also, in astrocytes there may be a transition from vimentin to GFAP during glial differentiation (Bignami and Dahl, 1985) or vimentin and GFAP may coexist in mature astrocytes (Fedoroff *et al.*, 1983). Vimentin and GFAP also coexist in radially oriented glia of the adults CNS, such as Bergmann fibers in the cerebellum (Björklund *et al.*, 1985), tanycytes of the ependyma (Shaw *et al.*, 1981), and Müller cells of the retina (Björklund and Dahl, 1985). In the pineal gland, both vimentin- and GFAP-containing cells are present, although vimentin seems to be more abundant (Schachner *et al.*, 1984; Schröder and Malhotra, 1986). Thus, it is apparent that astrocytes (and other cell types as well) may express GFAP as well as vimentin in the adult. The ciliated ependymal cells in adult do not express GFAP but they do express vimentin. In the human, these cells have been shown to express GFAP transiently during development (Roessmann *et al.*, 1980). It is difficult to ascribe a significance to these differential expressions of GFAP and/or vimentin in various situations reported in astrocytes, because it is not known what precise functional roles these proteins serve during development or in the adult. Now that cDNA clones specific for GFAP (Cowan *et al.*, 1985) and vimentin (Bloemendal *et al.*, 1985) are available, the potential exists that the gene expression can be studied in defined cell types.

Irrespective of the role of IF proteins in astrocytes, the presence of homopolymers and/or heteropolymers made up of GFAP and/or vimentin in situ needs to be examined. GFAP and vimentin have been shown to exist in the same IF filament in a glioma cell line (Quinlan and Franke, 1983; Wang, 1985). Vimentin also copolymerizes with desmin into heteropolymers (Quax *et al.*, 1985). Therefore, the existence of heteropolymers of GFAP and vimentin in astrocytes in situ could be a general phenomenon with varying contributions from the two proteins. The extent of contribution of each protein into heteropolymers could be variable and dependent on factors that determine the extent to which GFAP and vimentin may be present in the same cell. In respect to the presence of heteropolymers of IFs, it is recalled that in epithelial cells, the assembly of keratins into IFs necessarily requires acidic and basic keratins; for example, the IFs are heteropolymers. The three NF proteins may also associate into heteropolymers (Liem *et al.*, 1978).

It should be stressed that the expression of one or more core IF protein(s) does not necessarily lead to the formation of filaments. For example, Bignami *et al.* (1985) found evidence of transient expression of NF protein in the Purkinje cells of chick embryo by immunoreactivity; however, no 10-nm IFs could be detected. Therefore, the presence of other factors, such as IFAPs, may be required for the assembly of IFs in vivo, even though in vitro the core proteins are able to associate into filaments resembling IFs within cells.

9. FUNCTIONS ASSOCIATED WITH ASTROCYTES

An image of astrocytes as multifunctional cells is emerging as the nature, mechanism, and significance of their roles in the functioning of CNS are revealed (Figure 1). Astrocytes play a role in developing the nervous system, as they guide neurons to their destination. They play a role in the regulation of extracellular potassium levels in the brain (Bunge and Waksman, 1985; Goldstein and Betz, 1986; Newman, 1986). They contribute to the nodes of Ranvier (ffrench-Constant and Raff, 1986) and are associated with oligodendrocytes adjacent to the nodes by gap junctions (Waxman, 1986). They influence differentiation of endothelial cells (de Bault and Cancilla, 1980) and of oligodendrocytes by release of a growth factor(s) (Raff *et al.*, 1985). They are "coupled" to other astrocytes by gap junctions (Brightman and Reese, 1969). They react to injury of CNS (Duffy, 1983). Some of these functions are likely to be performed by a subpopulation(s) of astrocytes, but as pointed out on p. 282, no such subdivision has been attempted in this chapter. Some of the functions of astrocytes are depicted in a highly diagrammatic sketch in Figure 1, and some of the selected studies on these topics are elaborated below. [An elegant exposition of earlier studies on the role of astrocytes in particular and neuroglia in general is given by Kuffler and Nicholls (1976).] Astrocytes may also influence differentiation of endothelial cells. The enzyme γ-glutamyl transpeptidase (γ-GTP), which is involved in amino acid transport, appears to be maintained in high concentrations in endothelia of microvessels in mouse brain. An endothelial cell line, which is negative for γ-GTP, has been shown to be inducible for γ-GTP when cocultured with glial cells (glioma C6). In CNS, the endothelium is separated from the surrounding glial cells (astrocytes) by a basement membrane (Figure 1). It has been hypothesized that astrocytes provide the inductive influence to maintain high levels of γ-GTP activity, which may be necessary to transport amino acids across the blood–brain barrier.

Astrocytes are thought to participate in neurotransmitter uptake, metabolism, and release. For example, GABA could be taken up, metabolized to glutamate, which is transformed to glutamine; the latter could be released into the extracellular space from where it is taken up by neurons and metabolized to glutamate (Duffy, 1983).

Astrocytes have been implicated in the regulation of K^+ that transiently accumulates in the extracellular space during neuronal excitation. As a consequence of such an increase of K^+, the membrane potential of the glial cells becomes depolarized and returns to normal as the K^+ disappears by uptake and diffusion (Kuffler and Nicholls, 1976). Newman (1986) has demonstrated in dissociated astrocytes from optic nerve of salamander that the end feet of astrocytes have tenfold higher K^+ conductance than the conductance in other regions of these cells. Thus, K^+, taken up from the vicinity of neuronal processes, could be preferentially shunted out through the end feet; that is, K^+ could be siphoned off into the cerebrospinal fluid (CSF) and capillaries as the astrocytic end feet terminate on the pial surfaces of the brain and on capillaries. Goldstein and Betz (1986) have reported that the transport of K^+ from the brain into the endothelial cells is coupled with the transport of Na^+ out of the endothelial cells into the brain. This transport takes place through Na^+,K^+-ATPase pump, which is located asymmetrically on the plasma membrane of the endothelial cells, there being more Na^+,K^+-ATPase on the antiluminal plasma membrane (i.e., brain side) than on the luminal plasma membrane (capillary side). The K^+ accumulated in the endothelial cells is likely to move out into the capillary, although the specific nature of the K^+ channel seems not to be known. Newman (1986) has also demonstrated that the end feet of Müller cells in the retina of salamander are also specialized for K^+ conductance and thus could direct excess K^+ into the vitreous humor.

Gap junctions exist between astrocytes (Brightman and Reese, 1969) but their role, if any, in "cell coupling," whether ionic and/or metabolic, is not known. It should be of interest to determine how extensive these cell-to-cell contacts are among astrocytes in different regions of the CNS, which could be achieved by immunofluorescence microscopy using antisera to gap junction protein (27–28K). This protein has considerable homology in different tissues as determined by effective blocking of gap junctional conductance in different cell types by antibodies against the 27K liver gap junction protein (Hertzberg *et al.*, 1985). A considerable homology between human and rat liver 27K gap junction protein has been revealed by cDNA probes (Kumar and Gilula, 1986). It is recalled that gap junctions have recently been reported between perinodal astrocytic processes and the oligodendrocytes adjacent to the node of Ranvier (Waxman, 1986). Such cell contacts (gap junctions) are generally known to be associated with cell-to-cell coupling (Malhotra, 1983). The astrocytic processes could serve in providing nodes with an extracellular ion buffering system (Waxman, 1986).

In lesions of CNS produced experimentally (Miller *et al.*, 1986) or in disease states such as multiple sclerosis (MS), astrocytes proliferate. The factors that initiate this astrocytic response are not well understood. In this respect, it is of interest that cultured astrocytes, when stimulated by lipopolysaccharide or phytohemagglutinin, synthesize interleukin-3-like factor (30K; Frei *et al.*, 1985). This is relevant as astrocytes have recently been implicated in immune response (Wekerle *et al.*, 1986), such as presenting antigen to a T-lymphocyte cell line (Fontana *et al.*, 1984), production of interferon (Tedeschi *et al.*, 1986), and during development of inflammatory lesions of the brain (Bunge and Waksman, 1985). Several factors have been implicated in influencing the proliferation of astrocytes, such as myelin basic protein (MBP), or its fragments released by injury to myelin-forming cells, that is, oligodendrocytes in CNS (Bologa *et al.*, 1985), T-cell derived factors (Merrill *et al.*, 1984), and IL-1-like (18KD) peptide released by microglia (Giulian and Baker, 1985). Also, activated (but not resting) lymphocytes could cross the blood–brain barrier in the CNS and recognize their antigen on the surface of antigen-presenting cells, that is, astrocytes in this CNS (Wekerle *et al.*, 1986).

However, in respect to the viral etiology of MS it should be pointed out that evidence for such a pathological factor is not readily accepted (Hauser *et al.*, 1986; Karpas *et al.*, 1986) and the "factor" has thus far remained elusive. Recent attempts to detect the presence of specific RNA sequences for a suspect human T-lymphotropic virus (HTLV) by in situ hybridization in CNS tissues from MS patients have been negative (Hauser *et al.*, 1986). Also, attempts to find antibodies against HTLV in the sera or CSF of MS patients have not provided confirmatory evidence for involvement of HTLV in MS (Karpas *et al.*, 1986).

In respect to interactions between astrocytes and neurons, perhaps the best understood cell-to-cell adhesion molecules are represented by CAMs. The CAMs are cell surface glycoproteins that are present on the earliest embryonic cells as well as adult cells; they manifest developmental changes (Cunningham, 1986; Rutishauser, 1986). Relevant to astrocytes is a CAM termed Ng-CAM (neuron–glial cell adhesion molecule), which first appears on postmitotic neurons and mediates adhesion between neurons and between neurons and glia (Cunningham, 1986; Grumet *et al.*, 1984). The specific nature of molecules on glia to which Ng-CAM binds is not yet known.

Besides cell-to-cell adhesion molecules there are also cell-to-substrate adhesion molecules (SAMs; Cunningham, 1986), which are likely to serve essential roles in modulating cell movements during development and growth. SAMs are represented by laminin (Timpl *et al.*, 1983), fibronectin (Vartio and Vaheri, 1983), and cytotactin (Grumet *et al.*, 1985). There appears to be evidence that astrocytes produce fibronectin in vitro (Bunge and Waksman, 1985 vs. Liesi *et al.*, 1983) but lack receptors for its binding. Cytotactin has recently been isolated from 14-day embryonic chicken brains and immunofluorescence studies showed that this protein is found in extracellular spaces and in basement membranes of a variety of non-neuronal tissues. In the cerebellum, it was detected on the glial end feet, on Bergmann fibers, and in extracellular spaces. In cultures of brain, cytotactin was detectable on the surface of glial cells. In view of its association with Bergmann fibers, it has been suggested that cytotactin may play a part in providing a pathway created for migrating neurons (Grumet *et al.*, 1985). While cytotactin is of recent discovery, laminin has been known for several years (Timpl *et al.*, 1983) and is a matrix glycoprotein present in basement membranes. Using primary cultures from various regions of CNS (cerebellum, neostriatum, and substantia nigra) of rats (20–21-day fetal, newborn, and 5-day old), Liesi *et al.* (1983) have demonstrated that all GFAP-positive cells begin to show laminin in the cytoplasm detectable by double-label immunofluorescence microscopy. As the cultures aged, laminin began to disappear from GFAP-positive cells and there was no intracellular laminin in cells of age corresponding to mature star-shaped astrocytes. Such cultures showed laminin as a pericellular matrix on top of an astrocyte monolayer and between astrocytes and neuronlike cells present in culture. (No evidence was discerned that neurons produce laminin.) In the developing cerebellum, laminin has been seen in association with Bergmann fibers during the period neurons (granule cells) migrate from the external layer. No such staining is evident when the migration is complete (Bunge and Waksman, 1985). Thus, there is strong evidence that early astrocytes function in producing an extracellular matrix glycoprotein, which could serve in developing CNS in cell maturation, and migration along neuronal pathways. In support of this, retinal ganglion cells of embryonic chick (day 6) extend neurites on laminin substrate in culture, whereas those from embryonic chick (day 11) do not (Cohen *et al.*, 1986). Thus, neurites appear to undergo a change in their substrate requirement as the neurons mature. Furthermore, Liesi (1985) has found that laminin is continuously expressed by normal astrocytes of those adult CNS systems that support axonal growth and regeneration (e.g., goldfish retina) but absent from the CNS that fail to support growth and regeneration (e.g., optic nerve and optic tectum of the rat).

Laminin is known to have several domains with distinct activities, and it binds to cells as well as basement membranes, perhaps through a protein receptor (Roberts *et al.*, 1985). Laminin also binds to sulfated glycolipids, which could be involved in cell-to-cell interaction (Roberts *et al.*, 1985). It appears very early in development as an extracellular matrix protein, and it is demonstrable in places where basement membranes are subsequently laid down. It could serve a role in morphogenesis (Timpl *et al.*, 1983).

Yet another extracellular matrix constituent, relevant to CNS, is a brain-specific hyaluronate-binding protein (hyaluronectin 59K). It has been detected in the white matter of spinal cord and brain of the human and calf by indirect immunofluorescence microscopy using Mabs (Bignami and Dahl, 1986). With polyclonal antibodies to hyaluronectin, positive staining was seen more widely in human and calf brain, in embryonic and adult rat brain, as well as in mesenchyma. These results have been interpreted to imply that hyaluronectin exists in heterogeneous forms and astrocytes may produce some form of this extracellular matrix protein.

Astrocytes also have the ability to produce glycosaminoglycans and growth factors and the study of interactions between neurons and glia is an important aspect of neurobiology under active investigation (Hatten and Mason, 1986). Soluble factors released by glial cells (Lindner *et al.*, 1986) as well as cell contacts between neurons and glial cells (Hatten and Mason, 1986) appear to be involved in the migration of granule cells in cerebellum. In respect to migration of neurons, perhaps the best understood system is represented by the grasshopper, in which the pioneer neurons (discernible by Mabs to peroxidase) contact certain non-neuronal (epithelial) cells in their path, and an adhesive gradient of surface molecules along the surface of epithelial cells guides the pioneer neurons (Berlot and Goodman, 1984).

Janzer and Raff (1987) have reported that astrocytes can induce endothelial cells to form tight seals to produce nonleaky vessels reminiscent of blood–brain barrier in the CNS. The properties of the "barrier" were determined by using a fluorescent dye Evans blue, which could permeate through junctions formed by endothelial cells in the presence of purified astrocytes injected into the rat anterior eye chamber. Control experiments were carried out using purified meningeal cells and fibroblasts. The latter did not induce endothelial cells to form nonleaky vessels. These results provide a strong evidence that astrocytes that surround blood vessels in the CNS could be responsible for inducing blood–brain barrier, which is provided by tight junctions between endothelials in the CNS of higher vertebrates. The molecule(s) involved in this induction phenomenon remains to be determined (Janzer and Raff, 1987). It is also of interest that brain endothelial cells and blood cells have been reported to express a specific 74K protein on the cell surface, which appears around embryonic day 13 in chick and may play a role in the organization of the blood–brain barrier (Risau *et al.*, 1986).

10. FUTURE PROSPECTS

It is expected that further advances are likely to take place in the elucidation of additional proteins associated with the cytoskeletal framework in astrocytes such as spectrin, and this protein has recently been shown to be present as two subtypes in brain: one subtype, referred to as brain spectrin (240/235), is associated with the plasma membrane and cytoplasm of axons, while the other subtype, brain spectrin 240/235E), is associated with the plasma membrane and cytoplasmic organelles of perikarya and dendrites. The second subtype has also been detected in association with the plasma membrane and cytoplasmic organelles of the glial cell body (Zagon *et al.*, 1986). The second subtype brain spectrin (240/235E) reacts specifically with an antibody directed against erythrocyte spectrin, whereas the other brain subtype (240/235) is not detected with antibodies against erythrocyte spectrin (Riederer *et al.*, 1986). Further advances are also likely to take place in our understanding of the distribution of IFs and IFAPs, actin and actin-binding proteins, microtubules and MAPs, their developmental expression, and factors that influence assembly and disassembly of cytoskeletal structures. Such a progress should also facilitate further study of interactions between cytoskeletal components and other intracellular membranous organelles and between the cytoskeleton and plasma membrane. To understand the functioning of astrocytes, it will be important to reveal the nature and significance of interactions between astrocytes themselves, for example, through gap junctions (Brightman and Reese, 1969), between astrocytes and other cells in CNS, for example, neurons, oligodendrocytes, microglia, and endothelial cells, and between astrocytes and their matrix during maturation and their subsequent life. Such studies should facilitate an understanding of the factors such as glial-promoting factors that influence the growth of astrocytes (Giulian and Young, 1986; Giulian *et al.* 1986) and thus help understand the biology of astrocytes.

ACKNOWLEDGMENTS. I am most appreciative of my past and present colleagues, Ms. H. Good, Mr. F. Wong, Dr. R. Singh, Dr. B. Singh, Mr. R. Predy, Mr. R. Bhatnagar, and Ms. G. Sim, who have made contributions to my research included in this chapter. Mr. Predy has provided invaluable help in the completion of this chapter. Dr. K. G. Warren and T. K. Shnitka have kindly provided human autopsy samples of CNS, which have been used for raising and screening monoclonal antibodies in my research.

Mrs. K. Baert and Ms. J. Scheinas have taken great pains in typing the manuscript.

My research has been supported by grants awarded by the NSERC of Canada and the M.S.I. Foundation.

11. REFERENCES

Aebi, U., Cohn, J., Buhl, L., and Gerace, L., 1986, The nuclear lamina is a meshwork of intermediate-type filaments. *Nature* **323**:560–564.

Asai, D. J., Thompson, W. C., Wilson, L., Dresden, C. F., Schulman, H., and Purich, D. L., 1985, Microtubule-associated proteins (MAPs): A monoclonal antibody to MAP 1 decorates microtubules in vitro but stains stress fibers and not microtubules *in vivo*. *Proc. Natl. Acad. Sci. U.S.A.* **82**:1434–1438.

Asch, A. S., Leung, L. L. K., Shapiro, J., and Nachman, R. L., 1986, Human brain glial cells synthesize thrombospondin. *Proc. Natl. Acad. Sci. U.S.A.* **83**:2904–2908.

Bartlett, P. F., Noble, M. D., Pruss, R. M., Raff, M. C., Rattray, S., and Williams, C. A., 1981, Rat neural antigen-2 (ran-2): A cell surface antigen on astrocytes, ependymal cells, Müller cells and lepto-meninges defined by a monoclonal antibody. *Brain Res.* **204**:339–351.

Berlot, J., and Goodman, C. S., 1984, Guidance of peripheral pioneer neurons in the grasshopper: Adhesive hierarchy of epithelial and neuronal surfaces. *Science* **223**:493–496.

Bignami, A., and Dahl, D., 1973, Differentiation of astrocytes in the cerebellar cortex and the pyramidal tracts of the newborn rat. *Brain Res.* **49**:393–402.

Bignami, A., and Dahl, D., 1979, The radial glia of Müller in the rat retina and their response to injury. An immunofluorescence study with antibodies to the glial fibrillary acidic (GFA) protein. *Exp. Eye Res.* **28**:63–69.

Bignami, A., and Dahl, D., 1985, Glial filaments in brain differentiation and the transient expression of neurofilaments in Purkinje cell development. *Ann. N.Y. Acad. Sci.* **455**:563–574.

Bignami, A., and Dahl, D., 1986, Brain specific hyaluronate-binding protein: An immunohistological study with monoclonal antibodies of human and bovine central nervous system. *Proc. Natl. Acad. Sci. U.S.A.* **83**:3518–3522.

Bignami, A., Eng, L. F., Dahl, D., and Uyeda, C. T., 1972, Localization of the glial fibrillary acidic protein in astrocytes by immunofluorescence. *Brain Res.* **43**:429–435.

Bignami, A., Raju, T., and Dahl, D., 1982, Localization of vimentin, the nonspecific intermediate filament protein, in embryonal glia and in early differentiating neurons. *Dev. Biol.* **91**:286–295.

Binder, L. I., Frankfurter, A., Kim, H., Caceres, A., Payne, M. R., and Rebhun, L. I., 1984, Heterogeneity of microtubule-associated protein 2 during rat brain development. *Proc. Natl. Acad. Sci. U.S.A.* **81**:5613–5617.

Björklund, H., and Dahl, D., 1985, Glial fibrillary acidic protein (GFAP)-like immunoreactivity in the rodent eye. Comparison between peripheral glia of the anterior uvea and central glia of the retina. *J. Neuroimmunol.* **8**:331–345.

Björklund, H., Bignami, A., and Dahl, D., 1985, Immunohistochemical demonstration of glial fibrillary acidic protein in normal rat Müller glia and retinal astrocytes. *Neurosci. Lett.* **54**:363–368.

Bloemendal, H., Quax, W., Quax-Jeuken, Y., Van Den Heuvel, R., Egberts, W. V., and Van Den Broek, L., 1985, Vimentin and desmin cDNA clones: Structural aspects of corresponding proteins and genes. *Ann. N.Y. Acad. Sci.* **455**:95–105.

Bloom, G. S., Luca, F. C., and Vallee, R. B., 1985, Microtubule-associated protein 1B: Identification of a major component of the neuronal cytoskeleton. *Proc. Natl. Acad. Sci. U.S.A.* **82**:5404–5408.

Bologa, L., 1985, Oligodendrocytes, key cells in myelination and target in demyelinating diseases. *J. Neurosci. Res.* **14**:1–20.

Bologa, L., Deugnier, M.-A., Joubert, R., and Bisconte, J.-C., 1985, Myelin basic protein stimulates the proliferation of astrocytes: Possible explanation for multiple sclerosis plaque formation. *Brain. Res.* **346**:199–203.

Bovolenta, P., Liem, R. K. H., and Mason, C. A., 1984, Development of cerebellar astroglia: Transitions in form and cytoskeletal content. *Dev. Biol.* **102**:248–259.

Brenner, T., Lisak, R. P., Rostami, A., Pleasure, D. E., and Silberberg, D. H., 1986, Astrocytes, oligodendrocytes, and Schwann cells share a common antigenic determinant that cross-reacts with myelin basic protein: Identification with monoclonal antibody. *J. Neurosci.* **6**:1925–1933.

Brightman, M. W., and Reese, T. S., 1969, Junctions between intimately apposed cell membranes in the vertebrate brain. *J. Cell Biol.* **40**:648–677.

Bruni, J. E., Del Bigio, M. R., and Clattenburg, R. E., 1985, Ependyma: Normal and pathological. A review of the literature. *Brain Res.* **9**:1–19.

Bunge, R. P., and Waksman, B. H., 1985, Glial development and interactions. *Trends Neurosci.* **8**:424–427.

Burke, B., and Gerace, L., 1986, A cell free system to study reassembly of the nuclear envelope at the end of mitosis. *Cell* **44**:639–652.

Cohen, J., Burne, J. F., Winter, J., and Bartlett, P., 1986, Retinal ganglion cells lose response to laminin with maturation. *Nature* **322**:465–467.

Couchie, D., Fages, C., Bridoux, A. M., Rolland, B., Tardy, M., and Nunez, J., 1985, Microtubule-associated proteins and in vitro astrocyte differentiation. *J. Cell Biol.* **101**:2095–2103.

Cowan, N. J., Lewis, S. A., Balcarek, J. M., Krek, V., and Shelanski, M., 1985, Structural implications of cDNA clone encoding mouse glial fibrillary acidic protein. *Ann. N.Y. Acad. Sci.* **455**:575–582.

Cunningham, B. A., 1986, Cell adhesion molecules: A new perspective on molecular biology. *Trends Biochem. Sci.* **11**:423–429.

de Bault, L. E., and Cancilla, P. A., 1980, γ-Glutamyl transpeptidase in isolated brain endothelial cells: Induction by glial cells in vitro, *Science* **207**:653–655.

de Camilli, P., Harris, S. M., Jr., Huttner, W. B., and Greengard, P., 1983, Synapsin I (Protein I), a nerve terminal-specific phosphoprotein. II. Its specific association with synaptic vesicles demonstrated by immunocytochemistry in agarose-embedded synaptosomes. *J. Cell Biol.* **96**:1355–1373.

Dickson, J. G., Flanigan, T. P., Kemshead, J. T., Doherty, P., and Walsh, F. S., 1983, Identification of cell-surface antigens present exclusively on a sub-population of astrocytes in human foetal brain cultures. *J. Neuroimmunol.* **5**:111–123.

Drager, U. C., 1983, Coexistence of neurofilaments and vimentin in a neurone of adult mouse retina. *Nature* **303**:169–172.

Duffy, P. E., 1983, *Astrocytes: Normal, Reactive and Neoplastic*, Raven Press, New York.

Edwards, M. A., Yamamoto, M., Schwarting, G., and Caviness, V. S., Jr., 1986, Development of radial glia in the mouse: An immunohistochemical study with a cell-class specific monoclonal antibody. *Soc. Neurosci. Abstr.* **12**:182.

Eisenbarth, G. S., Walsh, F. S., and Nirenberg, M., 1979, Monoclonal antibody to a plasma membrane antigen of neurons. *Proc. Natl. Acad. Sci. U.S.A.* **76**:4913–4917.

Eng, L. F., Smith, M. E., de Vellis, J., and Skoff, R. P., 1985, Recent studies of the glial fibrillary acidic protein. *Ann. N.Y. Acad. Sci.* **455**:525–537.

Fallon, J. R., 1985, Preferential outgrowth of central nervous system neurites on astrocytes and Schwann cells as compared with nonglial cells in vitro. *J. Cell Biol.* **100**:198–207.

Fedoroff, S., White, R., Neal, J., Subrahmanyan, L., and Kalnins, V. I., 1983, Astrocyte cell lineage. II. Mouse fibrous astrocytes and reactive astrocytes in cultures have vimentin- and GFP-containing intermediate filaments. *Dev. Brain Res.* **7**:303–315.

ffrench-Constant, C., and Raff, M. C., 1986, The oligodendrocyte-type-2 astrocyte cell lineage is specialized for myelination. *Nature* **323**:335–338.

Flament-Durand, J., and Brion, J. P., 1985, Tanycytes: Morphology and functions: A review. *Int. Rev. Cytology* **96**:121–153.

Fontana, A., Fierz, W., and Wekerle, H., 1984, Astrocytes present myelin basic protein to encephalitogenic T-cell lines. *Nature* **307**:273–276.

Frei, K., Bodmer, S., Schwerdel, C., and Fontana, A., 1985, Astrocytes of the brain synthesize interleukin-3-like factors. *J. Immunol.* **135**:4044–4047.

Georgatos, S. D., and Marchesi, V. T., 1985, The binding of vimentin to human erythrocyte membranes: A model system for the study of intermediate filament–membrane interactions. *J. Cell Biol.* **100**:1955–1961.

Giulian, D., and Baker, T. J., 1985, Peptides released by ameboid microglia regulate astroglial proliferation. *J. Cell Biol.* **101**:2411–2415.

Giulian, D., and Young, D. G., 1986, Brain peptides and glial growth. II. Identification of cells that secrete glia-promoting factors. *J. Cell Biol.* **102**:812–820.

Giulian, D., Allen, R. L., Baker, T. J., and Tomozawa, Y., 1986, Brain peptides and glial growth. I. Glia-promoting factors as regulators of gliogenesis in the developing and injured central nervous system. *J. Cell Biol.* **102**:803–811.

Goldman, J. E., Geier, S. S., and Hirano, M., 1986, Differentiation of astrocytes and oligodendrocytes from germinal matrix cells in primary culture. *J. Neurosci.* **6**:52–60.

Goldstein, G. W., and Betz, A. L., 1986, The blood–brain barrier. *Sci. Am.* **255**:74–83.

Grumet, M., Hoffman, S., and Edelman, G. M., 1984, Two antigenically related neuronal cell adhesion molecules of different specificities mediate neuron–neuron and neuron–glia adhesion. *Proc. Natl. Adac. Sci. U.S.A.* **81**:267–271.

Grumet, M., Hoffman, S., Crossin, K. L., and Edelman, G. M., 1985, Cytotactin, an extracellular matrix protein of neural and non-neural tissues that mediates glia–neuron interaction. *Proc. Natl. Acad. Sci. U.S.A.* **82**:8075–8079.

Harris, J. R., and Marshall, P., 1981, The nuclear envelope and the nuclear pore complex: Some electron microscopic and biochemical considerations. In *Electron Microscopy of Proteins* (J. R. Harris, ed.), Vol. 1, pp. 39–124, Academic Press, New York.

Hatten, M. E., and Mason, C. A., 1986, Neuron–astroglia interactions *in vitro* and *in vivo*. *Trends Neurosci.* **9**:168–174.

Hauser, S. L., Aubert, C., Burks, J. S., Kerr, C., Lyon-Caen, O., de The, G., and Brahic, M., 1986, Analysis of human T-lymphotropic virus sequences in multiple sclerosis tissue. *Nature* **322**:176–177.

Huber, G., Alaimo-Beuret, D., and Matus, A., 1985, MAP3: Characterization of a novel microtubule-associated protein. *J. Cell Biol.* **100**:496–507.

Huber, G., Pehling, G., and Matus, A., 1986, The novel microtubule-associated protein MAP3 contributes to the *in vitro* assembly of brain microtubules. *J. Biol. Chem.* **261**:2270–2273.

Hertzberg, E. L., Spray, D. C., and Bennett, M. V. L., 1985, Reduction of gap junctional conductance by microinjection of antibodies against the 24-kDa liver gap junction polypeptide. *Proc. Natl. Acad. Sci. U.S.A.* **82**:2412–2416.

Ip, W., Hartzer, M. K., Pang, Y.-Y. S., and Robson, R. M., 1985, Assembly of vimentin *in vitro* and its implications concerning the structure of intermediate filaments. *J. Mol. Biol.* **183**:365–375.

Jacobs, J. R., and Stevens, J. K., 1986, Changes in the organization of the neuritic cytoskeleton during nerve growth factor—activated differentiation of PC12 cells: A serial electron microscopic study of the development and control of neurite shape. *J. Cell Biol.* **103**:895–906.

Jacque, C. M., Kujas, M., Poreau, A., Raoul, M., Collier, P., Racadot, J., Bauman, N., 1979, GFA and S100 protein levels as an index for malignancy in human gliomas and neurinomas. *J. Natl. Cancer Inst.* **62**:479–483.

Janzer, R. C., and Raff, M. C., 1987, Astrocytes induce blood–brain barrier properties in endothelial cells. *Nature* **325**:253–257.

Karpas, A., Kampf, U., Siden, Å., Koch, M., and Poser, S., 1986, Lack of evidence for involvement of known human retroviruses in multiple sclerosis. *Nature* **322**:177–178.

Kim, S. U., Moretto, G., Lee, V., and Yu, R. K., 1986, Neuroimmunology of gangliosides in human neurons and glial cells in culture. *J. Neurosci. Res.* **15**:303–321.

Köhler, G., and Milstein, C., 1975, Continuous cultures of fused cells secreting antibody of predefined specificity. *Nature* **256**:495–975.

Kubo, T., Fukuda, K., Mikami, A., Maeda, A., Takahashi, H., Mishina, M., Haga, T., Haga, K., Ichiyama, A., Kangawa, K., Kojima, M., Matsuo, H., Hirose, T., and Numa, S., 1986, Cloning, sequencing and expression of complementary DNA encoding the muscarinic acetylcholine receptor. *Nature* **232**:411–416.

Kuffler, S. W., and Nicholls, J. G., 1976, *From Neuron to Brain*, Sinauer Associates, Sunderland, MA.

Kumar, N. M., and Gilula, N. B., 1986, Cloning and characterization of human and rat liver cDNAs coding for a gap junction protein. *J. Cell Biol.* **103:**767–776.

Lagenaur, C., Sommer, I., and Schachner, M., 1980, Subclass of astroglia in mouse cerebellum recognized by monoclonal antibody. *Dev. Biol.* **79:**367–378.

Landis, D. M. D., 1985, Promise and pitfalls in immunocytochemistry. *Trends Neurosci.* **8:**312–317.

Lawson, D., 1983, Epinemin: A new protein associated with vimentin filaments in non-neural cells. *J. Cell Biol.* **97:**1891–1905.

Lawson, D., 1984, Distribution of epinemin in colloidal gold-labelled, quick-frozen, deep-etched cytoskeletons. *J. Cell Biol.* **99:**1451–1460.

Lee, V. M.-Y., and Page, C., 1984, The dynamics of nerve growth factor-induced neurofilament and vimentin filament expression and organization in PC12 cells. *J. Neurosci.* **4:**1705–1714.

Lemke, G., 1986, Molecular biology of the major myelin genes. *Trends Neurosci.* **9:**266–270.

Lemmon, V., 1986, Localization of a filamin-like protein in glia of the chick central nervous system. *J. Neurosci.* **6:**43–51.

Leterrier, J.-F., Liem, R. K. H., and Shelanski, M. L., 1982, Interactions between neurofilaments and microtubule-associated proteins: A possible mechanism for intraorganellar bridging. *J. Cell Biol.* **95:**982–986.

Levi, G., Gallo, V., and Ciotti, M. T., 1986, Bipotential precursors of putative fibrous astrocytes and oligodendrocytes in rat cerebellar cultures express distinct surface features and "neuron-like" γ-aminobutyric acid transport. *Proc. Natl. Acad. Sci. U.S.A.* **83:**1504–1508.

Liem, R. K. H., Yen, S.-H., Salomon, G. D., and Shelanski, M. L., 1978, Intermediate filaments in nervous tissues. *J. Cell Biol.* **79:**637–645.

Liesi, P., 1985, Laminin-immunoreactive glia distinguish regenerative adult CNS systems from non-regenerative ones. *EMBO J.* **4:**2505–2511.

Liesi, P., Dahl, D., and Vaheri, A., 1983, Laminin is produced by early rat astrocytes in primary culture. *J. Cell Biol.* **96:**920–924.

Lieska, N., Yang, H.-Y., and Goldman, R. D., 1985, Purification of the 300K intermediate filament-associated protein and its in vitro recombination with intermediate filaments. *J. Cell Biol.* **101:**802–813.

Lindner, J., Guenther, J., Nick, H., Zinser, G., Antonicek, H., Schachner, M., and Monard, D., 1986, Modulation of granule cell migration by a glia-derived protein. *Proc. Natl. Acad. Sci. U.S.A.* **83:**4568–4571.

McKeon, F. D., Kirschner, M. W., and Caput, D., 1986, Homologies in both primary and secondary structure between nuclear envelope and intermediate filament proteins. *Nature* **319:**463–468.

Malhotra, S. K., 1983, *The Plasma Membrane*, Wiley, New York.

Malhotra, S. K., Wong, F., Cumming, P., Ross, S. D., Shnitka, T. K., Manickavel, V., Warren, K. G., and Jeffrey, V., 1984a, A monoclonal antibody for cytoskeletal antigenic determinants(s) distinguishable from glial fibrillary acidic protein in astrocytes. *Microbios Lett.* **26:**151–157.

Malhotra, S. K., Wong, F., Sinha, N., Manickeval, V., Shnitka, T. K., and Warren, K. G., 1984b, A monoclonal antibody for astrocytes. *Ann. N.Y. Acad. Sci.* **436:**492–494.

Malhotra, S. K., Singh, R., Predy, R., Singh, B., Bhatnagar, R., Shnitka, T. K., and Warren, K. G., 1986, Characterisation of a 30K protein associated with cytoskeleton in astrocytes. *J. Cell Biol.* **103:**79a.

Mentaberry, A., Adesnik, M., Atchison, M., Norgård, E. M., Alvarez, F., Sabatini, D. D., and Colman, D. R., 1986, Small basic proteins of myelin from central and peripheral nervous systems are encoded by the same gene. *Proc. Natl. Acad. Sci. U.S.A.* **83:**1111–1114.

Merrill, J. E., Kutsunai, S., Mohlstrom, C., Hofman, F., Groopman, J., and Golde, D. W., 1984, Proliferation of astroglia and oligodendroglia in response to human T cell-derived factors. *Science* **224:**1428–1430.

Miller, R. H., Abney, E. R., David, S., Ffrench-Constant, C., Lindsay, R., Patel, R., Stone, J., and Raff, M. C., 1986, Is reactive gliosis a property of a distinct subpopulation of astrocytes? *J. Neurosci.* **6:**22–29.

Milstein, C., 1981, Monoclonal antibodies from hybrid myelomas. *Proc. R. Soc. Lond. [Biol.]* **211:**393–412.

Mishina, M., Takai, T., Imoto, K., Noda, M., Takahashi, T., and Numa, S., 1986, Molecular distinction between fetal and adult forms of muscle acetylcholine receptor. *Nature* **321:**406–411.

Misson, J. P., Edwards, M. A., Yamamoto, A., and Caviness, V. S., Jr., 1986, Origin of radial glial cells in mouse cerebral cortex: A study using combined immunohistochemical and [^{3}H]thymidine autoradiographical labelling. *Soc. Neurosci. Abstr.* **12:**1585.

Napolitano, E. W., Pachter, J. S., Chin, S. S. M., and Liem, R. K. H., 1985, β-Internexin, a ubiquitous intermediate filament-associated protein. *J. Cell Biol.* **101:**1323–1331.

Newman, E. A., 1986, High potassium conductance in astrocyte endfeet. *Science* **233:**453–454.

Noble, M., and Murray, K., 1984, Purified astrocytes promote the *in vitro* division of a bipotential glial progenitor cell. *EMBO J.* **3:**2243–2247.

Osborn, M., 1985, Summary: Intermediate filaments 1984, *Ann. N.Y. Acad. Sci.* **455:**669–681.

Pachter, J. S., and Liem, R. K., 1985, α-Internexin, a 66-kD intermediate filament-binding protein from mammalian central nervous tissues. *J. Cell Biol.* **101:**1316–1322.

Palade, G. E., 1967, Cell biology: Past and present, *Rockefeller Univ. Rev.*, 11–16.

Palay, S. L., and Chan-Palay, V., 1974, *Cerebellar Cortex: Cytology and Organization*, Springer-Verlag, New York.

Papasozomenos, S. C., and Binder, L. I., 1986, Microtubule-associated protein 2 (MAP2) is present in astrocytes of the optic nerve but absent from astrocytes of the optic tract. *J. Neurosci.* **6:**1748–1756.

Predy, R., Singh, D., Bhatnagar, R., Singh, R., and Malhotra, S. K., 1987, A new protein (J1-31 antigen, 30 kD) is expressed by astrocytes, Müller glia and ependyma. *BioSci. Rep.* **7:**491–502.

Pruss, R. M., Mirsky, R., Raff, M. C., Thorpe, R., Dowding, A. J., and Anderton, B. H., 1981, All classes of intermediate filaments share a common antigenic determinant defined by a monoclonal antibody. *Cell* **27:**419–428.

Quax, W., van den Broek, L., Vree Egberts, W., Ramaekers, F., and Bloemendal, H., 1985, Characterization of the hamster desmin gene: Expression and formation of desmin filaments in nonmuscle cells after gene transfer. *Cell* **43:**327–338.

Quinlan, R. A., and Franke, W. W., 1983, Molecular interactions in intermediate-sized filaments revealed by chemical cross-linking. Heteropolymers of vimentin and glial filament protein in cultured human glioma cells. *Eur. J. Biochem.* **132:**477–484.

Raff, M. C., Miller, R. H., and Noble, M., 1983, Glial cell lineages in the rat optic nerve. *Cold Spring Harbor Symp. Quant. Biol.* **48:**569–572.

Raff, M. C., Abney, E. R., and Fok-Seang, J., 1985, Reconstitution of a developmental clock in vitro: A critical role for astrocytes in the timing of oligodendrocyte differentiation. *Cell* **42:**61–69.

Raju, T., Bignami, A., and Dahl, D., 1981, *In vivo* and *in vitro* differentiation of neurons and astrocytes in the rat embryo. *Dev. Biol.* **85:**344–357.

Reyners, H., Gianfelici de Reyners, E., Regniers, L., and Maisin, J.-R., 1986, A glial progenitor cell in the cerebral cortex of the adult rat. *J. Neurocytol.* **15:**53–61.

Riederer, B. M., Zagon, I. S., and Goodman, S. R., 1986, Brain spectrin (240–235) and brain spectrin (240–235E): Two distinct spectrin subtypes with different locations within mammalian neural cells. *J. Cell Biol.* **102:**2088–2097.

Risau, W., Hallmann, R., Albrecht, U., and Henke-Fahle, S., 1986, Brain induces the expression of an early cell surface marker for blood brain barrier. *EMBO J.* **5**:3179–3183.
Roberts, D. D., Nageswara Rao, C., Magnani, J. L., Spitalink, S. L., Liotta, L. A., and Ginsburg, V., 1985, Laminin binds specifically to sulfated glycolipids. *Proc. Natl. Acad. Sci. U.S.A.* **82**:1306–1310.
Roessmann, U., Velasco, M. E., Sindely, S. D., and Gambetti, P., 1980, Glial fibrillary acidic protein (GFAP) in ependymal cells during development. An immunocytochemical study. *Brain Res.* **200**:13–21.
Rueger, D. C., Huston, J. S., Dahl, D., and Bignami, A., 1979, Formation of 100 Å filaments from purified glial fibrillary acidic protein *in vitro. J. Mol. Biol.* **135**:53–68.
Rutishauser, U., 1984, Developmental biology of a neural cell adhesion molecule. *Nature* **310**:549–554.
Rutishauser, U., 1986, Differential cell adhesion through spatial and temporal variations of NCAM. *Trends Neurosci.* **9**:374–378.
Schachner, M., 1982, Cell type-specific surface antigens in the mammalian nervous system. *J. Neurochem.* **39**:1–8.
Schachner, M., Hedley-Whyte, E. T., Hsu, D. W., Schoonmaker, G., and Bignami, A., 1977, Ultrastructural localization of glial fibrillary acidic protein in mouse cerebellum by immunoperoxidase labelling. *J. Cell Biol.* **75**:67–73.
Schachner, M., Huang, S.-K., Ziegelmuller, P., Bizzini, B., and Taugner, R., 1984, Glial cells in the pineal gland of mice and rats: A combined immunofluorescence and electron-microscopic study. *Cell Tissue Res.* **237**:245–252.
Schiffer, D., Giordana, M. T., Migheli, A., Giaccone, G., Pezzotta, S., and Mauro, A., 1986, Glial fibrillary acidic protein and vimentin in the experimental glial reaction the rat brain. *Brain Res.* **374**:110–118.
Schröder, H., and Malhotra, S. K., 1987, Characterization of rodent pineal astrocytes by immunofluorescence microscopy using a monoclonal antibody (J1-31). *Cell Tissue Res.* **248**:607–610.
Shaw, G., and Weber, K., 1983, The structure and development of the rat retina: An immunofluorescence microscopical study using antibodies specific for intermediate filament proteins. *Eur. J. Cell Biol.* **30**:219–232.
Shaw, G., Osborn, M., and Weber, K., 1981, An immunofluorescence microscopical study of the neurofilament triplet proteins, vimentin and glial fibrillary acidic protein within the adult rat brain. *Eur. J. Cell Biol.* **26**:68–82.
Singh, R., Singh, B., and Malhotra, S. K., 1986, A new "marker" protein for astrocytes. *BioSci. Rep.* **6**:73–80.
Soellner, P., Quinlan, R. A., and Franke, W. W., 1985, Identification of a distinct soluble subunit of an intermediate filament protein: Tetrameric vimentin from living cells. *Proc. Natl. Acad. Sci. U.S.A.* **82**:7929–7933.
Steinert, P. M., and Parry, D. A. D., 1985, Intermediate filaments: Conformity and diversity of expression and structure. *Annu. Rev. Cell Biol.* **1**:41–65.
Steinert, P. M., and Steven, A. C., 1985, Splitting hairs and other intermediate filaments. *Nature* **316**:767.
Steinert, P. M., Cantieri, J. S., Teller, D. C., Lonsdale-Eccles, J. D., and Dale, B. A., 1981, Characterization of a class of cationic proteins that specifically interact with intermediate filaments. *Proc. Natl. Acad. Sci. U.S.A.* **78**:4097–4101.
Tedeschi, B., Barrett, J. N., and Keane, R. W., 1986, Astrocytes produce interferon that enhances the expression of H-2 antigens on a subpopulation of brain cells. *J. Cell Biol.* **102**:2244–2253.
Timpl, R., Engel, J., and Martin, G. R., 1983, Laminin—a multifunctional protein of basement membranes. *Trends Biochem. Sci.* **8**:207–209.
Trojanowski, J. Q., Walkenstein, N., and Lee, V. M.-Y., 1986, Expression of neurofilament

subunits in neurons of the central and peripheral nervous system: An immunohistochemical study with monoclonal antibodies. *J. Neurosci.* **6**:650–660.

Valentino, K. L., Winter, J., and Reichardt, L. F., 1985, Applications of monoclonal antibodies to neuroscience research. *Annu. Rev. Neurosci.* **8**:199–232.

Vartio, T., and Vaheri, A., 1983, Fibronectin: Chains of domains with diversified functions. *Trends Biochem. Sci.* **8**:442–444.

Vaughn, J. E., and Peters, A., 1967, Electron microscopy of the early postnatal development of fibrous astrocytes. *Am. J. Anat.* **121**:131–152.

Wang, E., 1985, Intermediate filament associated proteins. *Ann. N.Y. Acad. Sci.* **455**:32–56.

Waxman, S. G., 1986, The astrocyte as a component of the node of Ranvier. *Trends Neurosci.* **9**:250–253.

Weber, K., and Geisler, N., 1985, Intermediate filaments: Structural conservation and divergence. *Ann. N.Y. Acad. Sci.* **455**:126–143.

Wekerle, H., Linington, C., Lassmann, H., and Meyermann, R., 1986, Cellular immune reactivity within the CNS. *Trends Neurosci.* **9**:271–277.

Wujek, J. R., and Akeson, R. A., 1986, A rat astrocyte-specific cell-surface antigen identified with a mouse monoclonal antibody. *Soc. Neurosci. Abstr.* **12**:1584.

Yen, S.-H., and Fields, K. L., 1985, A protein related to glial filaments in Schwann cells. *Ann. N.Y. Acad. Sci.* **455**:538–551.

Zagon, I. S., Higbee, R., Riederer, B. M., and Goodman, S. R., 1986, Spectrin subtypes in mammalian brain: An immunoelectron microscope study. *J. Neurosci.* **6**:2977–2986.

Chapter 9

Current Molecular Approaches to Experimental Thyroid Autoimmunity

Brian R. Champion, David C. Rayner, Patricia Hutchings, Anne Cooke, Kevin Page, Gopa Biswas-Hughes, Humberto de Assis Paiva, David Male, and Peter Byfield

1. INTRODUCTION

Autoimmune thyroiditis is characterized by mononuclear cell infiltration of the thyroid gland, with accompanying tissue damage, and associated cellular and humoral autoimmunity to thyroid antigens such as thyroglobulin (Tg) and microsomal antigen (for reviews see Bigazzi and Rose, 1985; Rose *et al.*, 1981; Weetman and McGregor, 1984; Weigle, 1980; Wick *et al.*, 1985). Since the seminal studies by Rose and Witebsky (1956) demonstrating the experimental induction of autoimmune thyroiditis, animal models have been used extensively, both in the search for clues to the etiology of human autoimmune thyroiditis and to investigate basic mechanisms in the regulation of autoimmunity. Thyroid

Brian R. Champion, Patricia Hutchings, Anne Cooke, Kevin Page, Gopa Biswas-Hughes, and Humberto de Assis Paiva Department of Immunology, Middlesex Hospital Medical School, London W1P 9PG, United Kingdom. **David Male** Institute of Psychiatry, Denmark Hill, London SE5 1JK, United Kingdom. **David C. Rayner** Department of Pathology, University of Manitoba, Winnipeg, Canada R3G 0W3. **Peter Byfield** Department of Endocrinology, MRC Clinical Research Centre, Harrow HA1 3UJ, United Kingdom. *Present address of B. R. C.:* Immunobiology Department, Glaxo Group Research Ltd., Greenford, UB6 0HE, United Kingdom.

autoimmunity arises spontaneously in a number of species (Bigazzi and Rose, 1975), but most work has concentrated on the obese strain chicken (Wick *et al.*, 1982, 1985), buffalo rats (Bigazzi and Rose, 1975), and the diabetes-prone BB/W rat (Sternthal *et al.*, 1981). Thyroiditis may also be induced experimentally in many species by injecting Tg and an appropriate adjuvant (Weigle, 1980). A different model of thyroiditis can be induced in rats by a combination of thymectomy and irradiation (Penhale *et al.*, 1973, 1975). Studies of these various models have provided valuable information regarding the mechanisms involved in thyroid autoimmunity. More recently, we and others have begun to use a variety of in vitro techniques to gain insight into the molecular basis of thyroid autoimmunity. In this chapter, we focus on three main areas in which we have been particularly interested. For a broader coverage of mechanisms in experimental autoimmune thyroiditis (EAT), readers are referred to the excellent reviews cited above.

2. NATURE OF THE AUTOANTIGENS

2.1. Thyroglobulin

2.1.1. Molecular Characterization

Thyroglobulin (Tg) is a high molecular weight (660-kD) glycoprotein that is synthesized by follicular epithelial cells in the thyroid and is stored in a colloidal form in the lumen of thyroid follicles. Tg functions as a precursor for the synthesis of the thyroid hormones thyroxine (T4) and triiodothyronine (T3), and it can thus be considered to be a prohormone. It consists of disulfide-linked homodimers of a 330-kD glycoprotein chain, of which approximately 10% is carbohydrate. It is relatively cysteine rich and has the potential to form as many as a hundred disulfide bonds, although how many are actually formed is not known (Van Herle *et al.*, 1979).

The formation of thyroid hormones involves extracellular iodination of tyrosine residues in Tg to form monoiodo- and diiodotyrosines (MIT and DIT) and the subsequent coupling of one of these residues to another DIT to form the iodothyronines T3 and T4, respectively (Figure 1). Both of these reactions are catalyzed by thyroid peroxidase at the apical membrane of the thyroid epithelial cell adjacent to the follicular lumen. Secretion of T4 and T3 involves the endocytic uptake of colloidal Tg by follicular cells, followed by lysosomal enzyme digestion in phagolysosomes. Thyroid hormones are thus released from the prohormone and pass into the extracellular space and thence into the circulation (Figure 2). This most likely occurs by diffusion of T4 and T3 through phagolysosomal and epithelial cell membranes but might also involve active ex-

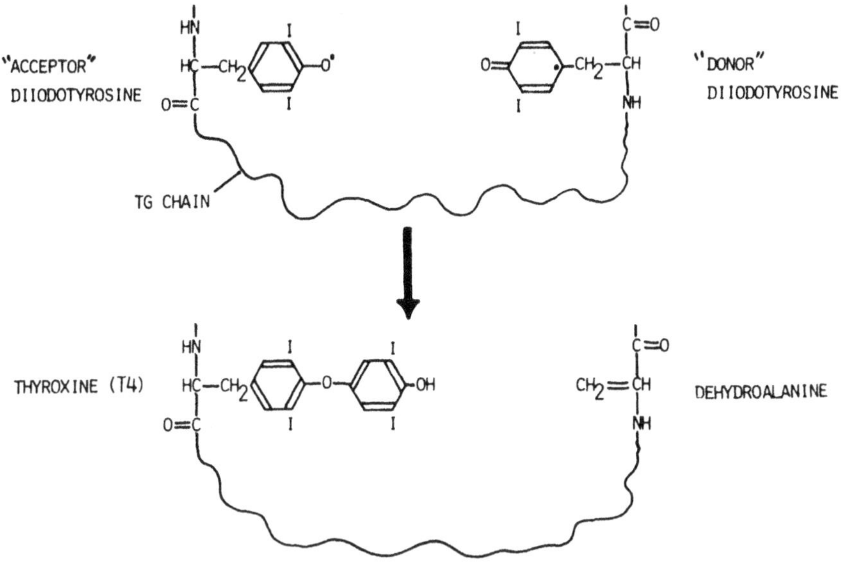

FIGURE 1. Formation of thyroid hormone residues within thyroglobulin. Iodination of tyrosine residues is catalyzed by the thyroid peroxidase to form mono- and diiodotyrosines (MIT and DIT, respectively). The same peroxidase also couples the diiodophenolic ring of a "donor" DIT to an "acceptor" DIT residue to form the ether tetraiodothyronine or thyroxine (T4). For triiodothyronine (T3) synthesis, a donor MIT is coupled to an acceptor DIT in the same way. As a result of these reactions, a dehydroalanine residue (which may also become hydrolyzed to form serine) is left in place of the donor iodotyrosine.

ocytosis. Intact Tg is also actively released into the circulation in low amounts (10–100 ng/mL). How this occurs is not yet clear, but it seems possible that undigested Tg, and perhaps fragments of partially digested Tg, might be released into the extracellular milieu by exocytosis.

The trapping of iodine and formation of iodothyronines in the thyroid are very efficient processes, enabling thyroid hormones to be made even under conditions of severe iodine deficiency. This efficiency is further enhanced by the salvage of iodine from free MIT and DIT by a deiodinase enzyme for reuse by the thyroid peroxidase.

With the recent advent of molecular biological technology, our knowledge of the structure of Tg has begun to advance rapidly. At least partial genetic analyses have been made on Tg from several species including rat, human, and cow (Di Lauro *et al.*, 1985; Malthiery and Lissitzky, 1985; Mercken *et al.*, 1985; Targovnik *et al.*, 1984). The complete organization of the rat Tg gene has been reported (Musti *et al.*, 1986). It is one of the largest genes ever described, being at least 170 kilobase pairs long and consisting of 42 exons of which all but two are 150–200 base pairs each; exons 9 and 10 are 1100 and 620 base pairs,

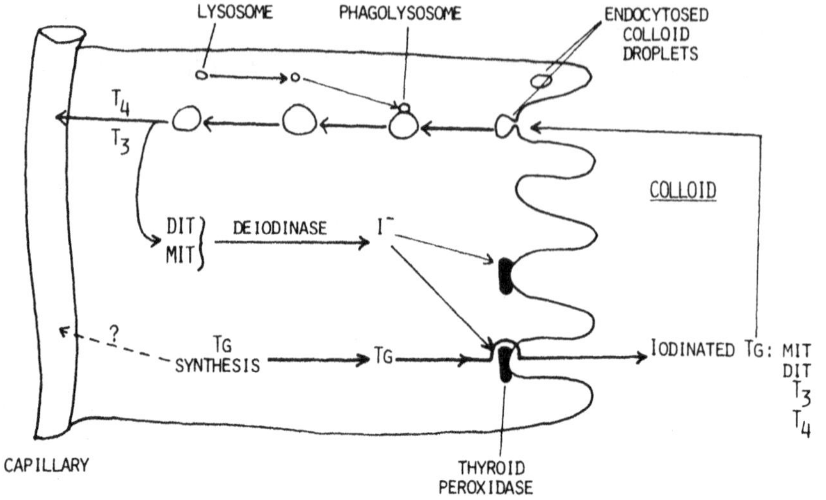

FIGURE 2. Thyroid hormone synthesis by the thyroid follicular epithelial cell. The newly synthesized Tg chain is glycosylated in the endoplasmic reticulum and then transported to the apical (internal) border of the thyroid cell, where it is released into the lumen of the follicle. Iodination and iodothyronine formation within Tg takes place at the extracellular surface of the membrane and is catalyzed by the thyroid peroxidase. This produces residues of MIT, DIT, T3, and T4 within the Tg molecule, which is stored in a colloidal form within the lumen of the follicle. To release free thyroid hormones into the circulation, droplets of colloid are endocytosed and the Tg is subjected to hydrolysis by lysosomal enzymes within phagolysosomes. The free T3 and T4 molecules are then released into the circulation, probably by diffusion although there may also be some active exocytosis. The iodine atoms are liberated from free MIT and DIT residues by a deiodinase enzyme and these iodines can be reutilized by the thyroid peroxidase. Although the transport of newly synthesized Tg is directed toward the luminal side of the thyroid cell, evidence that circulating Tg may on average have less iodine content than colloidal Tg suggests the possibility that some Tg may be released in a noniodinated form directly into the circulation.

respectively. About 95% of the Tg gene consists of introns of variable length, some consisting of several kilobases.

Most of the protein sequence information has been deduced from sequences of complementary and genomic DNA clones. Thus far, the full primary amino acid sequence has only been published for bovine Tg (Mercken *et al.*, 1985). Bovine Tg consists of 2769 amino acids, of which the first 19 constitute a hydrophobic signal peptide that is enzymatically removed in the endoplasmic reticulum. Four hormonogenic sites have been identified; one at tyrosine residue 5 in the N-terminal region and three in the C-terminal region at residues 2555, 2569, and 2748. The acceptor tyrosine (see Figure 1) at residue 2748 appears to be used primarily in the formation of T3, whereas the other sites represent locations for T4 synthesis (Marriq *et al.*, 1983). It is still unclear which tyrosine residues act as a donor during hormone formation, although they may reside within the same region of the molecule as the acceptor tyrosine. Indeed, there is a certain degree of clustering of tyrosine residues at either end of the protein chain where the hormonogenic sites are found (Mercken *et al.*, 1985). Of course, how many hormone residues are formed and at which of the potential sites depends largely on the level of iodine available.

The amino acid sequence surrounding the amino-terminal hormonogenic tyrosine shows remarkable evolutionary conservation, whereas the C-terminal hormonogenic domains show much less conservation and also differ in sequence from the N-terminal region (Vassart *et al.*, 1985). The structural conservation of the N-terminal region is consistent with observations that this area of the molecule contains the major hormonogenic site (Marriq *et al.*, 1984; Rawitch *et al.*, 1983), particularly under conditions of iodine deficiency.

The three-dimensional organization of the Tg molecule is still not known, although predictions of partial secondary structures have been made based on sequence information (Formisano *et al.*, 1985). However, about 75% of the molecule appears to be involved in imperfect repeats of cysteine-rich sequences, suggesting an organized domain structure. Interestingly, the C-terminal quarter of the molecule is free of repeat structures and this, together with the lack of sequence similarities in the hormonogenic domains, has led to the suggestion that the Tg gene may have evolved from the fusion of two different hormone-forming ancestral gene units (Musti *et al.*, 1986).

2.1.2. B-Cell Epitopes

Rose *et al.* (1982) and Kotani *et al.* (1985b) used a small panel of Tgs from different species in order to begin characterization of the epitopic specificities of monoclonal murine Tg autoantibodies. With the additional information from cross-blocking studies, one of these groups was able to define a minimum of eight autoantigenic epitopes on mouse Tg (Rose *et al.*, 1982). This is more than the apparent number of autoantigenic epitopes on human Tg (Nye *et al.*, 1980; Roitt *et al.*, 1958), perhaps reflecting differences between spontaneous and experimentally induced autoimmunity. We have recently established a large panel of Tgs from different mammalian orders and have used them to examine the specificities of autoantibodies and autoreactive T cells from experimental animals (B. R. Champion *et al.*, and H. de Assis Paiva *et al.*, unpublished observations). Serum autoantibodies from high responder mice (CBA, H-2^k) immunized with mouse Tg and adjuvant [either complete Freund's adjuvant (CFA) or lipopolysaccharide (LPS)] show preferential reactivity with mouse Tg but also cross-react quite strongly with rat Tg (Figure 3A). These EAT sera demonstrate only low (but still significant) reactivity with all other species tested. This specificity pattern of polyclonal murine Tg autoantibodies is clearly reflected in the dominant cross-reactivity profile of monoclonal antibodies from

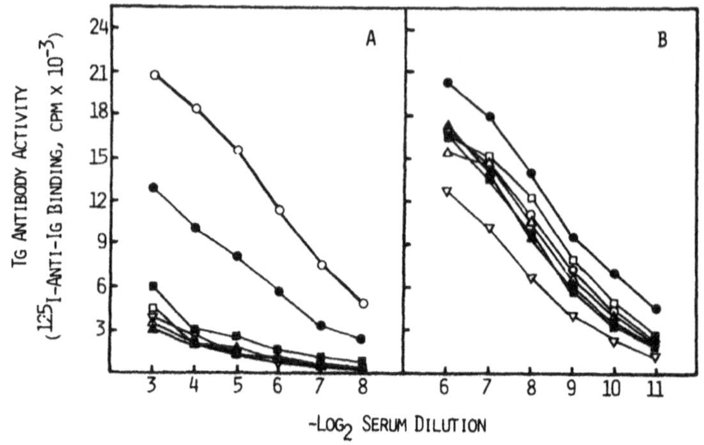

FIGURE 3. Cross-reactivity of mouse and rat serum Tg autoantibodies with Tgs from different species. Doubling dilutions of sera from mice (A) or rats (B) immunized with mouse or rat Tg (in CFA), respectively, were tested in a solid-phase radioimmunoassay for their ability to bind to Tgs prepared from different species. Antibody binding was detected with an ^{125}I-labeled affinity purified anti-immunoglobulin (Ig) second layer: sheep antimouse F(ab')$_2$ in (A) and rabbit antirat Ig in (B). ○——○, mouse TG; ●——●, rat Tg; ■——■, cow Tg; □——□, dog Tg; ▽——▽, human Tg; △——△, guinea pig Tg; ▲——▲, wallaby Tg. Other Tgs (see Figure 4) were also tested and bound autoantibodies to a similar degree as dog Tg (not shown).

similar mice. Most of the monoclonals we have obtained react preferentially with mouse Tg and only cross-react with rat Tg (Figure 4, group I). This agrees with the investigations of Rose *et al.* (1982). With our large panel of Tg species variants, we can demonstrate a minimum of four distinct epitopes (groups I–IV, Figure 4). The degree of cross-reactivity with rat Tg (not shown) suggests that the major group may be split into subgroups; cross-blocking studies are underway to investigate this possibility further.

In contrast to murine Tg autoantibodies, rat antibodies from animals immunized with rat Tg in CFA primarily recognize a determinant(s) shared by all species of Tg examined (Figure 3B). Thus, in the rat, highly conserved regions of Tg appear to be involved in its autoantigenicity and this suggests that the highly conserved N-terminal region of the Tg molecule may be involved. Thyroid hormone residues formed within all Tgs would be another candidate, but inhibition studies with T4 indicate that these epitopes are not of primary importance. Preliminary studies by others have indicated that human Tg autoantibodies show only a lower affinity cross-reactivity with other Tg species (Kohno *et al.*, 1985). Their data more closely resemble our findings with mouse autoantibodies, perhaps suggesting that mouse models of experimentally induced autoimmunity more closely resemble the human situation, at least at the B cell level.

A minor group of autoantibodies to Tg in some humans (Pearce *et al.*, 1981)

ORDER	SPECIES	AUTOANTIBODIES GROUP I (N=8)	GROUP II (N=1)	GROUP III (N=1)	GROUP IV (N=1)	AUTOREACTIVE T CELLS MTg9B3 (ADA2)	CH9	MTg12B	82H8	C4	PRIMED T CELLS
RODENTIA	MOUSE	+	+	+	+	+	+	+	+	+	+
	RAT	+	+/-	+	+	+	+	+	+	-	+
	FLYING SQUIRREL	-	-	-	+	+	+	-	NT	-	+
	GUINEA-PIG	-	-	-	+	+	+	+	NT	-	+
	BRAZILIAN TREE PORCUPINE	-	-	-	+	+	+	+	NT	-	NT
LAGOMORPHA	RABBIT	-	-	-	+	+	+	-	NT	-	+
INSECTIVORA	HEDGEHOG	-	-	-	+	+	+	-	NT	-	+/-
PRIMATES	MONKEY	-	-	-	+	+	+	-	NT	-	+
	PIG-TAILED MONKEY	-	-	-	+	+	+	-	NT	-	NT
	HUMAN	-	-	-	+	+	+	+/-	+	-	+
CARNIVORA	CAT	-	+/-	-	+	+	+	+	NT	-	+
	DOG	-	-	-	-	+	+	-	NT	-	+
	GIANT PANDA	-	-	-	-	+	+	-	NT	-	-
ARTIODACTYLA	SHEEP	-	-	-	-	+/-	-	-	NT	-	-
	PIG	-	-	-	-	+/-	-	+	NT	-	-
	MARKHOR	-	-	-	-	+/-	-	+/-	NT	-	+
	COW	-	-	-	-	+/-	-	+/-	NT	-	+/-
	REINDEER	-	-	-	-	+/-	-	-	NT	-	-
PERISSODACTYLA	ZEBRA	-	-	-	-	+	+	-	-	-	-
MARSUPIALIA	WOMBAT	-	-	+	+	-	+/-	-	NT	-	+/-
	WALLABY	-	-	+	+	-	+/-	-	NT	-	+
	BRUSH-TAILED OPOSSUM	-	-	+	+	+	+	-	NT	-	+
CLASS AVES	CHICKEN	-	-	-	-	-	-	-	NT	-	-

FIGURE 4. Reactivity of mouse autoantibodies and autoreactive T cells with a panel of Tgs prepared from different species. Reactivity is summarized in tabular form: +, positive; +/−, low reactivity only; −, negative; NT, not tested. Monoclonal autoantibodies (prepared from mice immunized with mouse Tg in CFA) could be separated into four groups (I–IV) on the basis of their cross-reactivity pattern, with most (group I) reflecting the predominant serum autoantibody reactivity (see Figure 3). Mouse Tg reactive T cell lines (MTg9B3, MTg12B, and 82H8), hybridomas (ADA2, CH9, and C4), and Tg-primed lymph node T cells were tested for their ability to respond to the different Tgs in the presence of irradiated syngeneic splenic accessory cells (ADA2 is a T cell hybridoma derived by fusion of the clone MTg9B3). The C-4 T cell hybridoma only reacted with mouse Tg and was subsequently found not to be autoreactive but to recognize an allelic Tg determinant. Four distinct T cell specificities are demonstrated: MTg9B3 (ADA2) and CH9, MTg12B, 82H8, and primed T cells. Major specificity differences distinguishing these T cells are boxed. The minor differences between MTg9B3 (ADA2) and CH9 may also reflect true epitopic differences (see Figure 6).

and obese strain chickens (Brown *et al.*, 1985) do appear to recognize T4 or T3 residues as haptenic groups within the Tg prohormone molecule. Such autoantibodies might conceivably exert physiological effects by depleting the available pool of circulating thyroid hormones, particularly if they are capable of reacting with hormones bound to carrier molecules such as albumin or thyroid binding globulin.

Studies with model globular proteins, such as lysozyme and myoglobin, have used species variants of known amino acid sequence to examine the structural basis of protein antigenicity (Benjamin *et al.*, 1984). Most antibodies appear to recognize conformational epitopes presented as particular electron cloud shapes on the surface of the molecule. It is likely that Tg autoantibodies also primarily recognize conformational epitopes on this globular protein. Unfortunately, the complexity of Tg has prevented the determination of amino acid sequences of most Tg species variants (with the exception of those for which molecular biological techniques have been applied: see Section 2.1.1) and knowledge of the three-dimensional structure of the molecule is still very limited. Because of the high number of disulfide bonds in Tg, it might be expected that at least some autoantibodies could recognize enzymatic fragments of Tg since these might be able to retain a conformation similar to that found in the native molecule. This is, in fact, true since tryptic (Rose *et al.*, 1982) and peptic (Stylos and Rose, 1977) fragments of Tg can interact with Tg autoantibodies and we have shown that larger fragments of human Tg, produced with staphylococcal V8 protease, do retain some reactivity with Tg antibodies (Male *et al.*, 1985a). Our recent studies of murine monoclonal autoantibodies have suggested the possibility that repeated epitopes may exist in the Tg molecule (B. R. Champion *et al.*, unpublished observations). It seems feasible that the repetitive structural domains in the Tg molecule (see Section 2.1.1) might be the site(s) of such epitopes.

2.1.3. T-Cell Epitopes

2.1.3.1. Cross-Reactivity of T-Cell Epitopes. Since the production of Tg autoantibodies is dependent on T cells (Rose *et al.*, 1981) and, at least in experimental models, T cells play a dominant role in establishing thyroid lesions (Kong *et al.*, 1986a), it is of great interest to examine the epitopic specificities of autoreactive Tg-specific T cells. T cells taken from animals immunized with Tg can transfer thyroiditis to naive syngeneic recipients if they are first activated with antigen or mitogen in vitro (Braley-Mullen *et al.*, 1981,1985; Okayasu, 1985; Simon *et al.*, 1986). In mice, the responsible T cells appear to recognize determinants on mouse Tg, which are shared with human and, to a lesser extent, pig and cow Tg (Kong *et al.*, 1986b; Romball and Weigle, 1984; Simon *et al.*, 1986). This also appears to be true for the T cells involved in generating cytotoxicity against mouse thyroid epithelial cells in vitro (Creemers *et al.*, 1983; Simon *et al.*, 1986) and supports the concept that cytotoxic T cells may play an important role in tissue damage in EAT. Furthermore, low responder mice (which develop Tg autoantibodies but not thyroiditis after challenge with Tg) fail to recognize such conserved epitopes (Simon *et al.*, 1985).

To analyze the role of T cells in thyroid autoimmunity more closely, we (Champion *et al.*, 1985, 1986a; Rayner *et al.*, 1987) and others (Maron *et al.*, 1983) have established T-cell lines and hybridomas autoreactive with mouse Tg. Analysis of the specificity profiles of our T cells using a panel of Tg species variants has revealed a minimum of four autoantigenic epitopes on mouse Tg, which can be recognized by T cells (summarized in Figure 4). As found with model protein antigens (Benjamin *et al.*, 1984), these T-cell epitopes all appear to differ from epitopes recognized by monoclonal Tg autoantibodies. Most of the T-cell lines we have obtained recognize determinants that are highly cross-reactive among Tg species variants. This supports the work of Kong *et al.* (1986a,b) indicating that cross-reactive T cells are important in thyroid autoimmunity in high responder mice. Preliminary observations have indicated that T cells from patients with autoimmune thyroiditis may also recognize epitopes shared between different Tg species (Shimojo *et al.*, 1986).

2.1.3.2. Allelic Tg Epitopes. The T-cell hybridoma C-4 (see Figure 4), which was derived from the autoreactive T cell line MTg12B, only recognized mouse Tg. Further analysis revealed that this T cell was responding to an allotypic determinant on Tg, since it was consistently unresponsive to syngeneic (CBA/Ca) Tg but responded well to Tg preparations from outbred mice (Rayner *et al.*, 1987). Previous studies have indicated that Tgs from allogeneic strains are better immunogens than syngeneic Tg (Tomazic and Rose, 1976; Vladutiu and Rose, 1972) and such Tgs have also been distinguished by monoclonal antibodies (Kong *et al.*, 1980; Kotani *et al.*, 1985a). The structural and genetic basis for the allelic T-cell epitope recognized by C-4 clearly requires further investigation. Nevertheless, the existence of such allelic epitopes on Tg is of considerable importance to the interpretation of experimental data where nonsyngeneic Tg is used, since such epitopes could allow the induction of autoimmune responses through a T-cell bypass mechanism (Weigle, 1980).

2.1.3.3. Role of Iodine in T-Cell Recognition of Tg. During studies of the epitopic specificities of our T cells, it became apparent that different preparations of human Tg varied in their ability to trigger the cells. When this phenomenon was examined more closely, it was found to be related to the degree of iodination (as measured by iodothyronine content) of the preparations (Champion *et al.*, 1986b); those with very low T4 content completely failed to trigger two independently derived T-cell populations. This finding was supported by the observation that murine Tg with undetectable T4 content (prepared from goitrous mice) was also unable to stimulate the same T cells (Figure 5). Thus, these

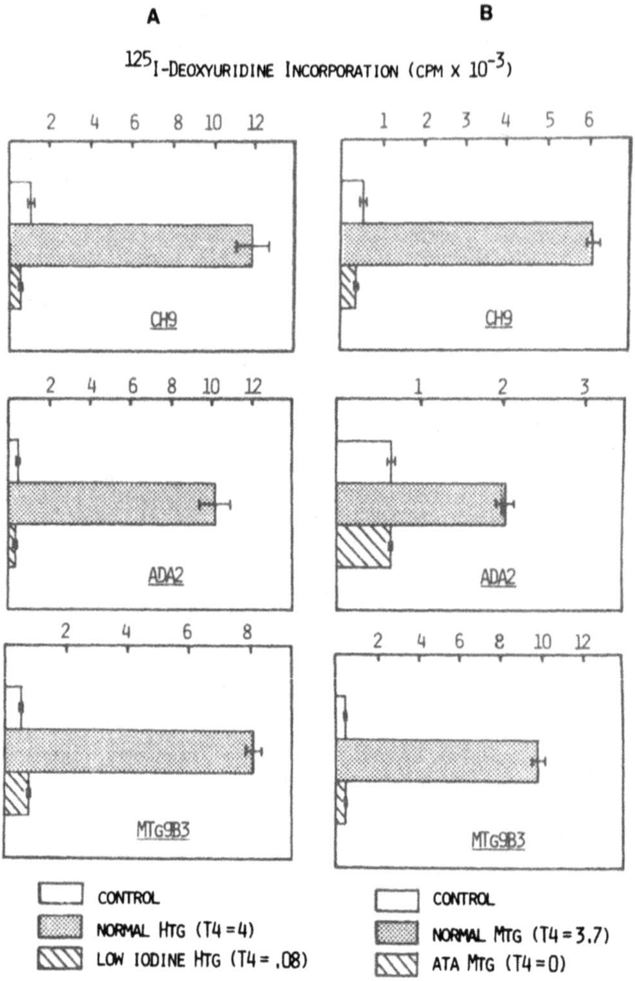

FIGURE 5. Effect of Tg iodination on recognition by autoreactive Tg-specific T cells. Reactivity of the independently derived T cell hybridomas, CH9 and ADA2, was assessed by their antigen-triggered release of interleukin-2 (IL-2) (measured by the ability of culture supernatants to maintain the proliferation of an IL-2-dependent CTLL line). The T cell clone MTg9B3 was tested in a direct proliferation assay. Proliferation of CTLL or line cells was assessed by the incorporation of ^{125}I-deoxyuridine. (A) The T cells were tested with a range of concentrations of human Tgs differing in their iodine content (as measured by their T4 level). Only the maximal response is illustrated. Normal human Tg (T4 = 4 residues/mole) stimulated the T cells well, but Tg of very low iodine content (T4 = 0.08 residues/mole) completely failed at any concentration tested. The use of other human Tgs with intermediate levels of iodine (not shown) showed that the level of T cell reactivity correlated well with Tg iodine content. (B) The importance of Tg iodination was confirmed with mouse Tg (MTg) prepared from mice treated with the goitrogen aminotriazole (ATA). Unlike normal MTg, this ATA MTg had undetectable T4 levels and was also unable to trigger the autoreactive T cells.

autoreactive T cells will only recognize Tg if it is sufficiently iodinated. In contrast, monoclonal and polyclonal Tg autoantibodies failed to discriminate Tgs of different iodine content. The structural basis for this Tg iodination requirement of T cells is the subject of current investigations. We could hypothesize that an iodinated epitope(s) (MIT, DIT, T3, or T4) might be recognized by the T cells, either by interacting directly with the T-cell receptor or by interacting with the MHC class II molecule (I-A^k) required for antigen presentation (Schwartz, 1985). Alternatively, iodination may allow the Tg to be processed correctly by antigen-presenting cells. A further possibility is that dehydroalanine residues, created at the site of donor tyrosines during hormonogenesis (see Figure 1), may be the important feature rather than iodinated tyrosines themselves. Whatever the molecular basis may be, iodination would seem to be important for Tg to act as a T-cell autoantigen in mice. The possible pathological importance of our findings is emphasized by a recent preliminary experiment, which indicated that noniodinated Tg may be unable to induce thyroiditis but will induce autoantibodies as efficiently as normally iodinated mouse Tg. This suggests the possibility that different epitopes are recognized by T cells that help autoantibody production and those that are involved in thyroid pathology. If confirmed, this would differ from experiments on chickens where highly iodinated Tg is a better inducer of autoantibodies than poorly iodinated Tg (Sundick *et al.*, 1986). It seems reasonable to suggest that the enhancing effects of dietary iodine on thyroid autoimmunity in humans (Beierwaltes, 1969; Harach *et al.*, 1985) and experimental animals (Allen *et al.*, 1986; Bagchi *et al.*, 1985) may be explained by the increased generation of iodination-related specific T-cell epitopes on Tg.

2.1.3.4. The Use of Tg Fragmentation to Define Further the T-Cell Epitopes. The characterization of T-cell epitopes on protein antigens has involved the production of defined antigen fragments and ultimately synthetic polypeptides corresponding to the active region of the antigen (Benjamin *et al.*, 1984; Schwartz, 1985). We have embarked on a similar line of investigation in order to localize the autoantigenic epitopes on Tg, which are recognized by our clonal T-cell populations. This is a difficult task because of the obvious complexity of the antigen (see Section 2.1.1). Nevertheless, early experiments using human Tg fragments produced with staphylococcal V8 protease demonstrated that Tg-reactive T cells could still recognize peptides of about 20 kD (Male *et al.*, 1985a). However, this is still larger than some intact protein antigens (e.g., cytochrome c and insulin). We have therefore turned to the use of tryptic peptides, which are much smaller and have proved successful with other protein antigens (Shimonkevitz *et al.*, 1983). Because of its high content of disulfide bonds, Tg must be reduced and alkylated before cleavage with trypsin. We have shown that the reduced and alkylated form of human Tg will still stimulate T cells that cross-react with human Tg. The T-cell hybridoma CH9 was also able to respond to unseparated tryptic peptides (Figure 6A) and we therefore hope to be able to isolate and characterize the critical Tg fragment. However, another

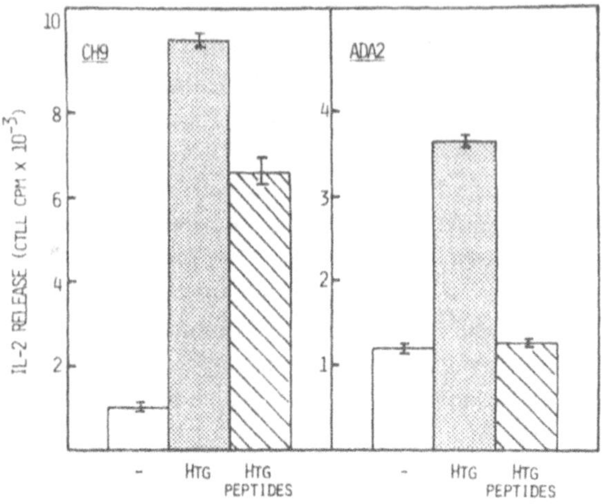

FIGURE 6. Reactivity of CH9 and ADA2 with tryptic peptides of human Tg. Normal human Tg was reduced and alkylated (HTg), to remove disulfide bonds, before being digested to small peptide fragments with trypsin (HTg peptides). These Tg preparations were then tested, in the presence of antigen-presenting cells, for their ability to trigger IL-2 release (measured with CTLL cells as described in Figure 5) from the T cell hybridomas CH9 and ADA2. These two hybridomas both reacted to reduced and alkylated Tg but could be differentiated by their reactivity to tryptic peptides of Tg.

clonotypically distinct T-cell hybridoma, ADA2 (Rayner *et al.*, 1987), appears to be unable to respond to tryptic peptides of human Tg (Figure 6B). This has also been confirmed using the T-cell clone, MTg9B3, from which ADA2 was derived. Thus, these two autoreactive T cells, which showed similar cross-reactivity patterns and iodination requirements (Figures 4 and 5), would appear to recognize distinct (but perhaps related) epitopes.

We are hopeful that knowledge of the structural basis for T-cell recognition of Tg will help in our understanding of thyroid autoimmunity and perhaps lead to novel approaches to the manipulation of the response, in humans as well as in experimental animals.

2.2. The Microsomal Antigen

Patients with autoimmune thyroid diseases often produce autoantibodies to a second thyroid antigen, the "microsomal antigen," so-called because of its localization in subcellular fractions of thyroid cells (Weetman and McGregor, 1984). This antigen is localized to the cytoplasmic and apical membranes of thyroid follicular epithelial cells and may play an important role in the pathogenesis of the disease. However, until recently the exact nature of the microsomal antigen, although subjected to biochemical analysis (Banga *et al.*, 1985), was not clear. It is perhaps because of this and the limited availability of antigen that only one study has reported the deliberate experimental induction of thyroid autoimmunity using thyroid microsomes (Mangkornkanok *et al.*, 1972).

The microsomal antigen has now been shown to be the thyroid peroxidase (TPO) enzyme responsible for hormonogenesis (see Figure 2). Antimicrosomal antibodies have been shown to bind well to purified TPO and also to be indistinguishable from anti-TPO antibodies (Czarnocka *et al.*, 1985, 1986; Mariotti *et al.*, 1986; Portmann *et al.*, 1985). Complementary DNA clones of TPO have also been produced recently (Dinsart *et al.*, 1986) and shown to code for products recognized by antimicrosomal autoantibodies.

Characterization of the microsomal/TPO antigen, both structurally and antigenically, should now proceed quite rapidly and perhaps experimental models can be established to investigate further the importance of this antigen in the initiation and pathogenesis of thyroid autoimmunity.

3. ANTIGEN PRESENTATION

3.1. Professional Antigen-Presenting Cells (APCs)

Macrophages and dendritic cells have sometimes been referred to as "professional APCs" because of their primary role in presenting foreign antigens to T cells in vitro. These cells, which must bear major histocompatibility complex (MHC) class II molecules (Ia) in order to present antigen to T cells, have also been shown to be important in thyroid autoimmunity. For example, dendritic cells enriched from the spleens of mice primed with mouse Tg in CFA can induce both Tg autoantibodies and thyroid lesions when transferred into naive syngeneic recipients (J. Farrant and S. Knight, personal communication). In the BB/W rat, spontaneous thyroiditis is associated with an increase in thyroidal dendritic APCs, and similar dendritic cells bearing thyroid antigens are found in the draining lymph node, where they may be involved in initiating the autoimmune response (Voorbij *et al.*, 1986). Whether or not Tg-bearing cells traveled in the afferent lymphatics to carry antigen from the thyroid to the local lymph node is not clear from these studies but seems likely in the light of evidence for this mode of antigen transport elsewhere in the body.

As might be expected, Ia-positive macrophages and/or dendritic cells are also able to function in vitro to present Tg to autoreactive T cells from animals with experimentally induced thyroid autoimmunity (Braley-Mullen *et al.*, 1983; Champion *et al.*, 1985; Weetman *et al.*, 1983).

3.2. B Cells

Recent evidence has indicated that B cells are also able to present antigens and some may be very efficient in this function by virtue of their ability to bind and concentrate specific antigen through the immunoglobulin receptor (Abbas *et al.*, 1985; Lanzavecchia, 1985; Rock *et al.*, 1984). This mode of antigen presentation may be particularly relevant in thyroid autoimmunity where the levels of circulating Tg are low (10–100 ng/mL). We have shown recently that murine B cells from mice primed with Tg are indeed highly efficient at presenting low antigen doses (as low as 100 ng/mL) to one of our Tg-specific T-cell hybridomas (Figure 7)(Hutchings *et al.*, 1987). This B cell presentation of Tg to T cells required Tg binding to the immunoglobulin receptor and was a productive collaboration since the activated T cells were able to provide help for Tg autoantibody production by the same B cells (see Figure 11). Whether or not similar T–B interactions occur in vivo awaits further investigation, but it is plausible that such

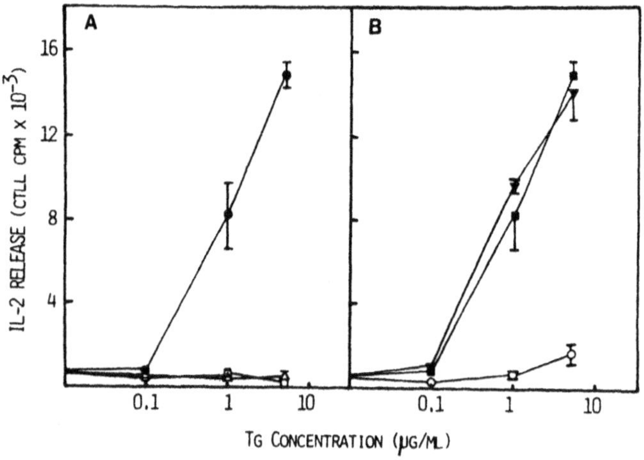

FIGURE 7. Highly efficient antigen presentation by Tg-specific B cells. Different antigen-presenting cell preparations were used to present low concentrations of Tg to the T cell hybridoma CH9. (A) Unseparated spleen cells from normal (△——△) or ovalbumin-primed (□——□) mice were only able to present Tg at high concentrations (not shown). However, cells from Tg-primed mice (●——●) were able to present low Tg concentrations (1–10 μg/mL). In some experiments, concentrations as low as 100 ng/mL could be presented by Tg-primed cells (not shown). (B) The efficient presentation by Tg-primed spleen cells (■——■) could be abrogated by the removal of B cells (○——○) but not T cells (▼——▼) from the antigen-presenting cell population. Blocking of B cell surface immunoglobulin receptors with anti-immunoglobulin also abrogated the efficient presentation of low Tg concentrations (not shown). This, and other evidence showing that dendritic cells were not involved, indicates that Tg-specific B cells can be very efficient at presenting antigen to T cells.

a mechanism may allow the maintenance of an autoimmune response to Tg following an initial triggering of B and T cells.

3.3. Thyroid Epithelial Cells

Since the demonstration of class II MHC antigen expression on thyroid epithelial cells from patients with autoimmune thyroiditis (Hanafusa *et al.*, 1983), there has been great interest in the possibility that, under pathological conditions, thyroid cells may assume an APC function and present their own (or extrinsic) antigens to T cells (Bottazzo *et al.*, 1983). Indeed, such Ia-positive thyroid cells can present foreign peptide antigens that do not require further processing (Londei *et al.*, 1984). While it is generally accepted that thyroid cells expressing class II antigens could play an exacerbating role in presenting antigens to the immune system, the possibility that such "aberrant" expression of class II molecules actually initiates autoimmune disease is a matter of controversy. In experimental and spontaneous autoimmune thyroiditis in rats, very few (if any) thyroid follicular cells appear to bear class II antigens, even in areas of dense lymphocytic infiltration (H. de Assis Paiva *et al.*, unpublished observations; Voorbij *et al.*, 1986). In the obese strain chicken, expression of Ia antigens on thyroid epithelial cells seems to be a secondary phenomenon resulting from lymphocytic infiltration (Wick *et al.*, 1985). Of course, how such observations relate to human thyroid disease is not yet clear.

Lymphocyte-mediated triggering of class II antigen expression is probably a result of the release of lymphokines, such as γ-interferon (γ-IFN), which has been shown to trigger Ia expression on cultured thyroid cells from humans and experimental animals (G. Biswas-Hughes *et al.*, unpublished observations; Salamero *et al.*, 1985; Todd *et al.*, 1985). In addition, we have shown that γ-IFN-stimulated cells also express increased levels of membrane Tg (as detected by indirect immunofluorescence staining) and release more Tg into the culture medium (Figure 8)(G. Biswas-Hughes *et al.*, unpublished observations). The mode of action of γ-IFN on Tg synthesis is not yet known, but if it occurs in vivo together with Ia expression it could lead to increased (or de novo) recognition by class II-restricted T cells and exacerbate the autoimmune pathology.

Antigen presentation to T helper cells by thyroid cells from experimental animals has not yet been directly demonstrated. The studies of Charreire and her colleagues have shown that primary T-cell sensitization can occur on thyroid epithelial cell monolayers (Yeni and Charreire, 1981). However, although it appears to be under I-A MHC subregion control (Salamero and Charreire, 1983), syngeneic sensitization seems to reflect the generation of class I-restricted cytotoxic T cells (Charreire and Salamero, 1984; Salamero and Charreire, 1985), which are thought to be responsible for the passive transfer of thyroid lesions in this system (Charreire and Michel-Bechet, 1982). Class I-restricted cytotoxic T

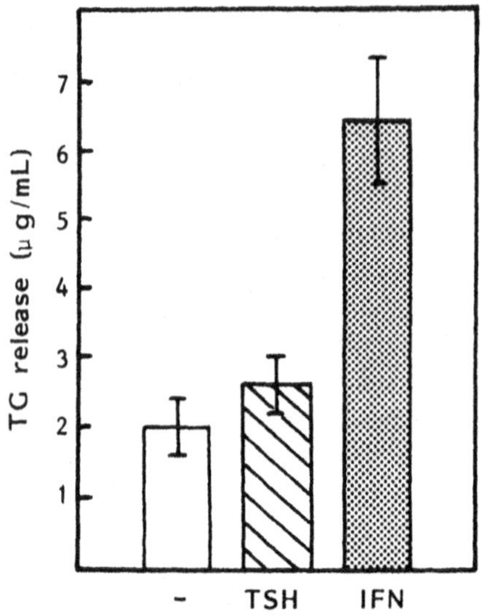

FIGURE 8. γ-Interferon triggers the release of thyroglobulin from cultured murine thyroid epithelial cells. Murine thyroid cells, prepared by collagenase digestion of pooled normal mouse thyroids, were cultured with or without (−) thyroid stimulating hormone (TSH) or γ-interferon (γ-IFN). Supernatants taken after 5 days of culture were tested for their Tg content by competition radioimmunoassay using solid-phase rabbit antimouse Tg and ^{125}I-labeled mouse Tg. The marked γ-interferon-induced increase in Tg release was confirmed in separate experiments showing increased surface membrane Tg (measured by indirect immunofluorescence) in the presence of this lymphokine (not shown).

cells, which recognize Tg on mouse thyroid cells, can also be generated by in vitro culture of Tg-primed lymph node cells (Creemers *et al.*, 1983). Thus, murine thyroid cells seem to be able to present Tg determinants to class I-restricted cytotoxic T cells but their ability to present antigen to class II-restricted T cells, when stimulated to express Ia molecules, awaits definitive experiments with T-cell clones.

The potential role of various APC types in presenting antigen (Tg) to class II-restricted T cells during the pathogenesis of autoimmune thyroid lesions is illustrated in Figure 9.

4. REGULATION OF AUTOIMMUNITY

4.1. Thyroglobulin Autoantibody Idiotypes

In recent years, evidence has indicated that cross-reactive or recurrent idiotypes (Ids) occur on autoantibodies and it has been proposed that such idiotypes may play a role in the regulation of the autoimmune response (Cooke, 1986). Matsuyama *et al.* (1983) provided evidence that such recurrent idiotypes occur on Tg autoantibodies from patients with Hashimoto's thyroiditis. A number of recent studies have attempted to assess the importance of idiotypic interactions in mice with EAT and buffalo rats with spontaneous autoimmune thyroiditis. A significant proportion of the Tg autoantibodies present in buffalo rats express a recurrent Id recognized by a polyvalent anti-idiotype (anti-Id) raised in rabbits.

This Id was found in up to 70% of the animals tested and in some cases constituted as much as 50% of the total Tg autoantibody. The Id was also found in the sera of other species having Tg antibodies, including those of patients with Hashimoto's thyroiditis (Zanetti *et al.*, 1983a). The anti-Id serum could inhibit the binding of Tg to the autoantibody and therefore appeared to be recognizing an

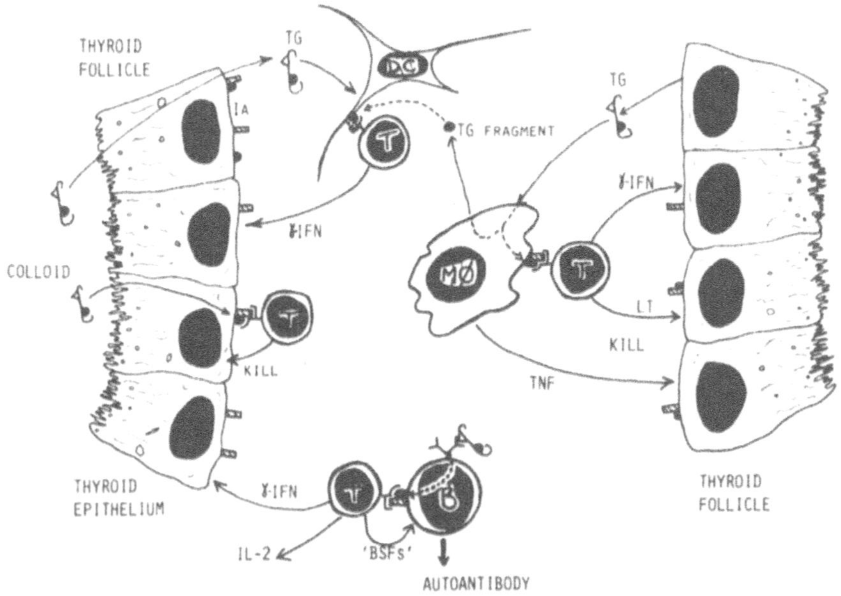

FIGURE 9. Antigen presentation in experimental thyroid autoimmunity. This figure illustrates some of the possible roles of antigen-presenting cell (APC) types in initiating and maintaining an autoimmune response to Tg. "Professional" APCs such as macrophages (Mϕ) and dendritic cells (DC) may process and present Tg or Tg fragments released from thyroid cells or colloidal stores of damaged follicles. Such APCs may also travel in afferent lymphatics, carrying Tg with them, and initiate autoimmune responses in the local lymph node. Activation of autoreactive T cells leads to the release of a variety of lymphokines, some of which may act on the thyroid cells. Lymphotoxin (LT) or tumor necrosis factor (TNF) may kill thyroid cells, thus contributing to the pathology and releasing more antigen to sustain the autoimmune response. γ-Interferon (γ-IFN) could trigger de novo expression of MHC class II molecules (Ia). Such Ia-positive thyroid cells might themselves present antigen to T helper cells or may act as targets for class II-restricted, Tg-specific cytotoxic T cells. Levels of MHC class I molecules are also increased on thyroid cells exposed to γ-IFN, perhaps also making them better targets for attack by class I-restricted cytotoxic T cells (not shown). B cells specific for Tg might concentrate and present their antigen to T cells, which then release various lymphokines, including those stimulating B cells ("BSFs"), which trigger the B cell to secrete specific autoantibody. Such autoantibodies might play a role in pathology by directing the antibody-dependent cellular cytotoxic (ADCC) mechanisms of cells such as macrophages or large granular lymphocytes (not shown).

antibody combining site-associated Id. In view of the wide distribution of this Id, the best explanation of the data may be that the paratope of Tg autoantibodies from different species is similar because it recognizes a conserved epitope on the Tg molecule. This Id is also detectable on spleen cells from rats with spontaneous or experimentally induced thyroiditis. This is reminiscent of the situation in mouse EAT where we have shown, using different adjuvants, that the spectrotype of induced autoantibodies is a characteristic of the strain, not of the method of EAT induction (D. Male *et al.*, unpublished observations).

Recurrent Ids are also found on Tg autoantibodies from mice with EAT. In CBA/Ca mice (H-2^k, EAT high responders), a site-nonassociated Id (D8.Id) was found in all animals tested, although it rarely constituted more than 5% of the total Tg autoantibodies (Male *et al.*, 1983). This Id was also detected on one of 14 other monoclonal Tg autoantibodies, indicating that recurrent Ids are not necessarily dominant in the autoimmune response. Similar levels of expression were found with the other site-nonassociated Ids G4 and P11, but the site-associated Id P4 was only detectable in two of the sera tested and was not present on any of the other monoclonal Tg autoantibodies. This contrasts with a study in BALB/c mice (H-2^d, low responders), which detected a site-associated Id (Id62) on up to 30% of the induced Tg autoantibodies (Zanetti *et al.*, 1983b).

Particular Ids are not uniquely associated with Tg autoantibodies since it is possible to detect Ids such as D8 on antibodies that do not bind Tg. It is also possible to increase the level of expression of recurrent Ids such as D8 by prior immunization with specific anti-Id, and in these circumstances the Id may constitute up to 30% of the induced Tg antibody (Male *et al.*, 1983). Spectrotypic examination of these antibodies from different strains of mice showed that a large proportion are identical to the prototype D8 Id-bearing monoclonal Tg autoantibody derived from CBA/Ca mice (Male *et al.*, 1985b). This means that a pool of Id-bearing B cells are available in this strain, which can be expanded selectively should a Tg autoantibody response be elicited. Presumably, the recurrent Ids that are seen in EAT are expressed on antibodies closest in shape to that required to form an anti-Tg paratope. However, since they can also be expressed on other antibodies, there would seem to be no specific link between the ability to express these particular Ids and the development of autoimmunity.

Although knowledge of the level of expression of different Ids allows some insight into how idiotypic interactions can affect different sets of autoantibodies, it does not show whether idiotypic interactions are of any significance in EAT. Evidence for an effective role of idiotypic interactions in these experimental conditions is still lacking. However, various experimental manipulations have been made, with the intention of modulating the Tg autoantibody response. Anti-Ids to both D8.Id and Id62 are able to induce or enhance Tg autoantibody production (Male *et al.*, 1983; Zanetti *et al.*, 1984). Polyvalent anti-Id62 was able to induce Tg antibodies in BALB/c mice (from which Id62 was originally isolated), buffalo rats, and the Fischer rat strain, which does not spontaneously develop thyroid autoimmunity. This further reinforces the view that Id62 is site-associated and that the anti-Id (or at least a proportion of it) may be an image of an epitope on Tg. In this case, the induction of Tg autoantibody is readily explained as being due to the direct stimulation of pools of autoreactive B cells recognizing this particular epitope. In the case of anti-D8.Id, evidence suggests that both B and T cells are stimulated in CBA/Ca mice, since T cells are more effective than B cells in transferring anti-Tg reactivity from animals primed with anti-Id to naive recipients. The observation that both T and B cells specific for a single antigen can share idiotopes has also been made on other systems. However, in the case of Tg the possibility is raised that autoreactive T and B cells could be controlled by a single idiotype-recognizing cell (Male *et al.*, 1986). Zanetti and colleagues have also suggested the possibility that Tg autoantibody is subject to idiotypic regulation, following their observation that Id62 (unlike many other defined idiotypes) is present on both the heavy and light chains of the monoclonal autoantibody (Zanetti *et al.*, 1985).

If idiotypic control of Tg autoantibody activity does occur, one might anticipate that, while anti-Id stimulates Tg antibodies, the corresponding Id might be suppressive. Experiments to examine this possibility have been carried out, both with Id62 in BALB/c mice and with D8.Id in CBA/Ca mice. Id62 did produce some suppression of the Tg autoantibody response when mice were subsequently immunized with low amounts of Tg (Zanetti, 1986). Pretreatment with D8.Id produced an almost total abrogation of the Tg antibody response in CBA/Ca mice (Male *et al.*, 1986). In these types of experiment, it is important to ensure that the apparent supression is not merely due to the sequestration of Tg by residual Id, thus preventing immunization (antibody-mediated immunosuppression). In our study using D8.Id, great care was taken to ensure that all detectable Id had been cleared from the circulation before immunizing the mice with Tg. Thus, in this case, it appeared that the Id was acting directly on a population of cells expressing anti-Id and, since both the Id^+ and the Id^- component of the response was suppressed, it appeared that the Id itself could activate a suppressor component of an immunoregulatory network (Male *et al.*, 1986).

In summary, idiotypic control of Tg autoantibody responses can be achieved experimentally but it is still not known whether such regulation occurs during the course of experimental or spontaneous thyroid autoimmunity. While attempts so far have concentrated on manipulating regulatory networks that involve recurrent Ids, there is still a reasonable possibility that physiologically active networks operate on private Ids present only within a particular individual.

4.2. T-Cell Lines and Hybridomas

As stated earlier, T cells from experimental animals primed with Tg can, after activation in vitro with antigen or mitogen, transfer thyroiditis to normal syngeneic recipients (Braley-Mullen *et al.*, 1981, 1985; Okayasu, 1985; Simon *et al.*, 1985, 1986). Long-term Tg-specific T-cell lines have also been shown to induce severe acute thyroid lesions in mice (Maron *et al.*, 1983). The great potential of such T-cell populations for studying the regulation of thyroid autoimmunity is well illustrated by the observation that the inactivated (irradiated) T-cell lines were able to vaccinate normal mice against the subsequent induction of experimental thyroid lesions. At present, the mechanism involved in the vaccination is uncertain, but it appears to act on the effector cells (since it is effective against line-mediated lesions) and may involve T-cell idiotypic interactions.

As discussed in Section 2.1.3, we have also established Tg-specific murine T-cell lines and clones (Champion *et al.*, 1985, 1986a). These cells are grown under different conditions than those reported by Maron *et al.* (1983) and are very inefficient at inducing significant thyroid lesions. These T cells bear most of the hallmarks of T helper cells; they are $Lyl^{+},2^{-}$, $L3T4^{+}$, I-A restricted and demonstrate activation-dependent secretion of a variety of factors that provide help for antigen-triggered B cells (bystander responses). However, the same T cells have profound suppressive effects on Tg autoantibody responses in vitro (Figure 10)(Champion *et al.*, 1986a). Recent evidence (B. R. Champion *et al.*, unpublished observations) has shown that these T cells are in fact indistinguishable from the recently described $L3T4^{+}$ class II-restricted, antigen-specific cytotoxic T cells (Golding *et al.*, 1985; Lukacher *et al.*, 1985; Morrison *et al.*, 1986; Tite and Janeway, 1984), which can also secrete nonspecific helper fac-

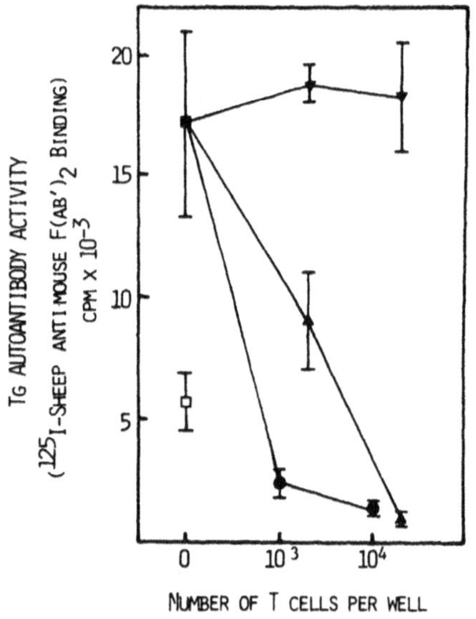

FIGURE 10. Suppression of Tg autoantibody responses in vitro by T cell lines. Autoantibody production by partially purified B cells (2 × 10^6 per well) from Tg-primed mice, cultured with Tg (5 μg/mL) in the presence or absence of T cells is shown: ●——●, the Tg-specific T cell clone MTg9B3; ▲——▲, the Tg-specific T cell line MTg12B; ▼——▼, the purified protein derivative (PPD)-specific T cell line PPD13P. B cells cultured alone without antigen are also shown (□). After 5 days, cultured cells were washed to remove antigen and the cells were recultured for a further 7 days before Tg antibody levels in culture media were determined by radioimmunoassay (as described in Figure 3). Low numbers of both the Tg-specific T cell populations, but not the PPD-specific T cell line, markedly suppressed the Tg autoantibody response. Subsequent experiments have indicated that the observed suppression is probably due to B cell killing, since both Tg-specific lines appear to be class II-restricted, cytotoxic T cells (not shown).

tors. Thus, our T-cell lines will help B cell responses through noncognate interactions but will kill Tg-specific B cells as a result of cognate interactions. Such class II-restricted Tg-specific T cells could conceivably play a role in thyroiditis by killing Ia-positive thyroid cells also expressing Tg antigenic determinants (see Figure 9). Although Ly2$^+$, class I-restricted cytotoxic T cells have been implicated in the pathogenesis of thyroid lesions (see Section 3.3), preliminary experiments by Kong and Waldmann (Y.-C. Kong, personal communication) have indicated that in vivo depletion of Ly2$^+$ T cells has little effect on the experimental induction of thyroiditis. It is interesting to speculate that, in this case, L3T4$^+$, class II-restricted cytotoxic T cells are activated in much greater numbers to compensate for the absence of Ly2$^+$, class I-restricted T cells. These same T cells might even secrete γ-IFN to induce Ia expression on thyroid cells, which could then act as targets for cytotoxicity. An alternative possibility is that class II-restricted cytotoxic T cells serve a protective function. By killing APCs expressing Tg determinants, they could down-regulate the response of true Tg-specific T helper cells, which are needed for the generation of class I-restricted cytotoxic cells (Creemers *et al.*, 1983).

As a different approach to the study of T cells in thyroid autoimmunity, we have established Tg-specific T-cell hybridomas (Rayner *et al.*, 1987). Because T-cell hybridomas can be obtained in large numbers, they have great potential for molecular studies of Tg-specific T cells. We have used T-cell hybridomas derived from our T-cell lines to study the epitopic specificities (see Section 2.1.3) and regulatory activities of autoreactive T cells. Two hybridomas (CH9 and ADA2), in contrast to the "suppressive" T cells from which they were derived, were able to provide specific help for Tg autoantibody responses (Figure 11). As

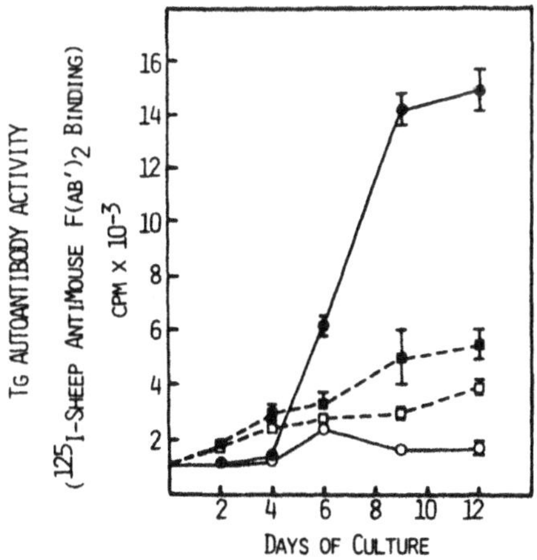

FIGURE 11. Help for Tg autoantibody responses in vitro by T cell hybridomas. The in vitro culture system described in Figure 10 was also used to test T cell hybridomas for their ability to help Tg autoantibody responses. The figure illustrates the antigen-dependent helper effect of CH9, a T cell hybridoma obtained from a fusion of the "suppressive" T cell line MTg12B. In this experiment, Tg was used at 0.1 μg/mL and the T cell hybridoma cells were irradiated to prevent their overgrowth in culture. □- - -□, B cells alone; ■- - -■, B cells + CH9; ○——○, B cells + Tg; ●——●, B cells + Tg + CH9. Another Tg-specific T cell hybridoma (ADA2) also helped B cell autoantibody responses but was somewhat less efficient than CH9 (not shown).

discussed in Section 3.2, this help was the result of specific antigen presentation by the Tg-specific B cells. Since, in at least one case (ADA2), the hybridoma was derived from a clonal T-cell line, it would seem that help can be observed because the hybridomas have lost their cytotoxic capability. This possibility is currently under investigation. Such functional losses are common with T-cell hybridomas that have to be cloned regularly to maintain their antigen specificity.

5. CONCLUDING REMARKS

In this chapter, we have used some of our experimental observations to highlight some specific aspects of experimental thyroid autoimmunity, which are beginning to be analyzed in detail at the molecular level. Clearly, other areas of research not covered here are also providing important insights. We hope that detailed molecular analyses, particularly utilizing monoclonal cell populations, will lead to a better understanding of the mechanisms involved in initiating and regulating thyroid autoimmunity as well as autoimmune responses in general.

6. REFERENCES

Abbas, A. K., Haber, S., and Rock, K. L., 1985, Antigen presentation by hapten-specific B lymphocytes. II. Specificity and properties of antigen-presenting B lymphocytes, and function of immunoglobulin receptors. *J. Immunol.* **135:**1661–1667.

Allen, E. M., Appel, M. C., and Braverman, L. E., 1986, The effect of iodide ingestion on the development of spontaneous lymphocytic thyroiditis in the diabetes-prone BB/W rat. *Endocrinology* **118:**1977–1981.

Bagchi, N., Brown, T. R., Urdanivia, E., and Sundick, R. S., 1985, Induction of autoimmune thyroiditis in chickens by dietary iodine. *Science* **230:**325–327.

Banga, J. P., Pryce, G., Hammond, L., and Roitt, I. M., 1985, Structural features of the autoantigens involved in thyroid autoimmune disease: Thyroid microsomal/microvillar antigen. *Mol. Immunol.* **22:**629–642.

Beierwaltes, W. H., 1969, Iodine and lymphocytic thyroiditis. *Bull. All. India Med. Sci.* **3:**145–146.

Benjamin, D. C., Berzofsky, J. A., East, I. J., Gurd, F. R. N., Hannum, C., Leach, S. J., Margoliash, E., Michael, J. G., Miller, A., Prager, E. M., Reichlin, M., Sercarz, E. E., Smith-Gill, S. J., Todd, P. E., and Wilson, A. C., 1984, The antigenic structure of proteins: A reappraisal. *Annu. Rev. Immunol.* **2:**67–101.

Bigazzi, P. E., and Rose, N. R., 1975, Spontaneous autoimmune thyroiditis in animals as a model of human disease. *Prog. Allergy* **19:**245–274.

Bigazzi, P. E., and Rose, N. R., 1985, Autoimmune thyroid disease. In *The Autoimmune Diseases* (N. R. Rose and I. R. Mackay, eds.), pp. 161–198, Academic Press, Orlando.

Bottazzo, G. F., Pujol-Borrel, R., Hanafusa, T., and Feldmann, M., 1983, Role of aberrant HLA-DR expression and antigen presentation in induction of endocrine autoimmunity. *Lancet* **2:**1115–1118.

Braley-Mullen, H., Thompson, J. G., Sharp, G. C., and Kyriakos, M., 1981, Adoptive transfer of experimental autoimmune thyroiditis (EAT) with *in vitro* activated lymph node cells from thyroglobulin sensitized guinea-pigs: Characterization of the cell that transfers EAT. *J. Immunol.* **127**:1767–1771.

Braley-Mullen, H., Thompson, J. G., Sharp, G. C., and Kyriakos, M., 1983, Adoptive transfer of experimental autoimmune thyroiditis (EAT) in guinea-pigs: Requirement for Ia-positive antigen-presenting cells for *in vitro* activation of effector T cells. *Cell. Immunol.* **81**:243–252.

Braley-Mullen, H., Johnson, M., Sharp, G. C., and Kyriakos, M., 1985, Induction of experimental autoimmune thyroiditis in mice with *in vitro* activated splenic T cells. *Cell. Immunol.* **93**:132–143.

Brown, T. R., Bagchi, N., and Sundick, R. S., 1985, Analysis of tri-iodothyronine and thyroxine-binding autoantibodies in chickens susceptible to autoimmune thyroiditis. *J. Immunol.* **134**:3845–3848.

Champion, B. R., Varey, A.-M., Katz, D., Cooke, A., and Roitt, I. M., 1985, Autoreactive T cell lines specific for mouse thyroglobulin. *Immunology* **54**:513–519.

Champion, B. R., Hutchings, P., Davies, S., Marshall-Clarke, S., Cooke, A., and Roitt, I. M., 1986a, Helper and suppressor activities of an autoreactive mouse thyroglobulin-specific T cell clone. *Immunology* **58**:51–56.

Champion, B., Hutchings, P., Rayner, D., Marshall-Clarke, S., Roitt, I. M., and Cooke, A., 1986b, Autoantigenicity of murine thyroglobulin. *J. Endocrinol. Invest.* **9**(Suppl.3):77.

Charreire, J., and Michel-Bechet, M., 1982, Syngeneic sensitization of mouse lymphocytes on monolayers of thyroid epithelial cells. III. Induction of thyroiditis by thyroid-sensitized T lymphoblasts. *Eur. J. Immunol.* **12**:420–425.

Charreire, J, and Salamero, J., 1984, Syngeneic sensitization of mouse lymphocytes on monolayers of thyroid epithelial cells. VI. Evaluation of its frequency using limiting dilution analysis. *Tissue Antigens* **23**:72–80.

Cooke, A., 1986, Idiotypes, anti-idiotypes and autoimmune disease. *Plasma Ther. Transfus. Technol.* **7**:161–170.

Creemers, P., Rose, N. R., and Kong, Y. M., 1983, Experimental autoimmune thyroiditis. *In vitro* cytotoxic effects of T lymphocytes on thyroid monolayers. *J. Exp. Med.* **157**:559–571.

Czarnocka, B., Ruf, J., Ferrand, M., Carrayon, P., and Lissitzky, S., 1985, Purification of the human thyroid peroxidase and its identification as the microsomal antigen involved in autoimmune thyroid diseases. *FEBS Lett.* **190**:147–152.

Czarnocka, B., Ruf, J., Ferrand, M., Lissitzky, S., and Carrayon, P., 1986, Evidence that thyroid peroxidase is the "microsomal" antigen. In *The Thyroid and Autoimmunity* (H. A. Drexhage and W. M. Wiersinga, eds.), pp. 84–85, Elsevier, Amsterdam.

Di Lauro, R., Obici, S., Condliffe, D., Ursini, V. M., Musti, A., Moscatelli, C., and Avvedimento, V. E., 1985, The sequence of 967 amino acids at the carboxyl-end of rat thyroglobulin. Location and surroundings of two thyroxine-forming sites. *Eur. J. Biochem.* **148**:7–11.

Dinsart, C., Libert, F., Ruel, J., Ludgate, M., Pommier, J., Dussault, J., and Vassart, G., 1986, Molecular cloning of a thyroid peroxidase cDNA fragment encoding epitopes involved in Hashimotos thyroiditis (HT). *J. Endocrinol. Invest.* **9**(Suppl.3):90.

Formisano, S., Moscatelli, C., Zarrilli, R., Di Ieso, B., Acquaviva, R., Obici, S., Palumbo, G., and Di Lauro, R., 1985, Prediction of the secondary structure of the carboxy-terminal third of rat thyroglobulin. *Biochem. Biophys. Res. Commun.* **133**:766–772.

Golding, H., Munitz, T. I., and Singer, A., 1985, Characterization of antigen-specific Ia-restricted, $L3T4^+$ cytolytic T lymphocytes and assessment of thymic influence on their self specificity. *J. Exp. Med.* **162**:943–961.

Hanafusa, T., Pujol-Borrel, R., Chiovata, R., Russel, R. C. G., Doniach, D., and Bottazzo, G. F., 1983, Aberrant expression of HLA-DR antigen on thyrocytes in Graves' disease: Relevance for autoimmunity. *Lancet* **2**:1111–1115.

Harach, H. R., Escalante, D. A., Onativia, A., Outes, J. L., Day, E. S., and Williams, E. D., 1985,

Thyroid carcinoma and thyroiditis in an endemic goitre region before and after iodine prophylaxis. *Acta Endocrinol.(Copenh.)* **108**:55–60.

Hutchings, P., Rayner, D. C., Champion, B. R., Marshall-Clarke, S., Macatonia, S., Roitt, I. M., and Cooke, A., 1987, High efficiency antigen presentation by thyroglobulin primed murine splenic B cells. *Eur. J. Immunol.* **17**:393–398.

Kohno, Y., Nakajima, H., and Tarutani, O., 1985, Interspecies cross-reactive determinants of thyroglobulin recognized by autoantibodies. *Clin. Exp. Immunol.* **61**:44–48.

Kong, Y. M., Rose, N. R., Elrehewy, M., Michaels, R., Giraldo, A. A., Accavitti, M. A., and Leon, M. A., 1980, Thyroid alloantigens in autoimmunity. *Transplant. Proc.* **12**:(Suppl-.1):129–134.

Kong, Y.-C. M., Bagnasco, M., and Canonica, G. W., 1986a, How do T cells mediate autoimmune thyroiditis? *Immunol. Today* **7**:337–339.

Kong, Y. M., Giraldo, A. A., Justin, J. M., Simon, L. L., and Fuller, B. E., 1986b, Are conserved regions on thyroglobulin thyroiditogenic? In *The Thyroid and Autoimmunity* (H. A. Drexhage and W. M. Wiersinga, eds.), pp. 151–152, Elsevier, Amsterdam.

Kotani, T., Maeda, M., Umeki, K., and Ohtaki, S., 1985a, Epitopic differences among rat thyroglobulins. *Immunol. Lett.* **9**:167–172.

Kotani, T., Maeda, M., Ohtaki, S., Kobayashi, S., and Okuyama, H., 1985b, Analyses of the specificity of mouse monoclonal antibodies to syngeneic and allogeneic thyroglobulin. *Clin. Immunol. Immunopathol.* **34**:147–157.

Lanzavecchia, A., 1985, Antigen-specific interaction between T and B cells. *Nature* **314**:537–538.

Londei, M., Lamb, J. R., Bottazzo, G. F., and Feldmann, M., 1984, Epithelial cells expressing aberrant MHC class II determinants can present antigen to cloned human T cells. *Nature* **312**:639–641.

Lukacher, A. E., Morrison, L. A., Braciale, V. L., Malissen, B., and Braciale, T. J., 1985, Expression of specific cytolytic activity by H-2I region restricted, influenza virus-specific T lymphocyte clones. *J. Exp. Med.* **162**:171–187.

Male, D., Pryce, G., Quartey-Papafio, R., and Roitt, I. M., 1983, The occurrence of defined idiotypes on autoantibodies to mouse thyroglobulin. *Eur. J. Immunol.* **13**:942–947.

Male, D. K., Champion, B. R., Pryce, G., Matthews, H., and Shepherd, P., 1985a, Antigenic determinants on human thyroglobulin differentiated using antigen fragments. *Immunology* **54**:419–427.

Male, D., Pryce, G., and Roitt, I. M., 1985b, Molecular analysis of induced idiotypes associated with autoanti-thyroglobulin. *Mol. Immunol.* **22**:255–263.

Male, D., Pryce, G., Cooke, A., Hutchings, P., Marshall-Clarke, S., and Roitt, I. M., 1986, T and B cell connections in experimentally induced autoimmunity. *Ann. N.Y. Acad. Sci.* **475**:94–105.

Malthiery, Y., and Lissitzky, S., 1985, Sequence of the 5′-end quarter of the human thyroglobulin messenger ribonucleic acid and of its deduced amino-acid sequence. *Eur. J. Biochem.* **147**:53–58.

Mangkornkanok, M., Marcowitz, A. S., and Battifora, H. A., 1972, Chronic thyroiditis in the rabbit induced with homologous thyroid microsomes. *Science* **178**:316–318.

Maron, R., Zerubavel, R., Friedman, A., and Cohen, I. R., 1983, T lymphocyte line specific for thyroglobulin produces or vaccinates against autoimmune thyroiditis in mice. *J. Immunol.* **131**:2316–2322.

Mariotti, S., Bechi, R., Anelli, S., Ruf, J., Lombardi, A., Chiovata, L., Czarnocka, B., Carayon, P., and Pinchera, A., 1986, Evidence for identity of serum thyroid microsomal and anti-human thyroid peroxidase autoantibodies. In *The Thyroid and Autoimmunity* (H. A. Drexhage and W. M. Wiersinga, eds.), pp. 86–87, Elsevier, Amsterdam.

Marriq, C., Rolland, M., and Lissitzky, S., 1983, Amino acid sequence of the unique 3,5,5′-

triiodothyronine-containing sequence from porcine thyroglobulin. *Biochem. Biophys. Res. Commun.* **112**:206–213.

Marriq, C., Lejeune, P. J., Malthiery, Y., Rolland, M., and Lissitzky, S., 1984, Precursor–product relationship between the 26kDa and 18kDa fragments formed by iodination of human thyroglobulin. *FEBS Lett.* **175**:140–146.

Matsuyama, T., Fukumori, J., and Tanaka, H., 1983, Evidence of unique idiotypic determinants and similar idiotypic determinants on human anti-thyroglobulin antibodies. *Clin. Exp. Immunol.* **51**:381–386.

Mercken, L., Simons, M.-J., Swillens, S., Massaer, M., and Vassart, G., 1985, Primary structure of bovine thyroglobulin deduced from the sequence of its 8431-base complementary DNA. *Nature* **316**:647–651.

Morrison, L. A., Lukacher, A. E., Braciale, V. L., Fan, D. P., and Braciale, T. J., 1986, Differences in antigen presentation to MHC class I- and class II-restricted influenza virus-specific cytolytic T lymphocyte clones. *J. Exp. Med.* **163**:903–921.

Musti, A. M., Avvedimento, E. V., Polistina, C., Ursini, V. M., Obici, S., Nitsch, L., Cocozza, S., and Di Lauro, R., 1986, The complete structure of the rat thyroglobulin gene. *Proc. Natl. Acad. Sci. U.S.A.* **83**:323–327.

Nye, L., Pontes de Carvalho, L. C., and Roitt, I. M., 1980, Restrictions in the response to autologous thyroglobulin in the human. *Clin. Exp. Immunol.* **41**:252–263.

Okayasu, I., 1985, Transfer of experimental autoimmune thyroiditis to normal syngeneic mice by injection of mouse thyroglobulin-sensitized T lymphocytes after activation with concanavilin A. *Clin. Immunol. Immunopathol.* **36**:101–109.

Pearce, C. J., Byfield, G. H., Edmonds, C. J., Lalloz, M. R. A., and Himsworth, R. L., 1981, Autoantibodies to thyroglobulin cross-reacting with iodothyronines. *Clin. Endocrinol.* **15**:1–10.

Penhale, W. J., Farmer, A., McKenna, R. P., and Irvine, W. J., 1973, Spontaneous thyroiditis in thymectomized and irradiated Wistar rats. *Clin. Exp. Immunol.* **15**:225–236.

Penhale, W. J., Farmer, A., and Irvine, W. J., 1975, Thyroiditis in T-cell depleted rats. Influence of strain, radiation dose, adjuvants and anti-lymphocyte serum. *Clin. Exp. Immunol.* **21**:362–375.

Portmann, L., Hamada, N., Heinrich, G., and De Groot, L. J., 1985, Anti-thyroid peroxidase antibody in patients with autoimmune thyroid disease: Possible identity with anti-microsomal antibody. *J. Clin. Endocrinol. Metab.* **61**:1001–1003.

Rawitch, A. B., Chernoff, S. B., Litwer, M. R., Rouse, J. B., and Hamilton, J. W., 1983, Thyroglobulin structure-function. The amino acid sequence surrounding thyroxine. *J. Biol. Chem.* **258**:2079–2082.

Rayner, D. C., Delves, P. J., Warren, D., Roitt, I. M., and Champion, B. R., 1987, T cell hybridomas specific for self and foreign thyroglobulins. *Immunology* **60**:231–237.

Rock, K. L., Benacerraf, B., and Abbas, A. K., 1984, Antigen presentation by hapten-specific B lymphocytes. I. Role of surface immunoglobulin receptors. *J. Exp. Med.* **160**:1102–1113.

Roitt, I. M., Campbell, P. N., and Doniach, D., 1958, The nature of the thyroid autoantibodies present in patients with Hashimoto's thyroiditis (lymphadenoid goitre). *Biochem. J.* **69**:248–257.

Romball, C. G., and Weigle, W. O., 1984, T cell competence to heterologous and homologous thyroglobulins during the induction of experimental autoimmune thyroiditis. *Eur. J. Immunol.* **10**: 887–893.

Rose, N. R., and Witebsky, E., 1956, Studies on organ specificity. V. Changes in the thyroid glands of rabbits following active immunization with rabbit thyroid extracts. *J. Immunol.* **76**:417–427.

Rose, N. R., Kong, Y. M., Okayasu, I., Giraldo, A. A., Beisel, K., and Sundick, R. S., 1981, T-cell regulation in autoimmune thyroiditis. *Immunol. Rev.* **55**:299–314.

Rose, N. R., Accavitti, M., Pydyn, E. F., Leon, M. A., and Brown, R. K., 1982, The use of hybridoma antibodies to probe the antigenic determinants of thyroglobulin. *Adv. Exp. Med. Biol.* **150**:23–35.

Salamero, J., and Charreire, J., 1983, Syngeneic sensitization of mouse lymphocytes on monolayers of thyroid epithelial cells. V. The primary syngeneic sensitization is under I-A subregion control. *Eur. J. Immunol.* **13**:948–951.

Salamero, J., and Charreire, J., 1985, Syngeneic sensitization of mouse lymphocytes on monolayers of thyroid epithelial cells (TEC). VII. Generation of thyroid specific cytotoxic effector cells. *Cell. Immunol.* **91**:111–118.

Salamero, J., Michel-Bechet, M., Wietzerbin, J., and Charreire, J., 1985, Syngeneic sensitization of mouse lymphocytes on monolayers of thyroid epithelial cells (TEC). VIII. γ-Interferon induced Ia antigen expression on TEC cultures. *Tissue Antigens* **25**:266–277.

Schwartz, R. H., 1985, T-lymphocyte recognition of antigen in association with gene products of the major histocompatibility complex. *Annu. Rev. Immunol.* **3**:237–261.

Shimojo, N., Kohno, Y., Tarutani, O., Sasaki, N., and Nakajima, H., 1986, Antigenic determinants on thyroglobulin recognized by T cells in patients with chronic thyroiditis. *Acta Endocrinol.* **111**:481–486.

Shimonkevitz, R., Kappler, J., Marrack, P., and Grey, H., 1983, Antigen recognition by H-2-restricted T cells. I. Cell-free antigen processing. *J. Exp. Med.* **158**:303–316.

Simon, L. L., Krco, C. J., David, C. S., and Kong, Y. M., 1985, Characterization of the in vitro T cell proliferative response to murine and human thyroglobulins in thyroiditis-susceptible and -resistant mice. *Cell. Immunol.* **94**:243–253.

Simon, L. L., Justen, J. M., Giraldo, A. A., Krco, C. J., and Kong, Y.-C. M., 1986, Activation of cytotoxic T cells and effector cells in experimental autoimmune thyroiditis by shared determinants of mouse and human thyroglobulins. *Clin. Immunol. Immunopathol.* **39**:345–356.

Sternthal, E., Like, A. A., Sarantis, K., and Braverman, L. E., 1981, Lymphocytic thyroiditis and diabetes in the BB/W rat. A new model of autoimmune endocrinopathy. *Diabetes* **30**:1058–1061.

Stylos, W. A., and Rose, N. R., 1977, Splitting of human thyroglobulin. VI. The antigenicity of the pepsin derived fragments. *Clin. Exp. Immunol.* **27**:245–253.

Sundick, R. S., Herdegen, D., Brown, T. R., and Bagchi, N., 1986, Thyroiditis induced by dietary iodine may be due to the increased immunogenicity of highly iodinated thyroglobulin. In *The Thyroid and Autoimmunity* (H. A. Drexhage and W. M. Wiersinga, eds.), pp. 213–214, Elsevier, Amsterdam.

Targovnik, H. M., Pohl, V., Christophe, D., Cabrer, B., Brocas, H., and Vassart, G., 1984, Structural organization of the 5′ region of the human thyroglobulin gene. *Eur. J. Biochem.* **141**:271–277.

Tite, J. P., and Janeway, C. A., Jr., 1984, Cloned helper T cells can kill B lymphoma cells in the presence of specific antigen: Ia restriction and cognate vs. non-cognate interactions in cytolysis. *Eur. J. Immunol.* **14**:878–886.

Todd, I., Pujol-Borrel, R., Hammond, L., Bottazzo, G. F., and Feldmann, M., 1985, Interferon-γ induces HLA-DR expression by thyroid epithelium. *Clin. Exp. Immunol.* **61**:265–273.

Tomazic, V., and Rose, N. R., 1976, Autoimmune thyroiditis. VIII. Role of different thyroid antigens in the induction of experimental autoimmune thyroiditis. *Immunology* **30**:63–68.

Van Herle, A. J., Vassart, G., and Dumont, J. E., 1979, Control of thyroglobulin synthesis and secretion. *N. Engl. J. Med.* **301**:239–249, 307–314.

Vassart, G., Bacolla, A., Brocas, H., Christophe, D., De Martynoff, G., Leriche, A., Mercken, L., Parma, J., Pohl, V., and Targovnik, H., 1985, Structure, expression and regulation of the thyroglobulin gene. *Mol. Cell. Endocrinol.* **40**:89–97.

Vladutiu, A. O., and Rose, N. R., 1972, Murine thyroiditis. III. Influence of syngeneic and allogeneic thyroid antigen on induction of the disease. *Clin. Exp. Immunol.* **11**:245–254.

Voorbij, H. A. M., Kabel, P. J., and Drexhage, H. A., 1986, Antigen presenting cells (APC's) in the thyroid autoimmune response in the BB/W rat. In *The Thyroid and Autoimmunity* (H. A. Drexhage and W. M. Wiersinga, eds.), pp. 166–169, Elsevier, Amsterdam.
Weetman, A. P., and McGregor, A. M., 1984, Autoimmune thyroid disease: Developments in our understanding. *Endocrine Rev.* **5**:309–355.
Weetman, A. P., McGregor, A. M., and Hall, R., 1983, Thyroglobulin uptake and presentation by macrophages in experimental autoimmune thyroiditis. *Immunology* **50**:315–318.
Weigle, W. O., 1980, Analysis of autoimmunity through experimental models of thyroiditis and allergic encephalomyelitis. *Adv. Immunol.* **30**:159–273.
Wick, G., Boyd, R., Hala, K., Thunold, S., and Kofler, H., 1982, Pathogenesis of spontaneous autoimmune thyroiditis in obese strain (OS) chickens. *Clin. Exp. Immunol.* **47**:1–18.
Wick, G., Most, J., Schauenstein, K., Kromer, K., Dietrich, H., Ziemiecki, A., Fassler, R., Schwarz, S., Neu, N., and Hala, K., 1985, Spontaneous autoimmune thyroiditis—a bird's eye view. *Immunol. Today* **6**:359–364.
Yeni, P., and Charreire, J., 1981, Syngeneic sensitization of mouse lymphocytes on monolayers of thyroid epithelial cells. I. Study of proliferative responses. *Cell. Immunol.* **62**:313–323.
Zanetti, M., Barton, R., and Bigazzi, P., 1983a, Anti-idiotypic immunity and autoimmunity. II. Idiotypic determinants of autoantibodies and lymphocytes in spontaneous and experimentally induced autoimmune thyroiditis. *Cell. Immunol.* **75**:292–301.
Zanetti, M., De Baets, M., and Rogers, J., 1983b, High degree of idiotypic cross-reactivity among murine antibodies to thyroglobulin. *J. Immunol.* **131**:2452–2457.
Zanetti, M., Rogers, J., and Katz, D. H., 1984, Induction of autoantibodies to thyroglobulin by anti-idiotypic antibodies. *J. Immunol.* **133**:240–243.
Zanetti, M., Liu, F. T., Rogers, J., and Katz, D. H., 1985, Heavy and light chains of the same idiotype. *J. Immunol.* **135**:1245–1251.
Zanetti, M., 1986, Idiotypic regulation of autoantibody production. *CRC Crit. Rev.* **6**:151–183.

Chapter 10

Biochemistry and Pathophysiology of the Molecular Forms of Cholinesterases

Zoltán Rakonczay and Stephen Brimijoin

1. INTRODUCTION

Acetylcholinesterase (acetylcholine acetylhydrolase, AChE: EC 3.1.1.7) and butyrylcholinesterase (acylcholine acylhydrolase, BuChE: EC 3.1.1.8) both possess the capacity to hydrolyze choline esters, although the latter accepts a much wider variety of substrates. These two enzymes are found in a large number of excitable and nonexcitable tissues in most species, including humans (for review see Silver, 1974). The two types of ChE (generic abbreviation for any cholinesterase) are readily distinguished not only by their substrate specificity but also by their response to selective inhibitors.

AChE is associated primarily with cholinergic neurons and cholinoceptive cells, but it is also found in other neurons and in non-neural cells like the erythrocyte. The importance of AChE in neurotransmission is well known. It is underscored by the toxicity associated with irreversible enzyme inhibition, which

Abbreviations used in this chapter: AChE, acetylcholinesterase; AD, Alzheimer's disease; ALS, amyotrophic lateral sclerosis; BuChE, butyrylcholinesterase; cDNA, complementary DNA; ChAT, choline acetyltransferase; ChE, cholinesterase; CSF, cerebrospinal fluid; EDL, extensor digitorum longus; FACS, fluorescence-activated cell sorter; McAb, monoclonal antibody; mdg, muscular dysgenesis; med, motor endplate disease; NTD, neural tube defects; PNH, paroxysmal nocturnal hemoglobinuria; PTA, phenyltrimethylammonium; RBC, red blood cell; SCG, superior cervical ganglion.

Zoltán Rakonczay Central Research Laboratory, Medical University, H-6720 Szeged, Hungary.
Stephen Brimijoin Department of Pharmacology, Mayo Clinic, Rochester, Minnesota 55905.

by a sustained parasympathomimetic action and disturbance of muscular contraction can rapidly lead to death.

BuChE is present in the plasma of many vertebrates (thus the term "plasma cholinesterase"). It is less well appreciated that this enzyme also occurs in tissues such as central and peripheral nervous system, muscle, liver, and various exocrine and endocrine glands (Goedde *et al.*, 1967; Silver, 1974). The function of BuChE is almost completely unknown, as are its natural substrates and inducers. Nevertheless, BuChE is of clinical importance in the hydrolysis of muscle relaxants (Brown *et al.*, 1981; Kalow and Genest, 1957), ester-type local anesthetics, and certain other drugs (La Du, 1971).

Both ChEs have unusually high turnover numbers. The great enzyme activity, along with the availability of specific enzyme inhibitors, has permitted extensive study of catalytic mechanisms and tissue distribution. One remarkable outcome has been the discovery that AChE and BuChE each exist in a series of molecular forms with close structural parallels (Brimijoin, 1983; Brzin *et al.*, 1983; Massoulié and Bon, 1982; Rakonczay, 1986; Rosenberry, 1985).

The large number of recent reviews on ChEs makes it unnecessary to cover all topics in detail. Classical reaction mechanism, kinetic properties, anatomical and histochemical analyses, and pharmacological studies are not considered in this chapter. Elsewhere, the present authors have reviewed the molecular biology and immunology of ChEs (Brimijoin and Rakonczay, 1986). Here, we concentrate on the molecular structure and biochemical properties of ChE molecular forms, the distribution of the enzyme species in different tissues, and the changes associated with some kinds of pathology.

2. MOLECULAR STRUCTURE AND BIOCHEMICAL PROPERTIES OF CHOLINESTERASES

Massoulié and Rieger (1969) were the first to demonstrate that fresh extracts of *Torpedo* electric organ contained multiple forms of AChE that could be separated by density gradient ultracentrifugation. These enzyme forms are stable entities with characteristic Stokes radii and sedimentation coefficients. Hall (1973) later found multiple molecular forms of AChE in extracts of rat diaphragm. One of these forms, a 16S species, was associated primarily with endplate-rich regions of the muscle. After denervation, this AChE form showed the largest proportional decrease in activity. Hall suggested that the various forms occupy different sites within the muscle. At least one of the forms appeared to be maintained by innervation, probably at the neuromuscular junction itself. Later work has borne out these conclusions.

AChE forms have been divided into two classes on the basis of their structural properties (Bon *et al.*, 1979). One class of forms shows a high degree of asymmetry by hydrodynamic criteria and possesses an elongated triple helical collagenlike "tail" structure in the electron micrographs (Dudai *et al.*, 1973; Rieger *et al.*, 1973). They are sensitive to collagenase and contain one, two, or three tetrameric arrays of subunits. These asymmetric forms are designated, respectively, A_4, A_8, and A_{12} (the latter corresponding to the 16S species). All other AChE forms are believed to be globular proteins that lack a collagenlike tail and contain one, two, and four subunits. These forms are designated, respectively, G_1, G_2, and G_4 (the latter corresponding to the tetrameric constituents of the asymmetric forms). This classification, while it does not describe the detailed structure of the molecules, is applicable to the molecular forms of BuChE as well.

Although ChEs are present in multiple forms, each form of a given enzyme appears to be enzymologically equivalent. Furthermore, no functionally significant subunit interactions are known. Vigny *et al.* (1978) used a stable, irreversible phosphorylating inhibitor to titrate the active sites of different forms of AChE from *Electrophorus, Torpedo,* rat, and chicken brain. They found that all molecular forms had the same catalytic activity per active site.

2.1. Heterogeneity and Quaternary Structure

The native molecular weights of AChE and BuChE are in the 70–85-kD range (Adamson *et al.*, 1975; Lockridge *et al.*, 1979; Muensch *et al.*, 1976; Rakonczay *et al.*, 1981a; Sørensen *et al.*, 1982) although estimates vary according to the species, sources, and methods. Each globular catalytic subunit possesses a single active site containing an "active serine." The enzymes are recognized as glycoproteins. In fact, the proportion of the molecular mass comprised by carbohydrate is 8% in the AChE from *Torpedo* electroplax (Taylor *et al.*, 1974), 13% in the AChE from human erythrocyte (Niday *et al.*, 1977), and 24% in the BuChE from human serum (Muensch *et al.*, 1976).

The function of the carbohydrate moieties is unknown. It can be speculated that the extensive carbohydrate content of BuChE has a role in protecting the enzyme from serum proteases. The smaller amounts of carbohydrate associated with AChE could have some effect on enzyme conformation. It is known that deglycosylated *Torpedo* AChE displays a molecular mass close to the figure of 66 kD deduced from the cDNA sequence (Schumacher *et al.*, 1986). As yet, however, no studies have been reported on the function or the three-dimensional structure of the enzyme in this state.

Based on in vitro solubility, three main types of AChE may be discerned: (1) high-salt-soluble, (2) low-salt-soluble, and (3) detergent-soluble. The high-salt-soluble forms are hydrophilic and asymmetric, with a collagenlike tail (i.e., exclusively A_4, A_8, or A_{12}). These forms occur mainly in the basal lamina of the neuromuscular junction. The low-salt-soluble (or "naturally soluble") AChEs are globular (G_1, G_2, or G_4). These are the secretory forms of the enzyme. The detergent-soluble AChEs are also globular but are membrane-associated entities. The proportions of the different globular forms vary significantly from tissue to tissue.

The globular forms may be separated further into subfractions according to electrophoretic mobility and lectin-binding properties. Nondenaturing electrophoresis revealed the heterogeneity of G_1 and G_4 AChE forms in rat sympathetic ganglia (Gisiger *et al.*, 1978). In the rat sciatic nerve, axonally transported G_4 was shown to differ in electrophoretic mobility from stationary forms of the same size (Kása and Rakonczay, 1982a). By comparing the lectin binding of AChEs from bovine erythrocyte, lymphocytes, brain, and muscle, Méflah *et al.* (1984) showed that otherwise identical molecular forms may differ in their glycosylation. For example, erythrocyte and lymphocyte membranes both contain a G_2 form, but only the former reacted with concanavalin A and *Ulex europeus* lectin. These results suggest that glycosylation contributes significantly to the polymorphism of AChE in a tissue-specific manner.

The structure of the tail component of asymmetric ChEs has been the subject of numerous studies but is still not well characterized. This component resembles collagen in its triple helical structure (Mays and Rosenberry, 1981), its sensitivity to collagenase, and its high content of hydroxyproline and hydroxylysine (Anglister *et al.*, 1979; Bon *et al.*, 1978; Silberstein *et al.*, 1982). There have been conflicting claims, however, regarding the immunological relation with collagen (Anglister *et al.*, 1979; Grassi *et al.*, 1983; Vigny *et al.*, 1983).

There is more agreement about the functional significance of the tail component. Several lines of evidence suggest that this component is responsible for anchoring asymmetric AChE in the basement membrane of the neuromuscular junction. For example, A_{12} AChE from eel electroplax binds to basement membrane constituents such as type V collagen and especially to heparin sulfate proteoglycans (Brandan and Inestrosa, 1984; Brandan *et al.*, 1985; Vigny *et al.*, 1983). This binding is dependent on the presence of the tail component (Vigny *et al.*, 1983). Purified A_{12} AChE from electroplax is found to contain a heparin-sulfate-like component (Vigny *et al.*, 1983). Finally, heparin specifically solubilizes A_8 and A_{12} AChE from endplate regions of the rat diaphragm.

The elements responsible for the plasmalemmal anchoring of membrane-bound globular ChEs are of equal interest. It is now established that the globular AChEs of *Torpedo* electroplax and human erythrocytes are anchored through a complex, inositol-containing glycolipid (Dutta-Choudhury and Rosenberry, 1984; Futerman *et al.*, 1985a, b; Haas *et al.*, 1986; Roberts and Rosenberry, 1985, 1986). In both cases, the linkage is to the C-terminal amino acid of the protein. The *Torpedo* glycolipid is highly sensitive to inositol-specific phospholipase C (Futerman *et al.*, 1985a). The mammalian anchor is much less sensitive to the lipases that have been tested (Haas *et al.*, 1986). It is not yet known whether the membrane anchoring of AChE in mammalian neurons is similar to either of the models elucidated so far.

Membrane anchoring by means of glycolipids is unusual but not unprecedented. The best documented case involves the variant surface glycoproteins of African trypanosomes. These proteins are attached through a glycophospholipid consisting of a chain of sugar residues coupled to dimyristyl-phosphatidyl inositol (Cardoso de Almeida and Turner, 1983; Ferguson *et al.*, 1985). When the latter component is cleaved by phospholipase C, the protein is released in a nonaggregating, hydrophilic form, and a formerly cryptic carbohydrate antigen is exposed (Turner *et al.*, 1985). Steiger *et al.* (1986) have now reported that the antigenic determinant common to solubilized tryposomal coat proteins is also found on membrane-bound AChEs from *Torpedo* electroplax and human red cell AChE, but only after digestion with an appropriate phospholipase. This finding underscores the commonality of the anchoring mechanisms involved.

2.2. Isolation and Purification

Because of the heterogeneous solubility properties of the various AChE species, sequential extraction is recommended. A significant amount of globular ChE (G_1, G_2, and G_4) can be solubilized in aqueous media without detergent. This procedure is partially selective in extracting hydrophilic forms, although some of the molecules do bind detergents (Gisiger et al., 1978; Lazar and Vigny, 1980; Rakonczay et al., 1981b; Rieger and Vigny, 1976). For *quantitative* extraction a detergent is necessary to solubilize more than minor amounts of the hydrophobic G_1, G_2, and G_4 forms (e.g., 0.5–1% Triton X-100). The asymmetric forms (A_4, A_8, and A_{12}) can be extracted with buffers of high ionic strength ($\geq$ 0.5 M NaCl), whether detergent is present or absent. These properties have made sequential extraction procedures popular for detailed analysis of ChE forms. For example, some investigators use salt in the initial homogenization buffer to extract hydrophilic globular and asymmetric forms; then they bring the membrane-associated forms into solution with detergent. Other investigators reverse the order of extractions.

Studying the solubilization of AChE from different species (*Torpedo*, trout, frog, chicken, rat, rabbit), Ramírez and his co-workers (1984) found that a major portion of the asymmetric forms could be solubilized in high salt (1 M NaCl). However, in most tissues, including electric organ, brain (retina), and muscle, there was a second pool that could only be solubilized in the presence of EDTA. These two pools were defined as class I and class II. Their proportions were found to be highly variable from tissue to tissue and species to species. It was theorized that class I forms are linked to tissue components by ionic forces, exclusively, while class II forms are additionally linked via bridges involving divalent cations, most likely Ca^{2+}. So far, however, clear-cut evidence of physiological differences between class I and class II forms is lacking.

The initial purification of AChE from electric organ of *Electrophorus* or *Torpedo* (the richest sources of the enzyme) or from brain involved conventional methods (Kremzner and Wilson, 1963; Lawler, 1964; Leuzinger and Baker, 1967). These methods consisted of ammonium sulfate precipitations and several laborious chromatographic steps with low yield of the pure product.

More recently, selective separation by ligand affinity chromatography has led to dramatic progress in the purifications of both AChE and BuChE (Chan *et al.*, 1972; Kalderon *et al.*, 1970; Lockridge and La Du, 1978; Massoulié and Bon, 1976; Rakonczay *et al.*, 1981a; Wenthold *et al.*, 1974). Affinity chromatography has been especially useful in the purification of AChE from sources such as human erythrocytes (Ott *et al.*, 1975) and fetal bovine serum (Ralston *et al.*, 1985). Generally, regardless of the enzyme source, AChE can now be isolated with high specific activity (3–10 mmole ACh hydrolyzed per minute per milligram of protein) and a purity of 90–98%, as judged by SDS–polyacrylamide gel electrophoresis. Even so, most of the purification procedures in current use are still lengthy, requiring several weeks to isolate microgram quantities of AChE from sources like brain, and they sometimes suffer from low yield.

The advent of monoclonal antibodies (McAbs) makes immunoaffinity purification a useful supplemental tool. Although AChE is a trace protein in brain, a single antibody with high affinity can selectivity and completely adsorb the enzyme from a crude extract. Recently, Brimijoin *et al.* (1983a) produced McAbs to human erythrocyte AChE. When conjugated to a solid matrix, one cross-reactive antibody ($A_2$45-16) was highly effective for initial purification of rabbit brain AChE. After elution from the immunadsorbent with pH 11 buffer (2000-fold enrichment, 50% yield), the preparation was purified further by affinity chromatography on phenyltrimethylammonium–Sepharose 4B (PTA–Sepharose) with decamethonium elution. The two-step procedure produced good yields of electrophoretically homogeneous AChE in fewer than 7 days (Mintz and Brimijoin, 1985a).

Another antibody ($F_4$61) is sufficiently salt-sensitive to release bound AChE at pH 7.4 with 2 M NaCl. The antibody was originally produced against rabbit brain AChE (Mintz and Brimijoin, 1985b) but it cross-reacts with human brain AChE. With the immunoaffinity procedure just described, we successfully purified AChE from the human erythrocyte (Brimijoin *et al.*, 1986) and from human brain (Rakonczay and Brimijoin, 1987. In both cases, the final affinity chromatography on PTA–Sepharose yielded an electrophoretically homogeneous preparation of globular AChE.

Asymmetric AChE has also been purified by immunoaffinity chromatography. In this case, the McAb (Nj-601) was raised against *Narke japonica* (Sakai *et al.*, 1985). From the Nj-601–Sepharose column, bound AChE was eluted by lowering the pH of the eluent to 2.8. Recovery was better than 45%. Similar yields were obtained from *Torpedo californica* AChE with the same procedure (Sakai *et al.*, 1985). These experiments forecast the possibility of purifying ChEs to homogeneity in one step.

2.3. Gene Cloning and Sequencing

For a detailed discussion of the molecular biology of the ChEs, based on work published through mid-1986, the reader is referred to a recent review (Brimijoin and Rakonczay, 1986). What follows is a briefer summary and update.

Recent rapid advancement in recombinant technology makes it feasible to elucidate the primary structure of large proteins by determining the base sequence of DNA complementary to a specific mRNA (cDNA). Taylor's group used this approach to characterize 11S *Torpedo* AChE. The results show that the enzyme is a 575 amino acid residue protein, with a calculated molecular weight of 65,612 daltons (Schumacher *et al.*, 1986). The slightly smaller size compared with the native catalytic subunits can be attributed to the absence of carbohydrate (7–8% by weight in the native enzyme). The enzyme contains eight cysteines, at least one of which is involved in intersubunit cross-linking. The 5′ region of the amino terminus encodes for 13 residues rich in hydrophobic amino acids, characteristic for the leader peptide of a membrane-associated or secreted protein. The active site serine is located at residue 200.

Schumacher *et al.* (1986) pointed out several interesting implications of the amino acid sequence of *Torpedo* AChE. First, comparisons with the four nicotinic acetylcholine receptor subunits show neither global nor local sequence homologies. Second, except for aliesterases and carboxylesterases, other serine hydrolases show only limited homologies with AChE, and these homologies are confined to the active center. Third, AChE bears a striking resemblance to thyroglobulin, another secreted glycoprotein (residue identity with the C-terminal portion of thyroglobulin is about 28%).

Progress has also been made in unraveling the structure of human plasma BuChE. Lockridge has completed the amino acid sequencing of this protein by classical Edman degradation procedures. Fragmentary information was published in early reports, including the sequence of a 29-residue peptide at the active site, a 6-residue peptide at the C-terminus, and a 22-residue peptide at the N-terminus (Lockridge, 1984; Lockridge and La Du, 1982). Comparison of these data with the amino acid sequence of *Torpedo* AChE (MacPhee-Quigley *et al.*, 1985; Schumacher *et al.*, 1986) shows a surprising degree of homology, given the evolutionary gulf. For example, the two ChEs are identical in 69% of the residues at the active site, with conservative replacement in another 17%. The C-termini are also strikingly similar; the N-termini are less so.

The complete amino acid sequence of human BuChE (Lockridge *et al.*, 1987) consists of 574 residues, 309 of which are identical to those in *Torpedo* AChE. It will be of the greatest interest to compare the BuChE sequence with that for human AChE, which is being pursued in other laboratories. The comparison must extend to the genomic level if we are to reach a proper understanding of the relation between these proteins. La Du and Lockridge (1986) experienced some difficulties with false positives in their original attempts to isolate a cDNA clone for human plasma BuChE by screening a human liver cDNA library. A specific clone has now been obtained after screening a brain cDNA library instead. The clone is still incomplete but does comprise 1400 bases, extending into the 3′ untranslated region (O. Lockridge, personal communication). Work on the genomic organization of AChE and BuChE continues intensively.

3. DISTRIBUTION OF CHOLINESTERASES

Much information has accumulated on the distribution of ChEs in nervous and non-nervous tissues of different species. It is generally assumed that AChE is the only type of ChE playing a functional role in nervous tissue. This assumption may or may not be justified, given the widespread distribution of BuChE. The latter enzyme has been detected in almost all major systems of the mammalian body (Silver, 1974), although there are considerable species differences in the levels and localization of activity. Despite the prevalence of BuChE, there have been few attempts to identify its natural substrate or its physiological significance in excitable tissues.

We have prepared four tables to summarize the types and abundance of the molecular forms of AChE and BuChE in a wide variety of tissues. Data are presented for several animal species, including some lower vertebrates and invertebrates. Most table entries are pairs of numbers. The first number in each pair is the sedimentation coefficient ($S_{20,w}$). The second number is the percentage abundance of the corresponding form (omitted when the data were insufficient). Relative abundance was usually determined from analyses of single-step extractions with salt and detergent. Data from sequential extractions (e.g., Toutant *et al.*, 1985) were combined except in the case of rat brain, where the properties of salt-soluble and detergent-soluble forms are shown separately.

3.1. Excitable Tissues

AChE is an essential part of the ACh system (together with acetylcholine receptors, AChR; the ACh-synthesizing enzyme choline acetyltransferase, ChAT, EC 2.3.1.6; and the membrane mechanism controlling choline uptake and release by neurons). Not surprisingly, AChE is found in highest concentrations at cholinergic synapses. However, there are neural regions that are high in AChE activity but low in ACh, AChR, and ChAT. Cerebellum is one example (Silver, 1974). The presence of AChE at a particular synapse therefore suggests, but does not necessarily prove, that the synapse is cholinergic.

3.1.1. Nervous Tissue

AChE is found widely throughout the central and peripheral nervous system of vertebrates (Brzin *et al.*, 1983; Massoulié and Bon, 1982; Massoulié *et al.*, 1980; Rakonczay, 1986; Silver, 1974). Generally, globular forms of AChE predominate, but small amounts of asymmetric species (mostly A_{12}) have also been documented (Massoulié and Bon, 1982; Rakonczay, 1986; Rakonczay *et al.*, 1981b; Rodriguez-Borrajo *et al.*, 1982) (Table I). It is interesting to note that the proportion of A_{12} AChE in brain seems to fall as the phylogenetic scale is ascended (Rodriguez-Borrajo *et al.*, 1982).

BuChE is also found in brain, typically at 1–10% of the levels of AChE (Silver, 1974). According to the older literature, the BuChE of brain is mostly associated with glial cells. This view is largely borne out by contemporary immunocytochemical studies (see Section 4.2).

3.1.2. Muscle Tissue

The predominant ChE in the skeletal muscle of vertebrates is AChE, but low amounts of BuChE activity are also present (Brimijoin, 1983; Brzin *et al.*, 1983; Massoulié and Bon, 1982) (Tables II and IV). Muscle AChE activity is highly concentrated at the neuromuscular junction, where phasic responses to motoneuron signals require the rapid inactivation of ACh. At the motor endplates, the very high concentrations of receptor molecules are associated with tailed asymmetric AChE, especially the most complex A_{12} form with its dense packing of active sites. Most of this enzyme is extracellular, localized within the synaptic basal lamina. Non-endplate regions in comparison contain primarily globular forms of AChE, both intra- and extracellular. Human muscle is unusual in that the pattern of molecular forms does not vary appreciably between endplate and non-endplate regions, although the total enzyme content is markedly lower outside the synapse (Carson *et al.*, 1979).

Reference to Table II will show that the relative abundance of the molecular forms of AChE in muscle depends both on the fiber type and on the animal species. There are few general rules, however. Commonly, the pattern of forms differs between muscles that are predominantly of the fast-twitch variety (e.g., rat EDL and chicken pectoralis) and muscles that are predominantly of the slow-twitch variety (e.g., rat soleus and chicken anterior latissimus dorsi). As Chaton-

Table I
Distribution of AChE Molecular Forms in Nervous Tissues

Source	Molecular form						References
	G_1	G_2	G_4	A_4	A_8	A_{12}	
	(sedimentation coefficient, $S_{20,w}$)/(percentage abundance)						
Human							
Brain, caudate	5/16	—	11/84	—	—	—	Sørensen *et al*. (1985)
Caudate	3.6/5	—	9.5/95	—	—	—	Atack *et al*. (1983,1986)
Parietal cortex	3.6/30	—	9.5/68	—	—	16/2	Atack *et al*. (1983,1986)
Cerebellum	3.6/14	—	9.5/86	—	—	—	Atack *et al*. (1983,1986)
Nucleus of Meynert	3.6/9	6.5/9	9.5/81	—	—	—	Atack *et al*. (1983,1986)
Dog							
Brain	4/5	—	10/95	—	—	—	Sørensen *et al*. (1985)
Rat							
Brain	4.4(s)/30[a]	—	10.7(s)/65[a]	—	—	16/trace	Rakonczay *et al*. (1981a,b)
	4(m)/5[b]	—	9.8(m)/95[b]	—	—	—	Rakonczay and Brimijoin (1985)
Spinal cord	4/?	—	10/60+	—	—	16/2	Fernandez *et al*. (1979b)
Sciatic nerve	4/7–28	6.5/7-9	10/63–81	—	—	16/2–3	Di Giamberardino and Couraud (1978); Kása and Rakonczay (1982a)
Phrenic nerve	4/?	6.5/?	10/60+	—	—	16/4.5	Fernandez *et al*. (1979b)
Hypoglossal nerve	4/?	6.5/?	10/60+	—	—	16/5.7	Fernandez *et al*. (1979b)
Obturator nerve	4/?	6.5/?	10/60+	—	—	16/2	Fernandez *et al*. (1979b)
Vagus nerve	—	6.5/30	10/58	—	—	16/4	Skau and Brimijoin (1980)
PC12 pheochromocytoma	[4 and	6]/61	10/28	—	—	16/10	Inestrosa *et al*. (1981)
Superior cervical ganglion	4/39	6.5/13	10/44	—	—	16/4	Gisiger *et al*. (1978)
Mouse							
Whole brain	3.8/12	—	9.5/88	—	—	—	Adamson *et al*. (1975); Rieger *et al*. (1980)
Cerebellum	3.8/22	—	9.5/75	—	—	16/3	Adamson *et al*. (1975); Rieger *et al*. (1980)

C-1300 neuroblastoma	4/54	6/28	9.6/18	—	—	—	Chang and Blume (1976)
T28 neuroblastoma	3.5/55	6/5	9.7/40	—	—	—	Lazar and Vigny (1980)
Nerve	—	5.5/20	10/77	—	—	16/3	Brimijoin and Schreiber (1982)
Rabbit							
Brain	4/14	—	10/80	—	—	16/3	Brimijoin and Mintz (unpublished observations)
Sciatic nerve	4/17	—	10/80	—	—	16/3	Brimijoin (1979)
Cow							
Brain	3.1/40	5.5/3	9.7/41	—	>13/4	>16/12	Grassi *et al.* (1982)
Superior cervical ganglion	4.8/30	7.8/13	10.5/7	8.7/35	13.0/6	17.1/9	Bon *et al.* (1979)
Horse							
Brain	4/5	—	10/95	—	—	—	Sørensen *et al.* (1985)
Pig							
Brain	5/10	—	10/90	—	—	—	Sørensen *et al.* (1985)
Sloth							
Whole brain	4/23	—	10/77	—	—	—	Rakonczay and Brimijoin (unpublished observations)
Chicken							
Brain	[4.7 and	7]/20	11/80	—	—	21.3/1	Rieger *et al.* (1980)
Retina	5/11	7/23	11/58	—	—	20/8	Gómez-Barriocanal *et al.* (1981)
Pigeon							
Ciliary ganglion	4/40	6.7/35	13.7/25	—	—	—	Paggi *et al.* (1977)
Frog							
Sciatic nerve	3.6/6	6/15	10.5/36	—	14/8	18/35	Couraud *et al.* (1985)
Torpedo marmorata							
Spinal chord	—	6-7/19	10-11/74	—	—	17-18/6	Toutant *et al.* (1985)
Electric nerve	—	6-7/80	10-11/20	—	—	—	Toutant *et al.* (1985)
Arthropods							
Brain (12 different species)	—	~5.6/100	—	—	—	—	Lenoir-Rousseaux (1985)

[a] s = soluble.
[b] m = membrane associated.

Table II
Distribution of AChE Molecular Forms in Muscle

Source	Molecular form						References
	G_1	G_2	G_4	A_4	A_8	A_{12}	
	(sedimentation coefficient, $S_{20,w}$)/(percentage abundance)						
Human							
Intercostal (e or n)[a,b]	3.5/27	6.0/29	11.0/6	9.5/4	13.3/9	16.7/25	Carson *et al.* (1979)
Psoas (n)[b]	4/43	—	10/14	—	13/21	16/22	Brimijoin (unpublished observations)
Dog							
Diaphragm (e)	3.1/32	5.4/25	—	—	13.5/17	16.7/25	Sketelj and Brzin (1985)
Rat							
Diaphragm (e)	4/32	6.5/9	10/26	8.5/4	12.5/9	16/20	Younkin *et al.* (1982)
EDL	4.4/50	—	10.3/35	—	—	16.1/15	Grosswald and Dettbarn (1983)
Soleus	4.4/25	—	10.3/20	—	12.3/25	16.1/30	Grosswald and Dettbarn (1983)
Left atrium	4/60	—	10/32	—	—	16/7	Skau and Brimijoin (1980)
Right atrium	4/36	—	10/56	—	—	16/7	Skau and Brimijoin (1980)
Cardiac ventricles	4/51	—	10/39	—	—	—	Skau and Brimijoin (1980)
Mouse							
EDL	3.6/24	5.5/12	10.6/31	8.2/1	12.7/10	16.2/22	Gisiger and Stephens (1983)
Soleus	3.6/29	5.5/18	10.6/6	8.2/4	12.7/18	16.2/28	Gisiger and Stephens (1983)
Diaphragm	4/26	—	10/27	—	—	16/32	Skau and Brimijoin (1981)
Cardiac atrium	—	5.5/32	10/48	—	—	16/19	Brimijoin and Schreiber (1982)
Guinea pig							
Diaphragm (e)	3.3/12	5.8/15	—	—	12.8/23	16.1/50	Sketelj and Brzin (1985)
Rabbit							
Diaphragm (e)	3.5/21	6.0/27	—	—	13/13	16/39	Sketelj and Brzin (1985)
Pig							
Diaphragm (e)	3.2/10	5.7/56	—	—	13.3/10	16.6/24	Sketelj and Brzin (1985)

Sloth							
Diaphragm	3.9/27	—	9.8/47	—	13.5/11	16.4/15	Rakonczay and Brimijoin (unpublished observations)
Chicken							
Posterior latissimus dorsi	—	6.5/75	11/15	—	—	19.5/10	Vigny *et al.* (1976)
Pectoralis	4.3/13	6.3/15	11.8/21	—	15.1/11	19.3/40	Chatonnet and Bacou (1983)
Anterior latissimus dorsi	4.3/15	6.3/34	11.8/32	—	15.1/5	19.3/14	Chatonnet and Bacou (1983)
Pigeon							
Pectoralis	4.3/42	—	13.5/35	—	16.2/7	20.8/16	Chatonnet and Bacou (1983)
Anterior latissimus dorsi	4.3/34	6.4/4	13.5/26	—	16.2/10	20.8/26	Chatonnet and Bacou (1983)
Zebra finch							
Syrinx	4.0/45	6.1/29	10.5/10	—	—	16.4/16	Bleisch *et al.* (1983)
Frog							
Cutaneous pectoris (e)	3.5/9	6/36	11/18	—	14/17	18/18	Nicolet *et al.* (1986)
Xenopus laevis							
Thigh muscle	3.5/10	5.5/38	10/8	—	13.5/16	17.5/19	Lappin and Rubin (1985)
Electrophorus electricus							
Electric organ	5.3[c]	7.7[c]	11.8[c]	9/1	14.2	18/4	Massoulié *et al.* (1980)
Torpedo marmorata							
Electric organ	—	7.1/55	—	9.1/2	13.4/11	17.4/31	Bon and Massoulié (1980)
Skeletal muscle	—	6–7/89	10–11/1	—	13–14/3	17–18/7	Toutant *et al.* (1985)
Narke japonica							
Electric organ	—	6/15	—	—	13/59	17/26	Abe *et al.* (1983)

[a]e = endplate rich.
[b]n = non-endplate.
[c]Lytic globular forms.

net and Bacou (1983) emphasize, the differences between the two varieties of muscle tend to hold regardless of the *metabolic* properties of the corresponding fibers. On the other hand, generalizations across animal species are risky, even when the species are closely related.

Many studies have focused on the AChE in the electric organs of *Electrophorus* and *Torpedo*. These organs possess a pure cholinergic innervation and are phylogenetically derived from muscle. They have lost their contractility but have amplified their electrical response to nerve stimulation. *Torpedo* electroplax contains a large proportion of globular, amphipathic G_2 AChE together with asymmetric forms (Massoulié and Bon, 1982). In contrast, the electric organ of *Electrophorus* contains only the asymmetric forms A_4, A_8, and A_{12} (Massoulié and Bon, 1982) (Table II).

3.2. Nonexcitable Tissues

AChE is not only associated with cholinergic nerve structures but is also found in nonexcitable tissues such as erythrocyte, liver, kidneys, and placenta (Silver, 1974). The high amounts of erythrocyte AChE and plasma BuChE have been known for some time. However, no physiological function has yet been ascribed to either enzyme in these locations.

Recently, several authors have confirmed the presence of AChE in the sera of some animals (Table III). The list includes humans, although it should be stressed that serum BuChE predominates overwhelmingly in this species. The AChE in most animal sera (e.g., rat, rabbit, dog, and cow) is primarily of the G_4 form. However, human serum contains G_2 AChE in addition, while mouse serum contains both G_1 and G_4 forms (Table IV).

The mammalian erythrocyte is particularly rich in AChE that is bound firmly to the external surface membrane. In most species, the erythrocyte AChE is exclusively of the dimeric globular form. The only known exception is the rat, whose erythrocytes contain monomeric AChE instead (i.e., the G_1 form).

As already mentioned, the characteristic esterase of serum in most species is BuChE (or an equivalent enzyme that prefers propionylcholine as a substrate). Of all bodily tissues, the serum is the richest source of BuChE. The circulating form of this enzyme is usually purely tetrameric. However, in chicken plasma there is a series of BuChE molecular forms (Table IV) whose structure remains to be analyzed in detail.

4. IMMUNOCHEMISTRY OF CHOLINESTERASES

Modern immunology has opened new approaches to the biology of cholinesterases. The last 5 years have seen particularly rapid progress in applying

Table III
Distribution of AChE Molecular Forms in Other Tissues

Source	Molecular form G_1	G_2	G_4	A_4	A_8	A_{12}	References
	(sedimentation coefficient, $S_{20,w}$)/(percentage abundance)						
Human							
Serum	—	7.6/57	10.9/47	—	—	—	Sørensen *et al.* (1986)
Red blood cell	—	6.5/100	—	—	—	—	Ott *et al.* (1975)
Normal amniotic fluid	3.5	7/90+	—	—	—	—	Bonham and Atack (1983)
Normal amniotic fluid	4.1/85	—	10.5/15	—	—	—	Rakonczay *et al.* (1985)
Amniotic fluid neural tube defect	3.5	7/40	10.1/60	—	—	—	Bonham and Atack (1983)
Cerebrospinal fluid	3.9/19	—	10.3/81	—	—	—	Rakonczay (unpublished observations)
Intestine (rectal biopsy)	4/82	—	10/18	—	—	—	Rakonczay and Németh (1984)
Dog							
Serum	—	—	10/100	—	—	—	Skau (1985)
Red blood cell	—	6.5/65	10/15	—	—	—	Skau (1985)
Rat							
Serum	—	—	10/100	—	—	—	Fernandez *et al.* (1979a); Skau (1985)
Red blood cell	4/100	—	—	—	—	—	Fernandez *et al.* (1979a); Rakonczay and Brimijoin (1985); Skau (1985)
Cerebrospinal fluid	—	—	10/95	—	—	—	Rakonczay and Brimijoin (unpublished observations)
Liver	4/92	—	10/8	—	—	—	Rakonczay and Brimijoin (unpublished observations)
Ileum	4/24	—	10/65	—	—	16/7	Skau and Brimijoin (1980)
Colon	4/25	—	10/68	—	—	16/4	Skau and Brimijoin (1980)
Mouse							
Serum	4/78	—	10/19	—	—	—	Skau (1985)
Red blood cell	—	6.5/100	—	—	—	—	Skau (1985)

(*continued*)

Table III (*Continued*)

Source	Molecular form						References
	G_1	G_2	G_4	A_4	A_8	A_{12}	
	(sedimentation coefficient, $S_{20,w}$)/(percentage abundance)						
Rabbit							
Serum	—	—	10/100	—	—	—	Rakonczay and Brimijoin (1985)
Red blood cell	—	6.5/100	—	—	—	—	Rakonczay and Brimijoin (1985)
Cerebrospinal fluid	—	—	10/100	—	—	—	Brimijoin (unpublished observations)
Submandibular salivary gland	4/20	—	10/64	—	—	16/16	Kása and Rakonczay (1982b)
Parotid gland	4/37	—	10/55	—	—	16/9	Kása and Rakonczay (1982b)
Cow							
Serum	—	—	10/90+	—	—	—	Rakonczay and Brimijoin (1985)
Red blood cell	—	6.5/100	—	—	—	—	Rakonczay and Brimijoin (1985)
Lymphocyte	—	6.0/100	—	—	—	—	Méflah *et al.* (1984)
Naja Naja oxiana							
Cobra venom	—	7.1/100	—	—	—	—	Raba *et al.* (1979)

Table IV
Distribution of BuChE Molecular Forms in Various Tissues

Source	Molecular form						References
	G_1	G_2	G_4	A_4	A_8	A_{12}	
	(sedimentation coefficient, $S_{20,w}$)/(percentage abundance)						
Human							
Brain caudate nucleus	3.8/30	—	10.4/70	—	—	—	Atack *et al.* (1986)
Cerebellum	3.8/39	—	10.4/61	—	—	—	Lockridge *et al.* (1979)
Parietal cortex	3.8/53	—	10.4/47	—	—	—	Lockridge *et al.* (1979)
Serum	—	—	11.1/100	—	—	—	Lockridge *et al.* (1979)
Intestine (rectal biopsy)	4/20	—	10/80	—	—	—	Rakonczay and Németh (unpublished observations)
Rat							
Superior cervical ganglion	4.2/85	—	10.8/15	—	—	—	Vigny *et al.* (1978)
Muscle (sternocleidomastoid)	4.2/50	7/22	10.8/25	—	—	17/trace	Vigny *et al.* (1978)
Horse							
Serum	—	—	9.9/100	—	—	—	Jansz and Cohen (1962)
Pig							
Parotid	—	—	9.7/100	—	—	—	Tucci and Seifter (1969)
Chicken							
Plasma	4.4/7	6.0/18	11.1/68	13.8/7	15.5/trace	—	Lyles *et al.* (1980)
Muscle	4.0/6	5.9/10	10.5/95	—	14.1/6	19.2/12	Lyles *et al.* (1980)
Torpedo marmorata							
Plasma	—	—	10–11/100	—	—	—	Toutant *et al.* (1985)
Spinal cord	—	—	10–11/100	—	—	—	Toutant *et al.* (1985)
Heart	4–5/56	6–7/31	10–11/8	—	[13–14	17–18]/5	Toutant *et al.* (1985)

immunochemistry, immunocytochemistry, and immunoassay to the study of AChE and BuChE. Again the reader is referred to Brimijoin and Rakonczay (1986) for a detailed review, but an outline of major results is given below.

Many polyclonal antibodies have been produced to AChE (Adamson, 1977; Gurari *et al.*, 1974; Rieger *et al.*, 1976; Williams, 1969; Zanetta *et al.*, 1981) and BuChE (Eckerson *et al.*, 1983; Hodkin *et al.*, 1965; Rubinstein *et al.*, 1970, 1976, 1978). There are disadvantages to polyclonal antisera (e.g., limited quantities, lack of reproducibility, different affinity to different epitopes). Nevertheless, such antisera have proved suitable for some interesting applications. A study localizing the AChE in *Electrophorus* electric organ (Benda *et al.*, 1970; Tsuji *et al.*, 1972) could be cited as one example among several. Monoclonal antibodies (McAbs) now offer a promising new means of gaining insight into the structure, function, and biosynthesis of ChEs (Brimijoin *et al.*, 1983a; Fambrough *et al.*, 1982; Mintz and Brimijoin, 1985b; Rakonczay and Brimijoin, 1986). A near-complete list of reported McAbs to AChE and BuChE can be found in the review by Brimijoin and Rakonczay (1986).

4.1. Species Cross-Reactivity

One of the potential applications of antibodies is for immunochemical studies of the relations among proteins from different animal species. Antibody cross-reactivity with ChEs has been found to depend markedly on evolutionary relations of the animals involved. For example, experiments with rabbit antisera to eel AChE and bovine caudate nucleus AChE revealed complete absence of cross-reactivity between the mammalian brain enzyme and either the globular or the elongated forms of the electric tissue enzyme (Greenberg *et al.*, 1977; Williams, 1969). Similarly, polyclonal antisera or McAbs to chicken brain AChE failed to recognize the ChE of any nonavian species (Rotundo, 1984). A rabbit antiserum against rat AChE did not react with the *Electrophorus, Torpedo,* chicken, or rabbit enzyme but did recognize AChE from various mammalian species. The order of affinities was: rat = mouse > bovine > human (Marsh *et al.*, 1984). Sørensen *et al.* (1985) studied the cross-reactivity of various enzymes with antisera to human AChE (RBC, brain) and bovine brain AChE. The anti-bovine AChE antiserum reacted in the order: bovine > horse > hog > dog > human.

Most McAbs to mammalian AChEs react only with comparable enzyme from other mammals (Brimijoin *et al.*, 1983a; Mintz and Brimijoin, 1985b; Rakonczay and Brimijoin, 1986). Fambrough *et al.* (1982) described five McAbs produced against human erythrocyte AChE. They reacted to a variable degree with most mammalian species tested, including the monkey, rabbit, dog, guinea pig, and calf, but excluding the rat. The reactivity of one antibody appeared to be restricted to primate AChE.

The narrow specificity of McAbs was used in one of our laboratories to investigate the phylogenetic relation among primate BuChEs (Mintz *et al.*, 1984). Some McAbs to human serum BuChE recognized the monkey, dog, cat, horse, and rabbit enzyme. However, one antibody ($B_2$12-1) cross-reacted only with BuChEs from higher primates (Brimijoin *et al.*, 1983a). Quantitative measurements established a series of affinities in the following order: human > chimpanzee = pigmy chimpanzee > gorilla >> orangutan >> gibbon > monkey (Mintz *et al.*, 1984). Unfortunately, the nature of the epitope recognized by $B_2$12-1 is unknown. The results are consistent with the idea that there have been significant changes in the conformation or structure of BuChE during the later stages of primate evolution.

Rodent brain AChE is particularly suitable for immunochemical study because of its ready availability. Recently, we produced seven McAbs to the rat enzyme (Rakonczay and Brimijoin, 1986). Figure 1 shows the cross-reactivities

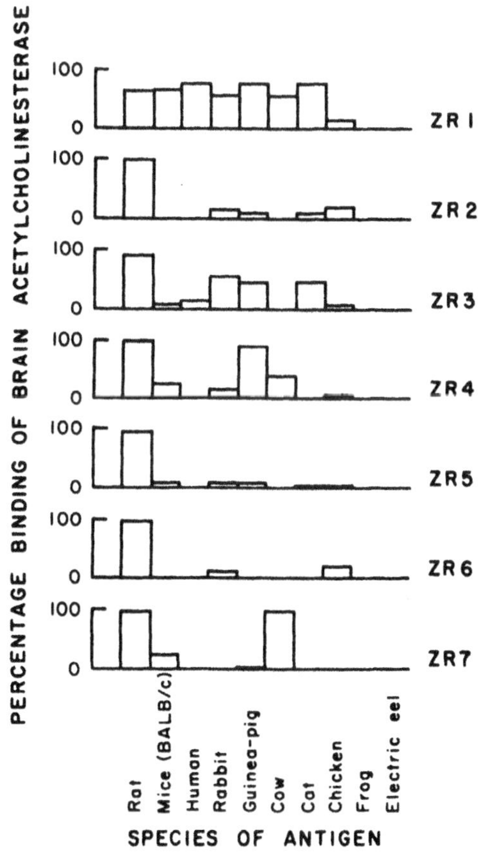

FIGURE 1. Cross-reactivity of antirat brain AChE monoclonal antibodies with brain AChE from various animal species. Each binding assay was performed with a variable amount of crude detergent extract of brain tissue with approximately 0.5 units of AChE activity. Antibody concentration was 10^{-6} M in all cases. (Reprinted from Rakonczay and Brimijoin, 1986, with permission from the *Journal of Neurochemistry*, Raven Press.)

of these McAbs against the brain AChEs of various animal species. None of the antibodies could bind frog or electric eel AChE to a measurable extent. Neither did they have much affinity for chicken brain enzyme. With mammalian AChEs, selective cross-reactivity was the rule. However, antibody ZR1 possessed essentially equivalent affinity for enzyme from each mammalian species tested. Since ZR1 bound mouse AChE as avidly as rat AChE, there must have been a breakdown of immune tolerance in the mouse from which the ZR1 hybridoma was derived. The broad species cross-reactivity of ZR1 suggests that at least one AChE epitope is shared by most mammals (including all inbred mouse strains tested). Unfortunately, among the series of antibodies to brain AChE, ZR1 has the lowest affinity, so its range of applications is limited.

4.2. Immunocytochemistry

The classical thiocholine method, based on ChE activity (Karnovsky and Roots, 1964; Koelle and Friedenwald, 1949), has been used with modifications for almost 40 years to identify cholinergic or cholinoceptive, cholinesterase-positive cells. These methods are quite sensitive. Also, with the aid of appropriate enzyme inhibitors, the two types of ChE can be well distinguished from each other. Nevertheless, there are advantages to immunocytochemistry that have motivated the production and characterization of antibodies for this purpose. Among these advantages are (1) increased specificity of staining (McAbs typically react only with AChE or with BuChE, not with both); (2) ability to localize inactive enzyme; (3) potential ability to localize a particular molecular form; and (4) ability to stain living cells.

The principles of immunocytochemistry require that a polyclonal or monoclonal antibody bind specifically to the target, in situ. Usually, this "primary antibody" is unlabeled. If the primary antibody remains bound to the antigen after rinsing and washing, it can be detected with a labeled secondary antibody (e.g., fluorescein- or peroxidase-conjugated anti-IgG from another animal species). The recently developed streptavidin–biotin system is suited for intensification of labeling in cases where extreme sensitivity is needed (Hsu *et al.*, 1981). Proper controls are always necessary because of the possibility of nonspecific deposition of IgG on tissue specimens. This problem is considered in more detail elsewhere (Brimijoin and Rakonczay, 1986).

Benda *et al.* (1970) and Tsuji *et al.* (1972) were among the first to report an immunocytochemical study of AChE. These workers examined the electroplax of the electric eel by indirect immunofluorescence with a polyclonal antiserum against AChE. They found intense fluorescence staining of the external membrane foldings of the innervated face. The distribution of this staining strikingly resembled the classical histochemical picture. A similar staining pattern was noted by Abe *et al.* (1983) in a study of McAbs against electric organ of *Narke japonica*. One of these McAbs subsequently proved to be directed toward AChE.

Fambrough *et al.* (1982) were the first to produce McAbs against human erythrocyte AChE. These antibodies gave specific immunofluroescence staining of neuromuscular junctions from several animals, including monkey, dog, guinea pig, and calf, but not rat, chicken, or frog. Two of the antibodies were used by Hedreen *et al.* (1984) to examine the morphology of AChE immunoreactivity in monkey visual cortex. In the latter study, AChE immunoreactive fibers were most dense in layers I, IIIB, IV, and VIB of Hassler and Wagner.

Recently, Rakonczay and Brimijoin (1986) obtained immunofluorescence staining of AChE in the large motor neurons of the rat spinal cord, and the smaller neurons in the rat caudate nucleus. This work utilized a mixture of McAbs against rat brain AChE. One of these antibodies (ZR1) has already been mentioned because of its broad species cross-reactivity. ZR1 displays an interesting sensitivity for AChE forms. In particular, this antibody prefers G_4 over G_1 and prefers membrane-bound forms over soluble ones (Rakonczay and Brimijoin, 1985). Unfortunately, the low affinity of the antibody makes it problematical for use in immunocytochemistry. Nevertheless, these observations foreshadow the possibility of form-specific tissue localization.

So far only one paper has been published on the immunocytochemical localization of BuChE in brain (Barth and Ghandour, 1983). This study, based on a polyclonal antiserum, showed specific staining of rat cerebellum in the Bergman glial fibers of the molecular layer and in glial cells of the granular layer. The results support the conclusion that BuChE is a marker of glial cells.

It is well known that AChE tends to be exposed on the external surface of cell membranes where, thanks to its unusual membrane anchoring, nearly the entire molecule is accessible to applied antibodies. This feature makes it possible to consider vital immunostaining (i.e., with live cells). If the immunostaining is strong enough, a fluorescence-activated cell sorter can separate AChE positive cells from unlabeled ones.

Human erythrocytes were probably the first live cells studied with McAbs to AChE. Indirect radiolabeling showed measurable and selective antibody binding to the cell surface (Brimijoin and Mintz, 1985). Further studies have been done with the aid of antihuman AChE McAbs, fluorescein-isothiocyanate labeled second antibodies, and a fluorescence-activated cell sorter. Using such a sytem, Brimijoin *et al.* (1986) and Dockter and Morrison (1986) identified separate populations of AChE-positive and AChE-negative RBCs in the blood of patients with paroxysmal nocturnal hemoglobinuria (see Section 5.4). It is not yet known whether this technique will provide a means for separating cholinergic neurons from other neurons in cell culture.

4.3. Immunoassays

Immunoassays can measure enzyme protein, independent of catalytic activity. Despite the increasing availability of antibodies to AChE and BuChE (surveyed by Brimijoin and Rakonczay, 1986), there have been few attempts to devise immunoassays for these proteins. One reason is that we can measure enzyme activity by sensitive spectrophotometric (Ellman *et al.*, 1961), radiometric (Johnson and Russell, 1975), or fluorometric (Parvari *et al.*, 1983) assays. However, there are four types of samples where the rationale for immunoassay is clear: (1) enzyme inactivated by an anticholinesterase; (2) enzyme inactivated by tissue proteases; (3) small amounts of one ChE to be measured in the presence of large amounts of another ChE; (4) individual molecular forms to be determined directly.

"Quantitative enzyme antigen immunoassay" was used first with AChE in amniotic fluid for screening neural tube defects (see Section 5.3) (Brock *et al.*, 1985; Hangaard *et al.*, 1984; Nørgaard-Pedersen *et al.*, 1983). In this procedure, polyclonal or monoclonal antibodies to human AChE, fixed to microtiter plates, were used to adsorb sample AChE. The intrinsic activity of the bound enzyme was then determined by conventional thiocholine-based AChE assay. A similar method was used to investigate possible changes in red cell and plasma AChE activity in Alzheimer's disease (St. Clair *et al.*, 1986). These solid-phase methods provide enhanced sensitivity and selectivity. However, they do not measure enzyme protein per se.

An immunodisplacement assay was developed by Brimijoin *et al.* (1986) to measure AChE protein in the erythrocyte membrane from normal subjects and patients with paroxysmal nocturnal hemoglobinuria (see Section 5.4). This assay, based on a specific, solid-phase monoclonal antibody, quantitates a sample in terms of its ability to reduce the antibody binding of a purified AChE standard. With the antibodies available, the assay's detection limit is about 10 ng. Control experiments show that immunodisplacement abilities of crude and purified preparations of AChE are identical when represented as a function of enzyme activity (Figure 2). This result leads to two conclusions: (1) the immunodisplacement assay is unaffected by irrelevant proteins of red blood cell and (2) crude and purified red cell AChE have the same catalytic properties.

A more sensitive, two-site immunoassay for AChE has also been developed (Brimijoin *et al.*, 1987). This assay has a detection limit of 100 pg. It is capable of measuring inhibited or partially proteolysed AChE in samples of low enrichment, such as muscle and peripheral nerve. Experiments with the two-site assay support the view that the marked loss of AChE activity from denervated rat skeletal muscle reflects a disappearance of enzyme protein (S. Brimijoin, Z. Rakonczay, and P. I. Hammond, unpublished observations).

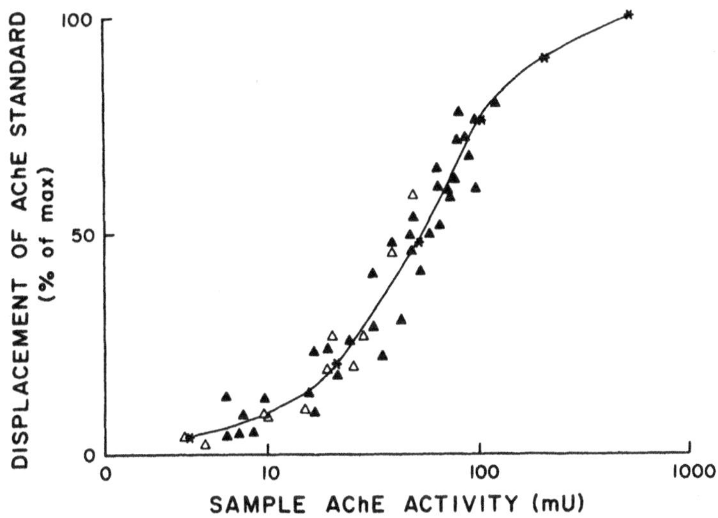

FIGURE 2. Immunodisplacement ability as a function of AChE activity in crude and pure enzyme samples. Asterisks and solid line = pure AChE; solid triangles = crude extracts of normal red cells; open triangles = crude extracts of red cells from paroxysmal nocturnal hemoglobinuria. (Reprinted from Brimijoin *et al.*, 1986, with permission from the Mayo Clinic Proceedings.)

4.4. Discrimination among AChE Molecular Forms

The work of B. P. Doctor and colleagues (Doctor *et al.*, 1983) originally demonstrated immunological differences between membrane-bound and hydrophilic (asymmetric) AChEs of *Torpedo* electroplax. This work has been amplified by Sakai *et al.* (1986) who immunized mice with *Narke japonica* asymmetric AChE. Several resulting McAbs recognized both asymmetric and detergent-solubilized enzyme. In addition, five McAbs were found that reacted only with asymmetric forms. Of these antibodies, two may react with the tail unit, but one may be directed toward the catalytic subunit. These McAbs will certainly be useful for studying the molecular structure and function of asymmetric AChE in the electric organ of electric rays.

The question of the immunological differences between soluble and membrane-associated forms of mammalian brain AChE has been debated by a number of authors. Some studies found that polyclonal antisera against purified detergent-solubilized G_4 AChE from rat brain (membrane-bound enzyme) did not recognize soluble AChE (Sørensen *et al.*, 1982; Zanetta *et al.*, 1981). Others have observed no difference in the immunoreactivity of the two enzyme fractions in rat and bovine brain (Marsh *et al.*, 1984) or in human brain (Gennari and Brodbeck, 1985).

Recently, Mintz and Brimijoin (1985b) characterized several McAbs in terms of their ability to bind soluble and membrane-associated AChE from rabbit brain. One of these antibodies ($F_4$43) proved unable to bind more than half of the soluble enzyme fraction. Furthermore, when the unbound material was isolated and retested, a fresh preparation of the same antibody bound little or none of it. In other words, antibody $F_4$43 discriminated against a subform of AChE that was particularly abundant in detergent-free, aqueous extracts. Other findings suggest that $F_4$43 partially distinguishes membrane-associated G_4 and G_1 AChE from the soluble counterparts (Rakonczay and Brimijoin, 1985). A similar property has been noted in the case of ZR1, an antirat brain AChE McAb. ZR1 also exhibits substantially higher avidity for tetrameric than for monomeric enzyme (Rakonczay and Brimijoin, 1985, 1986). The behavior of $F_4$43 and ZR1 should reawaken interest in the possibility of immunochemical differences between soluble and membrane-bound forms of mammalian brain AChE.

Monoclonal antibodies that could recognize tissue-specific types of AChE would be of considerable interest. One candidate for that purpose is again ZR1. This antibody binds brain and serum AChE from rat, rabbit, and cow but does not bind AChE from erythrocytes of the same species (Rakonczay and Brimijoin, 1985).

Attempts have been made to characterize the multiple allelic forms of human plasma BuChE by immunological methods. The "atypical" form of human plasma BuChE is associated with prolonged apnea following the administration of normal doses of succinylcholine (Davies *et al.*, 1960; Kalow and Genest, 1957). This feature reflects altered catalytic properties, presumably due to an altered active site with lowered affinity for substrate. It is naturally of interest to know whether the inherited variants of human BuChE can be distinguished immunologically. The answer appears to be "no, not readily." Eckerson *et al.* (1983) produced polyclonal antisera to BuChE. Brimijoin *et al.* (1983a) produced several McAbs against the "usual" form of the enzyme. Tests with all these antibodies demonstrated immunological similarity between the "atypical" and "usual" form of BuChE.

5. CHOLINESTERASE MOLECULAR FORMS IN PATHOLOGICAL CONDITIONS

The involvement of ChEs in certain pathological processes, especially the avian and murine muscular dystrophies, has been recognized for some time. Here we consider a few particularly interesting examples that connect AChE abnormalities with pathological states in animals and humans.

5.1. Alzheimer's and Parkinson's Diseases

Alzheimer's disease (AD) represents the largest number of presently incurable senile dementias. It is a neurodegenerative disorder characterized by progressive loss of memory and cognitive function, continuing without remission until death. The impairment of cognitive function appears to be correlated with an increased number of neuritic plaques and neurofibrillary tangles in the cerebral cortex, hippocampus, and amygdala (Hooper and Vogel, 1976; Tomlinson *et al.*, 1970).

Although several neurotransmitter systems may be affected in AD, the profound vulnerability of cholinergic neurotransmission in the disease process received considerable attention when first reported (Bowen *et al.*, 1976; Davies and Maloney, 1976; Perry *et al.*, 1977). Further cholinergic enzyme alterations have since been observed, especially losses of ChAT and AChE (Bowen *et al.*, 1976; Davies, 1979; Rossor *et al.*, 1980). The reductions of ChAT and AChE are commensurate with the severity of the disease, suggesting a selective central cholinergic lesion (Perry *et al.*, 1978). Postsynaptic ACh receptors, in contrast, appear to be normal (Davies and Verth, 1977; Reisine *et al.*, 1978), while BuChE has been reported to be elevated in cortical tissues (Atack *et al.*, 1983).

Some studies of AD have dealt with the ChEs of accessible tissues, primarily blood and CSF. Measurements in the CSF have given conflicting and ambiguous results. For example, Davies (1979) found a normal AChE content, while Soininen *et al.* (1981) found a reduced one (however, the direction of the change depended on whether the data were expressed per milliliter of CSF or per milligram of protein). To add to the confusion, there has been at least one report of elevated BuChE in the CSF of patients with AD (Arendt *et al.*, 1984).

Experiments on plasma AChE were hampered at first by the difficulties in measuring this enzyme in the presence of a large excess of BuChE. Accurate results can now be obtained by the judicious use of selective inhibitors or, as discussed earlier, by specific immunoadsorption methods. A careful recent study with selective inhibitors has documented a significant elevation in plasma AChE activity in AD (Atack *et al.*, 1985). This abnormality was not accompanied by any changes in erythrocyte AChE or plasma BuChE activity. The authors inferred that the increased levels of plasma AChE reflected excessive release of the enzyme from degenerating cholinergic neurons in the brain.

To study enzyme abnormalities in brain tissue is more difficult, since only postmortem samples are available. Fortunately, AChE is a stable enzyme that can be measured accurately in such material. It is well known that AChE is synthesized in the neuronal cell body and transported down the axon to the nerve terminal. In human cerebral cortex, many of the cholinergic nerve terminals are projections from large cholinergic neurons in the basal forebrain (Coyle *et al.*, 1983). The major component of the basal forebrain cholinergic system is the nucleus basalis of Meynert, where cholinergic neurons are substantially decreased in AD (Arendt *et al.*, 1985; Perry *et al.*, 1982; Whitehouse *et al.*, 1981, 1982). Lesions in these cell bodies may result in a loss of axonally transported "presynaptic" AChE at the nerve terminals in distant brain areas, especially cortex.

Recently, Atack *et al.* (1983) reported that the loss of AChE activity in the parietal cortex (Brodmann area 40) in AD is selective for the G_4 form of the enzyme, apparently sparing the G_1 form. The results have been confirmed and extended to cover more brain areas by Fishman *et al.* (1986), who found a decrease in G_4/G_1 ratio in Brodmann 9,10,11,21,40 and the amygdala. On the other hand, these investigators found no changes relative to age-matched controls in cortical areas 17 and 20, in the hippocampus, or in the cerebellum. The decrease in G_4/G_1 ratio does suggest a specific loss of the G_4 form, which is predominantly membrane bound. Fishman *et al.* (1986) proposed that the G_4 form is associated mainly with presynaptic structures (cholinergic axons), while the cortical G_1 form is associated with postsynaptic unaffected by AD.

Younkin *et al.* (1986) have focused on the asymmetric AChEs, which are barely detectable in normal human brain. Besides the previously reported decrease in membrane-associated G_4 AChE, they found *increases* in A_{12} and A_8 AChE in Brodmann area 21 of AD brains. The increases were small in absolute terms but large when expressed as percentages of control (354 and 345%, respectively). Younkin *et al.* (1986) proposed that the increase of asymmetric AChEs reflected an impairment of fast retrograde axonal transport in the affected neurons, leading to local deposition of transported protein. Confirmation of these findings will be eagerly awaited.

Fewer studies have been performed to characterize brain AChE in Parkinson's disease. The lack of emphasis is partly attributable to the focus on dopaminergic mechanisms in this movement disorder. It has recently been noted, however, that there is a significant loss of AChE activity in the frontal cortex of both demented and nondemented Parkinsonian subjects (Ruberg *et al.*, 1986). As in AD, the 10S or G_4 AChE was selectively involved, with a greater decrease in demented individuals. Cortical BuChE activity was decreased only in *nondemented* subjects. There was some tendency for AChE and BuChE levels in CSF to vary inversely with the levels in brain tissue. The apparent paradox might reflect changes in the activity of enzyme-release mechanisms in surviving cholinergic neurons.

5.2. Hirschsprung's Disease

Hirschsprung's disease, a disorder affecting bowel motility, appears to arise from the failure of ganglionic cells to migrate from the neural crest into the intramural plexus of the gut (Okamoto and Ueda, 1967). Histochemical studies have shown that the larger nerve bundles in the neuromuscular zone of aganglionic bowel segments are AChE-positive (Adams *et al.*, 1960; Kamijo *et al.*, 1953; Patrick *et al.*, 1980). Hirschsprung's disease is also associated with abnormal prominence of AChE-positive nerves in the circular muscle of the gut (Meier-Ruge, 1968).

Density gradient ultracentrifugation shows that Hirschsprung's disease not only involves a large elevation of AChE activity in gut tissue (sixfold or more) but also a change in the relative distribution of G_4 and G_1 forms (Figure 3).

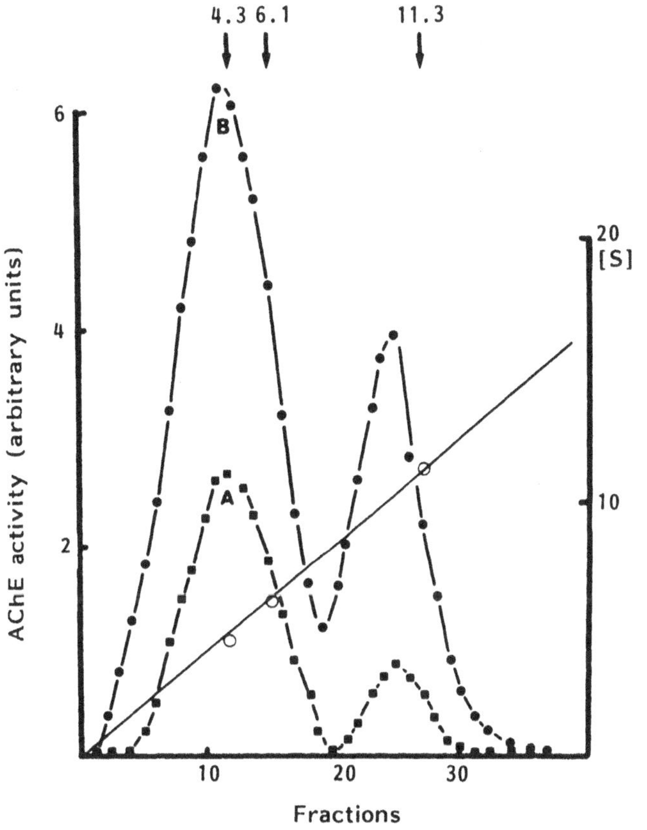

FIGURE 3. Sedimentation velocity profile of different molecular forms of AChE in human rectum: (A) normal and (B) Hirschsprung's disease. The suction biopsy samples were homogenized in 12.5 mM sodium phosphate buffer (pH 7.2), 0.4 M NaCl, 0.5% Triton X-100. After centrifugation at 100,000*g* for 60 min at 4°C, 200 μL of supernatant was mixed with markers, layered on a 5–20% sucrose gradient, and centrifuged at 107,000*g* for 32 h at 4°C; Beckman L8-55 ultracentrifuge, SW28.1 rotor. All fractions were assayed for AChE and enzyme markers. (Reprinted from Rakonczay and Németh, 1984, with permission.)

Thus, the G_4/G_1 ratio was 2.5 times higher in affected bowels than in normal ones (Rakonczay and Németh, 1984). The change in ratio could be ascribed to an excessive accumulation of the larger form. Since this form is a characteristic component of neurons, the accumulation of G_4 AChE could reflect the proliferation of nerve fibers in the aganglionic megacolon. The original results together with the supporting data of Bonham *et al.* (1985) suggest that the quantitative measurement of AChE molecular forms might serve as the basis of an improved diagnostic test for the disease.

5.3. Neural Tube Defects

Several reports have indicated that an elevated concentration of AChE in human amniotic fluid is an indicator of spina bifida and anencephaly, the two major types of neural tube closure defects (NTD) (Chubb *et al.*, 1979; Smith, 1979; Smith *et al.*, 1979). This finding is rational because most AChE is probably derived from the nervous system (Chubb *et al.*, 1979). The nervous tissues in NTD are either in direct contact with the amniotic fluid or just covered with a thin membrane (Warkany, 1971). In any event, AChE determination together with alpha-fetoprotein determination seems to permit the antenatal detection of NTD with fewer false positive results than are obtained with any other method (Milunsky and Sapirstein, 1982).

Electrophoretic analysis of AChE has proved most useful in the diagnosis of NTD. Consistently, in these cases, amniotic fluid contains an additional, faster migrating band of AChE (Brock and Hayward, 1980; Buamah *et al.*, 1980; Seller and Cole, 1980). Sedimentation studies show that normal amniotic fluids contain G_1 AChE (Bonham and Atack, 1983) and also G_4 AChE (Rakonczay *et al.*, 1985). Neural tube defects have been linked to a disproportionate increase in the G_4 form (Bonham and Atack, 1983). For example, in the amniotic fluid of the anencephalic fetus, Rakonczay *et al.* (1985) found a 3300% increase in G_4 AChE activity, with only a 196% increase in G_1 AChE activity (Figure 4). Polyacrylamide gel electrophoretic enzyme from this case showed a huge accumulation of AChE activity in the slower migrating band, with less in the faster migrating band (Figure 4).

Two bands of AChE activity were found in the electrophoresis gels of amniotic fluids from fetuses with pyloric stenosis, hydrocephaly spina bifida, multiple cyst syndrome, gastroschisis, and in one instance of dead twins (Elejalde *et al.*, 1986). Alpha-fetoprotein and BuChE were also elevated in the amniotic fluid of women with trisomies and in a broad range of abnormal fetuses (Elejalde *et al.*, 1986).

A model of NTD has been developed in chickens. Studies with this model showed that the amniotic fluid taken from affected embryos always contained a higher concentration of AChE than fluid from sham-operated but otherwise

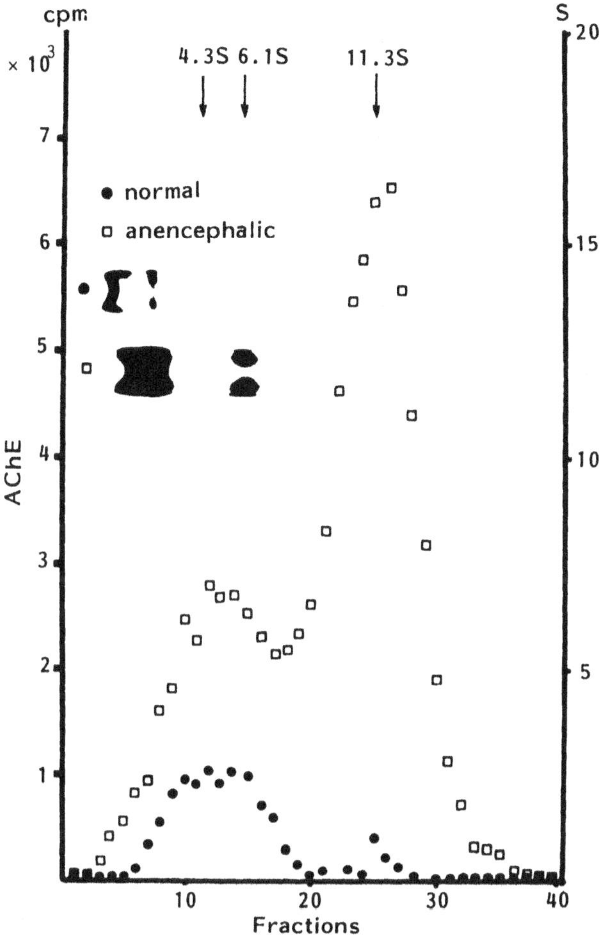

FIGURE 4. Sedimentation velocity profile of different molecular forms of AChE in amniotic fluid. Closed circles = normal; open squares = anencephalic. After centrifugation at 12,000*g* for 20 min at 4°C, 500 μL of supernatant was mixed with markers, layered on a 5–20% sucrose gradient, and centrifuged at 178,000*g* for 14 h at 4°C (Beckman SW41 rotor). All fractions were assayed for AChE by a radiometric assay (Johnson and Russell, 1975) and for enzyme markers. Polyacrylamide gel electrophoresis of amniotic fluids was carried out on 7.5% slab gels and the gels were stained for enzyme activity (Kása and Rakonczay, 1982a).

normal embryos (Pilowsky *et al.*, 1982). These results further support the use of AChE tests in the clinical detection of NTD.

Evidence has recently been presented that in mild malformations, namely, hydrocele (Verp *et al.*, 1984) or in a case of a fetal lymphangiectasia (Holzgreve *et al.*, 1986), the AChE test may be positive in the absence of NTD. Nevertheless, we conclude that the determination of ChE activity in amniotic fluid and its characterization by gel electrophoresis or density gradient ultracentrifugation may be of real help in the prenatal diagnosis of NTD.

5.4. Paroxysmal Nocturnal Hemoglobinuria

Paroxysmal nocturnal hemoglobinuria (PNH) is a rare, chronic, acquired, potentially life-threatening anemia characterized by the increased sensitivity of erythrocytes, platelets, and granulocytes to complement-mediated lysis (see review by Dacie, 1967). Dacie suggested that there are two distinct red cell populations in the blood of affected individuals: "normal" and "PNH."

An interesting feature of the red cells from this disease is their reduced content of AChE activity. Although the enzyme defect is more likely to be a consequence than a cause of membrane abnormality, PNH is nearly unique in this respect (Auditore and Hartmann, 1959; Metz *et al.*, 1961). As mentioned above, Brimijoin *et. al.* (1986) demonstrated that the reduction of AChE activity in PNH red cells reflects a loss of immunoreactive enzyme protein. In other words, this abnormality is not due to the presence of enzyme inhibitors or other factors modifying catalytic activity in the local environment.

Using flow-cytofluorimetric analysis of AChE immunoreactivity, Brimijoin *et al.* (1986) and Dockter and Morrison (1986) showed that the erythrocytes from normal donors belonged to one population while erythrocytes from PNH patients formed two distinct populations. In particular, the erythrocytes in PNH consisted of those containing a normal content of AChE and others containing no detectable AChE. These observations show that the abnormal PNH erythrocytes lack AChE molecules on their surface membranes, as suggested by the findings of Chow *et al.* (1985). The AChE-deficient cells are probably responsible for the characteristic hemolytic anemia (Rotoli *et al.*, 1984), although their complement sensitivity is probably related to a parallel deficiency of another protein, decay-accelerating factor (Nicholson-Weller *et al.*, 1983, 1985). Viewing AChE purely as a marker for the defective cells, it still seems that cell sorting may help in assessing clinical status, in quantitative evaluation of abnormal erythrocytes, and perhaps in developing new therapeutic modalities for PNH.

5.5. Muscular Dystrophy

Abnormalities affecting AChE in the human muscular dystrophies have received much attention because of their potential relation to the disturbed muscle function. Fortunately, a number of animal species are subject to muscle diseases that resemble the human dystrophies. The mechanisms of these disorders are doubtless different from one another and from their human counterpart, but they all involve characteristic abnormalities of AChE. Most studies in murine dystrophy have dealt with mice of the Jackson Laboratories 129/ReJ or C57BL/6J strains.

In the diaphragm of 129/ReJ dystrophic mice, Skau and Brimijoin (1981) found normal total AChE activity but greatly reduced activity of the G_4 molecular form. In the extensor digitorum longus (EDL), a fast-twitch hindlimb muscle, AChE was much reduced in total activity, but again the G_4 form was disproportionately affected or absent altogether. Gisiger and Stephens (1982) confirmed the loss of G_4 AChE in the dystrophic mouse EDL. They extended these studies to show similar biochemical as well as histochemical abnormalities in the distribution of AChE in other affected hindlimb muscles. However, the significance of the G_4 AChE deficit remained obscure until it was noted that murine dystrophy transformed the pattern of AChE forms in fast-twitch muscles into one resembling that of normal slow muscle (Gisiger and Stephens, 1983). Skau (1983) has interpreted this effect as a failure of maturation. In any case, it appears that the abnormal distribution of AChE forms in murine dystrophy is confined to skeletal muscle, since both brain and cardiac tissue exhibit normal proportions of G_4 AChE (Brimijoin and Schreiber, 1982).

In EDL of C57BL/6J dy^{2J}/dy^{2J} dystrophic mice, Lindenbaum and Livett (1983) again observed a decreased proportion of G_4 AChE activity. Sequential extraction with low salt, detergent, and high salt later revealed that this alteration affects only the membrane-bound forms (Poiana *et al.*, 1985). These observations further strengthen the idea of hampered assembly of G_4 AChE in the membrane of dystrophic muscle fiber. Such an effect could confine the cell to an immature developmental stage.

Another hereditary, autosomal recessive mutation affecting the murine peripheral neuromuscular system is "motor endplate disease" (Duchen *et al.*, 1967). The multiple molecular forms of AChE undergo abnormal development during the course of this disease. In gastrocnemius muscle, a decrease in the proportion of G_4 AChE activity is the major result (Rieger *et al.*, 1983). Similar changes occur after short-term denervation or tenotomy (Rieger *et al.*, 1983).

Dysgenic mice (mdg/mdg) suffer from a particularly severe genetic disorder of skeletal muscle. The single most striking feature of dysgenic mouse skeletal muscle is lack of contraction. Of course, mdg/mdg embryos die at birth (Bowden-Essien, 1972). Characteristic abnormalities in these animals affect the field of motor innervation and the pattern of motor nerve branching. AChE is also affected. Unlike the murine dystrophies and motor endplate disease, however, it is the A_{12} form of AChE that is disproportionately reduced (Rieger *et al.*, 1984).

In dystrophic hamsters (Bio-40.54), the result of electrophoretic experiments suggest that the affected muscles have an increase in the proportion of low molecular weight species of AChE (Henderson *et al.*, 1980). Apparently, this model has yet to be examined with sedimentation techniques.

Inherited muscular dystrophy of chicken is yet another animal model for human muscular dystrophies. In affected fast-twitch muscles of the dystrophic chicken, AChE is *overproduced* (Wilson *et al.*, 1968). Most of the AChE and BuChE molecular forms are elevated both in terms of the activity per unit weight of tissue and in an absolute sense (Lyles *et al.*, 1979, 1982; Silman *et al.*, 1979; Sketelj *et al.*, 1978; see also the review of Brimijoin, 1983).

In summary, the AChE abnormalities of the various animal dystrophies have little in common with each other, and their pathophysiologic relation to the human dystrophies is uncertain. Careful studies of the physiology and biochemistry of the neuromuscular junction in these systems, however, may clarify this picture. In addition, the animal models of neuromuscular disease have the potential to shed light on the functional role of AChE forms in normal tissue.

5.6. Amyotrophic Lateral Sclerosis

There is little direct information about AChE in human dystrophies or other neuromuscular disorders. Amyotrophic lateral sclerosis (ALS), a neurologic disease of unknown etiology and pathogenesis, is characterized by both upper and lower motor neuron degeneration (Hirano, 1973) as well as morphological changes in motor endplates and progressive failure of neuromuscular transmission (Bjornskov *et al.*, 1975). Festoff and Fernandez (1981) found that plasma AChE activity in an ALS patient was almost twice the control value, while red blood cell AChE activity showed no statistically significant difference from a control group. In comparison, in Duchenne muscular dystrophy neither the AChE of plasma (Sketelj *et al.*, 1983) nor that of erythrocytes (Igisu *et al.*, 1980) showed any change compared to the controls. Fernandez *et al.* (1986) found that total AChE activity in endplate-containing ALS muscle sections was reduced by an average of 65% below the level in control muscle. The G_4 and A_4 forms almost completely disappeared and the G_1, A_8, and A_{12} forms decreased significantly. The alterations in AChE species from ALS muscles may be related to disturbances in the normal "trophic" interactions between nerve and muscle.

6. CONCLUDING REMARKS AND PERSPECTIVES

Because of their importance from a clinical, pharmacologic, and toxicologic standpoint, ChEs have received close scrutiny for many years. The pace of investigation has probably accelerated since the initial characterization of the molecular forms of these enzymes. A number of important advances have resulted. Perhaps foremost is the derivation of the complete amino acid sequence for one member of the AChE family and one member of the BuChE family. We now are beginning to understand the molecular architecture of the individual forms of AChE in terms of the quaternary structure and the subunit composition as well. We have clearer pictures of the mechanisms involved in anchoring the asymmetric and hydrophobic globular forms in their typical locations. In addition to this progress, major gains have been made in the immunological analysis of ChEs.

It may soon be feasible to perform immunocytochemistry with monoclonal antibodies specific for certain of the molecular forms of AChE or BuChE. Such work would provide improved maps of the distribution of ChE forms within tissues and even within cells. Correlation of AChE antibody maps with ChAT immunocytochemistry, now also becoming practical, may provide additional insights into the functional significance of the localization of AChE molecular forms in the central and peripheral nervous system. The powerful methods of molecular biology will certainly help to solve many problems concerning the multiple variants of the molecular forms. These approaches will be most useful for clarifying the relation of these forms to one another and for defining the biochemical features that are crucial for their interaction with other tissue components, including the basal lamina and the surface membranes of cholinergic synapses.

The most important problems to be addressed, however, concern the physiological role of AChE forms and the implications of their involvement in neuromuscular pathology. Solving these problems remains a formidable challenge for the future.

ACKNOWLEDGMENTS. We thank Mrs. Éva Gál, Mrs. Mária Soós, and Mrs. Pamela Hammond for editorial assistance, and Mrs. Luanne Wussow for typing the manuscript. This work was supported by the Scientific Research Council, Ministry of Health, Hungary (06/4-20/457).

7. REFERENCES

Abe, T., Sakai, M., and Saisu, H., 1983, A monoclonal antibody against catalytic subunits of acetylcholinesterase in the electric organ of an electric ray, *Narke japonica*. *Neurosci. Lett.* **38**:61–66.

Adams, C. W., Marples, E. A., and Trounce, J. R., 1960, Achalasia of cardia and Hirschsprung's disease: The amount and distribution of cholinesterases. *Clin. Sci.* **19**:473–481.

Adamson, E. D., 1977, Acetylcholinesterase in mouse brain, erythrocytes and muscle. *J. Neurochem.* **28**:605–615

Adamson, E. C., Ayers, S. E., Deussen, Z. A., and Graham, C. F., 1975, Analysis of the forms of acetylcholinesterase from adult mouse brain. *Biochem. J.* **147**:205–214.

Anglister, L., Tarrab-Hazdai, R., Fuchs, S., and Silman, I., 1979, Immunological cross-reactivity between electric eel acetylcholinesterase and rat tail-tendon collagen. *Eur. J. Biochem.* **94**:25–29.

Arendt, T., Bigl, V., Tennstedt, A., and Arendt, A., 1985, Neuronal loss in different parts of the nucleus basalis is related to neuritic plaque formation in cortical target areas in Alzheimer's disease. *Neuroscience* **14:**1–14.

Arendt, T., Bigl, V., Walther, F., and Sonntag, M., 1984, Decreased ratio of CSF acetylcholinesterase to butyrylcholinesterase activity in Alzheimer's disease. *Lancet* **i:**173.

Atack, J. R., Perry, E. K., Bonham, J. R., Perry, R. H., Tomlinson, B. E., Blessed, G., and Fairbairn, A., 1983, Molecular forms of acetylcholinesterase in senile dementia of Alzheimer type: Selective loss of the intermediate (10S) form. *Neurosci. Lett.* **40:**199–204.

Atack, J. R., Perry, E. K., Perry, R. H., Wilson, I. D., Bober, M. J., Blessed, G., and Tomlinson, B. E., 1985, Blood acetyl- and butyrylcholinesterases in senile dementia of Alzheimer type. *J. Neurol. Sci.* **70:**1–12.

Atack, J. R., Perry, E. K., Bonham, J. R., Candy, J. M., and Perry, R. H., 1986, Molecular forms of acetylcholinesterase and butyrylcholinesterase in the aged human central nervous system. *J. Neurochem.* **47:**263–277.

Auditore, J. V., and Hartmann, R. C., 1959, Paroxysmal nocturnal hemoglobinuria. II. Erythrocyte acetylcholinesterase defect. *Am. J. Med.* **22:**401–410.

Barth, F., and Ghandour, M. S., 1983, Cellular localization of butyrylcholinesterase in adult rat cerebellum determined by immunofluorescence. *Neurosci. Lett.* **39:**149–153.

Benda, P., Tsuji, S., Daussant, J., and Changeux, J.-P., 1970, Localization of acetylcholinesterase by immunofluorescence in eel electroplax. *Nature* **225:**1149–1150.

Bjornskov, E. K., Dekker, N. D., Norris, F. H., and Stuart, M. E., 1975, Endplate morphology in amyotrophic lateral sclerosis. *Arch. Neurol.* **32:**711–712.

Bleisch, W. V., Luine, V. N., and McEwen, B. S., 1983, Molecular forms of acetylcholinesterase: Regulation in a testosterone-sensitive nerve-muscle axis. *J. Neurochem.* **40:**1523–1528.

Bon, S., and Massoulié, J., 1980, Collagen-tailed and hydrophobic components of acetylcholinesterase in *Torpedo marmorata* electric organ. *Biochemistry* **77:**4464–4468.

Bon, S., Cartaud, J., and Massoulié, J., 1978, The dependence of acetylcholinesterase aggregation at low ionic strength upon a polyanionic component. *Eur. J. Biochem.* **85:**1–14.

Bon, S., Vigny, M., and Massoulié, J., 1979, Asymmetric and globular forms of acetylcholinesterase in mammals and birds. *Proc. Natl. Acad. Sci. U.S.A.* **76:**2546–2550.

Bonham, J. R., and Atack, J. R., 1983, A neural tube defect specific form of acetylcholinesterase in amniotic fluid. *Clin. Chim. Acta* **135:**233–237.

Bonham, J. R., Dale, G., Scott, D., and Wagget, J., 1985, Molecular forms of acetylcholinesterase in Hirschsprung's disease. *Clin. Chim. Acta* **145:**297–305.

Bowden-Essien, F., 1972, An in vitro study of normal and mutant myogenesis in the mouse. *Dev. Biol.* **27:**351–364.

Bowen, D. M., Smith, C. B., White, P., and Davison, A. N., 1976, Neurotransmitter-related enzymes and indices of hypoxia in senile dementia and other abiotrophies. *Brain* **99:**459–496.

Brandan, E., and Inestrosa, N. C., 1984, Binding of the asymmetric forms of acetylcholinesterase to heparin. *Biochem. J.* **221:**415–422.

Brandan, E., Maldonado, M., Garrido, J., and Inestrosa, N. C., 1985, Anchorage of collagen-tailed acetylcholinesterase to the extracellular matrix is mediated by heparin sulfate proteoglycans. *J. Cell Biol.* **101:**985–992.

Brimijoin, S., 1979, Axonal transport and subcellular distribution of molecular forms of acetylcholinesterase in rabbit sciatic nerve. *Mol. Pharmacol.* **15:**641–648.

Brimijoin, S., 1983, Molecular forms of acetylcholinesterase in brain, nerve and muscle: Nature, localization and dynamics. *Prog. Neurobiol.* **21:**291–322.

Brimijoin, S., and Mintz, K. P., 1985, Human acetylcholinesterase: Immunochemical studies with monoclonal antibodies. *Biochim. Biophys. Acta* **828:**290–297.

Brimijoin, S., and Rakonczay, Z., 1986, Immunology and molecular biology of the cholinesterases: Current results and prospects. *Int. Rev. Neurobiol.* **28**:363–410.

Brimijoin, S., and Schreiber, P. A., 1982, Reduced axonal transport of 10S acetylcholinesterase in dystrophic mice. *Muscle Nerve* **5**:405–410.

Brimijoin, S., Mintz, K. P., and Alley, M. C., 1983a, Production and characterization of separate monoclonal antibodies to human acetylcholinesterase and butyrylcholinesterase. *Mol. Pharmacol.* **24**:513–520.

Brimijoin, S., Hammond, P. I., and Petitt, R. M., 1986, Paroxysmal nocturnal hemoglobinuria: Erythrocyte acetylcholinesterase deficit analyzed by immunoassay and fluorescence-activated sorting. *Mayo Clin. Proc.* **61**:522–529.

Brimijoin, S., Rakonczay, Z., and Hammond, P. I., 1987, Immunoassay of acetylcholinesterase. *Fed. Proc.* **46**:2557–2562.

Brock, D. J. H., and Hayward, C., 1980, Gel electrophoresis of amniotic fluid acetylcholinesterase as an aid to the prenatal diagnosis of fetal defects. *Clin. Chim. Acta* **108**:135–141.

Brock, D. J. H., Barron, L., and Heyningen, V., 1985, Prenatal diagnosis of neural-tube defects with a monoclonal antibody specific for acetylcholinesterase. *Lancet* **i**:5–8.

Brown, S. S., Kalow, W., Pilz, W., Whittaker, M., and Woronick, C. L., 1981, The plasma cholinesterases: A new perspective. *Adv. Clin. Chem.* **22**:1–123.

Brzin, M., Sketelj, J., and Klinar, B., 1983, Cholinesterases. In *Handbook of Neurochemistry* (A. Lajtha, ed.), Vol. 4, pp. 251–292, Plenum Press, New York.

Buamah, P. K., Evans, L., and Ward, M., 1980, Amniotic fluid acetylcholinesterase isoenzyme patterns in the diagnosis of neural tube defects. *Clin. Chim. Acta* **102**:147–151.

Cardoso de Almeida, M. L., and Turner, M. J., 1983, The membrane form of variant surface glycoproteins of *Trypanosoma brucei*. *Nature* **302**:349–352.

Carson, S., Bon, S., Vigny, M., Massoulié, J., and Fardeau, M., 1979, Distribution of acetylcholinesterase molecular forms in neural and non-neural sections of human muscle. *FEBS Lett.* **97**:348–352.

Chan, S. L., Shirachi, D. Y., Bhargava, H. N., Gardner, E., and Trevor, A. J., 1972, Purification and properties of multiple forms of brain acetylcholinesterase (EC 3.1.1.7). *J. Neurochem.* **19**:2747–2758.

Chang, C.-H., and Blume, A. J., 1976, Heterogeneity of acetylcholinesterase in neuroblastoma. *J. Neurochem.* **27**:1427–1435.

Chatonnet, A., and Bacou, F., 1983, Acetylcholinesterase molecular forms in the fast and slow muscles of the chicken and the pigeon. *FEBS Lett.* **161**:122–126.

Chow, F.-L., Telen, M. J., and Rosse, W. F., 1985, The acetylcholinesterase deficit in paroxysmal nocturnal hemoglobinuria: Evidence that the enzyme is absent from the cell membrane. *Blood* **66**:940–945.

Chubb, I. W., Pilowski, P. M., Springel, H. J., and Pollard, A. C., 1979, Acetylcholinesterase in amniotic fluid: An index of fetal neural development. *Lancet* **i**:688–690.

Couraud, J. Y., Nicolet, M., and Hässig, R., 1985, Rapid axonal transport of three molecular forms of acetylcholinesterase in the frog sciatic nerve. *Neuroscience* **14**:1141–1147.

Coyle, J. T., Price, D. L., and DeLong, M. R., 1983, Alzheimer's disease: A disorder of cortical cholinergic innervation. *Science* **219**:1184–1190.

Dacie, J. V., 1967, Paroxysmal nocturnal haemoglobinuria. In *The Haemolytic Anaemias, Congential and Acquired. IV. Drug-Induced Haemolytic Anemia, Paroxysmal Haemoglobinuria* (J. V. Dacie, ed.), pp. 1128–1160, Grune & Stratton, New York.

Davies, P., 1979, Neurotransmitter-related enzymes in senile dementia of the Alzheimer type. *Brain Res.* **171**:319–327.

Davies, P., and Maloney, A. J. F., 1976, Selective loss of central cholinergic neurons in Alzheimer's disease. *Lancet* **ii**:1403.

Davies, P., and Verth, A. H., 1977, Regional distribution of the muscarinic receptor in normal and Alzheimer's type dementia brain. *Brain Res.* **138**:385–392.

Davies, R. O., Marton, A. V., and Kalow, W., 1960, The action of normal and atypical cholinesterase of human serum upon a series of esters of choline. *Can. J. Biochem. Physiol.* **38**:545–551.

Di Giamberardino, L., and Couraud, J. Y., 1978, Rapid accumulation of high molecular weight acetylcholinesterase in transected sciatic nerve. *Nature* **271**:170–172.

Dockter, M. E., and Morrison, M., 1986, Paroxysmal nocturnal hemoglobinuria erythrocytes are of two distinct types: Positive or negative for acetylcholinesterase. *Blood* **67**:540–543.

Doctor, B. P., Camp, S., Gentry, M. K., Taylor, S. S., and Taylor, P., 1983, Antigenic and structural differences in the catalytic subunits of the molecular forms of acetylcholinesterase. *Proc. Natl. Acad. Sci. U.S.A.* **80**:5767–5771.

Duchen, L. W., Searle, A. G., and Strich, S. J., 1967, An hereditary motor end-plate disease in the mouse. *J. Physiol. (Lond.)* **189**:1–6P.

Dudai, Y., Herzberg, M., and Silman, I., 1973, Molecular structures of acetylcholinesterase from electric organ tissue of the electric eel. *Proc. Natl. Acad. Sci. U.S.A.* **70**:2473–2476.

Dutta-Choudhury, T. A., and Rosenberry, T. L., 1984, Human erythrocyte acetylcholinesterase is an amphipathic protein whose short membrane-binding domain is removed by papain digestion. *J. Biol. Chem.* **259**:5653–5660.

Eckerson, H. W., Oseroff, A., Lockridge, O., and La Du, B. N., 1983, Immunological comparison of the usual and atypical human serum cholinesterase phenotypes. *Biochem. Genet.* **21**:93–108.

Elejalde, B. R., Peck, G., and De Elejalde, M. M., 1986, Determination of cholinesterase and acetylcholinesterase in amniotic fluid. *Clin. Genet.* **29**:196–203.

Ellman, G. L., Courtney, D. K., Andres, V., and Featherstone, R. M., 1961, A new and rapid colorimetric determination of acetylcholinesterase activity. *Biochem. Pharmacol.* **7**:88–95.

Fambrough, D. M., Engel, A. G., and Rosenberry, T. L., 1982, Acetylcholinesterase of human erythrocytes and neuromuscular junctions: Homologies revealed by monoclonal antibodies. *Proc. Natl. Acad. Sci. U.S.A.* **79**:1078–1082.

Ferguson, M. A. J., Haldar, K., and Cross, G. A. M., 1985, *Trypanosoma brucei* variant surface glycoprotein has a *sn*-1,2-dimyristyl glycerol membrane anchor at its COOH terminus. *J. Biol. Chem.* **260**:4963–4968.

Fernandez, H. L., Duell, M. J., and Festoff, B. W., 1979a, Neurotrophic control of 16S acetylcholinesterase at the vertebrate neuromuscular junction. *J. Neurobiol.* **10**:441–454.

Fernandez, H. L., Duell, M. J., and Festoff, B. W., 1979b, Cellular distribution of 16S acetylcholinesterase. *J. Neurochem.* **32**:581–585.

Fernandez, H. L., Stiles, J. R., and Donoso, J. A., 1986, Skeletal muscle acetylcholinesterase molecular forms in amyotrophic lateral sclerosis. *Muscle Nerve* **9**:399–406.

Festoff, B. W., and Fernandez, H. L., 1981, Plasma and red blood cell acetylcholinesterase in amyotrophic lateral sclerosis. *Muscle Nerve* **4**:41–47.

Fishman, E. B., Siek, G. C., MacCallum, R. D., Bird, E. D., Volicer, L., and Marquis, J. K., 1986, Distribution of the molecular forms of acetylcholinesterase in human brain: Alterations in dementia of the Alzheimer type. *Ann. Neurol.* **19**:246–252.

Futerman, A. H., Low, M. G., Ackermann, K. E., Sherman, W. R., and Silman, I., 1985a, Identification of covalently bound inositol in the hydrophobic membrane-anchoring domain of *Torpedo* acetylcholinesterase. *Biochem. Biophys. Res. Commun.* **129**:312–317.

Futerman, A. H., Fiorini, R.-M., Roth, E., Low, M. G., and Silman, I., 1985b, Physicochemical behavior and structural characteristics of membrane-bound acetylcholinesterase from *Torpedo* electric organ. Effect of phosphatidylinositol-specific phospholipase C. *Biochem. J.* **226**:369–377.

Gennari, K., and Brodbeck, U., 1985, Molecular forms of acetylcholinesterase from human caudate nucleus: Comparison of salt-soluble and detergent-soluble tetrameric enzyme species. *J. Neurochem.* **44**:697–704.

Gisiger, V., and Stephens, H., 1982, Molecular forms and cytochemical distribution of acetylcholinesterase (AChE) in skeletal muscles of the 129/ReJ dystrophic mouse. *Can. J. Neurol. Sci.* **9**:361.

Gisiger, V., and Stephens, H. R., 1983, Asymmetric and globular forms of AChE in slow and fast muscles of 129/ReJ normal and dystrophic mice. *J. Neurochem.* **41**:919–929.

Gisiger, V., Vigny, M., Gautron, J., and Rieger, F., 1978, Acetylcholinesterase of rat sympathetic ganglion: Molecular forms, localization and effects of denervation. *J. Neurochem.* **30**:501–516.

Goedde, H. W., Doenicke, A., and Altland, K., 1967, *Pseudocholinesterasen. Pharmakogenetik, Biochemie, Klinik,* Springer, Berlin.

Gómez-Barriocanal, J., Barat, A., Escudero, E., Rodríguez-Borrajo, C., and Ramírez, G., 1981, Solubilization of collagen-tailed acetylcholinesterase from chick retina: Effect of different extraction procedures. *J. Neurochem.* **37**:1239–1249.

Grassi, J., Vigny, M., and Massoulié, J., 1982, Molecular forms of acetylcholinesterase in bovine caudate nucleus and superior cervical ganglion: Solubility properties and hydrophobic character. *J. Neurochem.* **38**:457–469.

Grassi, J., Massoulié, J., and Timpl, R., 1983, Relationship of collagen-tailed acetylcholinesterase with basal lamina components. Absence of binding with laminin, fibronectin, and collagen types IV and V, and lack of reactivity with different anti-collagen sera. *Eur. J. Biochem.* **133**:31–38.

Greenberg, A. J., Parker, K. K., and Trevor, A. J., 1977, Immunochemical studies of mammalian brain acetylcholinesterase. *J. Neurochem.* **29**:911–917.

Grosswald, D. E., and Dettbarn, W.-D., 1983, Nerve crush induced changes in molecular forms of acetylcholinesterase in soleus and extensor digitorum muscles. *Exp. Neurol.* **79**:519–531.

Gurari, D., Silman, I., and Fuchs, S., 1974, Immunochemical and enzymatic properties of the electric eel acetylcholinesterase–antiacetylcholinesterase system. *Eur. J. Biochem.* **43**:179–187.

Haas, R., Brandt, P. T., Knight, J., and Rosenberry, T. L., 1986, Identification of amine components in a glycolipid membrane-binding domain at the C-terminus of human erythrocyte acetylcholinesterase. *Biochemistry* **25**:3098–3105.

Hall, Z. W., 1973, Multiple forms of acetylcholinesterase and their distribution in endplate and nonendplate regions of rat diaphragm muscle. *J. Neurobiol.* **4**:343–361.

Hangaard, J., Sørensen, K., Brodbeck, U., and Nørgaard-Pedersen, B., 1984, Quantitative determination of acetylcholinesterase by enzyme antigen immunoassay: Methodological aspects and clinical use. *Scand. J. Clin. Lab. Invest.* **44**:717–724.

Henderson, N. S., Tweedle, C. D., and Kabara, J. J., 1980, Cholinesterase in muscle of dystrophic hamsters (Bio-40.54). *Neurochem. Res.* **5**:1221–1230.

Hedreen, J. C., Uhl, G. R., Bacon, S. J., Fambrough, D. M., and Price, D. L., 1984, Acetylcholinesterase-immunoreactive axonal network in monkey visual cortex. *J. Comp. Neurol.* **226**:246–254.

Hirano, A., 1973, Progress in the pathology of motor neuron diseases. *Prog. Neuropathol.* **2**:181–215.

Hodkin, W. E., Giblett, E. R., Levine, H., Bauer, W., and Matulsky, A. G., 1965, Complete pseudocholinesterase deficiency: Genetic and immunologic characterization. *J. Clin. Invest.* **44**:486–493.

Holzgreve, W., Rempen, A., and Beller, F. K., 1986, Fetal lymphangiectasia—another cause for a positive amniotic fluid acetylcholinesterase test. *Arch. Gynecol.* **239**:123–125.

Hooper, M. W., and Vogel, F. S., 1976, The limbic system in Alzheimer's disease. A neuropathologic investigation. *Am. J. Pathol.* **85**:1–20.

Hsu, S.-M., Raine, L., and Fanger, H., 1981, Use of avidin–biotin–peroxidase complex (ABC) in immunoperoxidase techniques: A comparison between ABC and unlabeled antibody (PAP) procedures. *J. Histochem. Cytochem.* **29:**577–580.

Igisu, H., Mawatari, S., Kuroiwa, Y., and Miyoshino, S., 1980, Erythrocyte acetylcholinesterase in Duchenne muscular dystrophy. *Clin. Chim. Acta* **105:**241–247.

Inestrosa, N. C., Reiness, C. G., Reichardt, L. F., and Hall, Z. W., 1981, Cellular localization of the molecular forms of acetylcholinesterase in rat pheochromocytoma PC12 cells treated with nerve growth factor. *J. Neurosci.* **1:**1260–1267.

Jansz, H. S., and Cohen, J. A., 1962, Pseudocholinesterase from horse serum. I. Purification and properties of the enzyme. *Biochim. Biophys. Acta* **56:**531–537.

Johnson, C. D., and Russell, R. L., 1975, A rapid, simple radiometric assay for cholinesterase suitable for multiple determinations. *Anal. Biochem.* **64:**229–238.

Kalderon, N., Silman, I., Blumberg, S., and Dudai, Y., 1970, A method for the purification of acetylcholinesterase by affinity chromatography. *Biochim. Biophys. Acta* **207:**560–562.

Kalow, W., and Genest, K., 1957, A method for the detection of atypical forms of human serum cholinesterase. Determination of dibucaine numbers. *Can. J. Biochem. Physiol.* **35:**339–346.

Kamijo, K., Hiatt, R. B., and Koelle, G., 1953, Congenital megacolon. A comparison of the spastic and hypertrophied segments with respect to cholinesterase activities and sensitivities to acetylcholines, DFP and the barium ion. *Gastroenterology* **24:**173–185.

Karnovsky, M. J., and Roots, L., 1964, A "direct-coloring" thiocholine method for cholinesterases. *J. Histochem. Cytochem.* **12:**219–221.

Kása, P., and Rakonczay, Z., 1982a, Histochemical and biochemical demonstration of the molecular forms of acetylcholinesterase in peripheral nerve of rat. *Acta Histochem.* **70:**244–257.

Kása, P., and Rakonczay, Z., 1982b, Biochemical and histochemical evidence of 16S acetylcholinesterase in salivary glands. *J. Neurochem.* **38:**278–280.

Koelle, G. B., and Friedenwald, J. S., 1949, A histochemical method for localizing cholinesterase activity. *Proc. Soc. Exp. Biol. Med.* **70:**617–622.

Kremzner, L. T., and Wilson, I. B., 1963, A chromatographic procedure for the purification of acetylcholinesterase. *J. Biol. Chem.* **238:**1714–1717.

La Du, B. N., 1971, Plasma esterase activity and the metabolism of drugs with ester groups. *Ann. N.Y. Acad. Sci.* **179:**684–694.

La Du, B. N., and Lockridge, O., 1986, Molecular biology of human serum cholinesterase. *Fed. Proc.* **45:**2965–2969.

Lappin, R. I., and Rubin, L. L., 1985, Molecular forms of acetylcholinesterase in *Xenopus* muscle. *Dev. Biol.* **110:**269–274.

Lawler, H. C., 1964, The preparation of a soluble acetylcholinesterase from brain. *Biochim. Biophys. Acta* **81:**280–288.

Lazar, M., and Vigny, M., 1980, Modulation of the distribution of acetylcholinesterase molecular forms in a murine neuroblastoma sympathetic ganglion cell hybrid cell line. *J. Neurochem.* **35:**1067–1079.

Lenoir-Rousseaux, J.-J. 1985, Les arthropodes possèdent une même forme moléculaire majoritaire d'acétylcholinestérase. *C.R. Soc. Biol.* **179:**741–747.

Leuzinger, W., and Baker, A. L., 1967, Acetylcholinesterase. I. Large-scale purification, homogeneity, and amino acid analysis. *Proc. Natl. Acad. Sci. U.S.A.* **57:**446–451.

Lindenbaum, M. H., and Livett, B. G., 1983, Acetylcholinesterase molecular forms in C57BL/6J dystrophic mice. *Muscle Nerve* **6:**638–645.

Lockridge, O., 1984, Amino acid composition and sequence of human serum cholinesterase: A progress report. In *Cholinesterases, Fundamental and Applied Aspects* (M. Brzin, E. A., Barnard, and D. Sket, eds.), pp. 5–20, Walter de Gruyter, Berlin.

Lockridge, O., and La Du, B. N. J., 1978, Comparison of atypical and usual human serum cholinesterase: Purification, number of active sites, substrate affinity, and turnover number. *J. Biol. Chem.* **253**:361–366.

Lockridge, O., and La Du, B. N., 1982, Loss of the interchain disulfide peptide and dissociation of the tetramer following limited proteolysis of the native human serum cholinesterase. *J. Biol. Chem.* **257**:12012–12018.

Lockridge, O., Eckerson, H. W., and La Du, B. N., 1979, Interchain disulfide bonds and subunit organisation in human serum cholinesterase. *J. Biol. Chem.* **254**:8324–8330.

Lockridge, O., Bartels, C. F., Vaughan, T. A., Wang, C. K., Norton, S. E., and Johnson, L. L., 1987, Complete amino acid sequence of human serum cholinesterase. *J. Biol. Chem.* **262**:549–557.

Lyles, J. M., Silman, I., and Barnard, E. A., 1979, Developmental changes in levels and forms of cholinesterases in muscles of normal and dystrophic chickens. *J. Neurochem.* **33**:727–738.

Lyles, J. M., Barnard, E. A., and Silman, I., 1980, Changes in the levels and forms of cholinesterases in the blood plasma of normal and dystrophic chickens. *J. Neurochem.* **34**:978–987.

Lyles, J. M., Silman, I., Di Giamberardino, L., Couraud, J.-Y., and Barnard, E. A., 1982, Comparison of the molecular forms of the cholinesterases in tissues of normal and dystrophic chickens. *J. Neurochem.* **38**:1007–1021.

MacPhee-Quigley, K., Taylor, P., and Taylor, S., 1985, Primary structures of the catalytic subunits from two molecular forms of acetylcholinesterase. A comparison of NH_2-terminal and active center sequences. *J. Biol. Chem.* **260**:12185–12189.

Marsh, D., Grassi, J., Vigny, M., and Massoulié, J., 1984, An immunological study of rat acetylcholinesterase: Comparison with acetylcholinesterases from other vertebrates. *J. Neurochem.* **43**:204–213.

Massoulié, J., and Bon, S., 1976, Affinity chromatography of acetylcholinesterase; the importance of hydrophobic interactions. *Eur. J. Biochem.* **68**:531–539.

Massoulié, J., and Bon, S., 1982, The molecular forms of cholinesterase and acetylcholinesterase in vertebrates. *Annu. Rev. Neurosci.* **5**:57–106.

Massoulié, J., and Rieger, F., 1969, L'acétylcholinestérase des organes électriques de poissons (Torpille et Gymnote). *Eur. J. Biochem.* **11**:441–455.

Massoulié, J., Bon, S., and Vigny, M., 1980, The polymorphism of cholinesterase in vertebrates. *Neurochem. Int.* **2**:161–184.

Mays, C., and Rosenberry, T. L., 1981, Characterization of pepsin-resistant collagen-like tail subunit fragments of 18S and 14S acetylcholinesterase from *Electrophorus electricus*. *Biochemistry* **20**:2810–2817.

Méflah, K., Bernard, S., and Massoulié, J., 1984, Interactions with lectins indicate differences in the carbohydrate composition of the membrane-bound enzymes acetylcholinesterase and 5′-nucleotidase in different cell types. *Biochimie* **66**:59–69.

Meier-Ruge, W., 1968, Das megacolon: seine Diagnose und Pathophysiologie, *Virchows Arch. (Pathol. Anat.)* **344**:67–85.

Metz, J., Stevens, K., van Rensburg, N. J., and Hart, D., 1961, Failure of in-vivo inhibition of acetylcholinesterase to affect erythrocyte life-span: The significance of the enzyme defect in paroxysmal nocturnal haemoglobinuria. *Br. J. Haematol.* **7**:458–463.

Milunsky, A., and Sapirstein, V. S., 1982, Prenatal diagnosis of open neural defects using the amniotic fluid acetylcholinesterase assay. *Obstet. Gynecol.* **59**:1–6.

Mintz, K. P., and Brimijoin, S., 1985a, Two-step immunoaffinity purification of acetylcholinesterase from rabbit brain. *J. Neurochem.* **44**:225–232.

Mintz, K. P., and Brimijoin, S., 1985b, Monoclonal antibodies to rabbit brain acetylcholinesterase: Selective enzyme inhibition, differential affinity for enzyme forms, and cross-reactivity with other mammalian cholinesterases. *J. Neurochem.* **45**:284–292.

Mintz, K. P., Weinshilboum, R. M., and Brimijoin, W. S., 1984, Evolution of butyrylcholinesterase in higher primates: An immunochemical study. *Comp. Biochem. Physiol.* **79C**:35–37.

Muensch, H., Goedde, H. W., and Yoshida, A., 1976, Human serum cholinesterase subunits and number of active sites of the major component. *Eur. J. Biochem.* **70**:217–223.

Nicholson-Weller, A., March, J. P., Rosenfield, S. I., and Austen, K. F., 1983, Affected erythrocytes of patients with paroxysmal nocturnal hemoglobinuria are deficient in the complement regulatory protein, decay accelerating factor. *Proc. Natl. Acad. Sci. U.S.A.* **80**:5066–5070.

Nicholson-Weller, A., Spicer, D. B., and Austen, K. F., 1985, Deficiency of the complement regulatory protein decay accelerating factor on membranes of erythrocytes, monocytes, and platelets in paroxysmal nocturnal hemoglobinuria. *N. Engl. J. Med.* **312**:1091–1097.

Nicolet, M., Pincon-Raymond, M., and Rieger, F., 1986, Globular and asymmetric acetylcholinesterase in frog muscle basal lamina sheaths. *J. Cell Biol.* **102**:762–768.

Niday, E., Wang, C. S., and Alaupovic, P., 1977, Studies on the characterization of human erythrocyte acetylcholinesterase and its interaction with antibodies. *Biochim. Biophys. Acta* **459**:180–193.

Nørgaard-Pedersen, B., Hangaard, J., and Bjerrum, O. J., 1983, Quantitative enzyme antigen immunoassay of acetylcholinesterase in amniotic fluid. *Clin. Chem.* **29**:1061–1064.

Okamoto, E., and Ueda, T., 1967, Embryogenesis of intramural ganglia of the gut and its relation to Hirschsprung's disease. *J. Pediatr. Surg.* **2**:437–443.

Ott, P., Jenny, B., and Brodbeck, U., 1975, Multiple molecular forms of purified human erythrocyte acetylcholinesterase. *Eur. J. Biochem.* **57**:469–480.

Paggi, P., Scarsella, G., and Toschi, G., 1977, Multiple forms of acetylcholinesterase in the ciliary ganglia and iris of the pigeon. *Neuroscience* **2**:1085–1093.

Parvari, R., Pecht, I., and Soreq, H., 1983, A microfluorometric assay for cholinesterases suitable for multiple kinetic determinations of picomoles of released thiocholine. *Anal. Biochem.* **133**:450–456.

Patrick, W. J. A., Besley, G. T. N., and Smith, I. I., 1980, Histochemical diagnosis of Hirschsprung's disease and a comparison of the histochemical and biochemical activity of acetylcholinesterase in rectal mucosal biopsies. *J. Clin. Pathol.* **33**:336–343.

Perry, E. K., Perry, R. H., Blessed, G., and Tomlinson, B. E., 1977, Necropsy evidence of central cholinergic deficits in senile dementia. *Lancet* **i**:189.

Perry, E. K., Tomlinson, B. E., Blessed, G., Bergmann, K., Gibson, P. H., and Perry, R. H., 1978, Correlation of cholinergic abnormalities with senile plaques and mental test scores in senile dementia. *Br. Med. J.* **2**:1457–1459.

Perry, R. H., Candy, J. M., Perry, E. K., Irving, D., Blessed, G., Fairbairn, A. F., and Tomlinson, B. E., 1982, Extensive loss of choline acetyltransferase activity is not reflected by neuronal loss in the nucleus of Meynert in Alzheimer's disease. *Neurosci. Lett.* **33**:311–315.

Pilowsky, P. M., Hodgson, A. J., and Chubb, I. W., 1982, Acetylcholinesterase in neural tube defects: A model using chick embryo amniotic fluid. *Neuroscience* **7**:1203–1214.

Poiana, G., Scarsella, G., Biagioni, S., Senni, M. I., and Cossu, G., 1985, Membrane acetylcholinesterase in murine muscular dystrophy in vivo and in cultured myotubes. *Int. J. Dev. Neurosci.* **3**:331–340.

Raba, R., Aaviksaar, A., Raba, M., and Siigur, J., 1979, Cobra venom acetylcholinesterase. Purification and molecular properties. *Eur. J. Biochem.* **96**:151–158.

Rakonczay, Z., 1986, Mammalian brain acetylcholinesterase. In *Neuromethods, Volume 5, Neurotransmitter Enzymes* (A. A. Boulton, G. B. Baker, and P. H. Yu, eds.), Chap. 9, pp. 319–360, Humana Press, Clifton, NJ.

Rakonczay, Z., and Brimijoin, S., 1985, Immunochemical differences among molecular forms of acetylcholinesterase in brain and blood. *Biochim. Biophys. Acta* **832**:127–134.

Rakonczay, Z., and Brimijoin, S., 1986, Monoclonal antibodies to rat brain acetylcholinesterase:

Comparative affinity for soluble and membrane-associated enzyme and for enzyme from different vertebrate species. *J. Neurochem.* **46**:280–287.

Rakonczay, Z., and Brimijoin, S., 1987, Monoclonal antibodies to human brain acetylcholinesterase: Properties and applications. *Cell. Mol. Neurobiol.* (in press).

Rakonczay, Z., and Németh, P., 1984, Change in the distribution of acetylcholinesterase molecular forms in Hirschsprung's disease. *J. Neurochem.* **43**:1194–1196.

Rakonczay, Z., Mallol, J., Schenk, H., Vincendon, G., and Zanetta, J.-P., 1981a, Purification and properties of the membrane-bound acetylcholinesterase from adult rat brain. *Biochim. Biophys. Acta* **657**:243–256.

Rakonczay, Z., Vincendon, G., and Zanetta, J.-P., 1981b, Heterogeneity of rat brain acetylcholinesterase: A study by gel filtration and gradient centrifugation. *J. Neurochem.* **37**:662–669.

Rakonczay, Z., Szüts, P., and Szabó, J., 1985, Significance of acetylcholinesterase molecular forms in the prenatal diagnosis of neural tube defects. *Acta Physiol. Hung.* **66**:280–281.

Ralston, J. S., Rush, R. S., Doctor, B. P., and Wolfe, A. D., 1985, Acetylcholinesterase from fetal bovine serum. Purification and characterization of soluble G_4 enzyme. *J. Biol. Chem.* **260**:4312–4318.

Ramírez, G., Gómez-Barriocanal, J., Barat, A., and Rodríguez-Borrajo, C., 1984, Two classes of collagen-tailed molecular forms of acetylcholinesterase. In *Cholinesterases, Fundamental and Applied Aspects* (M. Brzin, E. A. Barnard, and D. Sket, eds.), pp. 115–128, Walter de Gruyter, Berlin.

Reisine, T. D., Yamamura, H. I., Bird, E. D., Spokes, E. and Enna, S. J., 1978, Pre- and postsynaptic neurochemical alterations in Alzheimer's disease. *Brain Res.* **159**:477–481.

Rieger, F., and Vigny, M., 1976, Solubilization and physicochemical characterization of rat brain acetylcholinesterase: Development and maturation of its molecular forms. *J. Neurochem.* **27**:121–129.

Rieger, F., Bon, S., Massoulié, J., and Cartaud, J., 1973, Observation par microscopie électronique des formes allongées et globulaires de l'acétylcholinestérase de Gymnote (*Electrophorus electricus*). *Eur. J. Biochem.* **34**:539–547.

Rieger, F., Bon, S., Massoulié, Cartaud, J., Picard, B., and Benda, P., 1976, *Torpedo marmorata* acetylcholinesterase; a comparison with the *Electrophorus electricus* enzyme: Molecular forms, subunits, electron microscopy, immunological relationship. *Eur. J. Biochem.* **68**:513–521.

Rieger, F., Chételat, R., Nicolet, M., Kamal, L., and Poullet, M., 1980, Presence of tailed, asymmetric forms of acetylcholinesterase in the central nervous system of vertebrates. *FEBS Lett.* **121**:169–174.

Rieger, F., Shelanski, M. L., and Sidman, R. L., 1983, The multiple molecular forms of acetylcholinesterase in "motor end-plate disease" in the mouse (med^j and med allelic forms): Sensitivity of the 10S form to partial or total loss of muscle activity. *Exp. Neurol.* **79**:299–315.

Rieger, F., Powell, J. A., and Pincon-Raymond, M., 1984, Extensive nerve overgrowth and paucity of the tailed asymmetric form (16S) of acetylcholinesterase in the developing skeletal neuromuscular system of the dysgenic (mdg/mdg) mouse. *Dev. Biol.* **101**:181–191.

Roberts, W. L., and Rosenberry, T. L., 1985, Identification of covalently attached fatty acids in the hydrophobic membrane-binding domain of human erythrocyte acetylcholinesterase. *Biochem. Biophys. Res. Commun.* **133**:621–627.

Roberts, W. L., and Rosenberry, T. L., 1986, Selective radiolabelling and isolation of the hydrophobic membrane-binding domain of human erythrocyte acetylcholinesterase. *Biochemistry* **25**:3091–3098.

Rodriquez-Borrajo, C., Barat, A., and Ramirez, C., 1982, Solubilization of collagen-tailed molecular forms of acetylcholinesterase from several brain areas in different vertebrate species. *Neurochem. Int.* **4**:563–568.

Rosenberry, T. L., 1985, Structural distinctions among acetylcholinesterase forms. In *The Enzymes of Biological Membranes* (A. N. Martonosik, ed.), Vol. 3, pp. 403–429, Plenum Press, New York.

Rossor, M., Fahrenkrug, J., Emson, P., Mountjoy, C., Iversen, L., and Roth, M., 1980, Reduced cortical choline acetyltransferase activity in senile dementia of Alzheimer type is not accompanied by changes in vasoactive intestinal polypeptide. *Brain Res.* **201**:249–253.

Rotoli, B., Robledo, R., Scarpato, N., and Luzzatto, L., 1984, Two populations of erythroid cell progenitors in paroxysmal nocturnal hemoglobinuria. *Blood* **64**:847–851.

Rotundo, R. L., 1984, Synthesis, assembly, and processing of AChE in tissue cultured muscle. In *Cholinesterases, Fundamental and Applied Aspects* (M. Brzin, E. A. Barnard, and D. Sket, eds.), pp. 203–217, Walter de Gruyter, Berlin.

Ruberg, M., Rieger, F., Villageois, A., Bonnet, A. M., and Agid, Y., 1986, Acetylcholinesterase and butyrylcholinesterase in frontal cortex and cerebrospinal fluid of demented and non-demented patients with Parkinson's disease. *Brain Res.* **362**:83–91.

Rubinstein, H. M., Dietz, A. A., Hodges, L. K., Lubrano, T., and Czebotar, V. 1970, Silent cholinesterase gene: Variations in the properties of serum enzyme in apparent homozygotes. *J. Clin. Invest.* **49**:479–486.

Rubinstein, H. M., Dietz, A. A., Lubrano, T., and Garry, P. J., 1976, E_1^j, a quantitative variant at cholinesterase locus 1: Immunological evidence. *J. Med. Genet.* **13**:43–45.

Rubinstein, H. M., Dietz, A. A., and Lubrano, T., 1978, E_1^k, another quantitative variant at cholinesterase locus 1. *J. Med. Genet.* **15**:27–29.

Sakai, M., Saisu, H., and Abe, T., 1985, Comparison of asymmetric forms of acetylcholinesterase from the electric organ of *Narke japonica* and *Torpedo californica*. *Eur. J. Biochem.* **153**:497–502.

Sakai M., Koshigoe, N., Saisu, H., and Abe, T., 1986, Monoclonal antibodies specific to asymmetric forms of acetylcholinesterase from the electric organ of electric rays. *Biomed. Res.* **7**:1–5.

Schumacher, M., Camp, S., Maulet, Y., Newton, M., MacPhee-Quigley, K., Taylor, S. S., Friedmann, T., and Taylor, P., 1986, Primary structure of *Torpedo californica* acetylcholinesterase deduced from its cDNA sequence. *Nature* **319**:407–409.

Seller, M. J., and Cole, K. J., 1980, Polyacrylamide gel electrophoresis of amniotic fluid cholinesterases: A good prenatal test for neural tube defects. *Br. J. Obstet. Gynaecol.* **87**:1103–1108.

Silberstein, L., Inestrosa, N. C., and Hall, Z. W., 1982, Aneural muscle cell cultures make synaptic basal lamina components. *Nature* **295**:143–145.

Silman, I., Di Giamberardino, L., Lyles, J., Couraud, J.-Y., and Barnard, E. A., 1979, Parallel regulation of acetylcholinesterase and pseudocholinesterase in normal, denervated and dystrophic chicken skeletal muscle. *Nature* **280**:160–162.

Silver, A., 1974, *The Biology of Cholinesterases*, North-Holland, Amsterdam.

Skau, K. A., 1983, The acetylcholinesterase abnormality in dystrophic mice is a reflection of a maturational defect. *Brain Res.* **276**:192–194.

Skau, K. A., 1985, Acetylcholinesterase molecular forms in serum and erythrocytes of laboratory animals. *Comp. Biochem. Physiol.* **80C**:207–210.

Skau, K. A., and Brimijoin, S., 1980, Multiple molecular forms of acetylcholinesterase in rat vagus nerve, smooth muscle, and heart. *J. Neurochem.* **35**:1151–1154.

Skau, K. A., and Brimijoin, S., 1981, Abnormal distribution of skeletal muscle acetylcholinesterase molecular forms in dystrophic mice. *Exp. Neurol.* **74**:111–121.

Sketelj, J., and Brzin, M., 1985, Asymmetric molecular forms of acetylcholinesterase in mammalian skeletal muscles. *J. Neurosci. Res.* **14**:95–103.

Sketelj, J., McNamee, M. G., and Wilson, B. W., 1978, Effect of denervation on the molecular forms of acetylcholinesterase in normal and dystrophic chickens. *Exp. Neurol.* **60**:624–629.

Sketelj, J., Sasel, B., Zupancic, N., and Brzin, M., 1983, Plasma acetylcholinesterase in Duchenne muscular dystrophy. *Exp. Neurol.* **80:**329–336.

Smith, A. D., 1979, Amniotic fluid acetylcholinesterase and neural tube defects: Plea for standardization. *Lancet* **ii:**307–308.

Smith, A. D., Wald, N. J., Chuckle, H. S., Stirrat, G. M., Bobrow, M., and Lagercrantz, H., 1979, Amniotic fluid acetylcholinesterase as a possible diagnostic test for neural tube defects in early pregnancy. *Lancet* **i:**685–688.

Soininen, H., Halonen, T., and Riekkinen, P. J., 1981, Acetylcholinesterase activities in cerebrospinal fluid of patients with senile dementia of Alzheimer type. *Acta Neurol. Scand.* **64:**217–224.

Sørensen, K., Gentinetta, R., and Brodbeck, U., 1982, An amphiphile dependent form of human brain caudate nucleus acetylcholinesterase: Purification and properties. *J. Neurochem.* **39:**1050–1060.

Sørensen, K., Gennari, K., Brodbeck, U., Landauer, P., and Liefländer, M., 1985, Polymorphism and immunochemical cross-reactivity of acetylcholinesterases from the brains of human, dog, hog, bovine, and horse. *Comp. Biochem. Physiol.* **80C:**263–268.

Sørensen, K., Brodbeck, U., Rasmussen, A. G., and Nørgaard-Pedersen, B., 1986, Normal human serum contains two forms of acetylcholinesterase. *Clin. Chim. Acta* **158:**1–6.

St. Clair, D. M., Brock, D. J. H., and Barron, L., 1986, A monoclonal antibody assay technique for plasma and red cell acetylcholinesterase activity in Alzheimer's disease. *J. Neurol. Sci.* **73:**169–176.

Steiger, S., Cardoso de Almeida, M.-L., Blatter, M. C., Brodbeck, U., and Bordier, C., 1986, The membrane-anchoring systems of vertebrate acetylcholinesterase and variant surface glycoproteins of African trypanosomes share a common antigenic determinant. *FEBS Lett.* **199:**182–186.

Taylor, P., Jones, J. W., and Jacobs, N. M., 1974, Acetylcholinesterase from *Torpedo:* Characterization of an enzyme species isolated by lytic procedures. *Mol. Pharmacol.* **10:**78–92.

Tomlinson, B. E., Blessed, G., and Roth, M., 1970, Observations on the brains of demented old people. *J. Neurol. Sci.* **11:**205–242.

Toutant, J. P., Massoulié, J., and Bon, S., 1985, Polymorphism of pseudocholinesterase in *Torpedo marmorata* tissues. Comparative study of the catalytic and molecular properties of this enzyme with acetylcholinesterase. *J. Neurochem.* **44:**580–592.

Tsuji, S,, Rieger, F., Peltre, G., Massoulié, J., and Benda, P., 1972, Acetylcholinesterase du muscle, de la moelle épiniere et du cerveau de gymnote. *J. Neurochem.* **19:**989–997.

Tucci, A. F., and Seifter, S., 1969, Preparation and properties of porcine parotid butyrylcholinesterase. *J. Biol. Chem.* **244:**841–849.

Turner, M. J., Cardoso de Almeida, M. L., Gurnett, A. M., Raper, J., and Ward, J., 1985, Biosynthesis, attachment and release of variant surface glycoproteins of the African trypanosome. *Curr. Top. Microbiol. Immunol.* **117:**23–55.

Verp, M. S., Milunsky, A., Simpson, J. L., Graham, A., and Sabbagha, R., 1984, Elevated alphafetoprotein and acetylcholinesterase associated with hydrocele. *Am. J. Med. Genet.* **19:**651–652.

Vigny, M., Di Giamberardino, L., Couraud, J.-Y., Rieger, F., and Koenig, J., 1976, Molecular forms of chicken acetylcholinesterase: Effect of denervation. *FEBS Lett.* **69:**277–280.

Vigny, M., Bon, S., Massoulié, J., and Leterrier, F., 1978, Active site catalytic efficiency of acetylcholinesterase molecular forms in *Electrophorus, Torpedo,* rat and chicken. *Eur. J. Biochem.* **85:**317–323.

Vigny, M., Martin, G. R., and Grotendorst, G. R., 1983, Interactions of asymmetric forms of acetylcholinesterase with basement membrane components. *J. Biol. Chem.* **258:**8794–8798.

Warkany, J., 1971, *Congenital Malformations, Notes and Comments,* Year Book Medical Publishers, Chicago.

Wenthold, R. J., Mahler, H. R., and Moore, W. J., 1974, Properties of acetylcholinesterase from rat brain. *J. Neurochem.* **22:**945–949.

Whitehouse, P. J., Price, D. L., Clark, A. W., Coyle, J. T., and Delong, M. R., 1981, Alzheimer's disease: Evidence for selective loss of cholinergic neurons in the nucleus basalis. *Ann. Neurol.* **10:**122–126.

Whitehouse, P. J., Price, D. L., Struble, R. G., Coyle, J. T., and Delong, M. R., 1982, Alzheimer's disease and senile dementia—loss of neurons in the basal forebrain. *Science* **215:**1237–1239.

Williams, R. M., 1969, Antibodies to acetylcholinesterase. *Proc. Natl. Acad. Sci. U.S.A.* **62:**1175–1180.

Wilson, B. W., Montgomery, M. A., and Asmundson, R. V., 1968, Cholinesterase activity and inherited muscular dystrophy of the chicken. *Proc. Soc. Exp. Biol. Med.* **129:**199–206.

Younkin, S. G., Rosenstein, C., Collins, P. L., and Rosenberry, T. L., 1982, Cellular localization of the molecular forms of acetylcholinesterase in rat diaphragm. *J. Biol. Chem.* **257:**13630–13637.

Younkin, S. G., Goodridge, B., Katz, J., Lockett, G., Nafziger, D., Usiak, M. F., and Younkin, L. H., 1986, Molecular forms of acetylcholinesterase in Alzheimer's disease. *Fed. Proc.* **45:**2982–2988.

Zanetta, J.-P., Rakonczay, Z., Reeber, A., Kása, P., and Vincendon, G., 1981, Antibodies against the membrane-bound acetylcholinesterase from adult rat brain. *FEBS Lett.* **129:**293–296.

Index

GPSR Compliance

The European Union's (EU) General Product Safety Regulation (GPSR) is a set of rules that requires consumer products to be safe and our obligations to ensure this.

If you have any concerns about our products, you can contact us on ProductSafety@springernature.com

In case Publisher is established outside the EU, the EU authorized representative is:

Springer Nature Customer Service Center GmbH
Europaplatz 3
69115 Heidelberg, Germany

Batch number: 10761599

Printed by Printforce, the Netherlands